Die Grundlehren der mathematischen Wissenschaften

in Einzeldarstellungen
mit besonderer Berücksichtigung
der Anwendungsgebiete

Band 186

Handbook for Automatic Computation

Edited by

F. L. Bauer · A. S. Householder · F. W. J. Olver
H. Rutishauser † · K. Samelson · E. Stiefel

Volume II

J. H. Wilkinson · C. Reinsch

Linear Algebra

Chief editor
F. L. Bauer

Springer-Verlag Berlin Heidelberg New York 1971

Dr. J. H. Wilkinson, F.R.S.

National Physical Laboratory, Teddington, Middlesex

Dr. C. Reinsch

Mathematisches Institut der Technischen Universität
8 München 2, Arcisstr. 21

Geschäftsführende Herausgeber:

Prof. Dr. B. Eckmann

Eidgenössische Technische Hochschule Zürich

Prof. Dr. B. L. van der Waerden

Mathematisches Institut der Universität Zürich

AMS Subject Classifications (1971)
Primary   65 Fxx, 15-04
Secondary 65 G 05, 90 C 05

ISBN 3-540-05414-6 Springer-Verlag Berlin Heidelberg New York
ISBN 0-387-05414-6 Springer-Verlag New York Heidelberg Berlin

Preface

The development of the internationally standardized language ALGOL has made it possible to prepare procedures which can be used without modification whenever a computer with an ALGOL translator is available. Volume I a in this series gave details of the restricted version of ALGOL which is to be employed throughout the Handbook, and volume I b described its implementation on a computer. Each of the subsequent volumes will be devoted to a presentation of the basic algorithms in some specific areas of numerical analysis.

This is the first such volume and it was felt that the topic Linear Algebra was a natural choice, since the relevant algorithms are perhaps the most widely used in numerical analysis and have the advantage of forming a well defined class. The algorithms described here fall into two main categories, associated with the solution of linear systems and the algebraic eigenvalue problem respectively and each set is preceded by an introductory chapter giving a comparative assessment.

In spite of the self-contained nature of the linear algebra field, experience has shown that even here the preparation of a fully tested set of algorithms is a far greater task than had been anticipated. Almost all the algorithms presented here have received pre-publication in Numerische Mathematik and the need to check very carefully whenever an algorithm is modified, even in a minor detail, has meant that the preparation of a paper in the Handbook series has usually taken longer than would be required for a normal research paper. Failure to check may result in the introduction of an error into an algorithm which previously had enjoyed a long period of successful use.

It soon became obvious that it would be impractical, even if desirable, to aim at completeness in this volume. In general we have aimed to include only algorithms which at least in some limited area provide something approaching an optimum solution, this may be from the point of view of generality, elegance, speed or economy of storage. The omission of an algorithm should not be taken as indicating that it has been found wanting; in some cases it merely means that we were not fully satisfied with current implementations.

From its very nature a volume of this kind is essentially a joint effort. In many instances the basic mathematical concept of an algorithm has had a long period of evolution and many people have played a part in the progressive refinement of the original concept. Thanks are due to all who have thereby contributed indirectly to this volume but I would like to pay a special tribute to those who have submitted the ALGOL procedures published here and have assisted in testing them.

Professor F. L. Bauer and the late Professor H. Rutishauser, who are authors of the Handbook series, have been invaluable sources of information and have supervised much of the work in this volume. Three colleagues at the National

Physical Laboratory, Miss. H. Bowdler, Mr. R. S. Martin and Mrs. G. Peters have between them been responsible for the preparation of some half of the published algorithms, have made substantial contributions to them and have played a major part in their testing at all stages in their development. Special thanks are due to Mrs. E. Mann of the Technische Universität München, who tested a number of the earlier variants and virtually all of the final versions of the algorithms and gave valuable assistance in the proof reading.

The publication of a volume of this kind has special problems since a good deal of the typeset material has had to be retained in almost finished form over a long period. We are particularly grateful to the publishers for their patient courteous assistance in what must have been an unusually exacting task.

We are very much aware that the production of reliable algorithms is a continuing process and we shall appreciate it if users will inform us of failures or shortcomings in the performance of our procedures.

Teddington and München   J. H. Wilkinson
October 1970   C. H. Reinsch

Contents

General guide to the use of the volume

Most of the contributions have been subdivided into the following seven sections.

1. Theoretical Background
   briefly collecting the relevant formulae.

2. Applicability
   which sketches the scope of the algorithms.

3. Formal Parameter List
   with a specification of all input, output and exit parameters.

4. ALGOL Programs
   presenting the algorithms in the ALGOL 60 reference language.

5. Organisational and Notational Details
   explaining and commenting on special features, and showing why particular realizations of the formulae have been preferred to others.

6. Discussion of Numerical Properties
   usually referring to a published error analysis of the underlying algorithms.

7. Test Results and Examples of the Use of the Procedures
   giving the solutions obtained on a digital computer to a number of sample problems in order to facilitate the correct use of the procedures.

In some earlier contributions section 7 above was subdivided into two sections:

7. Examples of the Use of the Procedures.

8. Test Results.

List of Contributors

Bartels, R. H., Dr. Computer Sciences Department, Univ. of Texas, Austin, Texas 78712, U.S.A.

Barth, W., Dr. Rechenzentrum, Techn. Hochschule, 61 Darmstadt, Hochschulstr.2, Deutschland.

Bauer, F. L., Prof. Dr., Mathematisches Inst. der Techn. Univ., 8 München 2, Arcisstr. 21, Deutschland.

Boothroyd, J., Dr., Hydro-University Computing Centre, University of Tasmania, Hobart, Tasmania, Australia.

Bowdler, H. J., National Physical Laboratory, Division of Numerical & Applied Mathematics, Teddington, Middlesex, Great Britain.

Businger, P. A., Dr., Bell Telephone Laboratories, Mountain Avenue, Murray Hill, N.Y. 07974, U.S.A.

Dubrulle, A., Dr., I.B.M. Scientific Center, 6900 Fannin Street, Houston, Texas, U.S.A.

Eberlein, P. J., Prof. Dr., State University of New York at Buffalo, Amherst, New York, 14226, U.S.A.

Ginsburg, T., Dr., Inst. für angewandte Mathematik der Eidg. Techn. Hochschule Zürich, CH-8006 Zürich, Schweiz.

Golub, G. H., Prof. Dr., Computer Science Department, Stanford Univ., Stanford, California 94305, U.S.A.

Martin, R. S., National Physical Laboratory, Division of Numerical & Applied Mathematics, Teddington, Middlesex, Great Britain.

Parlett, B. N., Prof. Dr., Department of Computer Science, University of California, Berkeley, California 94720, U.S.A.

Peters, G., National Physical Laboratory, Division of Numerical & Applied Mathematics, Teddington, Middlesex, Great Britain.

Reinsch, C., Dr., Mathematisches Inst. der Techn. Univ., 8 München 2, Arcisstr. 21, Deutschland.

Rutishauser, H. †, Prof. Dr., Eidg. Techn. Hochschule Zürich, Fachgruppe Computer-Wissenschaften, CH-8006 Zürich, Schweiz. Clausiusstr. 55,

Schwarz, H. R., Dr., Inst. für Angewandte Mathematik, Eidg. Techn. Hochschule Zürich, CH-8006 Zürich, Schweiz.

Stoer, J., Prof. Dr., Inst. f. Angewandte Mathematik der Universität Würzburg, 87 Würzburg, Kaiserstr. 27, Deutschland.

Wilkinson F. R. S., J. H., Dr., National Physical Laboratory, Division of Numerical & Applied Mathematics, Teddington, Middlesex, Great Britain.

Zenger, Ch., Dr., Leibniz-Rechenzentrum der Bayerischen Akademie der Wissenschaften, 8 München 2, Barerstr. 21, Deutschland.

Introduction to Part I

Linear Systems, Least Squares and Linear Programming

by J. H. WILKINSON

1. Introduction

Algorithms associated with linear systems may be roughly classified under three headings:

i) Solution of non-singular systems of linear equations, matrix inversion and determinant evaluation.

ii) Linear least squares problems and calculation of generalized inverses.

iii) Linear programming

though the last two overlap to some extent with the first.

Nearly all the methods we describe for solving $n \times n$ systems of equations start by performing some factorization of the matrix A of coefficients and as a by-product the determinant of A is readily available. By taking the n columns of the unit matrix as right-hand sides, the inverse of A can be computed, though the volume of work can be reduced somewhat by taking account of the special nature of these n columns.

It was decided that iterative techniques for the solution of linear algebraic equations should be excluded from this volume since they have mainly been developed in connexion with the solution of partial differential equations.

It will be appreciated that the selection of the most efficient algorithms may be to some extent machine dependent. This is usually true, for example, when storage considerations become relevant. In the algorithms presented in this book no reference is made to the use of a backing store such as a magnetic tape or disc. For this reason the storage requirements in practice may be different from what they appear to be. For example, if iterative refinement of the solution of a system of linear equations is required, a copy of the original system must be retained. In practice this could be held in the backing store.

While it is hoped that many users will wish to take advantage of the tested algorithms in exactly the form in which they are published, others will probably wish to adapt them to suit their own special requirements. The descriptions accompanying the algorithms should make this easier to do.

2. List of Procedures

To facilitate reference to the procedures they are listed below in alphabetical order, together with the chapters in which they are described. Sets of related procedures are grouped together. We also include a number of basic procedures associated with complex arithmetic, etc. in this list.

3. Positive Definite Symmetric Matrices

The solution of an $n \times n$ system of linear equations with a positive definite symmetric matrix of coefficients is a simpler problem than that of a system with a matrix of general form or even with a non-positive definite symmetric matrix. Most of the algorithms we give for positive definite symmetric matrices are based on the Cholesky factorization $A = LL^T$, where L is a lower triangular matrix, or on the related factorization $A = LDL^T$ where L is unit lower triangular and D is a positive diagonal matrix. Such algorithms are numerically stable and very economical as regards the number of arithmetic operations. There is no significant difference between the effectiveness of the LL^T and LDL^T factorizations, but the former uses square roots and the latter does not. For some purposes the Cholesky factor L is specifically required.

3.1. Dense Positive Definite Matrices

a) Cholesky Factorizations

There are two main sets associated with the Cholesky factorization of a dense $n \times n$ positive definite matrix. They are *choldet1, cholsol1* and *cholinversion1*; and *choldet2, cholsol2* and *cholinversion2*. The second of these sets achieves economy of storage by working only with a linear array of $\frac{1}{2}n(n+1)$ elements consisting initially of the lower triangle of A which is later overwritten with L. This second set is marginally slower than the first, but should be used when storage considerations are paramount.

In each case *choldet* gives the LL^T factorization of A and computes the determinant as a by-product. After using *choldet*, systems of equations having A as the matrix of coefficients may be solved by using the corresponding *cholsol* and the inverse of A may be found by using the corresponding *cholinversion*.

Although the algorithms based on the Cholesky factorization are very accurate, indeed of optimal accuracy having regard to the precision of computation, the computed solutions or the computed inverse may not be of sufficient accuracy if the matrix A is ill-conditioned. The accuracy of the solutions or of the inverse can be improved while still using the original factorization of A. This is achieved in the algorithms *acc solve* and *acc inverse*. These algorithms enable us to determine by means of an iterative refinement process solutions or inverses which are "correct to working accuracy" provided A is not "singular to working accuracy". Iterative refinement ultimately gives a computed solution $\bar{x}$ such that

$$\|x - \bar{x}\|_\infty / \|x\|_\infty \leq k \times macheps,$$

where *macheps* is the machine precision and k is a constant of the order of unity.

In order to achieve this high accuracy these algorithms employ a procedure *innerprod* which calculates to double precision the inner-product of two single precision vectors. It is assumed that this is done in machine code. For consistency alternative versions of *choldet1* and *cholsol1* are given in which all inner-products are accumulated in double precision. For computers on which the accumulation of inner-products is inefficient, the standard versions of *choldet1* and *cholsol1* may be used, but the accumulation of inner-products in the computation of the residuals is an *essential* feature of the iterative refinement process.

The procedures *acc solve* and *acc inverse* are of great value even when more accurate solutions are not required. If $\bar{x}^{(1)}$ is the original solution and $\bar{x}^{(2)}$ is the first improved solution, then $\bar{x}^{(2)} - \bar{x}^{(1)}$ usually gives a very good estimate of the effect of end figure changes in the data (i.e. end figure changes from the point of view of *machine* word length) and enables one to assess the significance of the solution when the given matrix A is subject to errors. On a computer which accumulates double precision inner-products efficiently this important information is obtained very economically as regards computing time. However, it is necessary to retain A and the right-hand sides, though this could be done on the backing store.

b) The LDL^T Factorization

The LDL^T factorization is used in the set of procedures *symdet, symsol* and *syminversion*. The factorization of A is performed in *symdet* and $det(A)$ is produced as a by-product. After using *symdet*, systems of equations having A as the matrix of coefficients may be solved using *symsol* and the inverse of A may be found by using *syminversion*. It may be mentioned that algorithms analogous to *choldet2* etc. could be produced using the LDL^T decomposition and that versions of *acc solve* and *acc inverse* could be based on *symdet* and *symsol*. There is little to choose between corresponding algorithms based on the LL^T and LDL^T decompositions.

c) Gauss-Jordan Inversion

Two algorithms *gjdef1* and *gjdef2* are given for inverting a positive definite matrix *in situ*. The second of these economises in storage in much the same way as *choldet2*. The Gauss-Jordan algorithm should not be used for solving linear equations since it is less efficient than the *choldet-cholsol* combinations. For matrix inversion it provides an elegant algorithm of much the same accuracy as the *cholinversions*.

1*

3.2. Positive Definite Band Matrices

Many physical problems give rise to systems of equations for which the matrix is positive definite and of band form. For such matrices the Cholesky factorization is very effective since it preserves this band form. The procedure *chobanddet* gives the Cholesky factors and the determinant of such a matrix, A, and subsequent use of *chobandsol* provides the solution of systems of equations having A as the matrix of coefficients. It is rare for the inverse of such a matrix to be required (the inverse will be full even when A is of band form) and hence no procedure *choband inverse* is provided. Iterative refinement of the solution could be provided by a procedure analogous to *acc solve*. An alternative method for the solution of banded equations is discussed in the next section.

3.3. Sparse Positive Definite Matrices of Special Form

Many problems in mathematical physics give rise to sparse positive definite matrices in which the non-zero elements a_{ij} are defined in a systematic way as functions of i and j. This is particularly true of systems arising from finite difference approximations to partial differential equations. If the matrix is sufficiently sparse and the functions of i and j are sufficiently simple the procedure *cg* based on the conjugate gradient algorithm may be the most efficient method of solution.

Although the conjugate gradient method is sometimes thought of as an iterative method it would, with exact computation, give the exact solution in a finite number of steps. It has therefore been included, though *true* iterative methods have been excluded.

The *cg* algorithm is optimal in respect of storage since the matrix A is, in general, not stored at all. All that is needed is an auxiliary procedure for determining $A x$ from a given vector x. When A is of very high order the use of the conjugate gradient method (or of one of the iterative methods not included in this volume) may be mandatory since it may not be possible to store A even if it is of narrow band form. This is a common situation in connexion with systems arising from partial differential equations. They give rise to narrow banded matrices but in addition most of the elements within the band itself are zero and usually the procedure for computing $A x$ is very simple.

It cannot be too strongly emphasized that the procedure *cg* is seldom, if ever, to be preferred to Cholesky type algorithms on the grounds of accuracy. The *cg* method is less competitive in terms of speed if a number of different right-hand sides are involved since each solution proceeds independently and r solutions involve r times as much work. With Cholesky type methods the factorization of A is required only once, however many right-hand sides are involved.

4. Non-Positive Definite Symmetric Matrices

There are no algorithms in this volume designed to take advantage of symmetry in the solution of dense symmetric matrices and they must be solved using the procedures described in later sections. The two sets of procedures *bandet1*, *bansol1* and *bandet2*, *bansol2* may be used to deal with symmetric band matrices but they do not take advantage of symmetry and indeed this is destroyed during the factorization. Either of the procedures *bandet1, 2* may be used to factorize A and to find its determinant and subsequently *bansol1, 2* may be used to solve sets of equations having A as the matrix of coefficients. The first set of procedures is appreciably more efficient than the second, but *when A is symmetric, bandet2 also computes the number of positive eigenvalues*. It can therefore be used to find the number of eigenvalues greater than a given value p by working with $A - pI$.

5. Non-Hermitian Matrices

The solution of a system of equations having a dense non-Hermitian matrix of coefficients is a more severe problem than that for a symmetric positive definite system of equations. In particular the problem of scaling presents difficulties which have not been solved in an entirely satisfactory way. The procedures given here are in this respect something of a compromise between what is desirable and what is readily attainable.

5.1. Dense Matrices (Real and Complex)

The procedures depend on the LU factorization of the matrix A, where L is lower triangular and U is unit upper triangular. *Partial pivoting* is used in the factorization so that the algorithms effectively "correspond" to Gaussian elimination in which the pivotal element at each stage is chosen to be the largest element in the leading column of the remaining matrix, though this would give rise to a unit lower triangular L.

The procedure *unsymdet* produces the LU factorization of a real dense matrix A and produces the determinant as a by-product; subsequently any number of linear systems having A as the matrix of coefficients may be solved by using the procedure *unsymsol*. The procedures *compdet* and *compsol* perform the analogous operations for a complex matrix A.

To cover the cases when A is too ill-conditioned for the computed solution to be sufficiently accurate the procedures *unsym acc solve* and *cx acc solve* perform iterative refinement of the solution, using the initial factorization of A given by *unsymdet* or *compdet* respectively. Provided A is not almost singular to working accuracy iterative refinement will ultimately give a solution $\bar{x}$ which is correct to working accuracy, i.e. such that $\|x - \bar{x}\|_\infty / \|x\|_\infty \leq k \times macheps$, where *macheps* is the machine precision and k is a constant of order unity. Notice that "small" components of x may well have a low relative accuracy. We stress once again that one step of iterative refinement is of great value in giving an estimate of the sensitivity of the system of equations. If $\bar{x}^{(1)}$ and $\bar{x}^{(2)}$ are the first two solutions, then $\bar{x}^{(2)} - \bar{x}^{(1)}$ usually gives a good estimate of the effect of end figure changes in the data. (See earlier comments in 3.1 (a).)

5.2. Unsymmetric Band Matrices

The procedures *bandet1* and *bansol1* are designed to deal with real unsymmetric matrices of band form. They deal efficiently with banded matrices having a different number of non-zero diagonal columns on the two sides. An

LU decomposition with partial pivoting is used. (They can be used for symmetric band matrices; see 4.) *bandet1* gives the factorization of A and its determinant and subsequently *bansol1* may be used to give solutions corresponding to any number of right-hand sides.

bandet2 and *bansol2* can also be used for unsymmetric matrices but are less efficient; they were designed primarily for use with symmetric band matrices (see 4.).

6. Least Squares and Related Problems

a) The Standard Least Squares Problem

There are effectively two procedures designed for the solution of the least squares problem associated with a real $m \times n$ matrix, A, such that $m \geq n$ and $rank(A) = n$.

The first of these, *least squares solution*, is based on the determination of an orthogonal matrix Q such that $Q\tilde{A} = R$ where R is an $m \times n$ upper triangular matrix and $\tilde{A}$ is A with its columns suitably permuted as determined by a column pivoting strategy. The matrix Q is obtained as the product of n elementary orthogonal matrices of the type $I - 2ww^T$, where $\|w\|_2 = 1$. The vector $\tilde{x}$ such that $\|b - \tilde{A}\tilde{x}\|_2$ is a minimum is then determined from the equation $R\tilde{x} = Qb$. Any number of right-hand sides may be treated simultaneously. If the m columns of the $m \times m$ unit matrix are taken successively as right-hand sides the $n \times m$ matrix $(A^H A)^{-1} A^H$ is produced; this is the pseudo-inverse of A, the latter being assumed to be of rank n.

The procedure incorporates an iterative refinement process analogous to that described earlier for $n \times n$ systems of linear equations. For a compatible $m \times n$ system, i.e. a system for which $\min \|b - Ax\|_2 = 0$, iterative refinement leads to a solution which is "correct to working accuracy", but for an incompatible system the final accuracy may fall appreciably short of this. Least squares solutions are frequently required for systems which are far from compatible and for such systems iterative refinement serves little purpose.

The procedure can be used with $m = n$, in which case it gives the solution of a non-singular system of linear equations. It can therefore be used in the obvious way to perform matrix inversion. Although the orthogonal factorization is *slightly* superior to the LU factorization as regards numerical stability this superiority is hardly such as to justify the use of *least squares solution* in place of *unsymdet* and *unsymsol* when one is primarily interested in non-singular $n \times n$ systems.

The alternative to *least squares solution* is again based on an orthogonal triangularization of A, effectively using a variant of the Gram-Schmidt process. This gives an orthogonal basis for the columns of A and each right-hand side b is expressed in terms of this basis and a residual vector r which is orthogonal to it.

There is a set of four procedures based on this factorization. *ortholin1* gives the solution of the least squares problem $AX = B$ where B consists of k right-hand sides. *ortholin2* is provided for the special case when there is only one right-hand side, while *ortho1* gives the inverse of an $n \times n$ matrix. These three procedures are all provided with an iterative refinement section. The fourth procedure *ortho2* gives the inverse of an $n \times n$ matrix without iterative refinement and economises on storage by overwriting the inverse on the space originally holding A. The speed and performance of these procedures is roughly comparable with that of *least squares solution*. In particular iterative refinement does not generally give solutions correct to working accuracy for incompatible systems of $m \times n$ equations.

b) The General Least Squares Problem and Pseudo-Inverses

When the $m \times n$ matrix A is of rank less than n, the least squares problem does not have a unique solution and one is often interested in the solution x of the least squares problem such that $\|x\|_2$ is a minimum. This solution is given by $x = A^+ b$ where A^+ is the pseudo-inverse of A. There are two procedures associated with this problem, both based on the *singular value decomposition* of the matrix A. For the case $m \geq n$ we may express this decomposition in the form $A = U \Sigma V^T$ where $U^T U = V^T V = V V^T = I_n$ and $\Sigma = \mathrm{diag}\,(\sigma_1, \ldots, \sigma_n)$, the σ_i being the *singular values* of A. (U is an $m \times n$ matrix with orthogonal columns and V is an orthogonal matrix.) The corresponding expression for the pseudo-inverse is $V \Sigma^+ U^T$ where $\Sigma^+ = \mathrm{diag}\,(\sigma_1^+, \ldots, \sigma_n^+)$ and

$$\sigma_i^+ = \begin{cases} \sigma_i^{-1} & \text{if } \sigma_i > 0 \\ 0 & \text{if } \sigma_i = 0. \end{cases}$$

The singular value decomposition gives the most reliable determination of the "rank" of a matrix A, the elements of which are subject to errors. The computed singular values are always exact for some matrix which is equal to A almost to working accuracy and these values are of decisive importance in rank determination. Alternative simpler factorizations can in rare cases give an "incorrect" rank determination.

The procedure *svd* gives the set of singular values and the matrices U and V. The procedure *minfit* gives the solutions $A^+ b$ corresponding to each of p given right-hand sides b. The two procedures may be used to solve a variety of problems related to least squares problems and the calculation of the pseudo-inverse.

7. The Linear Programming Problem

Research in the field of linear programming has had surprisingly little contact with that in the main stream of numerical linear algebra, possibly for historical reasons. This is unfortunate since the two fields have a great deal in common. It is hoped that the inclusion of a linear programming procedure will serve to emphasize this fact and help to bring together the two groups of research workers.

In view of the fact that *lp* is the only procedure concerned with linear programming in this volume, it was felt that a fairly comprehensive treatment was warranted. *lp* may be used to solve the general problem:

Minimise

$$c_0 + c_{-m} x_{-m} + \cdots + c_{-1} x_{-1} + c_1 x_1 + \cdots + c_n x_n$$

subject to the relations

$$x_{-i} + \sum_{k=1}^{n} a_{ik} x_k = b_i, \qquad i = 1, 2, \ldots, m$$

$$x_i \geq 0 \quad (i \in I^+), \qquad x_i = 0 \quad (i \in I^0),$$

where I^+ and I^0 are disjoint subsets of $\{-m, \ldots, -1, 1, \ldots, n\}$.

The procedure is a variant of the simplex method based on a triangular factorization of the current basis A_J of the form $LA_J = R$, where R is a non-singular upper triangular matrix. Numerical stability is preserved by using a variant of the exchange algorithm independently suggested by Golub and Stoer, and developed by Bartels.

It is recognized that linear programming procedures are very machine dependent. Since very large problems are encountered, the use of the backing store is often mandatory and it is also important to deal efficiently with sparse matrices. These last two problems are to some extent bypassed by the use of two sub-procedures ap and p, the first of which deals with the handling of the matrix A and the second with operations on the matrices R and L and the various modified versions $\bar{B}$ of the right-hand sides. Specific versions of ap and p are presented to deal with the simple cases when the matrices $A, R, L, \bar{B}$ are stored as standard rectangular arrays.

Symmetric Decomposition of a Positive Definite Matrix[*]

by R. S. Martin, G. Peters, and J. H. Wilkinson

1. Theoretical Background

The methods are based on the following theorem due to Cholesky [1].

If A is a symmetric positive definite matrix then there exists a real non-singular lower-triangular matrix L such that

$$L L^T = A. \tag{1}$$

Further if the diagonal elements of L are taken to be positive the decomposition is unique.

The elements of L may be determined row by row or column by column by equating corresponding elements in (1). In the row by row determination we have for the i-th row

$$\sum_{k=1}^{j} l_{ik} l_{jk} = a_{ij} \text{ giving } l_{ij} = \left(a_{ij} - \sum_{k=1}^{j-1} l_{ik} l_{jk} \right) \Big/ l_{jj} \quad (j = 1, \ldots, i-1), \tag{2}$$

$$\sum_{k=1}^{i} l_{ik} l_{ik} = a_{ii} \text{ giving } l_{ii} = \left(a_{ii} - \sum_{k=1}^{i-1} l_{ik}^2 \right)^{\frac{1}{2}}. \tag{3}$$

There are thus n square roots and approximately $\frac{1}{6} n^3$ multiplications.

An alternative decomposition may be used in which the square roots are avoided as follows. If we define $\tilde{L}$ by the relation

$$L = \tilde{L} \operatorname{diag} (l_{ii}), \tag{4}$$

where L is the matrix given by the Cholesky factorization, then $\tilde{L}$ exists (since the l_{ii} are positive) and is a *unit* lower-triangular matrix. We have then

$$A = L L^T = \tilde{L} \operatorname{diag}(l_{ii}) \operatorname{diag}(l_{ii}) \tilde{L}^T = \tilde{L} \operatorname{diag}(l_{ii}^2) \tilde{L}^T = \tilde{L} D \tilde{L}^T, \tag{5}$$

where D is a positive diagonal matrix.

This factorization can be performed in n major steps in the i-th of which the i-th row of $\tilde{L}$ and d_i are determined. The corresponding equations are

$$\sum_{k=1}^{j} \tilde{l}_{ik} d_k \tilde{l}_{jk} = a_{ij} \text{ giving } \tilde{l}_{ij} d_j = a_{ij} - \sum_{k=1}^{j-1} \tilde{l}_{ik} d_k \tilde{l}_{jk} \quad (j = 1, \ldots, i-1), \tag{6}$$

$$\sum_{k=1}^{i} \tilde{l}_{ik} d_k \tilde{l}_{ik} = a_{ii} \text{ giving } d_i = a_{ii} - \sum_{k=1}^{i-1} \tilde{l}_{ik} d_k \tilde{l}_{ik} \tag{7}$$

[*] Prepublished in Numer. Math. 7, 362—383 (1965).

since $\tilde{l}_{jj}=1$. Expressed in this form the decomposition appears to take twice as many multiplications as that of CHOLESKY, but if we introduce the auxiliary quantities $\tilde{a}_{ij}$ defined by

$$\tilde{a}_{ij}=\tilde{l}_{ij}\,d_j \tag{8}$$

equations (6) and (7) become

$$\tilde{a}_{ij}=a_{ij}-\sum_{k=1}^{j-1}\tilde{a}_{ik}\,\tilde{l}_{jk}\qquad(j=1,\ldots,i-1), \tag{9}$$

$$d_i=a_{ii}-\sum_{k=1}^{i-1}\tilde{a}_{ik}\,\tilde{l}_{ik}. \tag{10}$$

We can therefore determine the $\tilde{a}_{ij}$ successively and then use them to determine the $\tilde{l}_{ij}$ and d_i. Notice that the $\tilde{a}_{ij}$ corresponding to the i-th row are not required when dealing with subsequent rows. The number of multiplications is still approximately $\frac{1}{6}n^3$ and there are no square roots.

Either factorization of A enables us to calculate its determinant since we have

$$\det(A)=\det(L)\det(L^T)=\left[\prod_{i=1}^{n}l_{ii}\right]^2 \tag{11}$$

and

$$\det(A)=\det(\tilde{L})\det(D)\det(\tilde{L}^T)=\prod_{i=1}^{n}d_i. \tag{12}$$

We can also compute the solution of the set of equations

$$A\,x=b \tag{13}$$

corresponding to any given right-hand side. In fact if

$$L\,y=b\quad\text{and}\quad L^T x=y \tag{14}$$

we have

$$A\,x=LL^T x=L\,y=b. \tag{15}$$

Equations (15) may be solved in the steps

$$y_i=\left(b_i-\sum_{k=1}^{i-1}l_{ik}\,y_k\right)\Big/l_{ii}\qquad(i=1,\ldots,n), \tag{16}$$

$$x_i=\left(y_i-\sum_{k=i+1}^{n}l_{ki}\,x_k\right)\Big/l_{ii}\qquad(i=n,\ldots,1) \tag{17}$$

involving n^2 multiplications in all and $2n$ divisions. Similarly if

$$\tilde{L}\,y=b,\qquad\tilde{L}^T x=D^{-1}y, \tag{18}$$

we have

$$A\,x=\tilde{L}D\tilde{L}^T x=\tilde{L}DD^{-1}y=b. \tag{19}$$

Equations (18) may be solved in the steps

$$y_i=b_i-\sum_{k=1}^{i-1}\tilde{l}_{ik}\,y_k\qquad(i=1,\ldots,n), \tag{20}$$

$$x_i = y_i/d_i - \sum_{k=i+1}^{n} \tilde{l}_{ki} x_k \qquad (i=n, \ldots, 1), \tag{21}$$

and there are again n^2 multiplications but only n divisions.

The inverse of A may be obtained by solving the equations

$$A x = e_i \qquad (i=1, \ldots, n), \tag{22}$$

the solution corresponding to the right-hand side e_i being the i-th column of A^{-1}. However, it is easier to take full advantage of symmetry and of the special form of the right-hand sides by observing that

$$A^{-1} = (L^T)^{-1} L^{-1} = (\tilde{L}^T)^{-1} D^{-1} \tilde{L}^{-1}. \tag{23}$$

With the Cholesky decomposition, denoting L^{-1} (a lower-triangular matrix) by P we have

$$p_{ii} = 1/l_{ii}, \tag{24}$$

$$p_{ji} = -\left(\sum_{k=i}^{j-1} l_{jk} p_{ki}\right)\bigg/ l_{jj} \qquad (j=i+1, \ldots, n). \tag{25}$$

In deriving A^{-1} from $P^T P$ we need compute only the lower triangle of elements and for these we have

$$(A^{-1})_{ji} = (P^T P)_{ji} = \sum_{k=j}^{n} p_{kj} p_{ki} \qquad (j=i, \ldots, n). \tag{26}$$

Notice that after $(A^{-1})_{ii}$ is computed p_{ji} is no longer required.

Similarly if we denote $\tilde{L}^{-1}$ (a *unit* lower-triangular matrix) by Q we have

$$q_{ii} = 1, \tag{27}$$

$$q_{ji} = -\sum_{k=i}^{j-1} \tilde{l}_{jk} q_{ki} \qquad (j=i+1, \ldots, n), \tag{28}$$

$$(A^{-1})_{ji} = \sum_{k=j}^{n} q_{kj} q_{ki}/d_k \qquad (j=i, \ldots, n) \tag{29}$$

and again after $(A^{-1})_{ii}$ is computed q_{ji} is no longer wanted.

As far as arithmetic operations are concerned there is marginally more work with the Cholesky decomposition but we shall see that it is marginally simpler from the point of view of the organization involved. Both decompositions are required in practice.

2. Applicability

The method of Cholesky may be used for positive definite matrices only. When the non-zero elements of A are in a narrow band centred on the diagonal, *chobanddet* and *chobandsol* [2] should be used. If A is symmetric but not positive definite no factorizations of types (1) and (5) may exist. An example is the matrix

$$\begin{bmatrix} 0 & 1 \\ 1 & 0 \end{bmatrix}; \tag{30}$$

notice that the absence of a factorization need not imply that A is ill-conditioned.

If A is non-negative semi-definite then real L exist satisfying (1) but will be singular and will not in general be unique. For example we have

$$\begin{bmatrix} 0 & 0 \\ 0 & 2 \end{bmatrix} = \begin{bmatrix} 0 & 0 \\ x & y \end{bmatrix} \begin{bmatrix} 0 & x \\ 0 & y \end{bmatrix} \tag{31}$$

provided only $x^2 + y^2 = 2$. Similarly

$$\begin{bmatrix} 0 & 0 \\ 0 & 2 \end{bmatrix} = \begin{bmatrix} 1 & 0 \\ x & 1 \end{bmatrix} \begin{bmatrix} 0 & \\ & 2 \end{bmatrix} \begin{bmatrix} 1 & x \\ 0 & 1 \end{bmatrix} \tag{32}$$

for all x.

If A has at least one negative eigenvalue then no real L can exist but there may be complex L with this property. When such a complex factorization exists the alternative decomposition exists with real $\tilde{L}$ and D but D has some negative diagonal elements. We have then

$$A = \tilde{L} D \tilde{L}^T = \tilde{L} D^{\frac{1}{2}} (\tilde{L} D^{\frac{1}{2}})^T \tag{33}$$

and since D has some negative elements $\tilde{L} D^{\frac{1}{2}} = L$ has some columns consisting of pure imaginary elements. The existence of the $\tilde{L} D \tilde{L}^T$ decomposition in the real field is an advantage in some cases. It should be emphasized, however, that it is only in the case when A is positive definite that the numerical stability of either decomposition is guaranteed.

Two complete sets, each consisting of three procedures, are given for the Cholesky factorization. In the first set (I) the original matrix A is retained while in the second set (II) L is overwritten on A as it is produced. The storage requirements of the second set are therefore about half those of the first, but since A is not retained we cannot compute the residuals, improve the computed solution or improve the computed inverse using the second set.

For the alternative factorization only the procedures corresponding to the first Cholesky set (III) are given. The details of the others should be obvious from these.

The procedures in each Cholesky set are

choldet. This produces L and $\det(A)$.

cholsol. Corresponding to r given right-hand sides b this produces the solutions of the equations $A x = b$. With either set *cholsol* must be preceded by *choldet* so that the Cholesky decomposition of A is already available. *cholsol* may be used any number of times after one use of *choldet*.

cholinversion. In both sets this procedure is independent of *choldet* and *cholsol*. It should be used only when *explicit* information on the inverse of A is required. When the only requirements are the solutions of $A x = b$ corresponding to a number of right-hand sides b it is more economical to use *choldet* followed by *cholsol*.

3. Formal Parameter List

3.I. Set I.

3.I.1 Input to procedure *choldet 1*

n order of the matrix A.

a elements of the positive definite matrix A. They must be given as described in Section 5.

Output of procedure *choldet 1*

a   elements of the lower triangle L of the Cholesky decomposition of A stored as described in Section 5.

p   a vector consisting of the reciprocals of diagonal elements of L.

d1 and *d2*   two numbers which together give the determinant of A.

fail   This is the exit used when A, possibly as the result of rounding errors, is not positive definite.

3.I.2 Input to procedure *cholsol 1*

n   order of the matrix A.

r   number of right-hand sides for which $A x = b$ is to be solved.

a   elements of the lower triangle L of the Cholesky decomposition of a positive definite matrix A as produced by procedure *choldet 1*.

p   the reciprocals of the diagonal elements of L as produced by procedure *choldet 1*.

b   the $n \times r$ matrix consisting of the r rigth-hand sides.

Output of procedure *cholsol 1*

x   the $n \times r$ matrix consisting of the r solution vectors.

3.I.3 Input to procedure *cholinversion 1*

n   order of the matrix A.

a   elements of the positive definite matrix A. They must be given as described in Section 5.

Output of procedure *cholinversion 1*

a   elements of A^{-1} stored as described in Section 5.

fail   This is the exit used when A, possibly as the result of rounding errors, is not positive definite.

3.II. Set II.

3.II.1 Input to procedure *choldet 2*

n   order of the matrix A.

a   elements of the positive definite matrix A stored as a linear array as described in Section 5.

Output of procedure *choldet 2*

a   elements of the lower triangle L of the Cholesky decomposition of A stored as described in Section 5.

d1 and *d2*   two numbers which together give the determinant of A.

fail   This is the exit used when A, possibly as the result of rounding errors, is not positive definite.

3.II.2 Input to procedure *cholsol 2*

n   order of the matrix A.

r   number of rigth-hand sides for which $A x = b$ is to be solved.

a   elements of the lower triangle L of the Cholesky decomposition of a positive definite matrix as produced by procedure *choldet 2*.

b   the $n \times r$ matrix consisting of the r right-hand sides.

Output of *cholsol 2*

b   the $n \times r$ matrix consisting of the *r* solution vectors.

3.II.3 Input to procedure *cholinversion 2*

n   order of the matrix *A*.
a   elements of the matrix *A* stored as a linear array as described in Section 5.

Output of procedure *cholinversion 2*

a   elements of A^{-1} stored as described in Section 5.
fail   This is the exit used when *A*, possibly as a result of rounding errors, is not positive definite.

3.III. Set III.

3.III.1 Input to procedure *symdet*

n   order of the matrix *A*.
a   elements of the positive definite matrix *A*. They must be given as described in Section 5.

Output of procedure *symdet*

a   elements of the lower triangle $\tilde{L}$ of the $\tilde{L}D\tilde{L}^{T}$ decomposition of *A* stored as described in Section 5.
p   a vector consisting of the reciprocals of the diagonal elements of *D*.
d1 and *d2*   two numbers which together give the determinant of *A*.
fail   This is the exit used when *A*, possibly as the result of rounding errors, is singular.

3.III.2 Input to procedure *symsol*

n   order of the matrix *A*.
r   number of right-hand sides for which $A x = b$ is to be solved.
a   elements of $\tilde{L}$ the lower triangle of the $\tilde{L}D\tilde{L}^{T}$ decomposition of a positive definite matrix *A* as produced by procedure *symdet*.
p   the reciprocals of the diagonal elements of *D* as produced by procedure *symdet*.
b   elements of the $n \times r$ matrix consisting of the *r* right-hand sides.

Output of procedure *symsol*

x   the $n \times r$ matrix consisting of the *r* solution vectors.

3.III.3 Input to procedure *syminversion*

n   order of the matrix *A*.
a   elements of the positive definite matrix *A*. This must be given as described in Section 5.

Output of procedure *syminversion*

a   elements of A^{-1} stored as described in Section 5.
fail   This is the exit used when *A*, possibly as the result of rounding errors, is singular.

4. ALGOL Programs

4.I. Set I.

procedure *choldet 1* (*n*) *trans*: (*a*) *result*: (*p, d1, d2*) *exit*: (*fail*);
value *n*; **integer** *d2, n*; **real** *d1*; **array** *a, p*; **label** *fail*;
comment The upper triangle of a positive definite symmetric matrix, *A*, is
stored in the upper triangle of an *n* × *n* array *a* [*i, j*], *i* = 1(1) *n*, *j* = 1(1) *n*.
The Cholesky decomposition *A* = *LU*, where *U* is the transpose of *L*,
is performed and stored in the remainder of the array *a* except for
the reciprocals of the diagonal elements which are stored in *p* [*i*],
i = 1(1) *n*, instead of the elements themselves. *A* is retained so that
the solution obtained can be subsequently improved. The determinant,
d1 × 2 ↑ *d2*, of *A* is also computed. The procedure will fail if *A*, modified
by the rounding errors, is not positive definite;

```
begin integer i, j, k; real x;
      d1 := 1;
      d2 := 0;
      for i := 1 step 1 until n do
      for j := i step 1 until n do
      begin  x := a[i, j];
             for k := i − 1 step − 1 until 1 do
             x := x − a[j, k] × a[i, k];
             if j = i then
             begin  d1 := d1 × x;
                    if x = 0 then
                    begin  d2 := 0;
                           go to fail
                    end;
      L1:      if abs(d1) ≥ 1 then
                    begin  d1 := d1 × 0.0625;
                           d2 := d2 + 4;
                           go to L1
                    end;
      L2:      if abs(d1) < 0.0625 then
                    begin  d1 := d1 × 16;
                           d2 := d2 − 4;
                           go to L2
                    end;
                    if x < 0 then go to fail;
                    p[i] := 1/sqrt(x)
             end
             else
             a[j, i] := x × p[i]
      end ij
end choldet 1;
```

procedure *cholsol 1* (*n*) *data*: (*r, a, p, b*) *result*: (*x*);
value *r, n*; **integer** *r, n*; **array** *a, b, p, x*;

comment Solves $A\,x=b$, where A is a positive definite symmetric matrix and
b is an $n\times r$ matrix of r right-hand sides. The procedure *cholsol 1*
must be preceded by *choldet 1* in which L is produced in $a[i,j]$ and
$p[i]$, from A. $A\,x=b$ is solved in two steps, $L\,y=b$ and $U\,x=y$. The
matrix b is retained in order to facilitate the refinement of x, but x
is overwritten on y. However, x and b can be identified in the call
of the procedure;

```
begin integer i, j, k; real z;
    for j:=1 step 1 until r do
    begin comment solution of L y=b;
        for i:=1 step 1 until n do
        begin z:= b[i,j];
            for k:=i−1 step −1 until 1 do
            z:=z−a[i, k]×x[k,j];
            x[i,j]:=z×p[i]
        end i;
        comment solution of U x=y;
        for i:=n step −1 until 1 do
        begin z:= x[i,j];
            for k:=i+1 step 1 until n do
            z:=z−a[k, i]×x[k,j];
            x[i,j]:=z×p[i]
        end i
    end j
end cholsol 1;
```

procedure *cholinversion 1* (n) *trans*: (a) *exit*: $(fail)$;
value n; **integer** n; **array** a; **label** *fail*;

comment The upper triangle of a positive definite symmetric matrix, A, is
stored in the upper triangle of an $(n+1)\times n$ array $a[i,j]$, $i=1(1)\,n+1$,
$j-1(1)\,n$. The Cholesky decomposition $A=LU$, where U is the transpose of L, is performed and L is stored in the remainder of the array a.
The reciprocals of the diagonal elements are stored instead of the
elements themselves. L is then replaced by its inverse and this in
turn is replaced by the lower triangle of the inverse of A. A is retained
so that the inverse can be subsequently improved. The procedure
will fail if A, modified by the rounding errors, is not positive definite;

```
begin integer i, j, k, i1, j1;
    real z, x, y;
    comment formation of L;
    for i:=1 step 1 until n do
    begin i1:=i+1;
        for j:=i step 1 until n do
        begin j1:=j+1;
            x:= a[i,j];
```

```
         for k := i−1 step −1 until 1 do
         x := x − a[j1, k] × a[i1, k];
         if j = i then
         begin if x ≤ 0 then go to fail;
                    a[i1, i] := y := 1/sqrt(x)
         end
         else
         a[j1, i] := x × y
    end j
end i;
comment inversion of L;
for i := 1 step 1 until n do
for j := i+1 step 1 until n do
begin z := 0;
      j1 := j+1;
      for k := j−1 step −1 until i do
      z := z − a[j1, k] × a[k+1, i];
      a[j1, i] := z × a[j1, j]
end ij;
comment calculation of the inverse of A;
for i := 1 step 1 until n do
for j := i step 1 until n do
begin z := 0;
      j1 := n+1;
      for k := j+1 step 1 until j1 do
      z := z + a[k, j] × a[k, i];
      a[j+1, i] := z
end ij
end cholinversion 1;
```

4.II. Set II.

procedure *choldet 2 (n) trans: (a) result: (d1, d2) exit: (fail);*

value n; **integer** $d2, n$; **real** $d1$; **array** a; **label** *fail*;

comment The lower triangle of a positive definite symmetric matrix, A, is stored row by row in a $1 \times n(n+1) \div 2$ array $a[i]$, $i = 1(1)n(n+1) \div 2$. The Cholesky decomposition $A = LU$, where U is the transpose of L, is performed and L is overwritten on A. The reciprocals of the diagonal elements are stored instead of the elements themselves. The determinant, $d1 \times 2 \uparrow d2$, of A is also computed. The procedure will fail if A, modified by the rounding errors, is not positive definite;

```
begin integer i, j, k, p, q, r;
      real x;
      d1 := 1;
      d2 := 0;
      p := 0;
      for i := 1 step 1 until n do
```

```
begin q := p+1;
        r := 0;
        for j := 1 step 1 until i do
        begin x := a[p+1];
                for k := q step 1 until p do
                begin r := r+1;
                        x := x - a[k] × a[r]
                end k;
                r := r+1;
                p := p+1;
                if i=j then
                begin d1 := d1×x;
                        if x=0 then
                        begin d2 := 0;
                                go to fail
                        end;
L1:             if abs(d1)≥1 then
                        begin d1 := d1×0.0625;
                                d2 := d2+4;
                                go to L1
                        end;
L2:             if abs(d1)<0.0625 then
                        begin d1 := d1×16;
                                d2 := d2-4;
                                go to L2
                        end;
                        if x<0 then go to fail;
                        a[p] := 1/sqrt(x)
                end
                else
                a[p] := x × a[r]
        end j
    end i
end choldet 2;


procedure cholsol 2 (n) data: (r, a) trans: (b);
value n, r; integer n, r; array a, b;
comment Solves A x=b, where A is a positive definite symmetric matrix and
        b is an n×r matrix of r right-hand sides. The procedure cholsol 2
        must be preceded by choldet 2 in which L is produced in a[i], from A.
        A x=b is solved in two steps, Ly=b and Ux=y. The matrices y
        and then x are overwritten on b;
begin integer i, j, k, p, q, s;
        real x;
        for j := 1 step 1 until r do
```

```
        begin comment solution of Ly=b;
            p:=1;
            for i:=1 step 1 until n do
            begin q:=i−1;
                x:=b[i,j];
                for k:=1 step 1 until q do
                begin x:=x−a[p]×b[k,j];
                    p:=p+1
                end k;
                b[i,j]:=x×a[p];
                p:=p+1
            end i;
            comment solution of Ux=y;
            for i:=n step −1 until 1 do
            begin s:=p:=p−1;
                q:=i+1;
                x:=b[i,j];
                for k:=n step −1 until q do
                begin x:=x−a[s]×b[k,j];
                    s:=s−k+1
                end k;
                b[i,j]:=x×a[s]
            end i
        end j
end cholsol 2;
```

procedure cholinversion 2 (n) trans: (a) exit: (fail);
value n; integer n; array a; label fail;
comment The lower triangle of a positive definite matrix, A, is stored row by
row in a $1 \times n(n+1) \div 2$ array $a[i]$, $i=1(1)n(n+1) \div 2$. The Cholesky
decomposition $A=LU$, where U is the transpose of L, is performed
and L is overwritten on A. The reciprocals of the diagonal elements
are stored instead of the elements themselves. L is then replaced by
its inverse and this in turn is replaced by the lower triangle of the
inverse of A. The procedure will fail if A, modified by the rounding
errors, is not positive definite;

```
begin integer i, j, k, p, q, r, s, t, u;
    real x, y;
    comment formation of L;
    p:=0;
    for i:=1 step 1 until n do
    begin q:=p+1;
        r:=0;
        for j:=1 step 1 until i do
        begin x:=a[p+1];
            for k:=q step 1 until p do
```

2*

```
            begin r := r + 1;
                  x := x − a[k] × a[r]
            end k;
            r := r + 1;
            p := p + 1;
            if i = j then
            begin if x ≤ 0 then go to fail;
                  a[p] := 1/sqrt(x)
            end
            else
            a[p] := x × a[r]
      end j
end i;
comment inversion of L;
p := r := t := 0;
for i := 2 step 1 until n do
begin p := p + 1;
      r := r + i;
      y := a[r + 1];
      for j := 2 step 1 until i do
      begin p := p + 1;
            s := t := t + 1;
            u := i − 2;
            x := 0;
            for k := r step − 1 until p do
            begin x := x − a[k] × a[s];
                  s := s − u;
                  u := u − 1
            end k;
            a[p] := x × y
      end j
end i;
comment calculation of the inverse of A;
p := 0;
for i := 1 step 1 until n do
begin r := p + i;
      for j := 1 step 1 until i do
      begin s := r;
            t := p := p + 1;
            x := 0;
            for k := i step 1 until n do
            begin x := x + a[s] × a[t];
                  s := s + k;
                  t := t + k
            end k;
            a[p] := x
```

```
            end j
        end i
end cholinversion 2;
```

4.III. Set III.

procedure symdet (n) trans: (a) result: (p, d1, d2) exit: (fail);
value n; **integer** d2, n; **array** a, p; **real** d1; **label** fail;
comment The upper triangle of a positive definite symmetric matrix, A, is
stored in the upper triangle of an $n \times n$ array $a[i, j], i=1(1)n, j=1(1)n$.
The decomposition $A = LDU$, where U is the transpose of L, which is
a unit lower triangular matrix, and D is a diagonal matrix, is performed and L is stored in the remainder of the array a, omitting the
unit diagonal, with the reciprocals of the elements of D stored in the
array $p[i], i=1(1)n$. A is retained so that the solution obtained can
be subsequently improved. The determinant, $d1 \times 2 \uparrow d2$, of A is also
computed. The procedure will fail if A, modified by the rounding
errors, is singular;

```
begin integer i, j, k:
      real x, y, z;
      d1 := 1;
      d2 := 0;
      for i := 1 step 1 until n do
      for j := 1 step 1 until i do
      begin x := a[j, i];
            if i=j then
            begin for k := j − 1 step − 1 until 1 do
                  begin y := a[i, k];
                        z := a[i, k] := y × p[k];
                        x := x − y × z
                  end k;
                  d1 := d1 × x;
                  if x=0 then
                  begin d2 := 0;
                        go to fail
                  end;
      L1:     if abs(d1) ≧ 1 then
                  begin d1 := d1 × 0.0625;
                        d2 := d2 + 4;
                        go to L1
                  end;
      L2:     if abs(d1) < 0.0625 then
                  begin d1 := d1 × 16;
                        d2 := d2 − 4;
                        go to L2
                  end;
                  p[i] := 1/x
            end
```

```
          else
          begin for k := j − 1 step − 1 until 1 do
                   x := x − a[i, k] × a[j, k];
                   a[i, j] := x
          end
     end ij
end symdet;
```

procedure symsol (n) data:(r, a, p, b) result:(x);
value n, r; **integer** n, r; **array** a, p, b, x;
comment Solves $A x = b$, where A is a positive definite symmetric matrix and
b is an $n \times r$ matrix of r right-hand sides. The procedure symsol must
be preceded by symdet in which L is produced in $a[i,j]$ and D is
produced in $p[i]$, from A. $A x = b$ is solved in two steps, $L y = b$ and
$DU x = y$. The matrix b is retained in order to facilitate the refinement
of x, but x is overwritten on y. However, x and b can be identified in
the call of the procedure;

```
begin integer i, j, k;
      real y;
      for j := 1 step 1 until r do
      begin comment solution of L y = b;
              for i := 1 step 1 until n do
              begin y := b[i, j];
                      for k := i − 1 step − 1 until 1 do
                      y := y − a[i, k] × x[k, j];
                      x[i, j] := y
              end i;
              comment solution of DU x = y;
              for i := n step − 1 until 1 do
              begin y := x[i, j] × p[i];
                      for k := i + 1 step 1 until n do
                      y := y − a[k, i] × x[k, j];
                      x[i, j] := y
              end i
      end j
end symsol;
```

procedure syminversion (n) trans:(a) exit:(fail);
value n; **integer** n; **array** a; **label** fail;
comment The upper triangle of a positive definite symmetric matrix, A, is
stored in the upper triangle of an $(n + 1) \times n$ array $a[i,j]$, $i = 1(1)n + 1$,
$j = 1(1)n$. The decomposition $A = LDU$, where U is the transpose of L,
which is a unit lower triangular matrix, and D is a diagonal matrix,
is performed and L and D together are stored in the remainder of the
array a, omitting the unit diagonal. The reciprocals of the elements
of D are stored instead of the elements themselves. L is then replaced
by its inverse and then both D and the inverse of L are replaced by

the lower triangle of the inverse of A. A is retained so that the inverse
can subsequently be improved. The procedure will fail if A, modified
by the rounding errors, is singular;

```
begin integer i, j, k, i1, j1;
      real x, y, z;
      comment formation of L;
      for i := 1 step 1 until n do
      begin i1 := i+1;
            for j := 1 step 1 until i do
            begin j1 := j+1;
                  x := a[j, i];
                  if i = j then
                  begin for k := j−1 step −1 until 1 do
                        begin y := a[i1, k];
                              z := a[i1, k] := y × a[k+1, k];
                              x := x − y × z
                        end k;
                        if x = 0 then go to fail;
                        a[i1, i] := 1/x
                  end
                  else
                  begin for k := j−1 step −1 until 1 do
                        x := x − a[i1, k] × a[j1, k];
                        a[i1, j] := x
                  end
            end j
      end i;
      comment inversion of L;
      for i := 2 step 1 until n do
      begin i1 := i+1;
            for j := 2 step 1 until i do
            begin j1 := j−1;
                  x := − a[i1, j1];
                  for k := i−1 step −1 until j do
                  x := x − a[i1, k] × a[k+1, j1];
                  a[i1, j1] := x
            end j
      end i;
      comment calculation of the inverse of A;
      for j := 1 step 1 until n do
      for i := j step 1 until n do
      begin i1 := i+1;
            x := a[i1, j];
            if i ≠ j then
            begin for k := i1 step 1 until n do
                  x := x + a[k+1, i] × a[k+1, j]
            end
```

```
    else
    for k := i1 step 1 until n do
    begin y := a[k+1, j];
           z := a[k+1, j] := a[k+1, k] × y;
           x := x + y × z
    end k;
    a[i1, j] := x
end ij
end syminversion;
```

5. Organisational and Notational Details

5.I. Set I.

In *choldet 1* the matrix A is an $n \times n$ array but only the upper-triangle of elements need to be correct since the sub-diagonal elements are not used. The matrix L, apart from its diagonal elements, is stored as the lower half of the square array A. In subsequent calculations it is the reciprocals of the l_{ii} which are required and hence the vector p with element $1/l_{ii}$ is stored separately. The storage is therefore in the form illustrated by

$$
\begin{bmatrix}
a_{11} & a_{12} & a_{13} & a_{14} \\
l_{21} & a_{22} & a_{23} & a_{24} \\
l_{31} & l_{32} & a_{33} & a_{34} \\
l_{41} & l_{42} & l_{43} & a_{44}
\end{bmatrix},
\begin{bmatrix}
l_{11}^{-1} \\
l_{22}^{-1} \\
l_{33}^{-1} \\
l_{44}^{-1}
\end{bmatrix} = p.
\tag{34}
$$

The determinant of A is computed whether it is required or not. Since determinants may assume a particularly wide range of values the determinant is stored as an ordered pair $d1$ and $d2$ such that $\det(A) = 2^{d2} d1$ where $d2$ is an integer and $\frac{1}{16} \le |d1| < 1$.

In *cholsol 1* the equations $AX = B$ are solved corresponding to the r different right-hand sides which form a matrix B. The $n \times r$ matrix of solutions X is stored separately so that B is available for the calculation of residuals if required.

In *cholinversion 1* a different form of storage from that in *choldet 1* is used in order that the complete lower triangle of A^{-1} should finally be available in the natural form. If we write

$$
P = L^{-1}, \qquad Q = A^{-1}
\tag{35}
$$

then the stored arrays at successive stages are illustrated by

$$
\begin{bmatrix}
a_{11} & a_{12} & a_{13} & a_{14} \\
l_{11}^{-1} & a_{22} & a_{23} & a_{24} \\
l_{21} & l_{22}^{-1} & a_{33} & a_{34} \\
l_{31} & l_{32} & l_{33}^{-1} & a_{44} \\
l_{41} & l_{42} & l_{43} & l_{44}^{-1}
\end{bmatrix},
\begin{bmatrix}
a_{11} & a_{12} & a_{13} & a_{14} \\
p_{11} & a_{22} & a_{23} & a_{24} \\
p_{21} & p_{22} & a_{33} & a_{34} \\
p_{31} & p_{32} & p_{33} & a_{44} \\
p_{41} & p_{42} & p_{43} & p_{44}
\end{bmatrix},
\begin{bmatrix}
a_{11} & a_{12} & a_{13} & a_{14} \\
q_{11} & a_{22} & a_{23} & a_{24} \\
q_{21} & q_{22} & a_{33} & a_{34} \\
q_{31} & q_{32} & q_{33} & a_{44} \\
q_{41} & q_{42} & q_{43} & q_{44}
\end{bmatrix}.
\tag{36}
$$

5.II. Set II.

In *choldet 2* the lower triangle of a symmetric matrix A is given as a linear array a of $\frac{1}{2}n(n+1)$ elements. Thus when $n=4$, for example we have

$$
\begin{bmatrix}
a_{11} & & & \\
a_{21} & a_{22} & & \\
a_{31} & a_{32} & a_{33} & \\
a_{41} & a_{42} & a_{43} & a_{44}
\end{bmatrix}
\rightarrow
\begin{bmatrix}
a_1 & & & \\
a_2 & a_3 & & \\
a_4 & a_5 & a_6 & \\
a_7 & a_8 & a_9 & a_{10}
\end{bmatrix}.
\tag{37}
$$

In general a_{ij} becomes a_k where $k=\dfrac{i}{2}(i-1)+j$. The elements of L are produced one by one and l_{ij} is overwritten on the element of A corresponding to a_{ij}, the reciprocal of l_{ii} being stored instead of the l_{ii}. The determinant is always computed and again the form $2^{d2}d1$ is used.

In *cholsol 2* the solution of $AX=B$ is overwritten on B; since we cannot compute the residuals there is no point in preserving B.

In *cholinversion 2* the linear array a representing A is overwritten by L as in *choldet 2*. L is then overwritten by L^{-1} which in turn is overwritten by the lower triangle of A^{-1} stored in the same form as the original A.

5.III. Set III.

The three procedures *symdet*, *symsol*, *syminversion* follow closely the corresponding procedures in the *chol 1* series, except that in general d_i^{-1} takes the place of l_{ii}^{-1}. However, in *symdet* the quantities $\tilde{a}_{ij}$ are required at intermediate stages in the computation. The elements $\tilde{l}_{ik}$ are produced row by row. During the determination of the fourth row of a 5×5 matrix the stored square arrays are successively

$$
\begin{bmatrix}
a_{11} & a_{12} & a_{13} & a_{14} & a_{15} \\
\tilde{l}_{21} & a_{22} & a_{23} & a_{24} & a_{25} \\
\tilde{l}_{31} & \tilde{l}_{32} & a_{33} & a_{34} & a_{35} \\
X & X & X & a_{44} & a_{45} \\
X & X & X & X & a_{55}
\end{bmatrix},
\begin{bmatrix}
a_{11} & a_{12} & a_{13} & a_{14} & a_{15} \\
\tilde{l}_{21} & a_{22} & a_{23} & a_{24} & a_{25} \\
\tilde{l}_{31} & \tilde{l}_{32} & a_{33} & a_{34} & a_{35} \\
\tilde{a}_{41} & \tilde{a}_{42} & \tilde{a}_{43} & a_{44} & a_{45} \\
X & X & X & X & a_{55}
\end{bmatrix},
\begin{bmatrix}
a_{11} & a_{12} & a_{13} & a_{14} & a_{15} \\
\tilde{l}_{21} & a_{22} & a_{23} & a_{24} & a_{25} \\
\tilde{l}_{31} & \tilde{l}_{32} & a_{33} & a_{34} & a_{35} \\
\tilde{l}_{41} & \tilde{l}_{42} & \tilde{l}_{43} & a_{44} & a_{45} \\
X & X & X & X & a_{55}
\end{bmatrix}
\tag{38}
$$

where the crosses denote quantities which may be arbitrary at the current stage. In addition the reciprocals of the d_i are stored as a separate vector in much the same way as in *choldet 1*. The relevant features of *cholinversion 1* and *symdet* are included in *syminversion*.

6. Discussion of the Numerical Properties

The Cholesky decomposition of a positive definite matrix has remarkable numerical stability. This is primarily because of the relation

$$
\sum_{j=1}^{i} l_{ij}^2 = a_{ii} \qquad (i=1, \ldots, n)
\tag{39}
$$

which ensures that the size of the l_{ij} is strictly limited by that of the a_{ij}. In fact if all a_{ij} satisfy the inequality $|a_{ij}| < 1$ then all l_{ij} satisfy the same inequality. The Cholesky decomposition therefore presents no difficulties even on a fixed-point computer. We have further (with exact computation)

$$\|L\|_2 = \|A\|_2^{\frac{1}{2}} = \lambda_1^{\frac{1}{2}}, \qquad \|L^{-1}\|_2 = \|A^{-1}\|_2^{\frac{1}{2}} = \lambda_n^{-\frac{1}{2}} \tag{40}$$

where the λ_i are the eigenvalues of A arranged in decreasing order. Although the alternative decomposition is not so convenient in fixed-point (since elements of $\tilde{L}$ can be appreciably larger than unity whatever bounds may be satisfied by the a_{ij}), in floating-point the upper bounds for the errors in both methods are virtually identical.

Bounds for the errors in the computed determinants, solution and inverse can be given in terms of the condition number $\varkappa$ defined by

$$\varkappa = \|A\|_2 \|A^{-1}\|_2 = \lambda_1/\lambda_n. \tag{41}$$

With standard floating-point computation and a t-digit mantissa the computed L satisfies the relation

$$LL^T = A + F, \tag{42}$$

where F satisfies both

$$\|F\|_2 \leqq k_1 n^{\frac{3}{2}} 2^{-t} \|A\|_2 \quad \text{and} \quad \|F\|_2 \leqq k_2 n^2 2^{-t} \max|a_{ij}|. \tag{43}$$

Here and elsewhere k_i denotes a constant of the order of unity which is dependent on the rounding procedure. Because of the statistical distribution of the rounding errors it is unlikely that $\|F\|_2$ will exceed $n^{\frac{3}{2}} 2^{-t} \|A\|_2$ or $n 2^{-t} \max|a_{ij}|$ in practice unless n is small.

If the eigenvalues of $A + F$ are λ_i' then from the first of relations (43)

$$- k_1 n^{\frac{3}{2}} 2^{-t} \lambda_1 \leqq \lambda_i - \lambda_i' \leqq k_1 n^{\frac{3}{2}} 2^{-t} \lambda_1. \tag{44}$$

Hence since $\det(LL^T) = \Pi l_{ii}^2 = \Pi \lambda_i'$ we have

$$\Pi(\lambda_i - k_1 n^{\frac{3}{2}} 2^{-t} \lambda_1) \leqq \det(LL^T) \leqq \Pi(\lambda_i + k_1 n^{\frac{3}{2}} 2^{-t} \lambda_1) \tag{45}$$

giving

$$(1 - k_1 n^{\frac{3}{2}} 2^{-t} \lambda_1/\lambda_i) \leqq \det(LL^T)/\det(A) \leqq (1 + k_1 n^{\frac{3}{2}} 2^{-t} \lambda_1/\lambda_i). \tag{46}$$

The bound for the relative error in the computed determinant is therefore primarily dependent on $k_1 n^{\frac{3}{2}} 2^{-t} \varkappa$. From (44) we see that if

$$\varkappa > 2^t/k_1 n^{\frac{3}{2}} \tag{47}$$

λ_n' can be negative, that is, rounding errors can destroy positive definiteness if $\varkappa$ is too large.

There are three sources of error in the computed inverse, the triangular factorization, the inversion of L and the calculation of $(L^{-1})^T L^{-1}$. Error bounds for the last two involve $\varkappa^{\frac{1}{2}}$ and are independent of $\varkappa$ respectively, so that when A is ill-conditioned the final error is dominated by the contribution from the triangular decomposition. If Q is the computed inverse we have certainly

$$\|A^{-1} - Q\|_2 / \|A^{-1}\|_2 \leq k_3 n^{\frac{3}{2}} 2^{-t} \varkappa \tag{48}$$

and in practice it is rare for the bound $n^{\frac{3}{4}} 2^{-t} \varkappa$ to be exceeded.

Similarly in the solution of $Ax = b$ the errors made in the solution of the triangular sets of equations are of minor importance and the computed solution satisfies the bounds

$$\|x - A^{-1} b\|_2 / \|A^{-1} b\|_2 \leq k_4 n^{\frac{3}{2}} 2^{-t} \varkappa. \tag{49}$$

It will be observed that we have organized the computation throughout so that each element of L, L^{-1}, A^{-1}, y and x is determined primarily as an inner-product. If it is possible to accumulate inner-products to double-precision then the bounds for the errors are much lower than those given above. We have, for example,

$$LL^T = A + F, \qquad \|F\|_2 \leq k_5 n^{\frac{1}{2}} 2^{-t} \|A\|_2 \tag{50}$$

and when A is ill-conditioned even this bound is usually an overestimate. The determination of these error bounds is given in [3] and [4]. In the second set of procedures based on the Cholesky decomposition a comparatively sophisticated inner-product procedure is required. Again if inner-products can be accumulated both the solution of $Ax = b$ and the computed inverse of A may be improved using the computed triangularizations. These modified procedures will be the subject of an additional note to be published later, (see Contribution I/2).

The significance of a "failure" when deriving a diagonal element l_{ii} depends very much on the context. If at the stage when the failure occurs we have $i < n$ and $-l_{ii}^2$ is appreciably greater than $2^{-t} \|A\|_2$, then usually the partial decomposition is useless. Often it will imply that the matrix A which should have been positive definite has been incorrectly computed or assigned. In any case the Cholesky decomposition cannot be completed without some modification. On the other hand the $\tilde{L} D \tilde{L}^T$ composition can be completed even if a d_i is negative, though numerical stability is guaranteed only if A is positive definite.

When *choldet* (or *symdet*) is used as a step in the process of inverse iteration for an eigenvector of a matrix B it is quite common for the matrix A to be defined by $B - \lambda I$ where λ is deliberately chosen to be very close to, but less than, the smallest eigenvalue of B. In this case rounding errors may well make $B - \lambda I$ non-positive definite and a failure may occur, most likely when $i = n$. It is permissable then to take $l_{nn} = 2^{-t} \|A\|_2^{\frac{1}{2}}$ and no harm will ensue. Generally as far as determinant evaluation is concerned a failure when computing l_{nn} is of minor importance and for this reason the factor l_{ii}^2 has been included before testing for failure.

7. Examples of the Use of the Procedures

The direct use of these procedures is obvious. On the whole the $\tilde{L} D \tilde{L}^T$ decomposition is very slightly superior. The LL^T decomposition is commonly used in connexion with RUTISHAUSER's LR algorithm [5] as applied to symmetric matrices, though usually this algorithm is applied only when A is of band form.

The Cholesky decomposition is also used in the solution of the eigenvalue problems

$$\det(A - \lambda B) = 0 \quad \text{and} \quad \det(AB - \lambda I) = 0 \tag{51}$$

when B is positive definite and A is symmetric. If $B=LL^T$ these problems reduce to

$$\det[L^{-1}A(L^{-1})^T - \lambda I]=0 \quad \text{and} \quad \det(L^TAL - \lambda I)=0 \tag{52}$$

respectively and both $L^{-1}A(L^{-1})^T$ and L^TAL are symmetric.

8. Test Results

The LL^T and LDL^T procedures have all been used in connexion with the leading principal minor of order seven of the infinite Hilbert matrix. In order to avoid rounding errors the matrix was scaled by the factor 360360 so that all coefficients were integers; hence A was given by

360360	180180	120120	90090	72072	60060	51480
180180	120120	90090	72072	60060	51480	45045
120120	90090	72072	60060	51480	45045	40040
90090	72072	60060	51480	45045	40040	36036.
72072	60060	51480	45045	40040	36036	32760
60060	51480	45045	40040	36036	32760	30030
51480	45045	40040	36036	32760	30030	27720

This matrix was chosen because it is ill-conditioned $(\varkappa \doteq \frac{1}{2}10^9)$ and also because the results will provide interesting comparisons with those obtained with the corresponding procedures of Contribution I/2 in which inner-products are accumulated.

The inverse of A was computed using *cholinversion 1 and 2* and *syminversion* and the solutions of $AX=I$ were obtained using *cholsol 1 and 2* and *symsol* so that in all six inverses were obtained. The computation was performed on KDF 9 which uses floating-point binary computation with 39 digits in the mantissa. All six inverses were remarkably accurate having regard to the condition number of A. Because of minor differences used in the counting techniques in the first and second sets of Cholesky procedures no two sets of answers were identical.

None of the inverses was significantly more accurate than any other. As typical of the results in general we give the two inverses obtained using *cholsol 1* and *cholinversion 1*. Both procedures gave identically the same L matrices of course, but the computation of the inverse is essentially different in the two cases. The procedure *cholsol 1* takes no advantage of the special form of the right-hand sides and naturally produces the full inverse without any regard for the fact that this should be symmetric; each column of the inverse is produced by a forward and a backward substitution. With *cholinversion 1* on the other hand, L^{-1} is computed and A^{-1} is then derived by matrix multiplication.

Inverse computed using choldet 1 followed by cholsol 1

Column 1	Column 2	Column 3
$+1.3597\,1314\,283_{10}-4$	$-3.2632\,5113\,753_{10}-3$	$+2.4474\,0603\,295_{10}-2$
$-3.2632\,5113\,738_{10}-3$	$+1.0442\,2850\,066_{10}-1$	$-8.8106\,0620\,311_{10}-1$
$+2.4474\,0603\,601_{10}-2$	$-8.8106\,0620\,282_{10}-1$	$+7.9295\,0934\,930_{10}+0$
$-8.1579\,3917\,875_{10}-2$	$+3.1326\,4074\,692_{10}+0$	$-2.9368\,4512\,786_{10}+1$
$+1.3460\,4954\,738_{10}-1$	$-5.3842\,0043\,662_{10}+0$	$+5.1919\,0848\,981_{10}+1$
$-1.0768\,3295\,031_{10}-1$	$+4.4304\,1069\,139_{10}+0$	$-4.3611\,9354\,458_{10}+1$
$+3.3330\,3738\,252_{10}-2$	$-1.3998\,8227\,242_{10}+0$	$+1.3998\,8675\,007_{10}+1$

Column 4

$-8.15793917156_{10}-2$
$+3.13264074649_{10}+0$
$-2.93684512777_{10}+1$
$+1.11879649598_{10}+2$
$-2.01907567906_{10}+2$
$+1.72294292795_{10}+2$
$-5.59956005459_{10}+1$

Column 5

$+1.34604954657_{10}-1$
$-5.38420043660_{10}+0$
$+5.19190848985_{10}+1$
$-2.01907567907_{10}+2$
$+3.70163885040_{10}+2$
$-3.19821597914_{10}+2$
$+1.04991937763_{10}+2$

Column 6

$-1.07683294804_{10}-1$
$+4.43041069058_{10}+0$
$-4.36119354451_{10}+1$
$+1.72294292794_{10}+2$
$-3.19821597912_{10}+2$
$+2.79117246791_{10}+2$
$-9.23930348819_{10}+1$

Column 7

$+3.33303738787_{10}-2$
$-1.39988227277_{10}+0$
$+1.39988675014_{10}+1$
$-5.59956005457_{10}+1$
$+1.04991937762_{10}+2$
$-9.23930348819_{10}+1$
$+3.07977132849_{10}+1$

Inverse computed using cholinversion 1 (Sub-diagonal elements only)

Row 1

$+1.35971314362_{10}-4$

Row 2

$-3.26325113762_{10}-3$
$+1.04422850063_{10}-1$

Row 3

$+2.44740603601_{10}-2$
$-8.81060620290_{10}-1$
$+7.92950934908_{10}+0$

Row 4

$-8.15793917877_{10}-2$
$+3.13264074697_{10}+0$
$-2.93684512785_{10}+1$
$+1.11879649600_{10}+2$

Row 5

$+1.34604954738_{10}-1$
$-5.38420043665_{10}+0$
$+5.19190848982_{10}+1$
$-2.01907567907_{10}+2$
$+3.70163885041_{10}+2$

Row 6

$-1.07683295031_{10}-1$
$+4.43041069138_{10}+0$
$-4.36119354457_{10}+1$
$+1.72294292795_{10}+2$
$-3.19821597913_{10}+2$
$+2.79117246790_{10}+2$

Row 7

$+3.33303738252_{10}-2$
$-1.39988227242_{10}+0$
$+1.39988675007_{10}+1$
$-5.59956005459_{10}+1$
$+1.04991937763_{10}+2$
$-9.23930348819_{10}+1$
$+3.07977132849_{10}+1$

The two inverses illustrate the points made in Section 6. All corresponding components of the two inverses agree to at least 9 significant decimal digits and nearly all agree to 11 significant decimal digits in spite of the fact that both inverses are, in general, in error in the fifth figure. This is because the main source of error is in the triangularization and this is identical in the two cases.

For a similar reason the inverse given by *cholsol 1* is almost exactly symmetric; this is because the errors arising from the forward and backward substitutions are comparatively unimportant and hence the computed inverse is very close to $(L^{-1})^T L^{-1}$ corresponding to the computed L; this last matrix is obviously *exactly* symmetric whatever errors L may have.

It cannot be too strongly emphasized that although the computed inverses have errors which approach 1 part in 10^4 *these errors are of the same order of* *magnitude as those which would be caused, for example, by a single perturbation of one part in 2^{39} in $a_{5,5}$ alone!*

The values obtained for the determinant were

choldet 1 $8.4741\,88295\,03 \times 10^{-2} \times 2^{52}$ i.e. $8.4741\,88295 \times 10^{-2} \times 2^{52}$

choldet 2 $8.4737\,13008\,72 \times 10^{-2} \times 2^{52}$

symdet $8.4731\,76059\,76 \times 10^{-2} \times 2^{52}$.

The exact determinant is $38161\,42770\,72600 = 8.47353\,9 \ldots \times 10^{-2} \times 2^{52}$, so that again the accuracy is high having regard to the condition of A.

A second test was performed on the computer TR 4 at the Mathematisches Institut der Technischen Hochschule, München, with the same matrix. The TR 4 effectively has a 38 binary digit mantissa but on the average the results were about a binary digit better than on KDF 9.

Acknowledgements. The authors wish to thank Mrs. E. MANN and Dr. C. REINSCH of the Technische Hochschule, Munich, for valuable assistance in the development of these algorithms. The work described here is part of the Research Programme of the National Physical Laboratory and this paper is published by permission of the Director of the Laboratory.

References

[1] Fox, L.: Practical solution of linear equations and inversion of matrices. Nat. Bur. Standards Appl. Math. Ser. **39**, 1—54 (1954).

[2] MARTIN, R. S., and J. H. WILKINSON: Symmetric decomposition of positive definite band matrices. Numer. Math. **7**, 355-361 (1965). Cf. I/4.

[3] WILKINSON, J. H.: Rounding errors in algebraic processes. Notes on Applied Science No. 32. London: Her Majesty's Stationery Office; New Jersey: Prentice-Hall 1963. German edition: Rundungsfehler. Berlin-Göttingen-Heidelberg: Springer 1969.

[4] WILKINSON, J. H.: The algebraic eigenvalue problem. London: Oxford University Press 1965.

[5] RUTISHAUSER, H.: Solution of eigenvalue problems with the LR-transformation. Nat. Bur. Standards Appl. Math. Ser. **49**, 47—81 (1958).

Contribution I/2

Iterative Refinement of the Solution
of a Positive Definite System of Equations*

by R. S. Martin, G. Peters, and J. H. Wilkinson

1. Theoretical Background

In an earlier paper in this series [1] the solution of a system of equations $A x = b$ with a positive definite matrix of coefficients was described; this was based on the Cholesky factorization of A. If A is ill-conditioned the computed solution may not be sufficiently accurate, but (provided A is not almost singular to working accuracy) it may be improved by an iterative procedure in which the Cholesky decomposition is used repeatedly.

This procedure may be defined by the following relations

$$A = LL^T, \quad x^{(0)} = 0,$$
$$r^{(s)} = b - A x^{(s)}, \quad (LL^T) d^{(s)} = r^{(s)}, \quad x^{(s+1)} = x^{(s)} + d^{(s)}. \tag{1}$$

In each iteration the residual $r^{(s)}$ corresponding to the current $x^{(s)}$ is computed and the correction $d^{(s)}$ to be added to $x^{(s)}$ is determined using the computed LL^T factorization.

The only reason that the first solution is not exact is because of the intervention of rounding errors. Since rounding errors are also made in the iterative refinement it is by no means obvious that the successive $x^{(s)}$ will be improved solutions. It has been shown by WILKINSON [2] that if A is not "too ill-conditioned" the $x^{(s)}$ will converge to the correct solution "to working accuracy" provided $r^{(s)}$ is computed to double-precision. Since the elements of A and of $x^{(s)}$ are single-precision numbers this may be achieved merely by accumulating inner-products. The remaining steps may be performed in single-precision.

Two terms used above call for some explanation. We say that $x^{(s)}$ is the correct solution "to working accuracy" if $\|x^{(s)} - x\|_\infty / \|x\|_\infty \leq 2^{-t}$ where the mantissae of our floating-point numbers have t binary digits. For the meaning of the term "too ill-conditioned" we must refer to the error analysis of the Cholesky method. It has been shown ([2, 3]) that the computed solution of $A x = b$ is the exact solution of a system $(A + \delta A) x = b$ where δA is dependent on b but is uniformly bounded. The upper bound m of $\|\delta A\|_2$ is dependent on the details of the arithmetic used, particularly upon whether or not inner-products are accumulated without intermediate rounding. We shall say that A is "too ill-conditioned" for the precision of computation that is being used if $m \|A^{-1}\|_2 \geq 1$. In this case $A + \delta A$ could be singular and in any case $(A + \delta A)^{-1} b$ may not

---

* Prepublished in Numer. Math. 8, 203—216 (1966).

agree with $A^{-1}b$ in any of its figures. If $m\|A^{-1}\|_2 = 2^{-p}$ ($p > 0$) then successive iterates certainly satisfy the relation

$$\|x - x^{(s+1)}\|_2 \leq \|x - x^{(s)}\|_2 2^{-p}/(1 - 2^{-p}).$$ (2)

Usually the statistical distribution of rounding errors will ensure that the $x^{(s)}$ will improve more rapidly than this.

A similar process may be used to refine an inverse derived from the Cholesky factorization of A as described in [1]. If we denote the first computed inverse by $X^{(1)}$ then we have the iterative refinement procedure defined by

$$\begin{aligned} X^{(s+1)} &= X^{(s)} + X^{(s)}(I - AX^{(s)}), \\ &= X^{(s)} + X^{(s)}B^{(s)}, \\ &= X^{(s)} + Z^{(s)}. \end{aligned}$$ (3)

In order to be effective it is essential that the right-hand side of (3) should not be expressed in the form $2X^{(s)} - X^{(s)}AX^{(s)}$. The "residual matrix" $I - AX^{(s)}$ must be computed using double-precision or accumulation of inner-products, each component being rounded only on completion. The conditions for convergence are the same as those for the iterative refinement of the solution of $Ax = b$, but if the condition is satisfied the convergence is effectively quadratic. Indeed ignoring rounding errors we have

$$I - AX^{(s+1)} = (I - AX^{(s)})^2.$$ (4)

In [1] several different procedures were given based on symmetric decompositions of A; here we give only the procedures related to *choldet 1, cholsol 1* and *cholinversion 1*. Analogous procedures could be made based on *symdet, symsol* and *syminversion*. Since A must be retained in order to form the residuals the decompositions in which the upper half of A is stored as a linear array of $\frac{1}{2}n(n+1)$ elements are not relevant.

2. Applicability

These procedures may be used to compute accurate solutions of systems of equations with positive definite matrices or to compute accurate inverses of such matrices. An essential feature of the algorithms is that in each iteration an accurate residual must be determined corresponding to the current solution. The i-th components r_i of a residual is given by

$$r_i = b_i - \sum_{j=1}^{n} a_{ij}x_j,$$ (5)

and since only the upper-half of A is stored this must be written in the form

$$r_i = b_i - \sum_{j=1}^{i-1} a_{ji}x_j - \sum_{j=i}^{n} a_{ij}x_j,$$ (6)

taking advantage of the symmetry of A. Similar remarks apply to the computation of $I - AX$ in the procedure for an accurate inverse.

In order to avoid an excessively complicated inner-product procedure it is convenient to compute the expression on the right-hand side of (6) in two stages. In the first stage $s_i = b_i - \sum\limits_{j=1}^{i-1} a_{ji} x_j$ is computed and in the second stage $\sum\limits_{j=i}^{n} a_{ij} x_j$ is subtracted from s_i. It is essential that s_i should not be rounded before subtracting the second inner-product since if this is done the refinement procedure is completely ineffective. This means that the inner-product procedure should be capable of starting from a previously computed double-precision inner-product when necessary.

Modified versions of *choldet 1, cholsol 1* and *cholinversion 1* are included in which the required inner-products are computed by means of the inner-product procedure used for the residuals. On many computers the accurate computation of inner-products may be very time consuming and in this case the versions of *choldet 1* etc. given in [1] may be used.

The procedures are

acc solve Corresponding to r given right-hand sides this either produces the correctly rounded solutions of the equation $A x = b$ or indicates that A is too ill-conditioned for this to be achieved without working to higher precision (or is possibly singular). *acc solve* must be preceded by *choldet 1* so that the Cholesky decomposition of A is available. *acc solve* may be used any number of times after one use of *choldet 1*.

acc inverse This either produces the correctly rounded inverse of a positive definite matrix or indicates that it is too ill-conditioned. The first approximation to the inverse is produced by *cholinversion 1*.

3. Formal Parameter List

3.1. Input to procedure *acc solve*.

n order of the matrix A.

r number of right-hand sides for which $A x = b$ is to be solved.

a an $n \times n$ array consisting of the upper-triangle of A and the subdiagonal elements of the lower-triangle L of the Cholesky decomposition of A produced by procedure *choldet 1*.

p the reciprocals of the diagonal elements of L as produced by procedure *choldet 1*.

b the $n \times r$ matrix consisting of the r right-hand sides.

eps the smallest number for which $1 + eps > 1$ on the computer.

        Output of procedure *acc solve*.

x the $n \times r$ matrix consisting of the r solution vectors.

l the number of iterations.

bb the $n \times r$ matrix consisting of the r residual vectors.

ill This is the exit which is used when there is no perceptible improvement from one iteration to the next.

3.2. Input to procedure *acc inverse*.

n order of the matrix A.

a   an $(n+1)\times n$ array the elements of the upper-triangle being those of A.

eps   the smallest number for which $1+eps>1$ on the computer.

Output of procedure *acc inverse*.

a   an $(n+1)\times n$ array the elements of the upper-triangle being those of A and the elements of the lower-triangle those of X, the accepted inverse.

l   the number of corrections added to the first inverse.

fail   the exit which is used if A, possibly as the result of rounding errors, is not positive definite.

ill   the exit used if there is no perceptible improvement from one iteration to the next. Both "*fail*" and "*ill*" indicate that A is too ill-conditioned for the algorithm to be successful without working to higher precision.

4. ALGOL Programs

procedure *innerprod* $(l, s, u, c1, c2, ak, bk)$ *bound variable*: (k) *result*: $(d1, d2)$;
value $l, s, u, c1, c2$;
integer l, s, u, k;
real $c1, c2, ak, bk, d1, d2$;
comment This procedure accumulates the sum of products $ak \times bk$ and adds it to the initial value $(c1, c2)$ in double precision. The bound variable k is to be used to indicate the subscript in the components of the vectors ak, bk over which a scalarproduct is to be formed.

Throughout the Handbook Series Linear Algebra, the actual parameters corresponding to ak and bk will be restricted to be real variables. If they are subscripted, all subscript expressions will be linear functions of the bound variable parameter k, or will be independent of k. This allows high efficiency handcoding of the subscript evaluation in the for statement (loop) of the procedure.

The body of this procedure cannot be expressed within ALGOL,

begin real $s1, s2$,
$\quad (s1, s2) := c1 + c2$,   **comment** dbl. pr. acc,
$\quad$**for** $k := l$ **step** s **until** u **do**
$\quad (s1, s2) := (s1, s2) + ak \times bk$,   **comment** dbl. pr. acc,
$\quad d1 := (s1, s2) rounded$,
$\quad d2 := ((s1, s2) - d1) rounded$ **comment** dbl. pr. acc,
end *innerprod*;

procedure *choldet 1* (n) *trans*: (a) *result*: $(p, d1, d2)$ *exit*: $(fail)$;
value n; **integer** $d2, n$; **real** $d1$; **array** a, p; **label** *fail*;
comment The upper triangle of a positive definite symmetric matrix, A, is stored in the upper triangle of an $n \times n$ array $a[i, j]$, $i = 1(1)n, j = 1(1)n$. The Cholesky decomposition $A = LU$, where U is the transpose of L, is performed and stored in the remainder of the array a except for the reciprocals of the diagonal elements which are stored in $p[i]$, $i = 1(1)n$, instead of the elements themselves. A is retained so that the solution obtained can be subsequently improved. The determinant, $d1 \times 2 \uparrow d2$, of A is also computed. The procedure will fail if A,

modified by the rounding errors, is not positive definite. Uses the procedure *innerprod*;

```
begin      integer i, j, k; real x, xx;
           d1 := 1;
           d2 := 0;
           for i := 1 step 1 until n do
           for j := i step 1 until n do
           begin innerprod (1, 1, i−1, −a[i, j], 0, a[i, k], a[j, k], k, x, xx);
                     x := −x;
                     if j=i then
                     begin
                                d1 := d1×x; if x≤0 then goto fail;
                     L1:    if abs (d1) ≥ 1 then
                                begin d1 := d1×0.0625;
                                         d2 := d2 +4;
                                         go to L1
                                end;
                     L2:    if abs (d1) < 0.0625 then
                                begin d1 := d1×16;
                                         d2 := d2 − 4;
                                         go to L2
                                end;
                                p [i] := 1/sqrt (x)
                     end
                     else
                     a[j, i] := x×p [i]
           end ij
end choldet 1;
```

```
procedure acc solve (n) data: (r, a, p, b, eps) result: (x, bb, l) exit: (ill);
value    n, r, eps;
integer n, r, l;
real      eps;
array    a, p, x, b, bb;
label     ill;
comment Solves A x=b where A is an n×n positive definite symmetric matrix
           and b is an n×r matrix of right hand sides, using the procedure
           cholsol 1. The procedure must be preceded by choldet 1 in which L is
           produced in a [i, j] and p [i]. The residuals bb=b − A x are calculated
           and A d=bb is solved, overwriting d on bb. The refinement is re-
           peated, as long as the maximum correction at any stage is less than
           half that at the previous stage, until the maximum correction is less
           than 2 eps times the maximum abs (x). Exits to label ill if the solution
           fails to improve. Uses the procedure innerprod which forms accurate
           innerproducts. l is the number of iterations;
```

begin

        procedure *cholsol 1* (n) *data*: (r, a, p, b) *result*: (x);

        value r, n; **integer** r, n; **array** a, b, p, x;

        comment Solves $A x = b$, where A is a positive definite symmetric matrix and b is an $n \times r$ matrix of r right-hand sides. The procedure *cholsol 1* must be preceded by *choldet 1* in which L is produced in $a[i, j]$ and $p[i]$, from A. $A x = b$ is solved in two steps, $L y = b$ and $U x = y$, and x is overwritten on y;

        begin   **integer** i, j, k; **real** y, yy;

                for $j := 1$ **step** 1 **until** r **do**

                begin comment solution of $L y = b$;

                      for $i := 1$ **step** 1 **until** n **do**

                      begin *innerprod* $(1, 1, i-1, b[i, j], 0, a[i, k],$

                                    $x[k, j], k, y, yy);$

                          $x[i, j] := -p[i] \times y$

                      end i;

                      comment solution of $U x = y$;

                      for $i := n$ **step** -1 **until** 1 **do**

                      begin *innerprod* $(i+1, 1, n, x[i, j], 0, a[k, i],$

                                    $x[k, j], k, y, yy);$

                          $x[i, j] := -p[i] \times y$

                      end i

              end j

        end *cholsol 1*;

        integer $i, j, k, d2$;

        real $d0, d1, c, cc, xmax, bbmax$;

        for $j := 1$ **step** 1 **until** r **do**

        for $i := 1$ **step** 1 **until** n **do**

        begin $x[i, j] := 0$;

                  $bb[i, j] := b[i, j]$;

        end;

        $l := 0$;

        $d0 := 3.0$;

  L3:   *cholsol 1* (n, r, a, p, bb, bb);

        $l := l + 1$;

        $d2 := 0$;

        $d1 := 0$;

        for $j := 1$ **step** 1 **until** r **do**

        for $i := 1$ **step** 1 **until** n **do**

        $x[i, j] := x[i, j] + bb[i, j]$;

        for $j := 1$ **step** 1 **until** r **do**

        begin $x max := bb max := 0$;

                  for $i := 1$ **step** 1 **until** n **do**

begin if $abs(x[i,j]) > xmax$ **then** $xmax := abs(x[i,j])$;
   if $abs(bb[i,j]) > bbmax$ **then** $bbmax := abs(bb[i,j])$;
   $innerprod(1, 1, i-1, -b[i,j], 0, a[k,i],$
   $\qquad\qquad\qquad\qquad x[k,j], k, c, cc)$;
   $innerprod(i, 1, n, c, cc, a[i,k], x[k,j], k, c, cc)$;
   $\qquad\qquad\qquad\qquad bb[i,j] := -c$

end;
if $bbmax > d1 \times xmax$ **then** $d1 := bbmax/xmax$;
if $bbmax > 2 \times eps \times xmax$ **then** $d2 := 1$;

end;
if $d1 > d0/2$ **then goto** ill;
$d0 := d1$; **if** $d2 = 1$ **then goto** $L3$;
end $acc\ solve$;

procedure $acc\ inverse(n)$ $data:(eps)$ $trans:(a)$ $result:(l)$ $exit:(fail, ill)$;
value n, eps;
integer n, l;
real eps;
array a;
label $fail, ill$;

comment The upper triangle of a positive definite symmetric matrix, A, is stored in the upper triangle of an $(n+1) \times n$ array $a[i,j]$, $i=1(1)n+1$, $j=1(1)n$. X, the inverse of A, is formed in the remainder of the array $a[i,j]$ by the procedure $cholinversion\,1$. The inverse is improved by calculating $X = X + Z$ until the correction, Z, is such that maximum $abs(Z[i,j])$ is less than $2eps$ times maximum $abs(X[i,j])$, where $Z = XB$ and $B = I - AX$. b is an $n \times n$ array and z is a $1 \times n$ array, X being overwritten a row at a time. Exits to the label fail if A is not positive definite and to the label ill if the maximum correction at any stage is not less than half that at the previous stage. l is the number of corrections applied. Uses the procedure $innerprod$;

begin

procedure $cholinversion\,1$ (n) $trans:(a)$ $exit:(fail)$;
value n; **integer** n; **array** a; **label** $fail$;
comment The upper triangle of a positive definite symmetric matrix, A, is stored in the upper triangle of an $(n+1) \times n$ array $a[i,j]$, $i=1(1)n+1$, $j=1(1)n$. The Cholesky decomposition $A = LU$, where U is the transpose of L, is performed and L is stored in the remainder of the array a. The reciprocals of the diagonal elements are stored instead of the elements themselves. L is then replaced by its inverse and this in turn is replaced by the lower triangle of the inverse of A. A is retained so that the inverse can be subsequently improved. The procedure will fail if A, modified by the rounding errors, is not positive definite. Uses the procedure $innerprod$;

```
begin    integer i, j, k, i1, j1;
         real x, xx, y;
         comment formation of L;
         for i := 1 step 1 until n do
         begin i1 := i+1;
               for j := i step 1 until n do
               begin j1 := j+1;
                     innerprod (1, 1, i−1, −a[i, j], 0,
                             a[i1, k], a[j1, k], k, x, xx);  x := −x;
                     if j=i then
                     begin if x≦0 then go to fail;
                           a[i1, i] := y := 1/sqrt (x)
                     end
                     else
                     a[j1, i] := x×y
               end j
         end i;
         comment inversion of L;
         for i := 1 step 1 until n do
         for j := i+1 step 1 until n do
         begin j1 := j+1; innerprod (i, 1, j−1, 0, 0, a[j1, k],
                                   a[k+1, i], k, x, xx);
               a[j1, i] := −a[j1, j]×x
         end ij;
         comment calculation of the inverse of A;
         for i := 1 step 1 until n do
         for j := i step 1 until n do
         innerprod (j+1, 1, n+1, 0, 0, a[k, j],
                                 a[k, i], k, a[j+1, i], xx)
end cholinversion 1;


integer i, j, k, j1;
real c, d, xmax, zmax, e;
array b[1:n, 1:n], z[1:n];
e := 1;
l := 0;
cholinversion 1 (n, a, fail);
for i := 1 step 1 until n do
for j := 1 step 1 until n do
begin j1 := j+1;
      if j≧i then
      begin innerprod (1, 1, i, if i=j then −1 else 0, 0,
                   a[k, i], a[j1, k], k, c, d);
            innerprod (i+1, 1, j, c, d, a[i, k], a[j1, k], k, c, d);
            innerprod (j1, 1, n, c, d, a[i, k], a[k+1, j], k, c, d)
      end
      else
```

Label: L1: (before `for i := 1 step 1 until n do`)

```
          begin innerprod (1, 1, j, 0, 0, a[k, i], a[j1, k], k, c, d);
                innerprod (j1, 1, i, c, d, a[k, i], a[k+1, j], k, c, d);
                innerprod (i+1, 1, n, c, d, a[i, k], a[k+1, j], k, c, d)
          end;
  b[i, j] := −c
  end;
  xmax := zmax := 0;
  for i := 1 step 1 until n do
  begin     for j := 1 step 1 until i do
            begin innerprod (1, 1, i, 0, 0, a[i+1, k], b[k, j], k, c, d);
                  innerprod (i+1, 1, n, c, d, a[k+1, i], b[k, j], k, z[j], d)
            end;
            for j := 1 step 1 until i do
            begin c := abs (a[i+1, j]);
                  d := abs (z[j]);
                  if c > xmax then xmax := c;
                  if d > zmax then zmax := d;
                  a[i+1, j] := a[i+1, j]+z[j]
            end
  end;
  l := l+1;
  d := zmax/xmax;
  if d > e/2 then goto ill;
  e := d;
  if d > 2×eps then goto L1;
end acc inverse;
```

5. Organisational and Notational Details

The details of *choldet 1*, *cholsol 1* and *cholinversion 1* are almost identical with those given in [1] except that the inner-products are all performed using the accurate inner-product procedure. This is not an essential change but it increases to some extent the range of matrices for which the iterative refinement will be successful.

In *acc solve* the first set of r solution vectors are taken to be null vectors so that the first residuals are the original right-hand sides. The successive residuals $r^{(s)}$ are stored in the array *bb* and are overwritten by the corresponding $d^{(s)}$. Iteration is terminated when

$$\|d^{(s)}\|_\infty \leq 2 \, eps \, \|x^{(s+1)}\|_\infty \tag{7}$$

for *all* right hand sides simultaneously, where *eps* is the relative machine precision. If in any iteration

$$\max \left(\|d^{(s)}\|_\infty / \|x^{(s+1)}\|_\infty \right) > \tfrac{1}{2} \max \left(\|d^{(s-1)}\|_\infty / \|x^{(s)}\|_\infty \right) \tag{8}$$

(where the maximization is over the r right-hand sides) the matrix is too ill-conditioned for reliable solutions to be obtained and a failure indication is given. The process may fail also at the stage when *choldet 1* is performed if A is almost singular to working accuracy.

In the inner-product procedure the parameters *c1* and *c2* are effectively the two halves of a double-precision number. They are both normalized single precision numbers and added together give an initial value to which the inner-product is added. Since the inner-product procedure must be performed in machine code, the means used to achieve the required effect will differ from one computer to another.

acc solve may be used any number of times after *choldet 1*. In practice if there are storage problems we may take $r=1$ and process one right-hand side at a time.

In *acc inverse* the lower triangles of successive approximations to the inverse are all overwritten on the original approximation produced by *cholinversion 1*. In each iteration the whole of the $n \times n$ matrix $B^{(s)} = I - A X^{(s)}$ is produced and stored in the array b. The matrix $Z^{(s)} = X^{(s)} B^{(s)}$ is produced a row at a time, remembering that since $Z^{(s)}$ and $X^{(s)}$ are symmetric we require only elements 1 to i of row i of $Z^{(s)}$. Each of the partial rows of $Z^{(s)}$ is overwritten in the $1 \times n$ array z. When each partial row is completed it can be added to $X^{(s)}$ since the corresponding row of $X^{(s)}$ is not required in the computation of the remaining rows of $X^{(s)} B^{(s)}$.

6. Discussion of the Numerical Properties

The error analysis of the procedures *choldet 1*, *cholsol 1* and *cholinversion 1* was discussed in [1]. When inner-products are accumulated the computed LL^T satisfies the relation

$$LL^T = A + F \quad \text{where} \quad \|F\|_2 \le k_1 n^{\frac{1}{2}} 2^{-t} \|A\|_2, \tag{9}$$

where here and later k_i is used to denote a constant of order unity. The computed solution $\bar{x}$ of $A x = b$ satisfies the relation

$$(A + G(b)) \bar{x} = b \quad \text{where} \quad \|G(b)\|_2 \le k_2 n^{\frac{1}{2}} 2^{-t} \|A\|_2. \tag{10}$$

Although $G(b)$ is dependent on b it is uniformly bounded. From (10) we have

$$\|\bar{x} - x\|_2 / \|x\|_2 < \alpha/(1 - \alpha), \tag{11}$$

where

$$\alpha = k_2 n^{\frac{1}{2}} 2^{-t} \|A\|_2 \|A^{-1}\|_2 = k_2 n^{\frac{1}{2}} 2^{-t} \varkappa(A). \tag{12}$$

The iterative refinement of a solution therefore converges if

$$\alpha/(1 - \alpha) < 1 \quad \text{i.e.} \quad \alpha < \tfrac{1}{2}.$$

For segments of the Hilbert matrix and similar matrices even the bound for $G(b)$ given above, satisfactory though it is, is a severe overestimate. This is well illustrated by the example in section 8.

The computed inverse X given by *cholinversion 1* with accumulation of inner-products satisfies the relation

$$\|X - A^{-1}\|_2 / \|A^{-1}\|_2 < \beta/(1 - \beta), \tag{13}$$

where

$$\beta = k_3 n^{\frac{1}{2}} 2^{-t} \varkappa(A) \tag{14}$$

and convergence is therefore guaranteed if $\beta < \tfrac{1}{2}$.

It is natural to ask whether *acc solve* can converge to a wrong solution. From the error analysis we know that this is impossible if $\alpha < \frac{1}{2}$. We observe that if this condition is not satisfied A is very ill-conditioned and we have an additional safeguard in *choldet 1*. It seems to be extremely difficult to construct an ill-conditioned matrix which does not reveal its shortcomings either in *choldet 1* or in the iterative refinement, and in practice it appears to be very safe to take the results of *acc solve* at their face value.

With *acc inverse* the situation is theoretically stronger since we have $I - AX$ for the alleged inverse X. If $I - AX = B$ and $\|B\| < 1$ in any norm then A is non-singular and

$$\|X - A^{-1}\| / \|A^{-1}\| \leq \|B\|. \tag{15}$$

If $\|B\| < \frac{1}{2}$ we certainly have a guarantee that the process will converge to the correctly rounded solution. When solutions corresponding to many different right-hand sides are required it seems attractive to compute X using *acc inverse* particularly since this takes full advantage of symmetry. If the iterative refinement converges and $\|I - AX\|_\infty$ for the accepted inverse is appreciably less than unity then we have a rigorous proof that X is the correctly rounded inverse. Unfortunately this does not mean that Xb is the correctly rounded solution corresponding to any right-hand side b. (For a discussion see [2], Chapter 3.) We would still have to perform an iterative process

$$x^{(s)} = 0, \qquad r^{(s)} = b - A x^{(s)}, \qquad x^{(s+1)} = x^{(s)} + X r^{(s)} \tag{16}$$

and its rate of convergence is no better than that of *acc solve*! When A is large it seems difficult to justify the use of *acc inverse* unless one is specifically interested in the inverse itself.

7. Examples of the Use of the Procedures

The uses of these procedures are self-evident.

8. Test Results

The procedure *acc solve* and *acc inverse* have been used on KDF 9 in connexion with the leading principal minor of order seven of the Hilbert matrix. KDF 9 has a 39 binary digit mantissa and is well designed for the accumulation of inner-products. As in [1], the matrix was scaled by the factor 360360 so that all coefficients were integers. In order to provide a comparison with the results given in [1] the accurate inner-product procedure was used in *choldet 1, cholsol 1* and *cholinversion 1* as well as in the calculation of the residuals.

acc solve was used with the seven right-hand sides given by the columns of the unit matrix. For economy of presentation we give only the results corresponding to the first column; the behaviour of the other columns is exactly analogous. The initial solution was correct to within 1 part in 10^7, most components being even more accurate than this. This is between two and three decimals better than the results obtained in [1] for *choldet 1* without accumulation of inner-products. The second solution was already correct to working accuracy and the third iterative merely served to show that the second was correct. (It should be appreciated that all published values were obtained by converting the values in the computer from binary to decimal. If the $d^{(1)}$ in the computer had been added to the $x^{(1)}$ in the computer without rounding, the result would have been correct to more than single-precision.) Notice that the first set of residuals is the smaller in spite of the fact that it corresponds to a less accurate solution. This is predictable from the error analysis. The modified *choldet 1* also gave a considerably more accurate determinant evaluation. Comparative results are

choldet 1 (without accumulation) $8.4741\,88295\,03_{10} - 2 \times 2^{52}$;

choldet 1 (with accumulation) $8.4735\,38254\,98_{10} - 2 \times 2^{52}$;

Correct value $8.4735\,3913\,\ldots_{10} - 2 \times 2^{52}$.

acc solve was also used with the seven right-hand sides given by the columns of ($360360 \times$ the unit matrix) for which the solutions are integers. As before the second iteration gave results which were "correct to working accuracy" but since the components are integers this means that the rounded components of $x^{(1)} + d^{(1)}$ were exact. Hence at the second iteration all residuals were exactly zero. For economy we present the results corresponding to the fifth column which has the solution with the largest norm.

acc inverse was also used in connexion with the scaled Hilbert matrix, the first solution being derived from the modified *cholinversion 1* with the accurate inner-product for comparison with the results in [1]. It will be seen that the inverse obtained using inner-product accumulation is much the more accurate. The second computed inverse was correct to working accuracy, the third inverse being necessary only to confirm this.

Example

"acc solve" with right-hand side $(1, 0, \ldots, 0)$

$d^{(0)}$	$x^{(1)} = 0 + d^{(0)}$	$r^{(1)} = b - A\,x^{(1)}$
$+1.3597\,5135\,069_{10} - 4$	$+1.3597\,5135\,069_{10} - 4$	$-4.9253\,7921\,559_{10} - 9$
$-3.2634\,0327\,676_{10} - 3$	$-3.2634\,0327\,676_{10} - 3$	$-5.8762\,2217\,552_{10} - 9$
$+2.4475\,5247\,690_{10} - 2$	$+2.4475\,5247\,690_{10} - 2$	$-7.9203\,1329\,411_{10} - 10$
$-8.1585\,0829\,840_{10} - 2$	$-8.1585\,0829\,840_{10} - 2$	$+2.1578\,9164\,315_{10} - 10$
$+1.3461\,5387\,384_{10} - 1$	$+1.3461\,5387\,384_{10} - 1$	$+2.4667\,5568\,860_{10} - 10$
$-1.0769\,2310\,159_{10} - 1$	$-1.0769\,2310\,159_{10} - 1$	$+1.6282\,6196\,970_{10} - 10$
$+3.3333\,3341\,518_{10} - 2$	$+3.3333\,3341\,518_{10} - 2$	$+1.1155\,1656\,801_{10} - 10$

$d^{(1)}$	$x^{(2)} = (x^{(1)} + d^{(1)})$ *rounded*	$r^{(2)} = b - A\,x^{(2)}$
$+9.0578\,0905\,471_{10} - 13$	$+1.3597\,5135\,975_{10} - 4$	$+5.4264\,0421\,486_{10} - 9$
$+1.3358\,5198\,681_{10} - 11$	$-3.2634\,0326\,340_{10} - 3$	$+4.8596\,3802\,532_{10} - 9$
$-2.9347\,7858\,424_{10} - 10$	$+2.4475\,5244\,755_{10} - 2$	$+4.3655\,7101\,580_{10} - 9$
$+1.3989\,3877\,246_{10} - 9$	$-8.1585\,0815\,850_{10} - 2$	$+3.9498\,6887\,642_{10} - 9$
$-2.7683\,6240\,838_{10} - 9$	$+1.3461\,5384\,615_{10} - 1$	$+3.6004\,6570\,336_{10} - 9$
$+2.4665\,7452\,585_{10} - 9$	$-1.0769\,2307\,692_{10} - 1$	$+3.3047\,1827\,681_{10} - 9$
$-8.1851\,5052\,054_{10} - 10$	$+3.3333\,3333\,333_{10} - 2$	$+3.0520\,9013\,618_{10} - 9$

"acc solve" with right-hand side (0, ..., 0, 360360, 0, 0)

$d^{(0)}$	$x^{(1)}=0+d^{(0)}$	$r^{(1)}=b-A\,x^{(1)}$
$+4.8510\,0010\,033_{10}+4$	$+4.8510\,0010\,033_{10}+4$	$-3.9145\,2026\,367_{10}+0$
$-1.9404\,0011\,270_{10}+6$	$-1.9404\,0011\,270_{10}+6$	$-6.4347\,3815\,918_{10}+0$
$+1.8711\,0013\,720_{10}+7$	$+1.8711\,0013\,720_{10}+7$	$-7.7243\,0419\,922_{10}-1$
$-7.2765\,0058\,982_{10}+7$	$-7.2765\,0058\,982_{10}+7$	$+2.4765\,7775\,879_{10}-1$
$+1.3340\,2511\,415_{10}+8$	$+1.3340\,2511\,415_{10}+8$	$+3.1143\,1884\,766_{10}-1$
$-1.1525\,9770\,196_{10}+8$	$-1.1525\,9770\,196_{10}+8$	$+2.9814\,1479\,492_{10}-1$
$+3.7837\,8034\,226_{10}+7$	$+3.7837\,8034\,226_{10}+7$	$+2.3474\,1210\,938_{10}-1$

$d^{(1)}$	$x^{(2)}=x^{(1)}+d^{(1)}$ (rounded)	$r^{(2)}=b-A\,x^{(2)}$
$-1.0032\,6554\,478_{10}-3$	$+4.8510\,0000\,000_{10}+4$	$+0.0000\,0000\,000$
$+1.1270\,1429\,771_{10}-1$	$-1.9404\,0000\,000_{10}+6$	$+0.0000\,0000\,000$
$-1.3720\,0943\,951_{10}-0$	$+1.8711\,0000\,000_{10}+7$	$+0.0000\,0000\,000$
$+5.8981\,9405\,087_{10}+0$	$-7.2765\,0000\,000_{10}+7$	$+0.0000\,0000\,000$
$-1.1415\,0403\,981_{10}+1$	$+1.3340\,2500\,000_{10}+8$	$+0.0000\,0000\,000$
$+1.0196\,2902\,556_{10}+1$	$-1.1525\,9760\,000_{10}+8$	$+0.0000\,0000\,000$
$-3.4226\,0782\,269_{10}+0$	$+3.7837\,8000\,000_{10}+7$	$+0.0000\,0000\,000$

acc inverse. First two solutions (lower triangle only)

1st inverse	2nd inverse
$+1.3597\,5135\,083_{10}-4;$	$+1.3597\,5135\,975_{10}-4;$
$-3.2634\,0327\,687_{10}-3;$	$-3.2634\,0326\,340_{10}-3;$
$+1.0442\,8907\,373_{10}-1;$	$+1.0442\,8904\,429_{10}-1;$
$+2.4475\,5247\,701_{10}-2;$	$+2.4475\,5244\,755_{10}-2;$
$-8.8111\,8918\,469_{10}-1;$	$-8.8111\,8881\,120_{10}-1;$
$+7.9300\,7038\,876_{10}-0;$	$+7.9300\,6993\,007_{10}+0;$
$-8.1585\,0829\,852_{10}-2;$	$-8.1585\,0815\,850_{10}-2;$
$+3.1328\,6729\,439_{10}+0;$	$+3.1328\,6713\,287_{10}+0;$
$-2.9370\,6313\,415_{10}+1;$	$-2.9370\,6293\,706_{10}+1;$
$+1.1188\,8120\,353_{10}+2;$	$+1.1188\,8111\,888_{10}+2;$
$+1.3461\,5387\,384_{10}-1;$	$+1.3461\,5384\,615_{10}-1;$
$-5.3846\,1569\,684_{10}+0;$	$-5.3846\,1538\,461_{10}+0;$
$+5.1923\,0807\,304_{10}+1;$	$+5.1923\,0769\,231_{10}+1;$
$-2.0192\,3093\,289_{10}+2;$	$-2.0192\,3076\,923_{10}+2;$
$+3.7019\,2339\,366_{10}+2;$	$+3.7019\,2307\,692_{10}+2;$
$-1.0769\,2310\,159_{10}-1;$	$-1.0769\,2307\,692_{10}-1;$
$+4.4307\,6950\,879_{10}+0;$	$+4.4307\,6923\,076_{10}+0;$
$-4.3615\,3880\,096_{10}+1;$	$-4.3615\,3846\,154_{10}+1;$
$+1.7230\,7706\,914_{10}+2;$	$+1.7230\,7692\,308_{10}+2;$
$-3.1984\,6182\,140_{10}+2;$	$-3.1984\,6153\,846_{10}+2;$
$+2.7913\,8486\,830_{10}+2;$	$+2.7913\,8461\,539_{10}+2;$
$+3.3333\,3341\,518_{10}-2;$	$+3.3333\,3333\,333_{10}-2;$
$-1.4000\,0009\,298_{10}+0;$	$-1.4000\,0000\,000_{10}+0;$
$+1.4000\,0011\,372_{10}+1;$	$+1.4000\,0000\,000_{10}+1;$
$-5.6000\,0048\,990_{10}+1;$	$-5.6000\,0000\,000_{10}+1;$
$+1.0500\,0009\,497_{10}+2;$	$+1.0500\,0000\,000_{10}+2;$
$-9.2400\,0084\,947_{10}+1;$	$-9.2399\,9999\,999_{10}+1;$
$+3.0800\,0028\,545_{10}+1;$	$+3.0800\,0000\,000_{10}+1;$

It should be appreciated that $-9.2399\,9999\,999_{10}+1$ is the correctly rounded decimal equivalent of the correctly rounded binary representation of $9.24_{10}+1$!

A second test was performed on the computer TR 4 at the Mathematisches Institut der Technischen Hochschule, München. The TR 4 has effectively a 38 binary digit mantissa; since it is not well adapted for the accumulation of inner-products, the latter were computed using true double-precision arithmetic. The results obtained generally confirmed those obtained on KDF 9 but as the rounding procedures are somewhat less satisfactory on TR 4 the accuracy of the first solutions was significantly poorer. However, in all cases the second solution was correct apart from the end figure and the third iteration served only to show that the second was correct. Because of the rounding procedures zero residuals are never obtained on TR 4 and this will be true of many other computers.

Acknowledgements. The authors wish to thank Dr. C. REINSCH of the Mathematisches Institut der Technischen Hochschule, München for a number of valuable comments and for his assistance in testing these algorithms. The work described here is part of the Research Programme of the National Physical Laboratory and this paper is published by permission of the Director of the Laboratory.

References

[1] MARTIN, R. S., G. PETERS, and J. H. WILKINSON: Symmetric decompositions of a positive definite matrix. Numer. Math. 7, 362-383 (1965). Cf. I/1.
[2] WILKINSON, J. H.: Rounding errors in algebraic processes. London: Her Majesty's Stationary Office; Englewood Cliffs, N.J.: Prentice-Hall 1963. German edition: Rundungsfehler. Berlin-Göttingen-Heidelberg: Springer 1969.
[3] — The algebraic eigenvalue problem. London: Oxford University Press 1965.

Inversion of Positive Definite Matrices by the Gauss-Jordan Method

by F. L. Bauer and C. Reinsch

1. Theoretical Background

Let A be a real $n \times n$ matrix and

$$y = A x \qquad (1)$$

the induced mapping $R_n \to R_n$. If $a_{1,1} \neq 0$, then one can solve the first of these equations for x_1 and insert the result into the remaining equations. This leads to the system

$$
\begin{aligned}
x_1 &= a'_{1,1} y_1 + a'_{1,2} x_2 + \cdots + a'_{1,n} x_n \\
y_2 &= a'_{2,1} y_1 + a'_{2,2} x_2 + \cdots + a'_{2,n} x_n \\
&\;\;\vdots \\
y_n &= a'_{n,1} y_1 + a'_{n,2} x_2 + \cdots + a'_{n,n} x_n
\end{aligned}
\qquad (2)
$$

with

$$
\begin{aligned}
& a'_{1,1} = 1/a_{1,1}, \quad a'_{1,j} = -a_{1,j}/a_{1,1}, \quad a'_{i,1} = a_{i,1}/a_{1,1}, \\
& a'_{i,j} = a_{i,j} - a_{i,1} a_{1,j}/a_{1,1} \quad (i, j = 2, \ldots, n).
\end{aligned}
\qquad (3)
$$

In the next step one may exchange the variables x_2 and y_2, provided that $a'_{2,2} \neq 0$. If n such steps are possible one arrives at a system

$$x = B y, \qquad (4)$$

whence $B = A^{-1}$, i.e., n steps of type (3) produce the entries of the inverse. At a typical stage we have the following situation. If after the k-th step the scheme of coefficients is divided into blocks

$$
\left(\begin{array}{c|c}
A_{11} & A_{21} \\
\hline
A_{12} & A_{22}
\end{array} \right)
\begin{array}{l} \} k \\ \\ \} n-k \end{array}
\qquad (5)
$$

then the $k \times k$ block A_{11} is the inverse of the corresponding block of A while A_{22} is the inverse of the corresponding block of A^{-1}. This is easily seen by putting $x_{k+1} = \cdots = x_n = 0$ in the first k equations of (1) or, respectively, $y_1 = \cdots = y_k = 0$ in the last $n - k$ equations of (4).

From this follows that A_{11} and A_{22} are both positive definite (for all k) if A is so. Consequently, the first diagonal entry of A_{22} is positive showing that for positive definite matrices the "next" step is always possible. By induction it also follows that

$$A_{21} = -A_{12}^T, \tag{6}$$

hence for positive definite matrices entries above the diagonal need not be computed.

2. Applicability

The two programs given below can be used to invert a positive definite matrix in situ. They should, however, not be used to solve a linear system $A x = b$ via $x = A^{-1} b$. Whenever A^{-1} is applied from left or right to a vector or a matrix, it is more economical to solve the corresponding linear system using the procedures of [3, 4]. This holds especially for quadratic forms

$$q^2 = b^T A^{-1} b$$

where the computed value, even under the influence of rounding errors, will always be positive if the triangular decomposition of A is used instead of A^{-1}:

$$\text{If} \quad A = L L^T \quad \text{then} \quad q = \|x\|_2 \quad \text{with } x \text{ from} \quad L x = b.\star$$

This is important for statistical applications.

3. Formal Parameter List

Input to procedure *gjdef1*:

n order of A,

$a\,[1:n,\,1:n]$ an array representing the matrix A. Values need be given only in the lower triangle including the diagonal.

Output of procedure *gjdef1*:

$a\,[1:n,\,1:n]$ an array representing the matrix A^{-1}. Only its lower triangle is computed whereas elements strictly above the diagonal retain their input values.

Input to procedure *gjdef2*:

n order of A,

$a\,[1:n\times(n+1)/2]$ entries of the lower triangle of A (including the diagonal) are stored row by row in the vector a, i.e., $a_{i,j}$ is given as $a\,[i\times(i-1)/2+j]$.

Output of procedure *gjdef2*:

$a\,[1:n\times(n+1)/2]$ entries of the lower triangle of A^{-1} (including the diagonal) are given row by row in the vector a, i.e., $(A^{-1})_{i,j}$ is given as $a\,[i\times(i-1)/2+j]$.

Moreover, both procedures provide an exit *fail* which is used if a pivot is not positive, i.e., if the matrix, possibly due to rounding errors, is not positive definite.

$\star$ The authors are indebted to Prof. H. Rutishauser for pointing out that fact.

4. ALGOL Programs

procedure $gjdef1(n)$ *trans*: (a) *exit*: $(fail)$;
value n; **integer** n; **array** a; **label** $fail$;
comment Gauss-Jordan algorithm for the in situ inversion of a positive definite
matrix $a[1:n, 1:n]$. Only the lower triangle is transformed;

begin integer i, j, k; **real** p, q; **array** $h[1:n]$;
   for $k := n$ **step** -1 **until** 1 **do**
   begin
      $p := a[1, 1]$; **if** $p \leq 0$ **then goto** $fail$;
      for $i := 2$ **step** 1 **until** n **do**
      begin
         $q := a[i, 1]$; $h[i] := ($**if** $i > k$ **then** q **else** $-q)/p$;
         for $j := 2$ **step** 1 **until** i **do**
            $a[i-1, j-1] := a[i, j] + q \times h[j]$
      end i;
      $a[n, n] := 1/p$;
      for $i := 2$ **step** 1 **until** n **do** $a[n, i-1] := h[i]$
   end k
end $gjdef1$;

procedure $gjdef2(n)$ *trans*: (a) *exit*: $(fail)$;
value n; **integer** n; **array** a; **label** $fail$;
comment Gauss-Jordan algorithm for the in situ inversion of a positive definite
matrix. Only the lower triangle is transformed which is stored row
by row in the vector $a[1:n \times (n+1)/2]$;

begin integer i, ii, ij, k, m; **real** p, q; **array** $h[1:n]$;
   $m := 0$;
   for $k := n$ **step** -1 **until** 1 **do**
   begin
      $p := a[1]$; **if** $p \leq 0$ **then goto** $fail$; $ii := 1$;
      comment use $p := a[s]$ and $ii := s$ if the lower triangle is stored in
         $a[s:s + n \times (n+1)/2 - 1]$;
      for $i := 2$ **step** 1 **until** n **do**
      begin
         $m := ii$; $ii := ii + i$; **comment** $m = i \times (i-1)/2$, $ii = i \times (i+1)/2$;
         $q := a[m+1]$; $h[i] := ($**if** $i > k$ **then** q **else** $-q)/p$;
         for $ij := m+2$ **step** 1 **until** ii **do**
            $a[ij-i] := a[ij] + q \times h[ij-m]$
      end i;
      $m := m - 1$; **comment** $m = n \times (n-1)/2 - 1$, $ii = n \times (n+1)/2$;
      $a[ii] := 1/p$;
      for $i := 2$ **step** 1 **until** n **do** $a[m+i] := h[i]$
   end k
end $gjdef2$;

5. Organisational and Notational Details

In both procedures a cyclic renumeration of the variables is applied in order that always the pair x_1 and y_1 is exchanged. This is achieved by a revolving storage arrangement with a shift of all entries by one place upwards and to the left per exchange step. Then, the corresponding algorithm is

$$a'_{n,n} = 1/a_{1,1}, \quad a'_{n,j-1} = -a_{1,j}/a_{1,1}, \quad a'_{i-1,n} = a_{i,1}/a_{1,1},$$
$$a'_{i-1,j-1} = a_{i,j} - a_{i,1} a_{1,j}/a_{1,1} \quad (i, j = 2, \ldots, n). \tag{7}$$

This simplifies the organisation since now all n steps are identical. On the other hand an auxiliary array of n components is necessary to hold the quantities $-a_{1,j}/a_{1,1}$. This device was first used by Rutishauser [5] and Petrie [6]. After the k-th step, A is now given by

$$\begin{pmatrix} A_{22} & -A_{12}^T \\ \hline A_{12} & A_{11} \end{pmatrix} \begin{matrix} \}n-k \\ \}k \end{matrix} \tag{8}$$

in lieu of (5) but with the same meaning of A_{ik}. Here, too, it is sufficient to compute the lower triangle though this does not correspond to the lower triangle of (5). According to this, the third equation in (7) is dropped while in the fourth j is restricted to values not larger than i. From (8) follows that for the $(k+1)$-st step

$$a_{1,j} = \begin{cases} a_{j,1} & \text{if } j \le n-k \\ -a_{j,1} & \text{if } j > n-k \end{cases} \tag{9}$$

and $n-k$ is to be replaced by k if counting of the exchange steps is done in backward direction.

Procedures *gjdef1* and *gjdef2* use the same algorithm and produce identical results even under the influence of roundoff errors. They differ only with respect to the storage arrangement.

6. Discussion of Numerical Properties

For positive definite matrices, the Gauss-Jordan process (in floating point arithmetic) is *gutartig* [1] in the sense that the absolute error in any entry of the computed inverse can be interpreted (up to a factor of order of magnitude 1) by an appropriate modification of the original matrix not larger than the machine precision [2].

7. Test Results

The following results were obtained using the TR 4 computer of the *Leibniz-Rechenzentrum der Bayerischen Akademie der Wissenschaften, München*. The computer has a floating point arithmetic with radix 16 and a 38 bit mantissa yielding a machine precision $\varepsilon = 2^{-35} \approx 2.9_{10} - 11$.

To test the formal correctness of the procedures, *gjdef1* and *gjdef2* were applied to the leading 7×7 segment of the Hilbert matrix scaled with 360360 in order to obtain precise input entries [3]. Output results are identical for both procedures and are listed in the Table with the first incorrect digit underlined.

Table. *Output of procedures gjdef1 and gjdef2*

1-st row
  $1.35976\,9220_{10}-4$

2-nd row
  $-3.26347\,64992_{10}-3$ $1.04431\,90465_{10}-1$

3-rd row
  $2.44762\,41626_{10}-2$ $-8.81148\,24518_{10}-1$ $7.93035\,72420_{10}\;0$

4-th row
  $-8.15878\,99658_{10}-2$ $3.13298\,24822_{10}\;0$ $-2.93717\,57800_{10}\;1$ $1.11892\,54334_{10}\;2$

5-th row
  $1.34620\,58996_{10}-1$ $-5.38482\,84021_{10}\;0$ $5.19251\,60558_{10}\;1$ $-2.01931\,25894_{10}\;2$
  $3.70207\,41379_{10}\quad 2$

6-th row
  $-1.07696\,83057_{10}-1$ $4.43095\,42898_{10}\;0$ $-4.36171\,94614_{10}\;1$ $1.72314\,79942_{10}\;2$
  $-3.19859\,27490_{10}\quad 2$ $2.79149\,85811_{10}\quad 2$

7-th row
  $3.33348\,24663_{10}-2$ $-1.40006\,10119_{10}\;0$ $1.40005\,96697_{10}\;1$ $-5.60023\,42886_{10}\;1$
  $1.05004\,32528_{10}\quad 2$ $-9.24037\,56739_{10}\quad 1$ $3.08012\,38345_{10}\;1$

References

1. Bauer, F. L., Heinhold, J., Samelson, K., Sauer, R.: Moderne Rechenanlagen. Stuttgart: Teubner 1965.
2. Bauer, F. L.: Genauigkeitsfragen bei der Lösung linearer Gleichungssysteme. Z. Angew. Math. Mech. **46**, 409–421 (1966).
3. Martin, R. S., Peters, G., Wilkinson, J. H.: Symmetric decomposition of a positive definite matrix. Numer. Math. **7**, 362–383 (1965). Cf. I/1.
4. — Iterative refinement of the solution of a positive definite system of equations. Numer. Math. **8**, 203–216 (1966). Cf. I/2.
5. Rutishauser, H.: Automatische Rechenplanfertigung für programmgesteuerte Rechenanlagen. Z. Angew. Math. Phys. **3**, 312–316 (1952).
6. Petrie, G. W., III: Matrix inversion and solution of simultaneous linear algebraic equations with the IBM 604 Electronic Calculating Punch. Simultaneous linear equations and eigenvalues. Nat. Bur. Standards Appl. Math. Ser. **29**, 107–112 (1953).

Symmetric Decomposition
of Positive Definite Band Matrices*

by R. S. MARTIN and J. H. WILKINSON

1. Theoretical Background

The method is based on the following theorem. If A is a positive definite matrix of band form such that

$$a_{ij}=0 \qquad (|i-j|>m) \tag{1}$$

then there exists a real non-singular lower triangular matrix L such that

$$LL^T=A, \quad \text{where} \quad l_{ij}=0 \qquad (i-j>m). \tag{2}$$

The elements of L may be determined row by row by equating elements on both sides of equation (2). If we adopt the convention that l_{pq} is to be regarded as zero if $q\leq 0$ or $q>p$, the elements of the i-th row are given by the equations

$$\sum_{k=i-m}^{i} l_{ik} l_{jk} = a_{ij} \text{ giving } l_{ij} = \left(a_{ij} - \sum_{k=i-m}^{i-1} l_{ik} l_{jk}\right)\bigg/ l_{jj} \qquad (j=i-m, \ldots, i-1), \tag{3}$$

$$\sum_{k=i-m}^{i} l_{ik}^2 = a_{ii} \text{ giving } l_{ii} = \left(a_{ii} - \sum_{k=i-m}^{i-1} l_{ik}^2\right)^{\frac{1}{2}}. \tag{4}$$

In general we shall be interested mainly in the case when m is much smaller than n. There are then approximately $\frac{1}{2}n(m+1)(m+2)$ multiplications and n square roots in the decomposition. As in the procedures for general symmetric matrices [1] there is an alternative decomposition of the form $\tilde{L} D \tilde{L}^T$ where $\tilde{l}_{ii}=1$ which requires no square roots but marginally more organisational instructions.

From the decomposition $\det(A)$ can be computed since we have

$$\det(A)=\det(LL^T)=\Pi \, l_{ii}^2. \tag{5}$$

The solution of the set of equations $Ax=b$ can be determined in the steps

$$Ly=b, \qquad L^Tx=y. \tag{6}$$

* Prepublished in Numer. Math. 7, 355−361 (1965).

Adopting a similar convention with respect to suffices outside the permitted ranges we have

$$y_i = \left(b_i - \sum_{k=i-m}^{i-1} l_{ik} y_k\right)\bigg/ l_{ii} \qquad (i=1, \ldots, n), \tag{7}$$

$$x_i = \left(y_i - \sum_{k=i+1}^{i+m} l_{ki} x_k\right)\bigg/ l_{ii} \qquad (i=n, \ldots, 1). \tag{8}$$

Approximately $2n(m+1)$ multiplications are involved in a solution and any number of right-hand sides can be processed when L is known.

The inverse of a band matrix is not of band form and will usually have no zero elements; hence when $m \ll n$, which we expect to be the case, it is undesirable to compute A^{-1}.

2. Applicability

The procedures should be used only when A is a positive definite symmetric band matrix and preferably only when $m \ll n$. As m approaches n it becomes less efficient to take advantage of the band form and *choldet* and *cholsol* [1] should be preferred. Band matrices arise most commonly in connexion with finite difference approximations to ordinary and partial differential equations and in such cases m is usually much smaller than n. There are two procedures in this set.

chobanddet. This produces L and $\det(A)$.

chobandsol. This must be preceded by *chobanddet* so that the Cholesky decomposition of A is already available. Corresponding to r right-hand sides b this produces the solutions of the equations $A x = b$. It may be used any number of times after one use of *chobanddet*.

Procedures analogous to *symdet* and *symsol* [1] can be constructed for band matrices but are not included here.

3. Formal Parameter List

3.1 Input to procedure *chobanddet*.

n order of the matrix A.

m number of lines of A on either side of the diagonal.

a elements of the positive definite band matrix A. They must be given as described in section 5.

Output of procedure *chobanddet*.

l the elements of the lower-triangle of the Cholesky decomposition of A stored as described in section 5.

d1 and *d2* two numbers which together give the determinant of A.

fail this is the exit used when A, possibly as a result of the rounding errors is not positive definite.

3.2 Input to procedure *chobandsol*.

n the order of the matrix A.

m the number of lines of A on either side of the diagonal.

r the number of right-hand sides for which $A x = b$ is to be solved.

4*

l   the elements of the lower-triangle of the Cholesky decomposition of the positive definite matrix *A* as produced by procedure *chobanddet*.

b   the $n \times r$ matrix formed by the matrix of right-hand sides.

Output of procedure *chobandsol*.

x   the $n \times r$ matrix formed by the solution vectors.

4. ALGOL Programs

procedure *chobanddet* (*n*, *m*) *data*: (*a*) *result*: (*l*, *d1*, *d2*) *exit*: (*fail*);
value *m*, *n*; **integer** *d2*, *m*, *n*; **real** *d1*; **array** *a*, *l*; **label** *fail*;
comment The lower half of a positive definite symmetric band matrix, *A*, with *m* lines on either side of the diagonal is stored as an $n \times (m+1)$ array, $a[i, k]$, $i = 1(1)n$, $k = 0(1)m$, $a[i, m]$ being the diagonal elements. The Cholesky decomposition $A = LU$, where *U* is the transpose of *L*, is performed and *L* is stored in $l[i, k]$ in the same form as *A*. The reciprocals of the diagonal elements are stored instead of the elements themselves. *A* is retained so that the solution obtained can subsequently be improved. However, *L* and *A* can be identified in the call of the procedure. The determinant, $d1 \times 2 \uparrow d2$, of *A* is also computed. The procedure will fail if *A*, modified by the rounding errors, is not positive definite;

```
begin integer i, j, k, p, q, r, s;
     real y;
     d1 := 1;
     d2 := 0;
     for i := 1 step 1 until n do
     begin p := (if i > m then 0 else m − i + 1);
           r := i − m + p;
           for j := p step 1 until m do
           begin s := j − 1;
                 q := m − j + p;
                 y := a[i, j];
                 for k := p step 1 until s do
                 begin y := y − l[i, k] × l[r, q];
                       q := q + 1
                 end k;
                 if j = m then
                 begin d1 := d1 × y;
                       if y = 0 then
                       begin d2 := 0;
                             go to fail
                       end;
           L1:       if abs (d1) ≥ 1 then
                       begin d1 := d1 × 0.0625;
                             d2 := d2 + 4;
                             go to L1
                       end;
```

```
L2:     if abs(d1) < 0.0625 then
            begin d1 := d1 × 16;
                  d2 := d2 − 4;
                  go to L2
            end;
            if y < 0 then go to fail;
            l[i, j] := 1/sqrt(y)
        end
        else
        l[i, j] := y × l[r, m];
        r := r + 1
    end j
  end i
end chobanddet;
procedure chobandsol (n, m) data:(r, l, b) result:(x);
value m, n, r; integer m, n, r; array l, b, x;
```

comment Solves $A x = b$, where A is a positive definite symmetric band matrix with m lines on either side of the diagonal and b is an $n \times r$ matrix of r right-hand sides. The procedure *chobandsol* must be preceded by *chobanddet* in which L is produced in $l[i, k]$, from A. $A x = b$ is solved in two steps, $L y = b$ and $U x = y$. The matrix b is retained in order to facilitate the refinement of x, but x is overwritten on y. However, x and b can be identified in the call of the procedure;

```
begin integer i, j, k, p, q, s;
      real y;
      s := m − 1;
      for j := 1 step 1 until r do
      begin comment solution of L y = b;
            for i := 1 step 1 until n do
            begin p := (if i > m then 0 else m − i + 1);
                  q := i;
                  y := b[i, j];
                  for k := s step −1 until p do
                  begin q := q − 1;
                        y := y − l[i, k] × x[q, j]
                  end k;
                  x[i, j] := y × l[i, m]
            end i;
            comment solution of U x = y;
            for i := n step −1 untii 1 do
            begin p := (if n − i > m then 0 else m − n + i);
                  y := x[i, j];
                  q := i;
                  for k := s step −1 until p do
```

```
                begin q := q + 1;
                        y := y − l[q, k] × x[q, j]
                end k;
                x[i, j] := y × l[i, m]
        end i
    end j
end chobandsol;
```

5. Organisational and Notational Details

To take full advantage both of symmetry and of the band form of the matrix an unconventional form of storage is used. We actually store a rectangular array A of dimension $n \times (m+1)$ and the relationship between this stored array and the conventional array in the case $n=8$, $m=3$ is illustrated by the arrays

		Stored array			*Lower-triangle of conventional array*			
x	x	x	a_{13}	a_{13}				
x	x	a_{22}	a_{23}	a_{22}	a_{23}			
x	a_{31}	a_{32} $\cdot$	a_{33}	a_{31}	a_{32}	a_{33}		
a_{40}	a_{41}	a_{42} $\cdot$	a_{43}	a_{40}	a_{41}	a_{42}	a_{43}	
a_{50}	a_{51}	a_{52}	a_{53}		a_{50}	a_{51}	a_{52}	a_{53}
a_{60} $\cdot$	a_{61}	a_{62}	a_{63}			a_{60}	a_{61}	a_{62}	a_{63}
a_{70} $\cdot$	a_{71}	a_{72}	a_{73}				a_{70}	a_{71}	a_{72}	a_{73}
a_{80}	a_{81}	a_{82}	a_{83}					a_{80}	a_{81}	a_{82}	a_{83}

(9)

Those elements of the stored array which correspond to the super-diagonal half of a row in the conventional array are indicated by the dotted lines. The elements in the upper left-hand corner of the stored array may be arbitrary since these are not used. Equations (3) and (4) are modified accordingly and the 'end effects' are taken care of in such a way as to ensure that no element is required which is not actually a true member of the conventional array. L is not overwritten on A (as it could be) so that A is still available for the computation of residuals if required. The reciprocals of the diagonal elements of L are stored instead of the elements themselves. The determinant of A is computed in *chobanddet* whether it is required or not and in order to avoid overspill and underspill the form $2^{d2}d1$ is used where $d2$ is an integer (a multiple of 4) and $d1$ is a real number such that $\frac{1}{16} \le |d1| < 1$.

In *chobandsol* the equations $Ax=b$ corresponding to r different right-hand sides are solved using equations (7) and (8). The vectors b are not overwritten but each x of equation (8) is overwritten on the corresponding y of equation (7). The vectors b are therefore still available for the computation of residuals.

6. Discussion of Numerical Properties

The reader is referred to the corresponding discussion given in connexion with the procedures *choldet*, *cholsol* etc. [1] much of which applies with obvious modifications to these procedures. With standard floating-point computation the computed L certainly satisfies the relation $LL^T = A + F$ where

$$\|F\|_2 \le k_1(m+1)^2\, 2^{-t} \max |a_{ii}| \le k_1(m+1)^2 2^{-t} \|A\|_2 \qquad (10)$$

though the error is usually appreciably smaller than is indicated by this bound. If inner-products can be accumulated then certainly

$$\|F\|_2 \leq k_2 (m+1) \, 2^{-t} \max |a_{i\,i}| \leq k_2 (m+1) \, 2^{-t} \|A\|_2 \tag{11}$$

and again this is usually an over estimate. (See, for example, [2]).

7. Examples of the Use of the Procedure

The Cholesky decomposition of a symmetric matrix is the essential requirement in the *LR* algorithm of RUTISHAUSER for symmetric band matrices [3].

When solutions of the equations $A\,x = b$ are required for a large number of different right-hand sides the procedure *Chobandsol* is very economical besides being very accurate. If inner-products can be accumulated then the correctly rounded solution can usually be obtained by the iterative procedure defined by

$$x^{(0)} = 0,$$
$$r^{(k)} = b - A\,x^{(k)}, \qquad A\,e^{(k)} = r^{(k)}, \tag{12}$$
$$x^{(k+1)} = x^{(k)} + e^{(k)}$$

iteration being terminated when $\|e^{(k)}\| \leq 2^{-t} \|x^{(k)}\|$. The accumulation of inner-products is essential only in the computation of the residual.

8. Test Results

Procedures *chobanddet* and *chobandsol* have been tested on the computer KDF 9 which uses a 39-binary digit mantissa. For comparison the same matrix was used as that employed by GINSBURG [4] to test the conjugate gradient algorithm. This is the matrix of order 40 given by

$$A =
\begin{bmatrix}
5 & -4 & 1 \\
-4 & 6 & -4 & 1 \\
1 & -4 & 6 & -4 & 1 \\
& 1 & -4 & 6 & -4 & 1 \\
& & & \cdot\,\cdot\,\cdot\,\cdot\,\cdot\,\cdot\,\cdot\,\cdot\,\cdot\,\cdot\,\cdot \\
& & & & 1 & -4 & 6 & -4 & 1 \\
& & & & & 1 & -4 & 6 & -4 \\
& & & & & & 1 & -4 & 5
\end{bmatrix}. \tag{13}$$

The computed solution and the correct solution corresponding to the right-hand side (1, 0, ..., 0) are given in the Table. The average error in the components of the solution is 2 parts in 10^7, a remarkably good result in view of the fact that $\varkappa(A) \approx 4.7 \times 10^5$. In general the results have two more correct decimals than those obtained by GINSBURG with the conjugate gradient method after 100 iterations, in spite of the fact that the KDF 9 works to a precision which is only half of a decimal greater than that used by GINSBURG. This confirms our general experience that the more stable of direct methods usually give results which are *more* rather than *less* accurate than those ultimately attained by iterative methods. There are often good reasons for using iterative methods, but accuracy of the results is not among these reasons.

Table

Computed Solution	Correct Solution	Computed Solution	Correct Solution
$1.31707304671_{10}1$	$1.31707317707_{10}1$	$1.04146325581_{10}2$	$1.04146341461_{10}2$
$2.53658511853_{10}1$	$2.53658536596_{10}1$	$1.01951203740_{10}2$	$1.01951219511_{10}2$
$3.66097524057_{10}1$	$3.66097560981_{10}1$	$9.92926673540_{10}1$	$9.92926829271_{10}1$
$4.69268243792_{10}1$	$4.69268292681_{10}1$	$9.61951066670_{10}1$	$9.61951219511_{10}1$
$5.63414573575_{10}1$	$5.63414634151_{10}1$	$9.26829119222_{10}1$	$9.26829268291_{10}1$
$6.48780415917_{10}1$	$6.48780487801_{10}1$	$8.87804733608_{10}1$	$8.87804878051_{10}1$
$7.25609673329_{10}1$	$7.25609756101_{10}1$	$8.45121812241_{10}1$	$8.45121951221_{10}1$
$7.94146248321_{10}1$	$7.94146341461_{10}1$	$7.99024257525_{10}1$	$7.99024390241_{10}1$
$8.54634043409_{10}1$	$8.54634146341_{10}1$	$7.49755971872_{10}1$	$7.49756097561_{10}1$
$9.07316961091_{10}1$	$9.07317073171_{10}1$	$6.97560857679_{10}1$	$6.97560975611_{10}1$
$9.52438903861_{10}1$	$9.52439024391_{10}1$	$6.42682817345_{10}1$	$6.42682926831_{10}1$
$9.90243774212_{10}1$	$9.90243902441_{10}1$	$5.85365753260_{10}1$	$5.85365853661_{10}1$
$1.02097547463_{10}2$	$1.02097560981_{10}2$	$5.25853567806_{10}1$	$5.25853658541_{10}1$
$1.04487790759_{10}2$	$1.04487804881_{10}2$	$4.64390163362_{10}1$	$4.64390243901_{10}1$
$1.06219497556_{10}2$	$1.06219512201_{10}2$	$4.01219441304_{10}1$	$4.01219512201_{10}1$
$1.07317058100_{10}2$	$1.07317073171_{10}2$	$3.36585307008_{10}1$	$3.36585365851_{10}1$
$1.07804862636_{10}2$	$1.07804878051_{10}2$	$2.70731659845_{10}1$	$2.70731707321_{10}1$
$1.07707301407_{10}2$	$1.07707317071_{10}2$	$2.03902403190_{10}1$	$2.03902439021_{10}1$
$1.07048764659_{10}2$	$1.07048780491_{10}2$	$1.36341439415_{10}1$	$1.36341463411_{10}1$
$1.05853642636_{10}2$	$1.05853658541_{10}2$	$6.82926708936_{10}0$	$6.82926829271_{10}0$

The computed value of the determinant was $4.104004559 \times 10^{-1} \times 2^{12}$ compared with the true value $1681 = 4.10400390 \ldots \times 10^{-1} \times 2^{12}$.

A second test was performed at the Mathematisches Institut der Technischen Hochschule München using the computer TR 4 which works to the same precision as that used by GINSBURG. The results were even better than those obtained on KDF 9, the errors in components of the solution being smaller by factors which varied from 0.49 to 0.039, thus illustrating even more conclusively the superior accuracy of the direct method.

Acknowledgements. The authors wish to thank Mrs. E. MANN and Dr. C. REINSCH for valuable assistance in the development of these algorithms. The work described here has been carried out as part of the Research Programme of the National Physical Laboratory and this paper is published by permission of the Director of the Laboratory.

References

[1] MARTIN, R. S., G. PETERS, and J. H. WILKINSON: Symmetric decompositions of a positive definite matrix. Numer. Math. 7, 362—383 (1965). Cf. I/1.

[2] WILKINSON, J. H.: The algebraic eigenvalue problem. London: Oxford University Press 1965.

[3] RUTISHAUSER, H.: Solution of the eigenvalue problem with the LR transformation. Nat. Bur. Standards Appl. Math. Ser. 49, 47—81 (1958).

[4] GINSBURG, T.: The conjugate gradient method. Numer. Math. 5, 191—200 (1963). Cf. I/5.

The Conjugate Gradient Method

by T. GINSBURG

1. Theoretical Background

The CG-algorithm is an iterative method to solve linear systems

$$(1) \qquad A x + b = 0$$

where A is a symmetric and positive definite coefficient matrix of order n. The method has been described first by Stiefel and Hesteness [1, 2] and additional information is contained in [3] and [4]. The notations used here coincide partially with those used in [5] where also the connexions with other methods and generalisations of the CG-algorithm are presented.

The CG-algorithm can be described in short as follows: Starting from an "initial guess" x_0 (which may be taken as $x_0 = 0$ if nothing is known about the solution) two sequences of vectors $x_0, x_1, x_2, \ldots$ and $r_0, r_1, r_2, \ldots$ are generated iteratively introducing two auxiliary vectors Δx_k and Δr_k and a series of numbers q_{k+1} and e_k as follows[*]:

$$(2) \qquad \Delta r_{-1} = \Delta x_{-1} = 0; \qquad r_0 = A x_0 + b;$$

$$(3) \qquad e_k = \begin{cases} 0 & \text{if } k = 0 \\ q_k \dfrac{(r_k, r_k)}{(r_{k-1}, r_{k-1})} & \text{if } k > 0, \end{cases}$$

$$(4) \qquad q_{k+1} = \frac{(r_k, A r_k)}{(r_k, r_k)} - e_k,$$

$$(5) \qquad \Delta x_k = \frac{1}{q_{k+1}} (-r_k + e_k \Delta x_{k-1}),$$

$$(6) \qquad x_{k+1} = x_k + \Delta x_k,$$

$$(7) \qquad \Delta r_k = \frac{1}{q_{k+1}} (-A r_k + e_k \Delta r_{k-1}),$$

$$(8) \qquad r_{k+1} = r_k + \Delta r_k.$$

---

[*] In contrast to [3] or [5] a substitution of the indices of the variables q_k and e_k has been made in order to be in agreement with their use in the QD-algorithm.

The main properties of the vectors x_k and r_k derived from the theory, are given by the relations

(9) $r_k = A\,x_k + b$,

(10) $(r_i, r_k) = 0 \quad \text{for} \quad i \neq k$,

(11) $x_n = x_\infty = \text{exact solution (i.e. } r_n = 0)$.

However, the roundoff-errors cause the actual computation to deviate from the theoretical course and therefore the relations (9), (10), (11) will not hold exactly. As a consequence x_n instead of being the exact solution may be—as the experiments in [5] have shown—quite far away from it.

It turned out, however, that this drawback can be corrected best simply by continuation of the algorithm beyond $k = n$ (see [5]). If this is done the residual vector r_k will sooner or later begin to decrease sharply★ such that ultimately x_k will be as good an approximation to x_∞ as it may be expected from the condition of the coefficient matrix★★.

An important problem is the determination of the moment when an iterative method does no longer improve the approximation (here the vector x) already obtained and therefore should be terminated. As indicated above, termination after n steps as given by theory is not feasible and therefore other conditions have been used for terminating the execution of procedure cg (cf. Section 6).

2. Applicability

The CG-algorithm can be used for the solution of linear systems (1) with positive definite symmetric coefficient matrices A. However, it is advantageous only for sparse matrices A, i.e. for matrices whose coefficients are mostly zero. In addition, the matrix A should not be too badly scaled, that is, the euclidian lengths of the n columns should not be of totally different order of magnitude. This requirement may ask for a transformation of the system to be solved; the safest way being to make all diagonal elements to 1 under preservation of symmetry.

Moreover, for its use in procedure cg, a coefficient matrix A must not be given as an array of $n \times n$ numbers, but as a rule (ALGOL-procedure) for computing from any given vector v the vector $A \times v$.

In addition to solving linear systems, the CG-algorithm can sometimes be used for investigating the spectrum of A. To this end we focus attention to the arrays q, e, which are formal output of procedure cg. Indeed, any sequence

$$Z_p = \{q[1], e[1], q[2], e[2], \ldots, e[p-1], q[p]\}$$

is a qd-line whose eigenvalues depend on the constant terms b, but are somehow related to the spectrum of A. Indeed, if the eigenvalues of Z_p are computed for increasing p, it may be observed that usually some of these tend to fixed values.

★ Experiments have shown that up to $3n$ steps may be needed in case of very bad condition (see Fig. 1).

★★ e.g. for 11-digit machine, hardly more then 5 digits may be expected for the solution of a system with a condition number 10^6 of the coefficient matrix.

If so, these "limits" may be considered as eigenvalues of A. It should be recognized however, that if p exceeds n, multipletts begin to appear which makes it sometimes difficult to identify eigenvalues of A. (See also Sections 7 and 8).

3. Formal Parameter List

Input to procedure cg:

n Order of the matrix A.

p Upper bound in the declaration of the arrays q, $e\,[1:p]$. For $k=1, 2, \ldots, p$ the values q_k, e_k occurring in formulae (3)–(7) are stored into $q\,[k]$, $e\,[k]$.

b **array** $b[1:n]$ containing the constant terms of the Eqs. (1).

op **procedure** $op\,(n, v)$ res: (av); **value** n; **integer** n; **array** v, av; **code** computing for an arbitrary vector $v\,[1:n]$ the product $av\,[1:n]=A \times v$ without changing v.

x In the normal case $(n>0)$ x is irrelevant at the beginning since the process then starts with $x=$null vector. However, n may be negative which serves as a flag that an initial guess for the solution x must be given. Of course, in this case $abs\,(n)$ is used internally instead of n.

nn Maximum number of CG-steps to be performed.

Output of procedure cg:

x **array** $x\,[1:n]=$ approximate solution of (1).

nn The number of CG-steps actually performed. If nn is equal to the originally given value, this indicates that more steps might have produced a more precise solution.

q, e **array** $q, e[1:p]$ containing the values q_k, e_k $(k=1, \ldots, p)$. If the final value of nn is below p, then these arrays are filled only up to $q[nn]$, $e[nn]$.

Exit from procedure cg:

$indef$ Exit used if for the first time a value q_k becomes non-positive. In this case the matrix A is either not positive definite or so illconditioned that its smallest eigenvalue is in the roundoff-error level. In either case no results are given.

4. ALGOL Program

Procedure cg uses two internally declared subprocedures $inprod$ and $lincom$. In cases where the execution of $op\,(n, v, av)$ requires not much computing time, transcription of these procedures into machine code may greatly improve the efficiency of procedure cg.

```
procedure cg (n, p, b, op) trans: (x, nn) result: (q, e) exit: (indef);
    value n, p;
    integer n, p, nn; array b, x, q, e; procedure op; label indef;
    begin
        integer k, j, sqn, ksqn;
        real rr, rar, oldrr, rfe, alpha, f, f2, ek, qk;
```

```
array r, dr, dx, ar [1 : if n > 0 then n else −n];
real procedure inprod (n, x, y); comment no globals;
    value n;
    integer n; array x, y;
    begin
        real s;
        integer j;
        s := 0;
        for j := 1 step 1 until n do s := s + x[j] × y[j];
        inprod := s;
end inprod;
procedure lincom (n, x, y, a) trans : (b, c); comment no globals;
    value n, x, y;
    real x, y; integer n; array a, b, c;
    begin
        integer j;
        for j := 1 step 1 until n do
        begin
            b[j] := (x × b[j] − a[j]) × y;
            c[j] := c[j] + b[j];
        end for j;
end lincom;
if n < 0 then
begin
    n := −n;
    op (n, x, ar);
    for j := 1 step 1 until n do r[j] := ar[j] + b[j];
end
else
    for j := 1 step 1 until n do
        begin x[j] := 0; r[j] := b[j] end;
rr := rfe := 0;
ksqn := sqn := sqrt (n);
alpha := 1;
f := exp (1/n);
f2 := f↑2;
for j := 1 step 1 until n do dx[j] := dr[j] := 0;
comment big loop for cg-step;
    for k := 0 step 1 until nn do
    begin
comment test for true residuals;
        if k = ksqn then
        begin
            ksqn := ksqn + sqn;
            alpha := alpha × f;
            f := f × f2;
```

```
            op (n, x, ar);
            rfe := 0;
            for j := 1 step 1 until n do
                    rfe := rfe + (ar[j] + b[j] − r[j])↑2;
            end if k = ksqn;
            op (n, r, ar);
            oldrr := rr;
            rr := inprod (n, r, r);
            rar := inprod (n, r, ar);
            ek := if k = 0 then 0.0 else qk × rr/oldrr;
            if k > 0 ∧ k ≤ p then begin q[k] := qk; e[k] := ek end;
            if rr ≤ alpha × rfe then nn := k;
            if k = nn then goto reg;
            qk := rar/rr − ek;
            if qk ≤ 0 then goto indef;
    comment computation of new r and x;
            lincom (n, ek, 1/qk, r, dx, x);
            lincom (n, ek, 1/qk, ar, dr, r);
        end for k;
reg:
end cg;
```

5. Organisational and Notational Details

For the sake of storage economy it is most important that the matrix is not defined in the usual way as a two-dimensional array of numbers but as a rule to compute the product Ax for any given vector x. For this reason procedure cg has no provision for the matrix A as an input parameter; instead of this it requires the above-mentioned rule as a procedure op.

According to the theory, the CG-algorithm generates sequences of vectors $r_k, x_k, \Delta r_k, \Delta x_k, A r_k, A x_k$, which are distinguished by the subscript k. This subscript gives the number of the CG-step in which these vectors are computed. The procedure cg however, does not distinguish these vectors explicitly with respect to k. In other words, in procedure cg the vectors $r, x, \Delta r, \Delta x, A r, A x$ occur as quantities taking in succession the vectors $r_k, x_k, \Delta r_k, \Delta x_k, A r_k, A x_k$ as actual values. For all these vectors the components are counted from 1 to n.

At the same time, the inner products (r_k, r_k), $(r_k, A r_k)$, (r_{k-1}, r_{k-1}) and the coefficients q_k, e_k, which are used in the CG-algorithm, are also not distinguished with respect to k and therefore appear as simple variables $rr, rar, oldrr, qk, ek$ respectively. On the other hand, at least for $k \leq p$ the q_k, e_k are stored into the formal arrays $q, e [1 : p]$ and thus are available after termination of procedure cg.

6. Discussion of Numerical Properties

Extensive studies of the numerical behaviour of the CG-method have been published in [5]. We therefore can confine here to the termination problem. Since roundoff errors prevent termination after n steps as stated by the theory and consequently the CG-method has to be treated as if it were an infinite iterative process, proper termination is not trivial:

1. Should the quantity $rr = \|r_k\|^2$ become precisely zero, then immediate termination is a must, because rr appears later as denominator.

2. Like for other iterative processes, we must—at least as a safeguard—prescribe an upper bound nn for the maximum number of CG-steps to be performed. Thus the process will terminate anyhow after nn steps, possibly without producing a sufficiently precise solution. It is difficult to give an advice for choosing nn, but if not other considerations suggest a smaller value, then $nn = 2 \times n$ seems appropriate for mildly illconditioned systems such as the plate problem in [5]. More critical cases may require higher values of nn, but if $5 \times n$ is not sufficient, we can hardly expect a solution at all. On the other hand, test example c) shows that a well-conditioned linear system can be solved in far less than n steps, and this is even more pronounced for big n.

3. In the ideal case, the computation would be discontinued as soon as continuation is no longer meaningful. For the latter, the true residuals give an indication: The vectors r_k occurring in the CG-algorithm are not computed as $A x_k + b$, but recursively from r_{k-1}, r_{k-2}. Because of the roundoff-errors, the r_k computed in this way (the so-called *recursive residuals*) are slightly different from the *true residuals* $\bar{r}_k$ obtained by computing $A x_k + b$, and this difference gives the desired information.

Thus we compute the quantity

$$(12) \qquad\qquad rfe = \|\bar{r}_k - r_k\|^2,$$

where r_k is calculated by (7), (8), $\bar{r}_k$ by (9). The value of rfe tends to increase with increasing k as may be seen from Fig. 1, where a typical behaviour of rr and rfe during execution of procedure cg is shown. As soon as rr and rfe become of the same order of magnitude, this indicates that $r_k, \bar{r}_k$ have reached the roundoff-error level.

However, we cannot use simply $rr < rfe$ as a termination criterion, because rr might forever remain slightly above rfe, but must use instead

$$(13) \qquad\qquad rr < alpha \times rfe,$$

where *alpha* is not a constant factor (this might fail again) but

$$(14) \qquad\qquad alpha = exp((k/n)\uparrow 2),$$

with k = number of current CG-step, n = order of the linear system. The steadily increasing factor *alpha* in (13) enforces finally termination also in cases of hopelessly bad condition of the coefficient matrix A, but will also cause swift termination when rfe and rr approach each other quickly.

4. Since the evaluation of rfe is time-consuming because it requires the computation of the true residual which involves a product $A x_k$, rfe is not computed

every step, but only every sqn^{th} step, where

(15) $$sqn = entier\,(sqrt\,(n))\,,$$

and is then kept constant until a new r/e is computed. On the other hand, (13) is tested every CG-step.

7. Examples of the Use of the Procedure

As an example for the procedure op defining the matrix A, we chose a problem from the elasticity theory (loaded bar) which leads to the following matrix:

(16)

$$\begin{Vmatrix}
5 & -4 & 1 \\
-4 & 6 & -4 & 1 \\
1 & -4 & 6 & -4 & 1 \\
 & 1 & -4 & 6 & -4 & 1 \\
 & & & & \cdots \\
 & & & & 1 & -4 & 6 & -4 & 1 \\
 & & & & & 1 & -4 & 6 & -4 & 1 \\
 & & & & & & 1 & -4 & 6 & -4 \\
 & & & & & & & 1 & -4 & 5
\end{Vmatrix}.$$

The procedure op for this matrix of order $n \geq 4$ has the following form:

```
procedure op (n, x) result: (ax);
value n; integer n; array x, ax;
begin integer j;
      ax[1]     := 5×x[1] −4×x[2] +x[3];
      ax[2]     := 6×x[2] −4×(x[1] +x[3]) +x[4];
      ax[n−1]   := 6×x[n−1] −4×(x[n−2] +x[n]) +x[n−3];
      ax[n]     := 5×x[n] −4×x[n−1] +x[n−2];
   for j:= 3 step 1 until n−2 do
      ax[j]     := 6×x[j] −4×(x[j−1] +x[j+1]) +x[j−2] +x[j+2];
end op;
```

A further example shows how to compute the eigenvalues of the partial qd-lines $Z_{10}, Z_{20}, Z_{30}, \ldots$, provided a procedure $qdalg$ for computing the eigenvalues of a given qd-line is available:

```
begin
     comment here n, nn, b[1:n] are assumed as being declared and having values;
     integer k, j;
     array z, q, e[1:nn], x[1:n];
     procedure qdalg (n, q, e) result: (z); value n; integer n; array q, e, z;
     code computing the eigenvalues z[1] through z[n] of the qd-line q[1], e[1],
          q[2], e[2] ... e[n−1], q[n] without changing these values and without
          changing nor making use of e[n];
```

```
procedure op (n, v) result: (av); value n; integer n; array v, av;
code as in the foregoing example;
cg (n, nn, b, op, x, nn, q, e, ind);
for k := 10 step 10 until nn do
begin
      outreal (2, k);
      qdalg (k, q, e, z);
      for j := 1 step 1 until k do outreal (2, z[j]);
end for k;
ind:
end;
```

8. Test Results

a) The plate problem as described in [5] was solved again with the same meshsize ($h = 1/8$, $n = 70$) and the same load. In this case the termination condition (13) was fulfilled after 94 steps. Table 1 exhibits the precise solution, the computed solution x_{94} and the residual vector r_{94}, the norm of the error vector (precise minus computed solution) being $1.11_{10} - 8$. Fig. 1 shows the quantities $rr = \|r_k\|^2$ and $rfe = \|A x_k + b - r_k\|^2$ as a function of the step number k (for this purpose the computation was extended over the termination point).

b) Again the plate problem with the finer mesh ($h = 1/16$, $n = 270$) and the same load. 320 steps were needed in this case until the criterion (13) caused termination. The computed solution has an error norm of $2.53_{10} - 7$ which is about 23 times bigger than for the coarse mesh, but can be explained by the approximately 16 times bigger condition number of the matrix for the fine mesh.

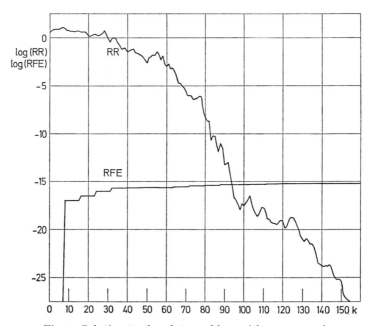

Fig. 1. Solution to the plate problem with coarse mesh

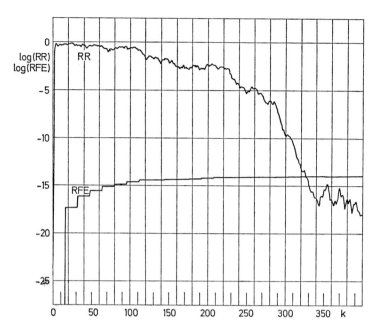

Fig. 2. Solution to the plate problem with fine mesh

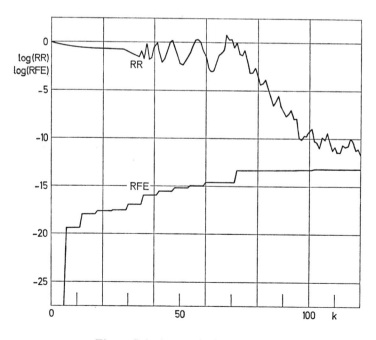

Fig. 3. Solution to the bar problem

Table 1

Precise solution, rounded to 12 figures after the decimal point:

0.6033 8853 4725	1.5228 5550 7082	2.3193 2738 0471	2.6546 1499 7594	2.4043 3919 1438	1.6553 3087 3833	0.5878 3872 0734
0.6670 5689 8255	1.6055 7661 6779	2.4028 0166 1686	2.7174 5920 6685	2.4340 2871 0056	1.6623 3954 8867	0.5879 1633 7364
0.7225 5578 0063	1.7271 3748 2651	2.5846 7695 7338	2.8906 1712 5814	2.5365 3271 1198	1.7067 8523 9250	0.5992 8507 2004
0.7584 7158 3441	1.8244 9251 6704	2.7736 9359 3777	3.0705 2030 7855	2.6127 1387 5655	1.7255 3547 6230	0.6009 0317 5172
0.7220 0835 4477	1.7381 4520 6849	2.6459 4659 6875	2.9261 5754 6494	2.4824 9508 2188	1.6360 5055 7261	0.5691 5878 0372
0.6117 0567 0108	1.4632 6029 7517	2.1931 1731 0241	2.4482 6108 1995	2.1385 4469 0028	1.4340 1328 1021	0.5026 6709 8980
0.4824 0458 8257	1.1554 8664 3324	1.7277 2851 9249	1.9536 8179 2094	1.7495 3332 9109	1.1951 6812 8218	0.4228 3423 7756
0.3684 1992 5831	0.8935 0297 7418	1.3483 3942 8830	1.5475 8812 9725	1.4129 1406 0923	0.9800 0585 7390	0.3494 8570 2869
0.2735 4272 5357	0.6904 4652 4090	1.0673 2152 3387	1.2486 0934 0653	1.1590 0824 8484	0.8136 0986 0644	0.2919 5212 8978
0.1752 9237 7877	0.5186 3748 8493	0.8514 4072 4312	1.0276 4122 8244	0.9722 9523 4249	0.6905 3327 6909	0.2491 7956 5887

Difference between precise solution and computed solution r_{94}:

$1.513_{10}-9$	$2.648_{10}-9$	$3.201_{10}-9$	$2.735_{10}-9$	$2.386_{10}-9$	$1.484_{10}-9$	$2.910_{10}-10$
$1.469_{10}-9$	$2.444_{10}-9$	$2.793_{10}-9$	$2.619_{10}-9$	$1.513_{10}-9$	$1.396_{10}-9$	$1.891_{10}-10$
$1.076_{10}-9$	$2.357_{10}-9$	$1.920_{10}-9$	$2.037_{10}-9$	$1.571_{10}-9$	$4.947_{10}-10$	$-2.910_{10}-11$
$9.895_{10}-10$	$1.688_{10}-9$	$1.571_{10}-9$	$1.804_{10}-9$	$1.105_{10}-9$	$3.492_{10}-10$	$-7.275_{10}-11$
$5.093_{10}-10$	$1.280_{10}-9$	$1.746_{10}-9$	$1.338_{10}-9$	$8.149_{10}-10$	$2.910_{10}-11$	$-2.764_{10}-10$
$5.384_{10}-10$	$1.513_{10}-9$	$1.746_{10}-9$	$1.979_{10}-9$	$8.731_{10}-10$	$-5.820_{10}-11$	$-7.275_{10}-11$
$3.929_{10}-10$	$1.047_{10}-9$	$1.600_{10}-9$	$1.280_{10}-9$	$7.566_{10}-10$	$-1.164_{10}-10$	$-9.458_{10}-11$
$4.365_{10}-10$	$8.585_{10}-10$	$1.426_{10}-9$	$1.222_{10}-9$	$4.656_{10}-10$	$7.275_{10}-11$	$-2.837_{10}-10$
$-1.018_{10}-10$	$2.037_{10}-10$	$4.074_{10}-10$	$3.492_{10}-10$	$-3.783_{10}-10$	$-5.529_{10}-10$	$-2.546_{10}-10$
$-6.184_{10}-10$	$-6.548_{10}-10$	$-6.402_{10}-10$	$-7.566_{10}-10$	$-1.018_{10}-9$	$-1.193_{10}-9$	$-4.474_{10}-10$

Recursive residual r_{94}:

$-4.035_{10}-11$	$5.043_{10}-10$	$-1.051_{10}-9$	$2.115_{10}-10$	$-2.105_{10}-10$	$1.350_{10}-10$	$-5.337_{10}-10$
$-1.699_{10}-9$	$7.846_{10}-11$	$4.932_{10}-10$	$-2.787_{10}-10$	$8.333_{10}-10$	$-3.302_{10}-9$	$3.838_{10}-9$
$3.704_{10}-9$	$-3.467_{10}-9$	$2.858_{10}-9$	$-2.549_{10}-9$	$-1.239_{10}-9$	$2.757_{10}-9$	$-3.407_{10}-10$
$-5.392_{10}-9$	$2.917_{10}-9$	$-3.454_{10}-10$	$1.409_{10}-10$	$1.664_{10}-9$	$-2.813_{10}-9$	$1.551_{10}-10$
$2.166_{10}-9$	$2.145_{10}-10$	$-2.098_{10}-9$	$1.955_{10}-9$	$-4.156_{10}-9$	$3.059_{10}-9$	$1.813_{10}-9$
$-1.508_{10}-9$	$-4.542_{10}-10$	$4.738_{10}-10$	$4.060_{10}-10$	$1.001_{10}-9$	$-9.757_{10}-10$	$-2.900_{10}-9$
$2.883_{10}-9$	$-8.830_{10}-10$	$-5.188_{10}-10$	$2.535_{10}-10$	$-1.926_{10}-9$	$4.039_{10}-9$	$-3.784_{10}-9$
$-2.177_{10}-9$	$1.846_{10}-9$	$-4.931_{10}-10$	$-1.649_{10}-9$	$5.157_{10}-9$	$-6.466_{10}-9$	$5.085_{10}-9$
$2.739_{10}-9$	$-1.919_{10}-9$	$3.814_{10}-10$	$1.427_{10}-10$	$-5.012_{10}-10$	$1.041_{10}-10$	$-1.173_{10}-9$
$-4.965_{10}-10$	$3.244_{10}-10$	$2.936_{10}-10$	$2.025_{10}-10$	$-1.882_{10}-10$	$2.675_{10}-10$	$1.627_{10}-10$

Note. The values are arranged in the same way as the gridpoints of the plate.

Table 2

Precise solution rounded to 12 digits after decimal point:

2.0089 7632 4128	2.4284 7688 1408	2.7280 7307 0141	2.8732 7686 7147	2.8510 2844 4043	2.6695 8508 8272
2.0603 5672 7229	2.4989 0971 1744	2.8078 2128 9988	2.9505 4878 8730	2.9144 2899 3084	2.7105 3453 5093
2.0682 5080 4542	2.5130 4773 9860	2.8243 0620 5059	2.9639 2293 3996	2.9196 1228 9814	2.7047 4424 9786
2.0187 0955 6465	2.4535 1923 7202	2.7574 4355 8675	2.8930 6873 6921	2.8484 7904 5959	2.6370 9550 9787
1.9110 8539 4290	2.3194 7268 5574	2.6062 3096 5286	2.7369 1485 2070	2.6999 7864 7649	2.5066 3601 8057
1.7582 2771 2072	2.1267 5166 8739	2.3889 0366 6823	2.5137 8863 5857	2.4902 3400 0853	2.3266 8960 8078
1.5847 8821 9746	1.9134 1818 7996	2.1501 3199 4977	2.2685 8666 2053	2.2579 8289 9341	2.1230 2200 3332

Difference between precise solution and computed solution x_{320}:

$-3.608_{10}-9$	$-5.937_{10}-9$	$-9.429_{10}-9$	$-8.905_{10}-9$	$-6.228_{10}-9$	$-6.402_{10}-10$
$-9.895_{10}-10$	$-2.095_{10}-9$	$-1.862_{10}-9$	$-2.211_{10}-9$	$1.513_{10}-9$	$4.598_{10}-9$
$3.492_{10}-10$	$9.313_{10}-10$	$1.222_{10}-9$	$2.328_{10}-9$	$5.238_{10}-9$	$7.217_{10}-9$
$5.413_{10}-9$	$5.762_{10}-9$	$6.170_{10}-9$	$5.995_{10}-9$	$6.402_{10}-9$	$8.905_{10}-9$
$8.352_{10}-9$	$7.916_{10}-9$	$8.789_{10}-9$	$9.313_{10}-9$	$9.255_{10}-9$	$1.228_{10}-8$
$1.079_{10}-8$	$1.036_{10}-8$	$1.187_{10}-8$	$1.327_{10}-8$	$1.525_{10}-8$	$1.542_{10}-8$
$1.362_{10}-8$	$1.356_{10}-8$	$1.606_{10}-8$	$1.897_{10}-8$	$1.973_{10}-8$	$1.897_{10}-8$

Residual vector r_{320}:

$-8.333_{10}-9$	$-4.623_{10}-9$	$9.244_{10}-9$	$-5.150_{10}-9$	$5.742_{10}-9$	$-7.984_{10}-9$
$-1.283_{10}-8$	$2.295_{10}-9$	$-7.899_{10}-9$	$-5.551_{10}-9$	$-7.240_{10}-9$	$3.274_{10}-9$
$1.831_{10}-8$	$1.783_{10}-9$	$-4.012_{10}-10$	$1.437_{10}-8$	$-9.816_{10}-9$	$-6.231_{10}-9$
$-8.628_{10}-9$	$-2.822_{10}-10$	$-1.747_{10}-9$	$-1.270_{10}-8$	$1.663_{10}-8$	$-5.121_{10}-9$
$-1.204_{10}-8$	$5.714_{10}-9$	$-7.385_{10}-9$	$3.274_{10}-9$	$7.968_{10}-9$	$1.052_{10}-8$
$4.860_{10}-9$	$3.493_{10}-9$	$5.895_{10}-9$	$1.016_{10}-8$	$-1.183_{10}-8$	$1.276_{10}-9$
$-8.608_{10}-9$	$1.054_{10}-8$	$-2.431_{10}-9$	$-9.483_{10}-9$	$-2.105_{10}-9$	$-7.817_{10}-9$

Note. These values pertain to only a section of the whole 18×15-grid for the plate problem (fine mesh), namely the section consisting of columns 5 through 10 and rows 5 through 11 of the grid.

Table 3. The twenty smallest and the six biggest eigenvalues of

Matrix A	Z_{40}	Z_{60}	Z_{80}	Z_{100}	Z_{120}
0.0000 3443 8091	0.0039 6253 2169	0.0003 8229 3787	0.0000 3443 8886	0.0000 3443 8352	0.0000 3443 8341
0.0005 4939 3867	0.0257 3089 3037	0.0027 2888 6134	0.0005 4939 4325	0.0005 4939 4111	0.0005 4939 4132
0.0027 6772 2900	0.0662 5501 4464	0.0086 8529 5482	0.0027 6772 3122	0.0027 6772 3195	0.0027 6772 3280
0.0086 8758 5094	0.1252 4226 721	0.0210 2353 1660	0.0086 8758 5533	0.0086 8758 5633	0.0085 8332 8578
0.0210 2357 8385	0.2035 1499 032	0.0431 2636 6755	0.0210 2357 8763	0.0210 2357 8852	0.0086 8758 5563
0.0431 2636 6245	0.2496 0614 456	0.0788 6278 1388	0.0431 2636 6583	0.0431 2636 6691	0.0209 9163 9660
0.0788 8278 1127	0.3130 6979 881	0.1325 9887 098	0.0788 8278 1527	0.0500 3849 5061	0.0210 2357 8886
0.1325 9887 068	0.4480 5654 163	0.2088 6878 021	0.1325 9887 117	0.0788 8278 1557	0.0431 2636 6985
0.2088 6877 975	0.6202 3551 143	0.3124 3271 906	0.2088 6878 049	0.0790 6426 8626	0.0431 2708 5742
0.3124 3271 857	0.8333 2244 463	0.4480 2754 836	0.2993 2867 525	0.1325 9887 159	0.0788 8278 1069
0.4480 2754 788	1.0911 063601	0.6202 3370 862	0.3124 3271 928	0.1326 6042 203	0.0788 8278 1770
0.6202 3370 787	1.3967 9950 29	0.8333 2232 619	0.4079 2533 515	0.2088 6878 025	0.1325 9887 073
0.8333 2232 565	1.7528 8661 33	1.0911 063524	0.4480 2754 838	0.2088 6893 312	0.1325 9887 158
1.0911 063519	2.1610 0876 89	1.3967 9950 29	0.6201 8506 860	0.3124 3271 944	0.2088 6878 044
1.3967 9950 23	2.6218 6622 50	1.7528 8661 41	0.6202 3370 856	0.3124 3271 990	0.2088 6878 082
1.7528 8661 31	3.1351 4174 94	2.1610 0876 91	0.8333 1123 139	0.4480 2754 880	0.3124 3271 915
2.1610 0876 85	3.6994 4652 82	2.2632 4606 45	0.8333 2232 633	0.4480 2754 920	0.3124 3271 947
2.6218 662245	4.3122 9026 73	2.6218 6622 56	1.0911 0635 22	0.6202 3370 921	0.4480 2754 931
3.1351 4174 91	4.9700 7655 52	2.9267 8944 25	1.0911 079342	0.6202 3371 054	0.4480 2754 939
3.6994 4652 81	5.6681 2397 14	3.1351 4174 96	1.3967 9950 23	0.8333 3232 647	0.6202 3370 953
14.3817 7527 8	15.1775 8177 9	15.5818 9390 2	15.8122 5529 1	15.8130 3616 7	15.8130 3617 1
14.8610 6279 6	15.2630 2985 6	15.5818 9450 3	15.8130 3617 2	15.8130 3617 0	15.8130 3617 1
15.2630 2985 8	15.5818 9450 3	15.8130 3615 0	15.8130 3617 3	15.9527 6362 1	15.9530 8724 9
15.5818 9450 5	15.8130 3617 2	15.8130 3617 2	15.9530 8659 7	15.9530 8725 0	15.9530 8725 1
15.8130 3617 4	15.9339 6715 0	15.9530 8725 3	15.9530 8725 5	15.9530 8725 2	15.9530 8725 2
15.9530 8725 7	15.9530 8725 5	15.9530 8725 4	15.9530 8725 5	15.9530 8725 3	15.9530 8725 4

Fig. 2 exhibits the quantities rr and rfe as a function of the step number k while Table 2 shows the solution, the error of the computed solution x_{320} and the residual r_{320} for a section of the whole 18×15 grid.

c) The Diriclet problem for a square mesh of 63×63 inner points. The boundary values are prescribed to be zero, while the load is 1 in the 37-th point of the 17-th row, 0 in all other points. Computing time was 9 min on the CDC 1604 computer, after procedures *lincom, inprod, op* had been transcribed into the assembly language of the 1604.

The execution of *cg* was terminated after 204 CG-steps, with

$$\|r_{204}\| = 2.18_{10} - 9, \quad \|A x_{204} + b\| = 3.15_{10} - 9, \quad \|x_{204} - x\| = 1.97_{10} - 8.$$

The component of x_{204} corresponding to the loadpoint is 0.7823 9244 099 (precise value $= \ldots 244145$).

d) Computation of the eigenvalues of the coefficient matrix of the bar problem (cf. p. 98 of [5]), with $n = 40$: First the system $A x = (1, 0, \ldots, 0)^T$ was solved and from the q- and e-values produced during this computation the eigenvalues of the qd-lines $Z_{40}, Z_{60}, Z_{80}, Z_{100}, Z_{120}$ were computed as indicated in "Examples of the use of the procedure". Table 3 compares the 20 smallest and the 6 biggest eigenvalues of every one of these qd-lines with the precise spectrum. Fig. 3 shows the course of rr and rfe (for the solution of the bar problem with $n = 40$) as a function of the step number k.

References

1. Hestenes, M. R., Stiefel, E.: Methods of Conjugate Gradients for Solving Linear Systems. Nat. Bur. Standards J. of Res. **49**, 409–436 (1952).
2. Stiefel, E.: Einige Methoden der Relaxationsrechnung. Z. Angew. Math. Phys. **3**, 1–33 (1952).
3. — Relaxationsmethoden bester Strategie zur Lösung linearer Gleichungssysteme. Comment. Math. Helv. **29**, 157–179 (1955).
4. — Kernel Polynomials in Linear Algebra and their Numerical Applications, Nat. Bur. Standards. Appl. Math. Ser. **49**, 1–22 (1958).
5. Engeli, M., Ginsburg, Th., Rutishauser, H., Stiefel, E.: Refined Iterative Methods for Computation of the Solution and the Eigenvalues of Self-adjoint Boundary Value Problems. Mitt. Inst. angew. Math. ETH Zürich, No. 8 (Basel: Birkhäuser 1959).

Solution of Symmetric and Unsymmetric Band Equations and the Calculations of Eigenvectors of Band Matrices[*]

by R. S. MARTIN and J. H. WILKINSON

1. Theoretical Background

In an earlier paper in this series [2] the triangular factorization of positive definite band matrices was discussed. With such matrices there is no need for pivoting, but with non-positive definite or unsymmetric matrices pivoting is necessary in general, otherwise severe numerical instability may result even when the matrix is well-conditioned.

In [1] the factorization of a general matrix using partial pivoting was described. This method can be applied to a general band matrix A of order n, advantage being taken of the band form to economize on arithmetic operations. For greater generality we shall assume that

$$a_{ij}=0 \quad \text{when} \quad j>i+m2 \quad \text{or} \quad i>j+m1, \tag{1}$$

so that the number of lines of non-zero elements above and below the diagonal can be different.

The modified algorithm has $n-1$ major steps, in the k-th of which the sub-diagonal elements in the k-th column of A are eliminated. The configuration at the beginning of the k-th major step is illustrated when $n=10$, $k=3$, $m1=2$, $m2=3$ by the array

$$
\begin{array}{l}
X \ X \ X \ X \ X \ X \\
X \ X \ X \ X \ X \ X \\
\text{Rows involved in} \quad \left\{ \begin{array}{l} X \ X \ X \ X \ X \\ X \ X \ X \ X \ X \\ X \ X \ X \ X \ X \ X \end{array} \right. \\
 X \ X \ X \ X \ X \ X \\
 X \ X \ X \ X \ X \ X \\
 X \ X \ X \ X \ X \\
 X \ X \ X \ X \\
 X \ X \ X
\end{array} \tag{2}
$$

Rows involved in 3rd major step

There are $m1$ non-zero sub-diagonal elements to be eliminated except of course in the last $m1$ major steps. The k-th major step is as follows.

(i) Determine $int[k]$ the smallest integer for which

$$|a_{int[k],k}| \ge |a_{i,k}| \qquad (i \ge k).$$

[*] Prepublished in Numer. Math. 9, 279—301 (1967).

(ii) Interchange rows k and $int[k]$.

(iii) Compute $m_{ik}=a_{ik}/a_{kk}$. (Note $|m_{ik}|\leqq 1$).

(iv) Replace row i by row $i - m_{ik}\times$row k for each value of i from $k+1$ to l where $l = min[n, k+m1]$.

In practice an economized method of storage is used as described in section 5. Because of interchanges there are up to $m1+m2+1$ elements in a typical final row.

If the multipliers m_{ik} and the integers $int[k]$ are retained the equations $Ax=b$ can be solved with any given right-hand side b. The vector b is first subjected to the same operations as A to give c and the system

$$Ux=c \tag{3}$$

is then solved, where U is the final upper-triangular matrix.

The determinant of A is given by the relation

$$det(A)=(-1)^s \prod_{i=1}^{n} u_{ii} \tag{4}$$

where s is the total number of row interchanges involved in the reduction of A to U. In general this reduction does not enable us to determine the leading principal minors of A.

There is an alternative method of pivoting which has the advantage of producing each of the leading principal minors p_r $(r=1, \ldots, n)$ of A as a by-product. In the case when A is defined by $B-\lambda I$ and B is symmetric the number of agreements in sign between consecutive p_r (p_0 being defined as $+1$) is the number of eigenvalues of B greater than λ.

For simplicity the alternative method is first described for a full matrix A. It consists of $n-1$ major steps. At the beginning of the k-th, only rows 1 to k of the original matrix have been modified and these have been reduced to the trapezoidal form illustrated when $n=6$, $k=3$ by the array

$$
\begin{matrix}
X & X & X & X & X & X \\
 & X & X & X & X & X \\
 & & X & X & X & X
\end{matrix} \tag{5}
$$

In the k-th major step, row $k+1$ of the original A is added to this array and the augmented matrix is reduced to trapezoidal form in k minor steps, of which the i-th is as follows.

(i) If $|a_{k+1,i}|>|a_{ii}|$ interchange row i and $k+1$.

(ii) Compute $m_{k+1,i}=a_{k+1,i}/a_{ii}$ (Note $|m_{k+1,i}|\leqq 1$).

(iii) Subtract $m_{k+1,i}\times$row i from row $(k+1)$.

The $(k+1)$-th leading principal minor is given by

$$p_{k+1}=(-1)^{s_{k+1}}a_{11}a_{22}\cdots a_{k+1,k+1} \tag{6}$$

where s_{k+1} is the total number of row interchanges in the first k major steps.

When A is of band form there are only $m1+1$ rows involved in each major step (fewer in the first and the last $m1$ steps), rows $1, 2, \ldots, k-m1$ playing no part in the $(k+1)$-th major step. Most commonly the algorithm is used when

$n \gg m1 + m2 + 1$ and since we are interested only in the signs of the p_i and the first $k - m1$ factors of p_{k+1} are the same as those of p_k, it is more economical to determine the sign of p_{k+1}/p_k in the k-th major step rather than to evaluate p_{k+1} explicitly.

As with the first method of factorization, if details of the row interchanges and the $m_{k+1,i}$ are retained the equation $A x = b$ with any right hand side b can be solved in the obvious way. We have also

$$det(A) = (-1)^{s_n} \prod_{i=1}^{n} u_{ii}. \tag{7}$$

The two methods of factorization involve the same number of arithmetic operations but the storage of the Boolean information giving the details of the interchanges in the second method is less economical in ALGOL on most computers than the storage of the integers $int[k]$ in the first method.

If either method of factorization is applied to the matrix $A - \lambda I$ the factors may be used to perform inverse iteration by solving the systems

$$(A - \lambda I) z_{r+1} = x_r, \qquad x_{r+1} = z_{r+1}/\alpha_{r+1} \tag{8}$$

where α_{r+1} is the element of z_{r+1} of maximum modulus. Each x_r therefore contains an element unity. If β_{r+1} is the element of z_{r+1} in the same position as the unit element of x_r then after each iteration $\lambda + \beta_{r+1}^{-1}$ gives an improved estimate of the eigenvalue. A convenient criterion for the termination of the iteration is

$$\|A x_r - (\lambda + \beta_r^{-1}) x_r\|_\infty \leq \text{tolerance}, \tag{9}$$

that is, that the residual vector corresponding to the current approximations to the eigenvector and eigenvalue should be smaller than some pre-assigned tolerance depending on $\|A\|_\infty$ and the machine precision. In this form the criterion involves a substantial amount of computation but using the relations (8) with $r = r - 1$ it can be recast in the form[*]

$$\begin{aligned} \text{tolerance} &\geq \|(A z_r - (\lambda + \beta_r^{-1}) z_r)/\alpha_r\|_\infty \\ &= \|(x_{r-1} - \beta_r^{-1} z_r)/\alpha_r\|_\infty. \end{aligned} \tag{10}$$

Remembering that z_r is merely the computed solution of $(A - \lambda I) z_r = x_{r-1}$ and that $A - \lambda I$ is normally very ill-conditioned, the use of (10) requires some justification. This is given in section 6. We shall assume that λ is a sufficiently good approximation to make a second factorization with an improved λ uneconomic.

Either of the factorizations may be used to solve the system

$$(A^T - \lambda I) y = b \tag{11}$$

and hence to perform inverse iteration to give the left-hand eigenvector corresponding to λ. Since the left hand eigenvector differs from the right-hand eigenvector only if A is unsymmetric and the second factorization has advantages only when A is symmetric, only the first factorization will be used for this purpose.

---

[*] The use of a criterion of this type was suggested by Dr. C. REINSCH of the Technische Hochschule München.

When A is symmetric an improved eigenvalue can be determined from an eigenvector x using the Rayleigh quotient defined by

$$x^T A x / x^T x = \lambda + x^T (A x - \lambda x) / x^T x, \qquad (12)$$

the second form being advantageous in practice (see for example [5], pp. 636—637). In the unsymmetric case an improved eigenvalue can be determined from left-hand and right-hand eigenvectors y and x using the generalized Rayleigh quotient

$$y^T A x / y^T x = \lambda + y^T (A x - \lambda x) / y^T x. \qquad (13)$$

2. Applicability

The procedures described in this paper may be applied to any band matrix whether symmetric or unsymmetric but will normally be used only when $m1 + m2 \ll n$. If A is a positive definite band matrix and eigenvectors are not required the procedures *chobandet* and *chobandsol* [2] should be used; however, when eigenvectors are to be calculated using inverse iteration *chobandsol* should not be used even when A is positive definite because $A - \lambda I$ will not in general have this property. The algorithms are

bandet 1. Gives the factorization of $A - \lambda I$, where A is a band matrix, using the first form of pivoting described in section 1. $det(A - \lambda I)$ is produced as a by-product in a form which avoids underspill or overspill.

bandet 2. Gives an alternative factorization to that provided by *bandet 1*. $det(A - \lambda I)$ is produced as in *bandet 1*, but in addition the number of agreements in sign between consecutive leading principal minors is given. When A is symmetric this gives the number $s(\lambda)$ of eigenvalues greater than λ. When this quantity is not required *bandet 1* is to be preferred.

bansol 1. May be used to solve $(A - \lambda I)X = B$ where A is a band matrix and B has r columns. It must be preceded by *bandet 1*. *bansol 1* may be used any number of times after one application of *bandet 1*.

bansol 2. Performs the same function as *bansol 1* but uses the alternative factorization. It must be preceded by *bandet 2*.

symray. May be used to compute r eigenvectors of a symmetric band matrix A corresponding to r approximate eigenvalues $\lambda_1, \ldots, \lambda_r$. Improved eigenvalues are computed from these eigenvectors using the Rayleigh quotient. The procedures *bandet 2* and *bansol 2* are used. The integers $s(\lambda_i)$ are also given.

unsray. May be used to compute the r left-hand and right-hand eigenvectors of a band matrix corresponding to r approximate eigenvalues $\lambda_1, \ldots, \lambda_r$. Improved eigenvalues are computed from the eigenvectors using the generalized Rayleigh quotient. The procedure *bandet 1* and *bansol 1* are used.

The procedures *symray* and *unsray* include a variety of features not all of which will be required in any specific application. The decision to make these procedures so comprehensive was taken so as to avoid providing a number of procedures having a good deal in common. Experience has shown that modifications which involve only the omission of unwanted features can be carried out accurately and quickly.

3. Formal Parameter List

3.1. Input to procedure *bandet 1*

n   order of the band matrix A.

m1   number of sub-diagonal lines of non-zero elements of A.

m2   number of super-diagonal lines of non-zero elements of A.

e   a parameter which influences the procedure to be adopted when a zero divisor occurs in the factorization as described in section "Organisational and notational details".

lambda   the factorization of $A - lambda\ I$ is performed rather than that of A itself.

macheps   the smallest number for which $1 + macheps > 1$ on the computer.

a   an $n \times (m1 + m2 + 1)$ array giving the elements of a band matrix A as described in section "Organisational and notational details".

Output of procedure *bandet 1*

a   an $n \times (m1 + m2 + 1)$ array giving the upper-triangle of the factorization of $A - \lambda I$ as described in section "Organisational and notational details".

d1, d2   two numbers which together give the determinant of $A - \lambda I$.

m   an $n \times m1$ array giving the lower-triangle of the factorization of $A - \lambda I$.

int   an $n \times 1$ integer array giving details of the pivoting used in the triangularization.

fail   failure exit, to be used when $e = 0$ and $A - lambda\ I$, modified by the rounding errors, is singular.

3.2. Input to procedure *bansol 1*

n   order of the band matrix $A - \lambda I$ which has been factorized using *bandet 1*.

m1   number of sub-diagonal lines in A.

m2   number of super-diagonal lines in A.

e   parameter which modifies the procedure for use in inverse iteration as described in section "Organisational and notational details".

r   number of right-hand sides.

a   upper-triangle of factorization of $A - \lambda I$ as given by *bandet 1*.

m   lower-triangle of factorization of $A - \lambda I$ as given by *bandet 1*.

int   an integer array giving details of pivoting as provided by *bandet 1*.

b   an $n \times r$ array consisting of the r right-hand sides.

Output of procedure *bansol 1*

b   an $n \times r$ array consisting of the r solutions.

3.3. Input to procedure *bandet 2*

n   order of the band matrix A.

m1   number of sub-diagonal lines of A.

m2   number of super-diagonal lines of A.

e   a parameter which influences the procedure to be adopted when a zero-divisor occurs in the factorization as described in section "Organisational and notational details".

lambda the factorization of $A - lambda\,I$ is performed rather than that of A itself.

macheps the smallest number for which $1 + macheps > 1$ on the computer.

a an $n \times (m1 + m2 + 1)$ array giving the elements of a band matrix A as described in section "Organisational and notational details".

Output of procedure *bandet 2*

a an $n \times (m1 + m2 + 1)$ array giving the upper-triangle of the factorization of $A - \lambda I$ as described in section "Organisational and notational details".

d1, d2 two numbers which together give the determinant of $A - \lambda I$.

c the number of eigenvalues of A greater than λ when A is symmetric. Otherwise irrelevant.

m an $n \times m1$ array giving the lower-triangle of the factorization of $A - \lambda I$.

int an $n \times m1$ Boolean array describing the interchanges used in the triangularization.

fail failure exit, to be used when $e = 0$ and $A - lambda\,I$, modified by the rounding errors, is singular.

3.4. Input to procedure *bansol 2*

n order of band matrix A.

m1 number of sub-diagonal lines of A.

m2 number of super-diagonal lines of A.

e parameter which modifies the procedure for use in inverse iteration as described in section "Organisational and notational details".

r number of right-hand sides.

a upper-triangle of factorization of $A - \lambda I$ as given by *bandet 2*.

m lower-triangle of factorization of $A - \lambda I$ as given by *bandet 2*.

int an $n \times m1$ Boolean array describing the interchanges in the factorization of $A - \lambda I$ as given by *bandet 2*.

b an $n \times r$ array consisting of r right-hand sides.

Output of procedure *bansol 2*

b an $n \times r$ consisting of the r solutions.

3.5. Input to procedure *symray*

n order of symmetric band matrix A.

m number of sub-diagonal or super-diagonal lines in A so that in *bandet 2* $m1 = m2 = m$.

r number of eigenvectors required.

macheps the smallest number for which $1 + macheps > 1$ on the computer.

a the $n \times (2m + 1)$ array giving the elements of A as described in section "Organisational and notational details".

la limit of the number of iterations permitted in the determination of any eigenvector.

lambda an $r \times 1$ array consisting of r approximate eigenvalues of A.

Output of procedure *symray*

lambda an $r \times 1$ array consisting of the r improved eigenvalues given by the Rayleigh quotients.

l an $r \times 1$ array giving the number of iterations taken for each eigen-
            vector.

c an $r \times 1$ array in which $c[i]$ gives the number of eigenvalues of A
            greater than $lambda[i]$.

z an $n \times r$ array giving the r eigenvectors "corresponding" to the
            given λ.

3.6. Input to procedure *unsray*

n order of band matrix A.

m1   number of lines of sub-diagonal elements of A.

m2   number of lines of super-diagonal element of A.

r number of eigenvectors required.

macheps the smallest number for which $1 + macheps > 1$ on the computer.

a an $n \times (m1 + m2 + 1)$ array giving the elements of A as described in
            section ,,Organisational and notational details".

la   limit of the number of iterations permitted in the determination of
            any eigenvector.

lambda   an $r \times 1$ array consisting of r approximate eigenvalues of A.

Output of procedure *unsray*

lambda   an $r \times 1$ array consisting of the r improved eigenvalues given by the
            Rayleigh quotients.

l an $r \times 1$ array giving the number of iterations taken for each right-
            hand eigenvector.

c an $r \times 1$ array giving the number of iterations taken for each left-
            hand eigenvector.

zr   an $n \times r$ array giving the right-hand eigenvectors.

zl   an $n \times r$ array giving the left-hand eigenvectors.

4. ALGOL Programs

procedure *symray* (n, m) *data*: $(r, macheps, a, la)$ *trans*: $(lambda)$
            result: (l, c, z) *exit*: $(fail)$;

value $n, m, r, macheps, la$; **integer** n, m, r, la; **real** *macheps*; **array** *lambda, a, z*;
integer array l, c; **label** *fail*;

comment Computes $z[i, j]$, the $n \times r$ array giving the r eigenvectors of the sym-
            metric band matrix, A, corresponding to r approximate eigenvalues
            given in the array $lambda[i]$, $i = 1(1)r$. Each eigenvector is computed
            by inverse iteration using *bandet 2* and *bansol 2*. In the first iteration
            for each eigenvector, a back-substitution only is performed using
            bansol 2 with $e = 1$ and a right-hand side consisting of unity elements.
            The integer array $l[i]$, $i = 1(1)r$, gives the number of iterations taken
            to determine each vector. A limit, *la*, is set to the maximum number
            of iterations permitted for any one eigenvector. The integer array
            $c[i]$, $i = 1(1)r$, is such that $c[i]$ is the number of eigenvalues of A
            greater than $lambda[i]$. From the r eigenvectors, r improved eigen-
            values are determined using the Rayleigh quotients and are over-
            written on the original eigenvalues in the array $lambda[i]$. The pro-
            cedure *innerprod* is used in the calculation of the Rayleigh quotients;

begin   **integer** $i, j, k, d2, lb, q$;
        real $eps, x, y, d1, r1$;
        array $aa[1:n, -m:m], v[1:n, 1:1], f[1:n, 1:m], u[1:n]$;
        Boolean array $int[1:n, 1:m]$;

        procedure $bandet\ 2\ (n, m1, m2)\ data:(e, lambda, macheps)$
                     $trans:(a)\ result:(d1, d2, c, m, int)\ exit:(fail)$;
        value $e, n, m1, m2, lambda, macheps$; **integer** $e, n, m1, m2, d2, c$;
        real $lambda, macheps, d1$; **array** a, m; **Boolean array** int;
        label $fail$;
        comment The band matrix, A, of order n with $m1$ sub-diagonal ele-
                ments and $m2$ super-diagonal elements in a typical row is
                stored as an $n \times (m1+m2+1)$ array, $a[1:n, -m1:m2]$.
                The matrix $(A - lambda \times I)$ is factorized into the product
                of a lower-triangular matrix and an upper-triangular ma-
                trix using binary row interchanges. The lower-triangle is
                stored as the $n \times m1$ array $m[i, j]$ and the upper-triangle is
                overwritten on A. Details of the binary row interchanges
                are stored in the Boolean $n \times m1$ array, $int[i, j]$. When A
                is symmetric, c gives the number of eigenvalues greater
                than $lambda$. The determinant, $d1 \times 2\uparrow d2$, is also computed.
                The parameter e influences the path to be taken when a
                zero pivot arises. The procedure fails when $e=0$ and
                $(A - lambda \times I)$, modified by the rounding errors, is sin-
                gular;
        begin   **integer** i, j, k, l, r, w;
                 real $norm, x$;
                 procedure $scaledet$;
                 if $d1 \neq 0$ **then**
                      begin for $d2 := d2+4$ **while** $abs(d1) \geq 1$ **do**
                          $d1 := d1 \times 0.0625$;
                          for $d2 := d2-4$ **while** $abs(d1) < 0.0625$ **do**
                          $d1 := d1 \times 16$
                    end $scaledet$;
                 $norm := 0$;
                 if $e=1$ **then**
                 for $i := 1$ **step** 1 **until** n **do**
                 begin $x := 0$;
                    for $j := -m1$ **step** 1 **until** $m2$ **do**
                    $x := x + abs(a[i, j])$;
                    if $norm < x$ **then** $norm := x$
                 end i;
                 for $i := 1$ **step** 1 **until** n **do**
                 $a[i, 0] := a[i, 0] - lambda$;
                 $l := m1$;
                 for $i := 1$ **step** 1 **until** $m1$ **do**

```
begin for j := 1−i step 1 until m2 do
        a[i, j−l] := a[i, j];
        l := l−1;
        for j := m2−l step 1 until m2 do
        a[i, j] := 0
end i;
d1 := 1;
d2 := 0;
c := if a[1, −m1] ≧ 0 then 1 else 0;
for k := 2 step 1 until n do
begin r := 1;
        for i := m1 step −1 until 1 do
        if k > i then
        begin w := k−i; l := m2−m1+i;
                if abs(a[k, −m1]) > abs(a[w, −m1]) then
                begin d1 := −d1;
                        for j := l step −1 until −m1 do
                        begin x := a[k, j]; a[k, j] := a[w, j];
                              a[w, j] := x
                        end j;
                        if a[k, −m1] ≧ 0 ≡ a[w, −m1] ≧ 0
                                then r := −r;
                        int[k, i] := true
                end
                else   int[k, i] := false;
                x := m[k, i] := if a[w, −m1] = 0 then 0
                        else a[k, −m1]/a[w, −m1];
                for j := 1−m1 step 1 until l do
                a[k, j−1] := a[k, j] − x×a[w, j];
                a[k, l] := 0
        end i;
        if k > m1 then
        begin x := a[k−m1, −m1]; d1 := d1×x;
                if x = 0 then
                begin d2 := 0;
                        if e = 1 then
                        a[k−m1, −m1] := norm×macheps
                        else go to fail
                end;
                scaledet
        end;
        if a[k, −m1] < 0 then r := −r;
        if r > 0 then c := c+1
end k;
for k := n−m1+1 step 1 until n do
begin x := a[k, −m1]; d1 := d1×x;
        if x = 0 then
```

$$\begin{aligned}
&\textbf{begin } d2 := 0; \\
&\qquad \textbf{if } e = 1 \textbf{ then } a[k, -m1] := norm \times macheps \\
&\qquad \textbf{else go to } fail \\
&\textbf{end}; \\
&scaledet \\
&\textbf{end } k
\end{aligned}$$

end *bandet 2*;

procedure *bansol 2* $(n, m1, m2)$ *data*: (e, r, a, m, int) *trans*: (b);
value $e, n, m1, m2, r$; **integer** $e, n, m1, m2, r$; **array** a, m, b;
Boolean array *int*;
comment When $e = 0$, this procedure solves $(A - lambda \times I) x = b$, where A is a band matrix of order n and $(A - lambda \times I)$ has previously been factorized using *bandet 2*. Here, b is an $n \times r$ matrix consisting of r right-hand sides. Each right-hand side requires one forward substitution and one back-substitution. When $e = 1$, a back-substitution only is used. This variant should be used as the first step in inverse iteration for the eigenvector corresponding to *lambda*;

begin **integer** i, j, k, l, w;
real y;
if $e = 0$ **then**
for $k := 2$ **step** 1 **until** n **do**
for $i := m1$ **step** -1 **until** 1 **do**
if $k > i$ **then**
begin $w := k - i$;
    if $int[k, i]$ **then**
    for $j := 1$ **step** 1 **until** r **do**
    begin $y := b[k, j]$; $b[k, j] := b[w, j]$; $b[w, j] := y$
    end j;
    $y := m[k, i]$;
    for $j := 1$ **step** 1 **until** r **do**
    $b[k, j] := b[k, j] - y \times b[w, j]$
end ik;
for $j := 1$ **step** 1 **until** r **do**
begin $l := -m1$;
    for $i := n$ **step** -1 **until** 1 **do**
    begin $y := b[i, j]$; $w := i + m1$;
        for $k := 1 - m1$ **step** 1 **until** l **do**
        $y := y - a[i, k] \times b[k + w, j]$;
        $b[i, j] := y / a[i, -m1]$;
        if $l < m2$ **then** $l := l + 1$
    end i
end j
end *bansol 2*;

```
eps := 0;
for i := 1 step 1 until n do
begin      x := 0;
           for j := − m step 1 until m do x := x + abs (a [i, j]);
           if eps < x then eps := x
end i;
eps := eps × macheps;
for k := 1 step 1 until r do
begin      for i := 1 step 1 until n do
           begin for j := − m step 1 until m do aa [i, j] : = a [i, j];
                 v [i, 1] := 1
           end i;
           lb := 0;
           bandet 2 (n, m, m, 1, lambda [k], macheps, aa, d1, d2, c [k], f,
                           int, fail);
           bansol 2 (n, m, m, 1, 1, aa, f, int, v);
           x := 0;
           for i := 1 step 1 until n do
           if abs (v [i, 1]) > abs (x) then
           begin x := v [i, 1]; j := i end;
           x := 1/x;
           for i := 1 step 1 until n do
           v [i, 1] := u [i] := x × v [i, 1];
L1:        d1 := 0; lb := lb + 1;
           bansol 2 (n, m, m, 0, 1, aa, f, int, v);
           y := 1/v [j, 1]; x := 0;
           for i := 1 step 1 until n do
           if abs (v [i, 1]) > abs (x) then
           begin x := v [i, 1]; j := i end;
           x := 1/x;
           for i := 1 step 1 until n do
           begin r1 := abs ((u [i] − y × v [i, 1]) × x);
                 if r1 > d1 then d1 := r1;
                 v [i, 1] := u [i] := x × v [i, 1]
           end i;
           if d1 > eps then
           begin if lb < la then go to L1 else lb := lb + 1 end;
           for i := 1 step 1 until n do
           z [i, k] := v [i, 1];
           l [k] := lb; d1 := lambda [k] := lambda [k] + y;
           x := 0; d2 := 1;
           for i := 1 step 1 until m do
           begin innerprod (i, 1, i, 0, 0, d1, v [j, 1], j, y, r1);
                 innerprod (1 − i, 1, m, − y, − r1, a [i, j], v [j + d2, 1],
                                                              j, y, r1);
                 x := x + y × v [i, 1]; d2 := d2 + 1
           end i;
```

```
              lb := n − m;
              for i := m +1 step 1 until lb do
              begin innerprod (i, 1, i, 0, 0, d1, v[j, 1], j, y, r1);
                    innerprod (− m, 1, m, − y, − r1, a[i, j], v[j+d2, 1],
                                                              j, y, r1);
                      x := x+y×v[i, 1]; d2 := d2+1
              end i;
              q := m −1;
              for i := lb +1 step 1 until n do
              begin innerprod (i, 1, i, 0, 0, d1, v[j, 1], j, y, r1);
                    innerprod (− m, 1, q, − y, − r1, a[i, j], v[j+d2, 1],
                                                              j, y, r1);
                      x := x+y×v[i, 1]; d2 := d2+1; q := q−1
              end i;
              y := 0;
              for i := 1 step 1 until n do
              y := y+v[i, 1]×v[i, 1];
              lambda[k] := lambda[k]+x/y
      end k
end symray;
```

procedure *unsray* (*n, m1, m2*) data:(*r, macheps, a, la*) trans:(*lambda*)
        result:(*l, c, zr, zl*) exit:(*fail*);
value *n, m1, m2, r, macheps, la*; **integer** *n, m1, m2, r, la*; **real** *macheps*;
array *lambda, a, zr, zl*; **integer array** *l, c*; **label** *fail*;
comment Computes $zr[i, j]$ and $zl[i, j]$, two $n×r$ arrays giving the r right-hand
        eigenvectors and the r left-hand eigenvectors of the band matrix, A,
        corresponding to the r approximate eigenvalues given in the array
        $lambda[i]$, $i=1(1)r$. The right-hand and left-hand eigenvectors cor-
        responding to $lambda[i]$ are both computed by inverse iteration using
        the decomposition of $(A − lambda ×I)$ given by *bandet 1*. In the first
        iteration for each right-hand eigenvector, a back-substitution only is
        performed using *bansol 1* with $e = 1$ and a right-hand side consisting
        of unity elements. Inverse iteration for the left-hand eigenvectors
        forms part of the main body of this procedure. The integer arrays $l[i]$
        and $c[i]$ give the number of iterations taken to determine each right-
        hand and left-hand eigenvector respectively. A limit, la, is set to the
        maximum number of iterations permitted for any one eigenvector.
        From the r eigenvectors, r improved eigenvalues are determined using
        the generalized Rayleigh quotients and are overwritten on the original
        eigenvalues in the array $lambda[i]$. The procedure *innerprod* is used
        in the calculation of the Rayleigh quotients;
begin   **integer** $i, j, k, d2, lb, mm, p, q$;
        real $eps, x, y, d1, r1$;
        array $aa[1:n, −m1:m2], v[1:n, 1:1], f[1:n, 1:m1], u[1:n]$;
        integer array $int[1:n]$;

procedure *bandet 1* $(n, m1, m2)$ *data*: $(e, lambda, macheps)$
                    trans: (a) *result*: $(d1, d2, m, int)$ *exit*: $(fail)$;
value $n, m1, m2, e, lambda, macheps$; **integer** $n, m1, m2, d2, e$;
real $d1, lambda, macheps$; **array** a, m; **integer array** *int*;
label *fail*;
comment The band matrix, A, of order n with $m1$ sub-diagonal ele-
                    ments and $m2$ super-diagonal elements in a typical row is
                    stored as an $n \times (m1+m2+1)$ array, $a[1:n, -m1:m2]$.
                    The matrix $(A - lambda \times I)$ is factorized into the product
                    of a lower-triangular matrix and an upper-triangular ma-
                    trix using partial pivoting. The lower triangle is stored as
                    the $n \times m1$ array $m[i, j]$ and the upper-triangle is over-
                    written on A. Details of the interchanges are stored in the
                    array *int*$[i]$, $i = 1(1) \ n$. The determinant, $d1 \times 2 \uparrow d2$, is also
                    computed. The parameter e influences the path to be taken
                    when a zero pivot arises. The procedure fails when $e = 0$
                    and $(A - lambda \times I)$, modified by the rounding errors, is
                    singular;
begin   **integer** i, j, k, l;
                    real $x, norm$;
                    $norm := 0$;
                    if $e = 1$ **then**
                    for $i := 1$ **step** 1 **until** n **do**
                    begin $x := 0$;
                            for $j := -m1$ **step** 1 **until** $m2$ **do**
                            $x := x + abs(a[i, j])$;
                            if $norm < x$ **then** $norm := x$
                    end i;
                    for $i := 1$ **step** 1 **until** n **do**
                    $a[i, 0] := a[i, 0] - lambda$;
                    $l := m1$;
                    for $i := 1$ **step** 1 **until** $m1$ **do**
                    begin for $j := 1-i$ **step** 1 **until** $m2$ **do**
                            $a[i, j-l] := a[i, j]$;
                            $l := l-1$;
                            for $j := m2-l$ **step** 1 **until** $m2$ **do**
                            $a[i, j] := 0$
                    end i;
                    $d1 := 1$; $d2 := 0$; $l := m1$;
                    for $k := 1$ **step** 1 **until** n **do**
                    begin $x := a[k, -m1]$; $i := k$;
                            if $l < n$ **then** $l := l+1$;
                            for $j := k+1$ **step** 1 **until** l **do**
                            if $abs(a[j, -m1]) > abs(x)$ **then**
                            begin $x := a[j, -m1]$; $i := j$ **end** j;
                            $int[k] := i$; $d1 := d1 \times x$;
                            if $x = 0$ **then**

```
                begin d2 := 0;
                    if e = 1 then a[k, − m1] := norm × macheps
                    else go to fail
                end;
                if d1 ≠ 0 then
                begin for d2 := d2 + 4 while abs (d1) ≧ 1 do
                    d1 := d1 × 0.0625;
                    for d2 := d2 − 4 while abs (d1) < 0.0625 do
                    d1 := d1 × 16
                end;
                if i ≠ k then
                begin d1 := − d1;
                    for j := − m1 step 1 until m2 do
                    begin x := a[k, j]; a[k, j] := a[i, j];
                        a[i, j] := x
                    end j
                end;
                for i := k + 1 step 1 until l do
                begin x := m[k, i − k] := a[i, − m1]/a[k, − m1];
                    for j := 1 − m1 step 1 until m2 do
                    a[i, j − 1] := a[i, j] − x × a[k, j];
                    a[i, m2] := 0
                end i
            end k
    end bandet 1;


procedure bansol 1 (n, m1, m2) data : (e, r, a, m, int) trans : (b);
value n, m1, m2, e, r; integer n, m1, m2, e, r; array a, m, b;
integer array int;
comment When e = 0 this procedure solves (A − lambda × I) x = b,
            where A is a band matrix of order n and (A − lambda × I)
            has previously been factorized using bandet 1. Here, b is an
            n × r matrix consisting of r right-hand sides. Each right-
            hand side requires one forward substitution and one back-
            substitution. When e = 1, a back-substitution only is used.
            This variant should be used as the first step in inverse
            iteration for the eigenvector corresponding to lambda;
begin      integer i, j, k, l, w;
            real x;
            l := m1;
            if e = 0 then
            for k := 1 step 1 until n do
            begin i := int[k];
                if i ≠ k then
                for j := 1 step 1 until r do
                begin x := b[k, j]; b[k, j] := b[i, j]; b[i, j] := x
                end j;
```

```
                    if l < n then l := l + 1;
                    for i := k + 1 step 1 until l do
                    begin x := m[k, i − k];
                            for j := 1 step 1 until r do
                            b[i, j] := b[i, j] − x × b[k, j]
                    end i
            end k;
            for j := 1 step 1 until r do
            begin l := − m1;
                    for i := n step − 1 until 1 do
                    begin x := b[i, j];  w := i + m1;
                            for k := 1 − m1 step 1 until l do
                            x := x − a[i, k] × b[k + w, j];
                            b[i, j] := x/a[i, − m1];
                            if l < m2 then l := l + 1
                    end i
            end j
    end bansol 1;


    mm := m1 + m2 + 1;  eps := 0;
    for i := 1 step 1 until n do
    begin       x := 0;
                for j := − m1 step 1 until m2 do
                x := x + abs(a[i, j]);
                if eps < x then eps := x
    end i;
    eps := eps × macheps;
    for k := 1 step 1 until r do
    begin       for i := 1 step 1 until n do
                begin for j := − m1 step 1 until m2 do
                        aa[i, j] := a[i, j];
                        v[i, 1] := 1
                end i;
                lb := 0;
                bandet 1 (n, m1, m2, 1, lambda[k], macheps, aa, d1, d2, f, int,
                                                                        fail);
                bansol 1 (n, m1, m2, 1, 1, aa, f, int, v);
                x := 0;
                for i := 1 step 1 until n do
                if abs(v[i, 1]) > abs(x) then
                begin x := v[i, 1];  p := i end;
                x := 1/x;
                for i := 1 step 1 until n do
                v[i, 1] := u[i] := x × v[i, 1];
L1:             d1 := 0;  lb := lb + 1;
                bansol 1 (n, m1, m2, 0, 1, aa, f, int, v);
                y := 1/v[p, 1];  x := 0;
```

```
for i := 1 step 1 until n do
if abs (v [i, 1]) > abs (x) then
begin x := v [i, 1]; p := i end;
x := 1/x;
for i := 1 step 1 until n do
begin r1 := abs ((u [i] − y × v [i, 1]) × x);
            if r1 > d1 then d1 := r1;
            v [i, 1] := u [i] := x × v [i, 1]
end i;
if d1 > eps then
begin if lb < la then go to L1 else lb := lb + 1 end;
for i := 1 step 1 until n do
u [i] := zr [i, k] := v [i, 1];
l [k] := lb; lambda [k] := lambda [k] + y; lb := 0;
d1 := 0; d2 := 1; lb := lb + 1;
for j := 1 step 1 until n do
begin x := v [j, 1]; q := j − m1;
            for i := j − 1 step − 1 until d2 do
            x := x − aa [i, q − i] × v [i, 1];
            v [j, 1] := x/aa [j, − m1];
            if j ≥ mm then d2 := d2 + 1
end j;
for j := n step − 1 until 1 do
begin x := 0;
            d2 := if j + m1 > n then n else j + m1;
            for i := j + 1 step 1 until d2 do
            x := x − f [j, i − j] × v [i, 1];
            v [j, 1] := v [j, 1] + x; i := int [j];
            if i ≠ j then
            begin x := v [j, 1]; v [j, 1] := v [i, 1]; v [i, 1] := x
            end
end j;
y := 1/v [p, 1]; x := 0;
for i := 1 step 1 until n do
if abs (v [i, 1]) > abs (x) then
begin x := v [i, 1]; p := i end;
x := 1/x;
for i := 1 step 1 until n do
begin r1 := abs ((u [i] − y × v [i, 1]) × x);
            if r1 > d1 then d1 := r1;
            v [i, 1] := u [i] := x × v [i, 1]
end i;
if d1 > eps then
begin if lb < la then go to L2 else lb := lb + 1 end;
c [k] := lb; d1 := lambda [k]; x := 0; d2 := 1;
for i := 1 step 1 until m1 do
```
L2:

```
begin innerprod (i, 1, i, 0, 0, d1, zr [j, k], j, y, r1);
      innerprod (1 − i, 1, m2, − y, − r1,
      a [i, j], zr [j + d2, k], j, y, r1);
      x := x + y × v [i, 1]; d2 := d2 + 1
end i;
lb := n − m2;
for i := m1 + 1 step 1 until lb do
begin innerprod (i, 1, i, 0, 0, d1, zr [j, k], j, y, r1);
      innerprod (− m1, 1, m2, − y, − r1, a [i, j],
                                    zr [j + d2, k], j, y, r1);
      x := x + y × v [i, 1]; d2 := d2 + 1
end i;
q := m2 − 1;
for i := lb + 1 step 1 until n do
begin innerprod (i, 1, i, 0, 0, d1, zr [j, k], j, y, r1);
      innerprod (− m1, 1, q, − y, − r1, a [i, j],
                                    zr [j + d2, k], j, y, r1);
      x := x + y × v [i, 1]; d2 := d2 + 1; q := q − 1
end i;
y := 0;
for i := 1 step 1 until n do
begin y := y + zr [i, k] × v [i, 1]; zl [i, k] := v [i, 1]
end;
lambda [k] := lambda [k] + x/y
                              end k
end unsray;
```

5. Organisational and Notational Details

In both *bandet 1* and *bandet 2* the matrix A is stored as an $n \times (m1 + m2 + 1)$ array. Since symmetry, when it exists, is destroyed, no advantage of symmetry is taken when storing A. For convenience the column suffices run from $-m1$ to $m2$, the diagonal elements having the suffix 0 in the original array. The storage is adequately described by the array for the case $m1 = 2$, $m2 = 1$, $n = 7$ which is

$$\begin{bmatrix} 0 & 0 & a_{11} & a_{12} \\ 0 & a_{21} & a_{22} & a_{23} \\ a_{31} & a_{32} & a_{33} & a_{34} \\ a_{42} & a_{43} & a_{44} & a_{45} \\ a_{53} & a_{54} & a_{55} & a_{56} \\ a_{64} & a_{65} & a_{66} & a_{67} \\ a_{75} & a_{76} & a_{77} & 0 \end{bmatrix} \tag{14}$$

Notice that zeros are required in the initial and final rows. In all cases a value of λ is subtracted before triangularization, so that λ must be set equal to zero if A itself is required.

During the reduction a realignment is necessary so that coefficients involving the variable which is about to be eliminated form a column. This is illustrated in 15 (a) and 15 (b) where the storage immediately before the elimination of x_1 and x_2 respectively in *bandet 1* are given.

$$
\text{(a)} \qquad\qquad\qquad\qquad \text{(b)}
$$

$$
\left.\begin{bmatrix}
a_{11} & a_{12} & 0 & 0 \\
a_{21} & a_{22} & a_{23} & 0 \\
a_{31} & a_{32} & a_{33} & a_{34} \\
a_{42} & a_{43} & a_{44} & a_{45} \\
a_{53} & a_{54} & a_{55} & a_{56} \\
a_{64} & a_{65} & a_{66} & a_{67} \\
a_{75} & a_{76} & a_{77} & 0
\end{bmatrix}\right\} \qquad
\left.\begin{bmatrix}
a_{11} & a_{12} & a_{13} & a_{14} \\
a_{22} & a_{23} & a_{24} & 0 \\
a_{32} & a_{33} & a_{34} & 0 \\
a_{42} & a_{43} & a_{44} & a_{45} \\
a_{53} & a_{54} & a_{55} & a_{56} \\
a_{64} & a_{65} & a_{66} & a_{67} \\
a_{75} & a_{76} & a_{77} & 0
\end{bmatrix}\right\} \qquad (15)
$$

In each case the brackets indicate the rows involved in the current reduction. Apart from the first stage the realignment takes place quite naturally in the reduction.

When performing inverse iteration or evaluating a determinant the emergence of a zero pivotal element is in no sense a failure of the routine. On the contrary inverse iteration is particularly effective when λ is almost an eigenvalue (in which case zero pivots are quite likely to arise) while a zero value of $det\,(A - \lambda I)$ is obviously not a "failure". When solving equations on the other hand, the emergence of a zero pivot means that $A - \lambda I$, possibly modified by rounding errors, is singular and a failure indication is essential. The marker e determines whether or not *bandet 1* or *bandet 2* should fail. It is set equal to 1 when the determinant or inverse iteration are required and to zero for the solution of equations. When $e = 1$ a zero pivot is replaced by *macheps* $\|A\|_\infty$. In order to avoid overspill and underspill $det\,(A - \lambda I)$ is stored as an ordered pair *d1* and *d2* such that $det\,(A - \lambda I) = 2^{d2}\,d1$ where d2 is an integer and $\frac{1}{16} \le |d1| < 1$.

In inverse iteration a limit *la* is placed on the number of iterations which are permitted. If this is exceeded the number of iterations is given as $la + 1$. After each iteration an improved approximation for λ is available. If the initial λ were a very poor approximation it might be more economical to perform a further factorization of $A - \lambda I$ using this improved value. The procedures given here should make it obvious how to do this.

The choice of initial vector in inverse iteration presents some difficulty. For the right-hand eigenvector we have started with a back-substitution and solved

$$U x = e. \qquad (16)$$

(See the discussion given by WILKINSON [5] in connexion with tri-diagonal matrices.) Since *bansol 1* (or *bansol 2*) is used as a procedure in inverse iteration, this procedure has a parameter e which determines whether or not the solution starts at the back-substitution stage. For left-hand eigenvectors such a stratagem is inappropriate. Instead we have taken the computed right-hand vector as initial approximation. If A is symmetric this gives immediate convergence; in the general case the right-hand vector cannot be orthogonal to the left-hand vector unless the relevant eigenvalues correspond to a non-linear divisor.

When calculating improved eigenvalues by means of the Rayleigh quotient using Eqs. (12) or (13) it is essential that $A x - \lambda x$ should be computed accurately. This has been done by using a machine code procedure in which inner-products are accumulated. The details of the procedure have been given in [3]. The i-th element of $A x - \lambda x$ should be computed from

$$- \lambda x_i + a_{i1} x_1 + a_{i2} x_2 + \cdots + a_{in} x_n \tag{17}$$

and not from

$$a_{i1} x_1 + a_{i2} x_2 + \cdots + (a_{ii} - \lambda) x_i + \cdots + a_{in} x_n. \tag{18}$$

Each element of $A x - \lambda x$ is rounded to single precision before computing $x^T (A x - \lambda x)$ in (12) or $y^T (A x - \lambda x)$ in (13); inner product accumulation is not needed in the latter computations or in the computation of $x^T x$ or $y^T x$. When the Rayleigh quotients are computed in this way the exact sum of the computed correction (a single-precision number) and the approximate λ is usually correct to more than single-precision, though the addition must be executed outside the ALGOL scheme if this accuracy is to be realised.

6. Discussion of Numerical Properties

The use of interchanges ensures that both decompositions are, in general, numerically stable. Each gives an exact factorization of a matrix $A + E$ rather than A where the bounds for E are dependent upon the machine precision. The matrix E is itself of band form and has $m1$ non-zero sub-diagonal lines, but because of interchanges it has up to $m1 + m2$ lines of non-zero super-diagonal elements. An upper bound for E is given by

$$\|E\|_\infty \leq 2^{-t}(m1 + m2 + 1) g \tag{19}$$

where g is the maximum element which arises in $|A|$ at any stage of its reduction. In practice g is seldom greater than $\max |a_{ij}|$ in the original matrix.

Discussions of the accuracy of the computed determinant or solutions therefore follow very closely those given in [2].

The accuracy attainable by inverse iteration is also very high and in general each computed eigenvector x_r is an exact eigenvector of some matrix $A + E_r$, where E_r is bounded as in relation (19). For a full discussion of the accuracy of inverse iteration see WILKINSON [5].

The use of the criterion (10) calls for special comment. When λ is close to an eigenvalue (which will normally be the case) the computed solution of $(A - \lambda I) z_r = x_{r-1}$ will differ substantially from the true solution. However, the analysis in [5] shows that the computed solution satisfies exactly an equation

$$(A + F_r - \lambda I) z_r = x_{r-1}, \tag{20}$$

where $\|F_r\|$ is small. Hence when

$$\|(x_{r-1} - \beta_r^{-1} z_r)\|_\infty / \|z_r\|_\infty \leq \text{tolerance} \tag{21}$$

we have

$$\|(A + F_r) z_r - (\lambda + \beta_r^{-1}) z_r\|_\infty / \|z_r\|_\infty \leq \text{tolerance} \tag{22}$$

giving

$$\|(A+F_r) x_r - (\lambda+\beta_r^{-1}) x_r\|_\infty \leq \text{tolerance} \tag{23}$$

and x_r is a normalized vector. This implies that x_r and $\lambda+\beta_r^{-1}$ are an exact eigenvalue and eigenvector of a matrix $A+G_r$ with

$$\|G_r\|_\infty \leq \|F_r\|_\infty + \text{tolerance}. \tag{24}$$

The error in the Rayleigh quotient for λ_p is unlikely to exceed

$$a + 2^{-2t} (m1+m2+1) \max |a_{ij}|/b \cos \vartheta_p \tag{25}$$

where a is max $|\lambda_p - \lambda_i|$ for those λ_i for which $|\lambda_p - \lambda_i| < 2^{-t}$ and b is min $|\lambda_p - \lambda_i|$ for the remaining λ_i, while ϑ_p is the angle between the right-hand and left-hand eigenvectors and is therefore zero when A is symmetric. Notice this means that in the symmetric case well separated roots will be obtained correct almost to double-precision although true double-precision work is not required at any stage. In the unsymmetric case this will again be true provided $\cos \vartheta_p$ is not small.

7. Test Results

The procedures have been tested in connexion with two band matrices using the computer KDF 9 which has a 39 binary digit mantissa. The first is the matrix of order seven given by

$$A_1 = \begin{bmatrix} 5 & -4 & 1 & & & & \\ -4 & 6 & -4 & 1 & & & \\ 1 & -4 & 6 & -4 & 1 & & \\ & 1 & -4 & 6 & -4 & 1 & \\ & & 1 & -4 & 6 & -4 & 1 \\ & & & 1 & -4 & 6 & -4 \\ & & & & 1 & -4 & 5 \end{bmatrix}$$

This was used merely to test the formal correctness of the procedures. The determinant of $(A_1 - \lambda I)$ was evaluated for $\lambda = 3.75$, 4.00, and 4.25. Both procedures *bandet 1* and *bandet 2* gave identical results. (The two methods give identical results only when the pivotal row at the r-th stage is in row r or row $r+1$ throughout.) The computed values were

$$det (A_1 - 3.75\, I) = -5.8757\,8997\,014_{10} - 1 \times 2^{12}$$
$$det (A_1 - 4.0\, I) = \text{zero},$$
$$det (A_1 - 4.25\, I) = +7.2967\,0271\,0273_{10} - 1 \times 2^{12}.$$

The value zero is, of course, exact; the other two values are correct to within two in the last figure. The procedure *bandet 2* gave 4, 4, 3 for the number of eigenvalues greater than 3.75, 4.0, 4.25 respectively.

The solution corresponding to $\lambda=0$ and the right-hand side

$$(0, 0, 0, 1, 0, 0, 0)$$

was computed using *bansol 1* and *bansol 2*; again both procedures gave identical results. These are given in Table 1.

Table 1

Computed solution of $A_1 x = e_4$	Exact solution
$3.9999\,9999\,938_{10}+0$	4.0
$7.4999\,9999\,884_{10}+0$	7.5
$9.9999\,9999\,843_{10}+0$	10.0
$1.0999\,9999\,983_{10}+1$	11.0
$9.9999\,9999\,837_{10}+0$	10.0
$7.4999\,9999\,876_{10}+0$	7.5
$3.9999\,9999\,934_{10}+0$	4.0

The second test matrix used was that derived from the usual five-point approximation to the Laplace operator on the unit square with a 10×10 mesh. The matrix A_2 is of order 81 and is of the form

$$A_2 = \begin{bmatrix} B & -I & & & \\ -I & B & -I & & \\ & & \cdots & & \\ & & -I & B & -I \\ & & & -I & B \end{bmatrix}, \quad \text{where} \quad B = \begin{bmatrix} 4 & -1 & & & \\ -1 & 4 & -1 & & \\ & & \cdots & & \\ & & -1 & 4 & -1 \\ & & & -1 & 4 \end{bmatrix}$$

and I is the identity matrix of order 9. The eigenvalues λ_{pq} are given by

$$\lambda_{pq} = 4 - 2\cos p\,\pi/10 - 2\cos q\,\pi/10 \qquad (p, q = 1, \ldots, 9).$$

Obviously $\lambda_{pq} = \lambda_{qp}$ so that all eigenvalues are double except those for which $p = q$.

As a test of the sensitivity of the *bandet* procedures three approximations

$$\mu_1 = 1.6488\,5899\,080, \qquad \mu_2 = 1.6488\,5899\,083, \qquad \mu_3 = 1.6488\,5899\,085$$

were given to the root $\lambda_{33} = 4 - 4\cos 54°$. (The second approximation μ_2 is correct to working accuracy.) *bandet 2* gave as the number of roots greater than μ_1, μ_2, and μ_3 the values 71, 71 and 70 respectively showing a correct changeover. The values given for the determinants were

	bandet 1	*bandet 2*
$det(A - \mu_1 I)$	$+4.4669\,5394\,056_{10} - 1 \times 2^{40}$	$+4.6168\,6362\,427_{10} - 1 \times 2^{40}$
$det(A - \mu_2 I)$	$+9.5140\,3176\,005_{10} - 2 \times 2^{40}$	$+1.1065\,2103\,301_{10} - 1 \times 2^{40}$
$det(A - \mu_3 I)$	$-3.7243\,9535\,521_{10} - 1 \times 2^{40}$	$-3.4721\,8668\,597_{10} - 1 \times 2^{40}$

For values so near to a root a low relative error cannot be expected, but the sensitivity is such that linear interpolation (or extrapolation) in any pair of values gives the zero correct to the working accuracy.

Inverse iteration was performed with all three values of μ using *symray* and *unsray*. One back-substitution and one iteration were needed in every case for right-hand vectors and one iteration for left-hand vectors. All vectors were of much the same accuracy. Since A is symmetric the nine vectors, three from *symray* and six from *unsray*, should all have been the same. In Table 2 a result obtained with *unsray* is given. The first column gives the first 9 elements, the second column the next 9 etc. The true value of the r-th element of the s-th column should be $\sin 3s\pi/10 \sin 3r\pi/10$. In general the components are accurate to about eleven decimals.

Table 2

Computed eigenvector corresponding to $\lambda_{3,3}$		
$+6.5450\,8497\,195_{10}-1$	$+7.6942\,0884\,297_{10}-1$	$+2.4999\,9999\,990_{10}-1$
$+7.6942\,0884\,308_{10}-1$	$+9.0450\,8497\,191_{10}-1$	$+2.9389\,2626\,128_{10}-1$
$+2.5000\,0000\,020_{10}-1$	$+2.9389\,2626\,159_{10}-1$	$+9.5491\,5028\,008_{10}-2$
$-4.7552\,8258\,126_{10}-1$	$-5.5901\,6994\,361_{10}-1$	$-1.8163\,5632\,012_{10}-1$
$-8.0901\,6994\,355_{10}-1$	$-9.5105\,6516\,282_{10}-1$	$-3.0901\,6994\,388_{10}-1$
$-4.7552\,8258\,134_{10}-1$	$-5.5901\,6994\,364_{10}-1$	$-1.8163\,5632\,010_{10}-1$
$+2.5000\,0000\,005_{10}-1$	$+2.9389\,2626\,146_{10}-1$	$+9.5491\,5028\,083_{10}-2$
$+7.6942\,0884\,290_{10}-1$	$+9.0450\,8497\,180_{10}-1$	$+2.9389\,2626\,143_{10}-1$
$+6.5450\,8497\,180_{10}-1$	$+7.6942\,0884\,284_{10}-1$	$+2.4999\,9999\,997_{10}-1$
$-4.7552\,8258\,165_{10}-1$	$-8.0901\,6994\,384_{10}-1$	$-4.7552\,8258\,145_{10}-1$
$-5.5901\,6994\,395_{10}-1$	$-9.5105\,6516\,293_{10}-1$	$-5.5901\,6994\,361_{10}-1$
$-1.8163\,5632\,015_{10}-1$	$-3.0901\,6994\,367_{10}-1$	$-1.8163\,5631\,978_{10}-1$
$+3.4549\,1502\,800_{10}-1$	$+5.8778\,5252\,299_{10}-1$	$+3.4549\,1502\,831_{10}-1$
$+5.8778\,5252\,277_{10}-1$	$+9.9999\,9999\,998_{10}-1$	$+5.8778\,5252\,299_{10}-1$
$+3.4549\,1502\,801_{10}-1$	$+5.8778\,5252\,288_{10}-1$	$+3.4549\,1502\,814_{10}-1$
$-1.8163\,5632\,002_{10}-1$	$-3.0901\,6994\,369_{10}-1$	$-1.8163\,5631\,992_{10}-1$
$-5.5901\,6994\,366_{10}-1$	$-9.5105\,6516\,278_{10}-1$	$-5.5901\,6994\,355_{10}-1$
$-4.7552\,8258\,139_{10}-1$	$-8.0901\,6994\,359_{10}-1$	$-4.7552\,8258\,127_{10}-1$
$+2.4999\,9999\,992_{10}-1$	$+7.6942\,0884\,279_{10}-1$	$+6.5450\,8497\,180_{10}-1$
$+2.9389\,2626\,146_{10}-1$	$+9.0450\,8497\,175_{10}-1$	$+7.6942\,0884\,290_{10}-1$
$+9.5491\,5028\,228_{10}-2$	$+2.9389\,2626\,144_{10}-1$	$+2.5000\,0000\,000_{10}-1$
$-1.8163\,5631\,994_{10}-1$	$-5.5901\,6996\,381_{10}-1$	$-4.7552\,8258\,142_{10}-1$
$-3.0901\,6994\,380_{10}-1$	$-9.5105\,6516\,305_{10}-1$	$-8.0901\,6994\,370_{10}-1$
$-1.8163\,5632\,013_{10}-1$	$-5.5901\,6994\,383_{10}-1$	$-4.7552\,8258\,139_{10}-1$
$+9.5491\,5027\,994_{10}-2$	$+2.9389\,2626\,141_{10}-1$	$+2.5000\,0000\,012_{10}-1$
$+2.9389\,2626\,153_{10}-1$	$+9.0450\,8497\,189_{10}-1$	$+7.6942\,0884\,299_{10}-1$
$+2.5000\,0000\,009_{10}-1$	$+7.6942\,0884\,291_{10}-1$	$+6.5450\,8497\,188_{10}-1$

The improved values using the Rayleigh correction were correct in all cases to not less than 22 decimals though, of course, to achieve the full potential accuracy the single-precision Rayleigh correction must be added to the single-precision λ using full double-precision addition. To check the use of condition (10) iteration was performed using $\lambda=1.67$, convergence to the true vector required 15 iterations.

As a final check the sensitivity of the eigenvalue count in *bandet 2* was tested in the neighbourhood of the double root $\lambda_{4,5}$ of A_2 given by

$$\lambda_{5,4} = \lambda_{4,5} = 4 - 2\cos\pi/2 - 2\cos\pi/5 = 2.3819\,6601\,125\ldots.$$

The two values $\mu_1 = 2.3819\,6601\,100$ and $\mu_2 = 2.3819\,6601\,150$ were taken. The procedure *bandet 2* gave 64 eigenvalues greater than μ_1 and 62 greater than μ_2, thus correctly detecting the presence of the double root. Since there are two independent eigenvectors corresponding to $\lambda_{4,5}$ it was not expected that inverse iteration would give the same vector. In each case the computed vector corresponding to μ_1 was substantially different from that corresponding to μ_2 *and between them they gave almost complete digital information on the relevant invariant 2-space.* The accuracy of the Rayleigh quotient is not impaired by the presence of coincident roots though close roots affect it adversely. In fact the Rayleigh values were all correct to 22 decimals.

Tests were also performed on the computer TR 4 at the Mathematisches Institut der Technischen Hochschule München with each of the above examples and generally confirmed the results obtained on KDF 9.

Acknowledgements. Once again the authors wish to thank Dr. C. Reinsch of the Technische Hochschule, München, who not only performed the second test runs but suggested a number of important improvements in the algorithms. The work described here is part of the Research Programme of the National Physical Laboratory and is published by permission of the Director of the Laboratory.

References

[1] Bowdler, H. J., R. S. Martin, G. Peters, and J. H. Wilkinson: Solution of Real and Complex Systems of Linear Equations. Numer. Math. **8**, 217—234 (1966). Cf. I/7.

[2] Martin, R. S., and J. H. Wilkinson: Symmetric decomposition of positive definite band matrices. Numer. Math. **7**, 355—361 (1965). Cf. I/4.

[3] —, G. Peters, and J. H. Wilkinson: Iterative refinement of the solution of a positive definite system of equations. Numer. Math. **8**, 203—216 (1966). Cf. I/2.

[4] Peters, G., and J. H. Wilkinson: The calculation of specified eigenvectors by inverse iteration. Cf. II/18.

[5] Wilkinson, J. H.: The algebraic eigenvalue problem. London: Oxford University Press 1965.

Solution of Real and Complex Systems of Linear Equations*

by H. J. Bowdler, R. S. Martin, G. Peters, and J. H. Wilkinson

1. Theoretical Background

If A is a non-singular matrix then, in general, it can be factorized in the form $A = LU$, where L is lower-triangular and U is upper-triangular. The factorization, when it exists, is unique to within a non-singular diagonal multiplying factor.

If we write $LU = (LD)(D^{-1}U)$, two choices of the diagonal matrix D are of particular interest, those for which LD is *unit* lower-triangular and $D^{-1}U$ is *unit* upper-triangular respectively. These two factorizations are the ones provided by the Doolittle and Crout algorithms respectively, though they are usually described in different terms. The Doolittle factorization is basically that given by Gaussian elimination and might for this reason be preferred but the Crout factorization has slight advantages for our purpose.

The elements of the matrices L and U of the Crout factorization may be determined in the order r-th column of L, r-th row of U $(r = 1, 2, \ldots, n)$ by means of the equations

$$\sum_{k=1}^{r-1} l_{ik} u_{kr} + l_{ir} = a_{ir} \qquad (i = r, \ldots, n), \tag{1}$$

$$\sum_{k=1}^{r-1} l_{rk} u_{ki} + l_{rr} u_{ri} = a_{ri} \qquad (i = r+1, \ldots, n). \tag{2}$$

The process fails at the r-th stage if $l_{rr} = 0$. If we write

$$A = \begin{bmatrix} A_{11} & A_{12} \\ \hline A_{21} & A_{22} \end{bmatrix}, \qquad L = \begin{bmatrix} L_{11} & 0 \\ \hline L_{21} & L_{22} \end{bmatrix}, \qquad U = \begin{bmatrix} U_{11} & U_{12} \\ \hline 0 & U_{22} \end{bmatrix}, \tag{3}$$

where A_{11}, L_{11}, U_{11} are $r \times r$ matrices then

$$det(A_{11}) = det(L_{11}) \, det(U_{11}) = \prod_{i=1}^{r} l_{ii}. \tag{4}$$

Failure therefore occurs at the r-th stage if $l_{ii} \neq 0$ $(i = 1, \ldots, r-1)$ but $l_{rr} = 0$. If all the leading principal minors of A are non-singular, the Crout factorization can be completed and is obviously, by construction, unique.

---

* Prepublished in Numer. Math. **8**, 217−234 (1966).

However, the process can break down even when A is very well-conditioned, for example for the matrix

$$A = \begin{bmatrix} 0 & 1 \\ 1 & 0 \end{bmatrix}. \tag{5}$$

Even more serious (because it happens much more frequently) it may be numerically unstable when A is well-conditioned, for example when

$$A = \begin{bmatrix} \varepsilon & a \\ b & c \end{bmatrix}, \tag{6}$$

where ε is small and a, b, c are not. In fact the process is "unnecessarily" unstable whenever a leading principal sub-matrix of A is much more ill-conditioned than A itself.

In Gaussian elimination "unnecessary" failures and instability may usually be avoided by *partial pivoting*. (See, for example, WILKINSON [7, 8].) In terms of the Doolittle factorization partial pivoting has the effect of re-ordering the rows of A to give an $\tilde{A}$ such that the latter may be factorized in the form $\tilde{A} = L^{(1)} U^{(1)}$ where $L^{(1)}$ is unit lower-triangular with $|l_{ij}^{(1)}| \leq 1$. If we write

$$\tilde{A} = L^{(1)} U^{(1)} = (L^{(1)} D)(D^{-1} U^{(1)}) = L^{(2)} U^{(2)} \tag{7}$$

and choose D so that $D^{-1} U^{(1)}$ is unit upper-triangular we obtain the equivalent Crout factorization. In the latter factorization neither the relations $|l_{ij}^{(2)}| \leq 1$ nor the relations $|u_{ij}^{(2)}| \leq 1$ are true in general though $|l_{ij}^{(2)}| \leq |l_{jj}^{(2)}|$ $(i > j)$. However, an elementary error analysis shows that this Crout factorization has the same numerical stability as the corresponding Doolittle factorization.

The algorithm which gives the required factorization directly is most simply described in the usual algorithmic terms including the storage arrangement. There are n major steps and at the beginning of the r-th step the stored array is illustrated when $n = 5$, $r = 3$ by

$$\begin{bmatrix} l_{11} & u_{12} & u_{13} & u_{14} & u_{15} \\ l_{21} & l_{22} & u_{23} & u_{24} & u_{25} \\ l_{31} & l_{32} & a_{33} & a_{34} & a_{35} \\ l_{41} & l_{42} & a_{43} & a_{44} & a_{45} \\ l_{51} & l_{52} & a_{53} & a_{54} & a_{55} \\ int_1 & int_2 \end{bmatrix}. \tag{8}$$

The r-th step is as follows.

(i)   Compute $l_{ir} = a_{ir} - \sum\limits_{k=1}^{r-1} l_{ik} u_{kr}$ and overwrite on a_{ir} $(i = r, \ldots, n)$.

(ii)   If int_r is the smallest integer for which $|l_{int_r, r}| = \max\limits_{i \geq r} |l_{ir}|$, then store the integer int_r and interchange the whole of rows r and int_r in the current array.

(iii)   Compute $u_{ri} = \left(a_{ri} - \sum\limits_{k=1}^{r-1} l_{rk} u_{ki}\right) \Big/ l_{rr}$ and overwrite on a_{ri} $(i = r+1, \ldots, n)$.

After n such steps A is replaced by L and U and the n integers $int_1, \ldots, int_n$ give the details of the interchanges. If P_r denote the permutation matrix, premultiplication by which gives an interchange of rows r and int_r, then

$$P_n \ldots P_2 P_1 A = \tilde{A} = LU, \tag{9}$$

in other words we have the orthodox Crout decomposition of $\tilde{A}$ defined by Eq. (9).

The triangles L and U and the array of integers int_r enable us to solve the equations $Ax = b$ corresponding to any right-hand side b. For we have

$$P_n \ldots P_2 P_1 A x = P_n \ldots P_2 P_1 b$$

or $\hspace{10cm} (10)$

$$LU x = P_n \ldots P_2 P_1 b.$$

We have only to subject b to the appropriate interchanges to give $\tilde{b}$ (say) and then solve

$$L y = \tilde{b}, \quad \text{and} \quad U x = y. \tag{11}$$

In general the computed solution of $A x = b$ can be refined, using the Crout factorization, by means of the iterative procedure

$$x^{(0)} = 0, \qquad r^{(s)} = b - A x^{(s)},$$
$$LU d^{(s)} = P_n \ldots P_2 P_1 r^{(s)}, \qquad x^{(s+1)} = x^{(s)} + d^{(s)}. \tag{12}$$

If A is not too ill-conditioned this refinement procedure ultimately gives the correctly rounded solution provided the residual vector is computed using accumulated inner-products, or double precision if this is not practicable. (See the corresponding discussion in [5].)

Pivoting was introduced to avoid the obvious shortcomings of the orthodox procedure as exposed by the matrix (6). If the columns of A are scaled by powers of two no additional rounding errors are involved and the pivotal selection is unaffected. In fact the final answers are identical even as regards rounding errors if the appropriate de-scaling is performed on completion. However, scaling by rows does affect the pivotal selection. If, for example, the first row of matrix (6) is multiplied by the appropriate power of two, a row interchange will not be required for the modified matrix and instability again results. Clearly our pivotal strategy implicitly involves some assumption about the scaling of rows since almost any pivotal selection could be achieved with the above algorithm merely by multiplying rows by the appropriate powers of two.

The problems involved in scaling a matrix so as to justify the use of the pivotal strategy we have described are surprisingly complex. Such scaling is usually referred to as *equilibration* and it has been discussed by BAUER [1], and FORSYTHE and STRAUS [2], and an examination of the practical implications given by WILKINSON [7, 8].

In the algorithm in this paper we have chosen the scaling factors $k(i)$ so that each row of A is of unit length. If the rows are actually multiplied by the appropriate factors rounding errors are introduced *ab initio*. Further, when solving $A x = b$ using the LU factorization, the elements of b must be multiplied by the same factors. It is more accurate and economical to leave A unscaled and choose int_r so that

$$\left| k(int_r) \, l_{int_r, r} \right| = \max_{i \geq r} \left| k(i) \, l_{i r} \right| \tag{13}$$

at each stage.

The discussion applies equally whether A and b are real or complex, though in the latter case complex arithmetic is involved.

2. Applicability

unsymdet may be used to give the Crout factorization of a square matrix A and it produces $det(A)$ as a by-product in a form which avoids overspill or underspill.

unsymsol may be used to solve $AX=B$ where A is an $n \times n$ matrix and B an $n \times r$ matrix. It must be preceded by *unsymdet*. *unsymsol* may be used any number of times after one application of *unsymdet* so that we may solve, for example, the iterative system $A x^{(s+1)} = x^{(s)}$ $(s = 1, 2, ...)$.

unsym acc solve may be used to give correctly rounded solutions of $AX=B$ provided A is not too ill-conditioned. It must be preceded by *unsymdet* applied to A and it uses procedure *unsymsol*. If A is too ill-conditioned for X to be determined, an indication is given to this effect. *unsym acc solve* uses the iterative procedure defined by Eq. (12) and it is essential that the residual vectors be determined using accumulation of inner-products or double-precision computation.

compdet, *compsol* and *cx acc solve* perform the corresponding functions for complex A and complex right-hand sides.

3. Formal Parameter List

3.1. Input to procedure *unsymdet*.

n order of the matrix A.

a elements of the matrix A stored as an $n \times n$ array.

eps   smallest number for which $1 + eps > 1$ on the computer.

       Output of procedure *unsymdet*

a an $n \times n$ matrix the lower triangle of which is L and the upper-triangle is U of the Crout factorization. Since U is unit upper-triangular its diagonal elements are not stored.

d1, d2 two numbers which together give the determinant of A.

int   a set of n integers the i-th of which gives the number of the i-th pivotal row.

fail   the exit used if A, possibly as the result of rounding errors, is singular or almost singular.

3.2. Input to procedure *unsymsol*

n order of the matrix A.

r number of right-hand sides for which $A x = b$ is to be solved.

a elements of L and U, the Crout factorization of A produced by *unsymdet*.

int   a set of n integers produced by *unsymdet* as described above.

b the $n \times r$ matrix consisting of the r right-hand sides.

       Output of procedure *unsymsol*

b the $n \times r$ matrix consisting of the r solution vectors.

3.3. Input to procedure *unsym acc solve*.

n	order of the matrix A.
r	number of right-hand sides for which $A x = b$ is to be solved.
a	elements of the matrix A stored as an $n \times n$ array.
aa	the $n \times n$ array consisting of L and U as produced by *unsymdet* for the matrix A.
p	a set of integers describing the interchanges as produced by *unsymdet* for the matrix A.
b	the $n \times r$ matrix consisting of the r right-hand sides.
eps	the smallest number for which $1 + eps > 1$ on the computer.

Output of procedure *unsym acc solve*

x	the $n \times r$ matrix formed by the r solution vectors.
bb	the $n \times r$ matrix formed by the r residual vectors.
l	the number of iterations.
ill	is the exit which is used when there is no perceptible improvement from one iteration to the next.

3.4. Input to procedure *compdet*

n	order of the complex matrix A.
a	an $n \times 2n$ array such that the (p, q) element of A is $a[p, 2q-1] + i a[p, 2q]$.

Output of procedure *compdet*

a	an $n \times 2n$ array of complex numbers. The lower triangle consists of the elements of L and the upper-triangle of the elements of U without the unit diagonal elements.

detr, deti, dete
three numbers which together give the complex determinant of A.

int	a set of n integers, the i-th of which gives the number of the i-th pivotal row.
fail	the exit used if A, possibly as the result of rounding errors, is singular or almost singular.

3.5. Input to procedure *compsol*

n	order of the complex matrix A.
r	number of right-hand sides for which $A x = b$ is to be solved.
a	an $n \times 2n$ array formed by the elements of L and U, the Crout factorization of A produced by *compdet*.
int	a set of n integers produced by *compdet* as described above.
b	an $n \times 2r$ array formed by the r right-hand sides. The p-th element of the q-th right-hand side is $b[p, 2q-1] + i b[p, 2q]$.

Output of procedure *compsol*.

b	an $n \times 2r$ array formed by r solution vectors.

3.6. Input to procedure *cx acc solve*.

n	order of complex matrix A.
r	number of right-hand sides for which $A x = b$ is to be solved.

a   an $n \times 2n$ array consisting of the complex elements of A in the same
        form as in *compdet*.
aa   an $n \times 2n$ array consisting of L and U of the Crout decomposition of A
        as provided by *compdet*.
p   a set of integers as produced by *compdet* for the matrix A.
b   the $n \times 2r$ matrix consisting of the r complex right-hand sides.
eps   the smallest number for which $1 + eps > 1$ on the computer.

        Output of procedure *cx acc solve*.

x   the $n \times 2r$ matrix formed by the r complex solution vectors.
bb   the $n \times 2r$ matrix formed by the r residual vectors.
l   the number of iterations.
ill   is the exit used if there is no perceptible improvement from one iteration
        to the next.

4. ALGOL Programs

procedure *innerprod* $(l, s, u, c1, c2, ak, bk)$ *bound variable*: (k) *result*: $(d1, d2)$;
value $l, s, u, c1, c2$;
integer l, s, u, k;
real $c1, c2, ak, bk, d1, d2$;
comment This procedure accumulates the sum of products $ak \times bk$ and adds
            it to the initial value $(c1, c2)$ in double precision. The bound vari-
            able k is to be used to indicate the subscript in the components of
            the vectors ak, bk over which a scalarproduct is to be formed.
            Throughout the Handbook Series Linear Algebra, the actual para-
            meters corresponding to ak and bk will be restricted to be real variables.
            If they are subscripted, all subscript expressions will be linear func-
            tions of the bound variable parameter k, or will be independent of k.
            This admits high efficiency handcoding of the subscript evaluation in
            the for statement (loop) of the procedure.
            The body of this procedure cannot be expressed within ALGOL.
begin real $s1, s2$,
        $(s1, s2) := c1 + c2$,   **comment** dbl. pr. acc,
        for $k := l$ **step** s **until** u **do**
        $(s1, s2) := (s1, s2) + ak \times bk$,   **comment** dbl. pr. acc,
        $d1 := (s1, s2) rounded$,
        $d2 := ((s1, s2) - d1) rounded$ **comment** dbl. pr. acc.
end *innerprod*;

procedure *cx innerprod* $(l, s, u, cr, ci, akr, aki, bkr, bki)$ *bound variable*: (k)
            result: (dr, di);
value l, s, u, cr, ci;
integer l, s, u, k;
real $cr, ci, akr, aki, bkr, bki, dr, di$;
comment *cx innerprod* computes
            $(dr, di) := (cr, ci)$ — sum over k $(akr, aki) \times (bkr, bki)$ using the pro-
            cedure *innerprod*. (cr, ci) denotes a complex number with realpart cr
            and imaginary part ci;

begin real h, hh;

Real part: $innerprod\,(l, s, u, -cr, 0, akr, bkr, k, h, hh)$;

                    $innerprod\,(l, s, u, -h, -hh, aki, bki, k, dr, hh)$;

Imaginary part: $innerprod\,(l, s, u, -ci, 0, aki, bkr, k, h, hh)$;

                    $innerprod\,(l, s, u, h, hh, akr, bki, k, h, hh)$;

                    $di := -h$

end *cx innerprod*;

procedure *unsymdet* (n, eps) *trans*: (a) *result*: $(d1, d2, int)$ *exit*: $(fail)$;

value n, eps; **integer** $d2, n$; **real** $d1, eps$; **array** a, int; **label** *fail*;

comment The unsymmetric matrix, A, is stored in the $n \times n$ array $a[i, j]$, $i = 1(1)n$, $j = 1(1)n$. The decomposition $A = LU$, where L is a lower triangular matrix and U is a unit upper triangular matrix, is performed and overwritten on A, omitting the unit diagonal of U. A record of any interchanges made to the rows of A is kept in $int[i]$, $i = 1(1)n$, such that the i-th row and the $int[i]$-th row were interchanged at the i-th step. The determinant, $d1 \times 2 \uparrow d2$, of A is also computed. The procedure will fail if A, modified by the rounding errors, is singular or almost singular. Uses the procedure *innerprod*;

begin   **integer** i, j, k, l;

           real x, y, yy;

           for $i := 1$ **step** 1 **until** n **do**

           begin $innerprod\,(1, 1, n, 0, 0, a[i, j], a[i, j], j, y, yy)$; $int[i] := 1/sqrt\,(y)$

           end i;

           $d1 := 1$;

           $d2 := 0$;

           for $k := 1$ **step** 1 **until** n **do**

           begin $l := k$;

               $x := 0$;

               for $i := k$ **step** 1 **until** n **do**

               begin $innerprod\,(1, 1, k-1, -a[i, k], 0, a[i, j], a[j, k], j, y, yy)$;

                    $a[i, k] := -y$; $y := abs\,(y \times int[i])$;

                    if $y > x$ **then**

                    begin $x := y$;

                         $l := i$

                    end

               end i;

               if $l \neq k$ **then**

               begin $d1 := -d1$;

                    for $j := 1$ **step** 1 **until** n **do**

                    begin $y := a[k, j]$;

                         $a[k, j] := a[l, j]$;

                         $a[l, j] := y$

                    end j;

                    $int[l] := int[k]$

               end;

```
                int [k] := l;
                d1 := d1 × a[k, k];
                if x < 8 × eps then goto fail;
        L1:     if abs (d1) ≥ 1 then
                begin d1 := d1 × 0.0625;
                      d2 := d2 + 4;
                      go to L1
                end;
        L2:     if abs (d1) < 0.0625 then
                begin d1 := d1 × 16;
                      d2 := d2 − 4;
                      go to L2
                end;
                x := −1/a[k, k];
                for j := k + 1 step 1 until n do
                begin innerprod (1, 1, k−1, −a[k, j], 0, a[k, i], a[i, j], i, y, yy);
                      a[k, j] := x × y end j
          end k
end unsymdet;


procedure unsym acc solve (n) data: (r, a, aa, p, b, eps) result: (x, bb, l) exit: (ill);
value    n, r, eps;
integer  n, r, l;
array    a, aa, p, x, b, bb;
real     eps;
label    ill;
```

comment Solves $A x = b$ where A is an $n \times n$ unsymmetric matrix and b is an $n \times r$ matrix of right hand sides, using the procedure *unsymsol*. The procedure must be preceded by *unsymdet* in which L and U are produced in $aa[i, j]$ and the interchanges in $p[i]$.
The residuals $bb = b − A x$ are calculated and $A d = bb$ is solved, over-writing d on bb. The refinement is repeated, as long as the maximum correction at any stage is less than half that at the previous stage, until the maximum correction is less than $2 eps$ times the maximum x. Exits to label *ill* if the solution fails to improve. Uses the procedure *innerprod* which forms accurate innerproducts. l is the number of iterations;

begin

```
                procedure unsymsol (n) data: (r, a, int) trans: (b);
                value r, n; integer r, n; array a, b, int;
```

                comment Solves $A x = b$, where A is an unsymmetric matrix and b is an $n \times r$ matrix of r right-hand sides. The procedure *unsymsol* must be preceded by *unsymdet* in which L and U are produced in $a[i, j]$, from A, and the record of the interchanges is produced in $int[i]$. $A x = b$ is solved in three steps, interchange the elements of b, $Ly=b$ and $Ux=y$. The matrices y and then x are overwritten on b;

begin   **integer** i, j, k;

real x, xx;

comment interchanging of elements of b;

for $i := 1$ **step** 1 **until** n **do**

if $int[i] \neq i$ **then**

for $k := 1$ **step** 1 **until** r **do**

begin $x := b[i, k]$;
           $b[i, k] := b[int[i], k]$;
           $b[int[i], k] := x$

end;

for $k := 1$ **step** 1 **until** r **do**

begin comment solution of $Ly = b$;

    for $i := 1$ **step** 1 **until** n **do**

    begin $innerprod(1, 1, i-1, b[i, k], 0, a[i, j]$,
        $b[j, k], j, x, xx)$; $b[i, k] := -x/a[i, i]$ **end** i;

    comment solution of $Ux = y$;

    for $i := n$ **step** -1 **until** 1 **do**

    begin $innerprod(i+1, 1, n, b[i, k], 0, a[i, j]$,
        $b[j, k], j, x, xx)$; $b[i, k] := -x$ **end** i

end k

end $unsymsol$;

integer $i, j, k, d2$;

real $d0, d1, c, cc, xmax, bbmax$;

for $j := 1$ **step** 1 **until** r **do**

for $i := 1$ **step** 1 **until** n **do**

begin $x[i, j] := 0$;
           $bb[i, j] := b[i, j]$;

end;

$l := 0$;

$d0 := 0$;

$L3$: $unsymsol(n, r, aa, p, bb)$;

$l := l+1$;

$d2 := 0$;

$d1 := 0$;

for $j := 1$ **step** 1 **until** r **do**

for $i := 1$ **step** 1 **until** n **do**

$x[i, j] := x[i, j] + bb[i, j]$;

for $j := 1$ **step** 1 **until** r **do**

begin $xmax := bbmax := 0$;

        for $i := 1$ **step** 1 **until** n **do**

$\quad\quad\quad\quad$ **begin if** $abs(x[i,j]) > xmax$ **then** $xmax := abs(x[i,j])$;
$\quad\quad\quad\quad\quad$ **if** $abs(bb[i,j]) > bbmax$ **then** $bbmax := abs(bb[i,j])$;
$\quad\quad\quad\quad\quad$ $innerprod(1, 1, n, -b[i,j], 0, a[i,k], x[k,j], k, c, cc)$;
$\quad\quad\quad\quad\quad$ $bb[i,j] := -c$
$\quad\quad\quad$ **end**;
$\quad\quad\quad$ **if** $bbmax > d1 \times xmax$ **then** $d1 := bbmax/xmax$;
$\quad\quad\quad$ **if** $bbmax > 2 \times eps \times xmax$ **then** $d2 := 1$;
$\quad\quad$ **end**;
$\quad\quad$ **if** $d1 > d0/2 \wedge l \neq 1$ **then goto** ill;
$\quad\quad$ $d0 := d1$;
$\quad\quad$ **if** $d2 = 1$ **then goto** $L3$;
end $unsym\ acc\ solve$;

procedure $compdet(n)$ $trans: (a)$ $result: (detr, deti, dete, int)$ $exit: (fail)$;
value n; **integer** $dete, n$; **real** $detr, deti$; **array** a, int; **label** $fail$;
comment The complex unsymmetric matrix, A, is stored in the $n \times 2n$ array
$\quad\quad\quad$ $a[i,j]$, $i = 1(1)n$, $j = 1(1)2n$, so that the general element is $a[i, 2j-1]$
$\quad\quad\quad$ $+i \times a[i, 2j]$. The decomposition $A = LU$, where L is a lower triangu-
$\quad\quad\quad$ lar matrix and U is a unit upper triangular matrix, is performed
$\quad\quad\quad$ and overwritten on A in the same form as A, omitting the unit
$\quad\quad\quad$ diagonal of U. A record of any interchanges made to the rows of A
$\quad\quad\quad$ is kept in $int[i]$, $i = 1(1)n$, such that the i-th row and the $int[i]$-th
$\quad\quad\quad$ row were interchanged at the i-th step. The determinant, $(detr + i \times$
$\quad\quad\quad$ $deti) \times 2 \uparrow dete$, of A is also computed. The procedure will fail if A,
$\quad\quad\quad$ modified by the rounding errors, is singular. Uses the procedures
$\quad\quad\quad$ $innerprod$ and $cx\ innerprod$;
begin $\quad$ **integer** i, j, k, l, p, pp;
$\quad\quad\quad$ **real** w, x, y, z, v;
$\quad\quad\quad$ **for** $i := 1$ **step** 1 **until** n **do**
$\quad\quad\quad$ $innerprod(1, 1, n+n, 0, 0, a[i,j], a[i,j], j, int[i], w)$;
$\quad\quad\quad$ $detr := 1$;
$\quad\quad\quad$ $deti := 0$;
$\quad\quad\quad$ $dete := 0$;
$\quad\quad\quad$ **for** $k := 1$ **step** 1 **until** n **do**
$\quad\quad\quad$ **begin** $\quad l := k$;
$\quad\quad\quad\quad\quad\quad$ $p := k+k$;
$\quad\quad\quad\quad\quad\quad$ $pp := p-1$;
$\quad\quad\quad\quad\quad\quad$ $z := 0$;
$\quad\quad\quad\quad\quad\quad$ **for** $i := k$ **step** 1 **until** n **do**
$\quad\quad\quad\quad\quad\quad$ **begin** $cx\ innerprod(1, 1, k-1, a[i, pp], a[i, p]$,
$\quad\quad\quad\quad\quad\quad\quad\quad$ $a[i,j+j-1], a[i,j+j], a[j,pp], a[j,p], j, x, y)$;
$\quad\quad\quad\quad\quad\quad\quad\quad$ $a[i, pp] := x$;
$\quad\quad\quad\quad\quad\quad\quad\quad$ $a[i, p] := y$;
$\quad\quad\quad\quad\quad\quad\quad\quad$ $x := (x \uparrow 2 + y \uparrow 2)/int[i]$;
$\quad\quad\quad\quad\quad\quad\quad\quad$ **if** $x > z$ **then**

```
                    begin z := x;
                           l := i
                    end
            end i;
            if l ≠ k then
            begin detr := − detr;
                    deti := − deti;
                    for j := n+n step −1 until 1 do
                    begin z := a[k, j];
                            a[k, j] := a[l, j];
                            a[l, j] := z
                    end j;
                    int[l] := int[k]
            end;
            int[k] := l;
            x := a[k, pp];
            y := a[k, p];
            z := x↑2 + y↑2;
            w := x × detr − y × deti;
            deti := x × deti + y × detr;
            detr := w;
            if abs(detr) < abs(deti) then w := deti;
            if w = 0 then
            begin dete := 0;
                    go to fail
            end;
L1:         if abs(w) ≧ 1 then
            begin w := w × 0.0625;
                    detr := detr × 0.0625;
                    deti := deti × 0.0625;
                    dete := dete + 4;
                    go to L1
            end;
L2:         if abs(w) < 0.0625 then
            begin w := w × 16;
                    detr := detr × 16;
                    deti := deti × 16;
                    dete := dete − 4;
                    go to L2
            end;
            for j := k+1 step 1 until n do
```

```
            begin p := j+j;
                  pp := p − 1;
                  cx innerprod (1, 1, k−1, a[k, pp], a[k, p],
                  a[k, i+i−1], a[k, i+i], a[i, pp], a[i, p], i, v, w);
                  a[k, pp] := (v×x + w×y)/z;
                  a[k, p] := (w×x − v×y)/z
            end j
      end k
end compdet;
```

procedure cx acc solve (n) data: (r, a, aa, p, b, eps) result: (x, bb, l) exit: (ill);
value   n, r, eps;
integer n, r, l;
real   eps;
array   a, aa, p, x, b, bb;
label   ill;
comment Solves $A x = b$ where A is an $n \times 2n$ complex unsymmetric matrix
          and b is an $n \times 2r$ complex matrix of r right-hand sides. The general
          elements are $a[i, 2j − 1] + i \times a[i, 2j]$ and $b[i, 2j − 1] + i \times b[i, 2j]$.
          The procedure must be preceded by compdet in which L and U are
          produced in $aa[i, j]$ and the interchanges in $p[i]$. The residuals
          $bb = b − A x$ are calculated and $A d = bb$ is solved, overwritting d on bb.
          The refinement is repeated, as long as the maximum correction at
          any stage is less than half that at the previous stage, until the maxi-
          mum correction is less than $2 eps$ times the maximum x for each
          right hand side. Exits to label ill if the solution fails to improve.
          l is the number of iterations. Uses the procedures innerprod and
          cx innerprod which form accurate inner products;
begin

          procedure compsol (n) data: (r, a, int) trans: (b);
          value r, n; **integer** r, n; **array** a, int, b;
          comment Solves $A x = b$, where A is a complex unsymmetric matrix
                    and b is an $n \times 2r$ complex matrix of r right-hand sides,
                    so that the general element is $b[i, 2j − 1] + i \times b[i, 2j]$.
                    The procedure compsol must be preceded by compdet in
                    which L and U are produced in $a[i, j]$, from A, and the
                    record of the interchanges is produced in $int[i]$. $A x = b$
                    is solved in three steps, interchange elements of b, $L y = b$
                    and $U x = y$. The matrices y and then x are overwritten
                    on b;
          begin   **integer** i, j, p, k, pp, kk;
                    real x, y, z;
                    comment interchanging of elements of b;
                    for i := 1 **step** 1 **until** n **do**
                    if int[i] ≠ i **then**
                    for j := r + r **step** −1 **until** 1 **do**
```
```

```
begin x := b[i, j];
      b[i, j] := b[int [i], j];
      b[int [i], j] := x
end j;
for k := r + r step − 2 until 2 do
begin comment solution of L y = b;
      kk := k − 1;
      for i := 1 step 1 until n do
      begin cx innerprod (1, 1, i−1, b[i, kk], b[i, k],
            a[i, j+j−1], a[i, j+j], b[j, kk],
                        b[j, k], j, x, y);
            p := i + i;
            pp := p − 1;
            z := a[i, pp] ↑ 2 + a[i, p] ↑ 2;
            b[i, kk] := (x × a[i, pp] + y × a[i, p])/z;
            b[i, k] := (y × a[i, pp] − x × a[i, p])/z
      end i;
      comment solution of U x = y;
      for i := n step −1 until 1 do
      cx innerprod (i+1, 1, n, b[i, kk], b[i, k],
      a[i, j+j−1], a[i, j+j], b[j, kk], b[j, k], j,
                  b[i, kk], b[i, k])
            end k
end compsol;

integer i, j, k, d2, c, cc;
real e, d0, d1, xmax, bbmax;
for j := 1 step 1 until r + r do
for i := 1 step 1 until n do
begin    x[i, j] := 0;
         bb[i, j] := b[i, j];
end;
l := 0;
d0 := 0;
```
L3:
```
         compsol (n, r, aa, p, bb);
         l := l + 1;
         d2 := 0;
         d1 := 0;
for j := 1 step 1 until r + r do
for i := 1 step 1 until n do
x[i, j] := x[i, j] + bb[i, j];
for j := 1 step 1 until r do
begin    xmax := bbmax := 0;
         c := j + j;
         cc := c − 1;
         for i := 1 step 1 until n do
```

$$\textbf{begin } e := x[i, cc] \uparrow 2 + x[i, c] \uparrow 2;$$
$$\textbf{if } e > xmax \textbf{ then } xmax := e;$$
$$e := bb[i, cc] \uparrow 2 + bb[i, c] \uparrow 2;$$
$$\textbf{if } e > bbmax \textbf{ then } bbmax := e;$$
$$cx\, innerprod\, (1, 1, n, b[i, cc], b[i, c],$$
$$a[i, k+k-1], a[i, k+k], x[k, cc], x[k, c], k,$$
$$bb[i, cc], bb[i, c])$$
$$\textbf{end};$$
$$\textbf{if } bbmax > d1 \times xmax \textbf{ then } d1 := bbmax/xmax;$$
$$\textbf{if } bbmax > (2 \times eps) \uparrow 2 \times xmax \textbf{ then } d2 := 1;$$
$$\textbf{end};$$
$$\textbf{if } d1 > d0/4 \;\wedge\; l \neq 1 \textbf{ then goto } ill;$$
$$d0 := d1;$$
$$\textbf{if } d2 = 1 \textbf{ then goto } L3;$$
$$\textbf{end } cx\, acc\, solve;$$

5. Organisational and Notational Details

In both *unsymdet* and *compdet* the Euclidean norm of each row is stored in the array $int[i]$. These values are overwritten with the integers specifying the interchanges. In *compdet* the squares of the Euclidean norms are stored to avoid an excessive use of square roots. In *unsymdet* $det(A)$ is given by $2^{d2} \cdot d1$ where $\frac{1}{16} \leq |d1| < 1$ or $d1 = 0$. In *compdet* $det(A)$ is given by $2^{dete}[detr + i\, deti]$ where $\frac{1}{16} \leq \max(|detr|, |deti|) < 1$ or $detr = deti = 0$.

In *unsym acc solve* and *cx acc solve* it is necessary to retain the original A and B matrices. The factorization of A is performed in the array aa and the array bb is used to store both the successive residual vectors and the successive corrections. An accurate inner-product procedure is required in computing residuals and for simplicity only one inner-product procedure (real or complex as appropriate) has been used throughout. If the accumulation of inner-products or double-precision computation is inefficient a simple inner-product procedure may be used except when computing the residuals. The procedure *innerprod* is identical with that used in [5] and adds the inner-product to a double-precision initial value given by $c1$ and $c2$. The procedure *cx innerprod* accepts a single precision complex number as initial value.

Iteration is terminated when

$$\|d^{(s)}\|_\infty \leq 2\, eps\, \|x^{(s+1)}\|_\infty \tag{14}$$

for all right hand sides. In the complex case this test is carried out in the form

$$\|d^{(s)}\|_\infty^2 \leq 4\, eps^2\, \|x^{(s+1)}\|_\infty^2. \tag{15}$$

If in any iteration

$$\max(\|d^{(s)}\|_\infty/\|x^{(s+1)}\|_\infty) > \tfrac{1}{2}\max(\|d^{(s-1)}\|_\infty/\|x^{(s)}\|_\infty) \tag{16}$$

(where the maximization is over the r right-hand sides) the matrix is too ill-conditioned for reliable solutions to be obtained and a failure indication is given. Again in the complex case the squares of both sides of (16) are used.

6. Discussion of the Numerical Properties

The computed matrices LU are such that

$$LU = \tilde{A} + \tilde{E} \tag{17}$$

where $\tilde{A}$ is the permuted form of A and $\tilde{E}$ may be regarded as the matrix of rounding errors. The upper bound for $\tilde{E}$ depends on the nature of the inner-product procedure used in the factorization and on the size of the maximum element g in the reduced matrices which would have arisen in Gaussian elimination of A with partial pivoting. When inner-products are accumulated we have

$$|\tilde{e}_{ij}| \leq g\, 2^{-t} \tag{18}$$

and it is uncommon with partial pivoting for g to be much greater than $\max\limits_{i,j} |a_{ij}|$. The determinant of a matrix which is very close to A is therefore obtained.

The rounding errors made in the forward and backward substitutions are comparatively unimportant and the computed solution of $A x = b$ is effectively

$$x = (A+E)^{-1} b \tag{19}$$

where E is the perturbation of A corresponding to the perturbation $\tilde{E}$ of $\tilde{A}$. Hence we have

$$\|x - A^{-1}b\|_2 / \|A^{-1}b\|_2 \leq \|E\|_2 \|A^{-1}\|_2 / (1 - \|E\|_2 \|A^{-1}\|_2), \tag{20}$$

so that if $\|E_2\| \|A^{-1}\|_2 = 2^{-p}$ $(p > 0)$

$$\|x - A^{-1}b\|_2 / \|A^{-1}b\|_2 \leq 2^{-p} / (1 - 2^{-p}). \tag{21}$$

If the residuals are computed using accumulation of inner-products the rounding errors made in that part are comparatively unimportant and in the iterative refinement procedure we have

$$\|x^{(s+1)} - A^{-1}b\|_2 \leq \|x^{(s)} - A^{-1}b\|_2\, 2^{-p} / (1 - 2^{-p}). \tag{22}$$

For detailed discussions see [7] or [8].

7. Examples of the Use of the Procedures

The uses of the procedures are in general self-evident. In later procedures in this series *compdet* will be used to evaluate $det(A_r \lambda^r + \cdots + A_1 \lambda + A_0)$ for complex values of λ in the solution of the generalized eigenvalue problem.

An interesting application of *unsym acc solve* and *cx acc solve* is in the determination of accurate eigenvectors. Here, given λ an approximate eigenvalue of A and an arbitrary vector $x^{(0)}$ the system of vectors defined by $(A - \lambda I) x^{(s+1)} = x^{(s)}$ is determined *using the relevant accurate solve procedure in each step*. (For this it is essential that λ is not too close to an eigenvalue, or the accurate solve procedure will fail.) In this way correctly rounded eigenvectors may be obtained even when A is quite ill-conditioned from the point of view of its eigenproblem. (Notice that this is a refinement of Wielandt-iteration).

8. Test Results

The procedures for real matrices have been used on KDF 9 in connexion with the leading principal minor of order seven of the Hilbert matrix. (KDF 9 has a 39 binary digit mantissa.) The scaled version of this matrix has been used as in [4] and [5]. Inner-products were accumulated throughout since this gives no loss of speed on KDF 9. On some computers it may be advisable to restrict the use of an accurate inner-product procedure to the computation of residuals only.

unsymdet gave the value $8.4735\,33760\,12_{10} - 2 \times 2^{52}$ for the determinant, which was somewhat less accurate than that given by *choldet 1* with accumulation of inner-products [5] but appreciably more accurate than that given by *choldet 1* without accumulation [4]. The correct value is $8.4735\,3913\ldots_{10} - 2 \times 2^{52}$

unsym acc solve was used with the seven columns of $(360360 \times \text{unit matrix})$ as right-hand sides. After the second iteration the solutions were correct to working accuracy and as the elements are integers this meant that they were exact. The corresponding residuals were therefore exactly zero and the third iteration confirmed that the second was correct. For economy of presentation and for comparison with [5] we give in Table 1 only the results for the fifth column. Again the results of the first iteration were slightly inferior to those obtained with *choldet 1* and accumulation of inner-products [5].

Table 1

"unsym acc solve" with right-hand side $(0, \ldots, 0, 360360, 0, 0)$

$d^{(0)}$	$x^{(1)} = 0 + d^{(0)}$	$r^{(1)} = b - A\,x^{(1)}$
$+4.8510\,0315\,047_{10} + 4$	$+4.8510\,0315\,047_{10} + 4$	$+7.2853\,1169\,891_{10} + 0$
$-1.9404\,0128\,440_{10} + 6$	$-1.9404\,0128\,440_{10} + 6$	$+7.9840\,9938\,812_{10} - 1$
$+1.8711\,0125\,732_{10} + 7$	$+1.8711\,0125\,732_{10} + 7$	$-1.4319\,4198\,608_{10} - 2$
$-7.2765\,0495\,356_{10} + 7$	$-7.2765\,0495\,356_{10} + 7$	$+1.2209\,0101\,242_{10} - 1$
$+1.3340\,2591\,837_{10} + 8$	$+1.3340\,2591\,837_{10} + 8$	$+3.7329\,6737\,671_{10} - 2$
$-1.1525\,9840\,116_{10} + 8$	$-1.1525\,9840\,116_{10} + 8$	$+2.8892\,9939\,270_{10} - 2$
$+3.7837\,8265\,214_{10} + 7$	$+3.7837\,8265\,214_{10} + 7$	$-6.4476\,6807\,556_{10} - 1$
$d^{(1)}$	$x^{(2)} = x^{(1)} + d^{(1)}$	$r^{(2)} = b - A\,x^{(2)}$
$-3.1504\,7711\,930_{10} - 2$	$+4.8510\,0000\,000_{10} + 4$	$+0.0000\,0000\,000$
$+1.2843\,9797\,744_{10} + 0$	$-1.9404\,0000\,000_{10} + 6$	$+0.0000\,0000\,000$
$-1.2573\,2505\,215_{10} + 1$	$+1.8711\,0000\,000_{10} + 7$	$+0.0000\,0000\,000$
$+4.9535\,6773\,435_{10} + 1$	$-7.2765\,0000\,000_{10} + 7$	$+0.0000\,0000\,000$
$-9.1837\,4631\,470_{10} + 1$	$+1.3340\,2500\,000_{10} + 8$	$+0.0000\,0000\,000$
$+8.0116\,5081\,013_{10} + 1$	$-1.1525\,9760\,000_{10} + 8$	$+0.0000\,0000\,000$
$-2.6521\,3798\,520_{10} + 1$	$+3.7837\,8000\,000_{10} + 7$	$+0.0000\,0000\,000$

The procedures for complex matrices were tested using the same scaled Hilbert matrix pre-multiplied by a diagonal complex matrix with elements

$$(1+i,\ 1-i,\ 1+2i,\ 1-2i,\ 1+3i,\ 1-3i,\ 1+4i)$$

compdet gave the value

$$(3.3099\,7558\,401_{10} - 2 + i\,1.3239\,9014\,944_{10} - 1)\,2^{60}$$

for the determinant compared with the true value

$$(3.3099\,7622\ldots_{10}-2+i\,1.3239\,9048\ldots_{10}-1)\,2^{60}.$$

It is perhaps worth remarking that the error is less than can be caused by a change of 1 part in 2^{39} in elements of A.

cx acc solve was used with the seven right-hand sides consisting of the columns of the identity matrix. Again after the second iteration the results were correct to working accuracy. In Table 2 we show the progress of the solution corresponding to the first right-hand side only. All right-hands sides behaved in the same manner. The corrections $d^{(s)}$ are not given. The solution is $\frac{1}{2}(1-i)$ times the solution given in [5].

Table 2

"cx acc solve" with right-hand side $(1, 0, \ldots, 0)$

Real part of $x^{(1)}$	Imag. part of $x^{(1)}$
$+6.7987\,5827\,723_{10}-5$	$-6.7987\,5818\,339_{10}-5$
$-1.6317\,0224\,716_{10}-3$	$+1.6317\,0221\,003_{10}-3$
$+1.2237\,7684\,112_{10}-2$	$-1.2237\,7680\,564_{10}-2$
$-4.0792\,5656\,083_{10}-2$	$+4.0792\,5642\,388_{10}-2$
$+6.7307\,7390\,627_{10}-2$	$-6.7307\,7365\,673_{10}-2$
$-5.3846\,1951\,681_{10}-2$	$+5.3846\,1930\,230_{10}-2$
$+1.6666\,6804\,923_{10}-2$	$-1.6666\,6797\,911_{10}-2$

Real part of $r^{(1)}$	Imag. part of $r^{(1)}$
$-3.2900\,0471\,311_{10}-\ 9$	$-3.8102\,8542\,051_{10}-11$
$-5.5630\,1671\,395_{10}-11$	$-1.5073\,4891\,236_{10}-\ 9$
$-9.6781\,2496\,810_{10}-11$	$-1.0084\,5554\,130_{10}-10$
$+7.1188\,8326\,066_{10}-11$	$+3.7296\,3437\,911_{10}-10$
$-4.6139\,0925\,466_{10}-11$	$-1.1232\,2595\,669_{10}-10$
$-3.7410\,0750\,378_{10}-11$	$+4.9302\,7840\,776_{10}-11$
$-2.7174\,1917\,857_{10}-\ 9$	$-1.6891\,9278\,565_{10}-\ 9$

Real part of $x^{(2)}$	Imag. part of $x^{(2)}$
$+6.7987\,5679\,876_{10}-5$	$-6.7987\,5679\,876_{10}-5$
$-1.6317\,0163\,170_{10}-3$	$+1.6317\,0163\,170_{10}-3$
$+1.2237\,7622\,378_{10}-2$	$-1.2237\,7622\,378_{10}-2$
$-4.0792\,5407\,925_{10}-2$	$+4.0792\,5407\,925_{10}-2$
$+6.7307\,6923\,076_{10}-2$	$-6.7307\,6923\,076_{10}-2$
$-5.3846\,1538\,462_{10}-2$	$+5.3846\,1538\,462_{10}-2$
$+1.6666\,6666\,667_{10}-2$	$-1.6666\,6666\,667_{10}-2$

Real part of $r^{(2)}$	Imag. part of $r^{(2)}$
$+5.4264\,0421\,486_{10}-9$	$+0.0000\,0000\,000$
$+0.0000\,0000\,000$	$-4.8596\,3802\,532_{10}-9$
$+6.5483\,5652\,369_{10}-9$	$+2.1827\,8550\,790_{10}-9$
$-1.9749\,3443\,821_{10}-9$	$-5.9248\,0331\,463_{10}-9$
$+7.2009\,3140\,671_{10}-9$	$+3.6004\,6570\,336_{10}-9$
$-3.3047\,1827\,681_{10}-9$	$-6.6094\,3655\,362_{10}-9$
$+7.6302\,2534\,045_{10}-9$	$+4.5781\,3520\,427_{10}-9$

A second test was performed on the computer TR 4 at the Mathematisches Institut der Technischen Hochschule, München. The TR 4 has effectively a 38 binary digit mantissa; since it is not well adapted for the accumulation of inner products the latter were computed using true double-precision arithmetic. The results obtained generally confirmed those on KDF 9 but as the rounding procedures are somewhat less satisfactory on TR 4 a third iteration was required in all cases to attain final accuracy and a fourth to confirm that the third was correct.

Acknowledgements. The authors wish to thank Dr. C. REINSCH of the Mathematisches Institut der Technischen Hochschule, München for several valuable comments and for his assistance in testing these algorithms. The work described here is part of the Research Programme of the National Physical Laboratory and this paper is published by permission of the Director of the Laboratory.

References

[1] BAUER, F. L.: Optimally scaled matrices. Numer. Math. **5**, 73—87 (1963).

[2] FORSYTHE, G. E.: Crout with pivoting. Comm. ACM **3**, 507—508 (1960).

[3] —, and E. G. STRAUS: On best conditioned matrices. Proc. Amer. Math. Soc. **6**, 340—345 (1955).

[4] MARTIN, R. S., G. PETERS, and J. H. WILKINSON: Symmetric decompositions of a positive definite matrix. Numer. Math. **7**, 362—383 (1965). Cf. I/1.

[5] — — — Iterative refinement of the solution of a positive definite system of equations. Numer. Math. **8**, 203—216 (1966). Cf. I/2.

[6] McKEEMAN, W. M.: Crout with equilibration and iteration. Comm. ACM **5**, 553—555 (1962).

[7] WILKINSON, J. H.: Rounding errors in algebraic processes. London: Her Majesty's Stationery Office; Englewood Cliffs, N. J.: Prentice-Hall 1963.
German edition: Rundungsfehler. Berlin-Göttingen-Heidelberg: Springer 1969.

[8] — The algebraic eigenvalue problem. London: Oxford University Press 1965.

Contribution I/8

Linear Least Squares Solutions
by Housholder Transformations* **

by P. Businger and G. H. Golub

1. Theoretical Background

Let A be a given $m \times n$ real matrix with $m \geq n$ and of rank n and b a given vector. We wish to determine a vector $\hat{x}$ such that

$$\|b - A\hat{x}\| = \min.$$

where $\|...\|$ indicates the euclidean norm. Since the euclidean norm is unitarily invariant

$$\|b - Ax\| = \|c - QAx\|$$

where $c = Qb$ and $Q^T Q = I$. We choose Q so that

$$QA = R = \begin{pmatrix} \tilde{R} \\ ... \\ 0 \end{pmatrix} \ \}(m-n) \times n \tag{1}$$

and $\tilde{R}$ is an upper triangular matrix. Clearly,

$$\hat{x} = \tilde{R}^{-1} \tilde{c}$$

where $\tilde{c}$ denotes the first n components of c.

A very effective method to realize the decomposition (1) is via Householder transformations [1]. Let $A = A^{(1)}$, and let $A^{(2)}, A^{(3)}, ..., A^{(n+1)}$ be defined as follows:

$$A^{(k+1)} = P^{(k)} A^{(k)} \qquad (k = 1, 2, ..., n).$$

$P^{(k)}$ is a symmetric, orthogonal matrix of the form

$$P^{(k)} = I - \beta_k u^{(k)} u^{(k)T}$$

* Reproduction in Whole or in Part is permitted for any Purpose of the United States government. This report was supported in part by Office of Naval Research Contract Nonr-225(37) (NR 044-11) at Stanford University.

** Prepublished in Numer. Math. 7, 269−276 (1965).

where the elements of $P^{(k)}$ are derived so that

$$a_{i,k}^{(k+1)} = 0$$

for $i = k+1, \ldots, m$. It can be shown, cf. [2], that $P^{(k)}$ is generated as follows:

$$\sigma_k = \left(\sum_{i=k}^{m} (a_{i,k}^{(k)})^2 \right)^{\frac{1}{2}},$$

$$\beta_k = [\sigma_k (\sigma_k + |a_{k,k}^{(k)}|)]^{-1},$$

$$u_i^{(k)} = 0 \quad \text{for} \quad i < k,$$

$$u_k^{(k)} = \operatorname{sgn} (a_{k,k}^{(k)}) (\sigma_k + |a_{k,k}^{(k)}|),$$

$$u_i^{(k)} = a_{i,k}^{(k)} \quad \text{for} \quad i > k.$$

The matrix $P^{(k)}$ is not computed explicitly. Rather we note that

$$A^{(k+1)} = (I - \beta_k u^{(k)} u^{(k)T}) A^{(k)}$$
$$= A^{(k)} - u^{(k)} y_k^T$$

where

$$y_k^T = \beta_k u^{(k)T} A^{(k)}.$$

In computing the vector y_k and $A^{(k+1)}$, one takes advantage of the fact that the first $(k-1)$ components of $u^{(k)}$ are equal to zero.

At the kth stage the column of $A^{(k)}$ is chosen which will maximize $|a_{k,k}^{(k+1)}|$. Let

$$s_j^{(k)} = \sum_{i=k}^{m} (a_{i,j}^{(k)})^2 \qquad j = k, k+1, \ldots, n.$$

Then since $|a_{k,k}^{(k+1)}| = \sigma_k$, one should choose that column for which $s_j^{(k)}$ is maximized. After $A^{(k+1)}$ has been computed, one can compute $s_j^{(k+1)}$ as follows:

$$s_j^{(k+1)} = s_j^{(k)} - (a_{k,j}^{(k+1)})^2$$

since the orthogonal transformations leave the column lengths invariant.

Let $\bar{x}$ be the initial solution obtained, and let $\hat{x} = \bar{x} + e$. Then

$$\|b - A\hat{x}\| = \|r - Ae\|$$

where

$$r = b - A\bar{x}, \quad \text{the residual vector.}$$

Thus the correction vector e is itself the solution to a linear least squares problem. Once A has been decomposed, and if the transformations have been saved, then it is a simple matter to compute r and solve for e. The iteration process is continued until convergence. There is no assurance, however, that all digits of the final solution will be correct.

2. Applicability

The algorithm *least squares solution* may be used for solving linear least squares problems, systems of linear equations where A is a square matrix, and thus also for inverting matrices.

3. Formal Parameter List

The matrix A is a given matrix of an overdetermined system of m equations in n unknowns. The matrix B contains p right hand sides, and the solution is stored in X. If no solution can be found then the problem is left unsolved and the emergency exit *singular* is used. The termination procedure is dependent upon *eta* which is the smallest number such that $fl(1+eta) > 1$ where $fl(...)$ indicates the floating point operation.

4. ALGOL Program

procedure *least squares solution* (a, x, b, m, n, p, eta) *e xit*: (*singular*);
    value m, n, p, eta;
    array a, x, b; **integer** m, n, p; **real** *eta*; **label** *singular*;
  comment The array $a[1:m, 1:n]$ contains the given matrix of an overdetermined
            system of m linear equations in n unknowns $(m \geq n)$. For the p right
            hand sides given as the columns of the array $b[1:m, 1:p]$, the least
            squares solutions are computed and stored as the columns of the
            array $x[1:n, 1:p]$. If $rank(a) < n$ then the problem is left unsolved
            and the emergency exit *singular* is used. In either case a and b
            are left intact. *eta* is the relative machine precision;
begin
      real procedure *inner product* (i, m, n, a, b, c);
        value m, n, c;
        real a, b, c; **integer** i, m, n;
      comment The body of this inner product routine should preferably be
                replaced by its double precision equivalent in machine code;
      begin
        for $i := m$ **step** 1 **until** n **do** $c := c + a \times b$;
        inner product $:= c$
      end *inner product*;
      procedure *decompose* $(m, n, qr, alpha, pivot)$ *exit*: (*singular*);
        value m, n;
        integer m, n; **array** $qr, alpha$; **integer array** *pivot*;
        label *singular*; **comment** nonlocal real procedure *inner product*;
  comment Decompose reduces the matrix given in the array $qr[1:m, 1:n]$
            $(m \geq n)$ to upper right triangular form by means of n elementary
            orthogonal transformations $(I-beta\ uu')$. The diagonal elements of
            the reduced matrix are stored in the array $alpha[1:n]$, the off diagonal
            elements in the upper right triangular part of qr. The nonzero com-
            ponents of the vectors u are stored on and below the leading diagonal
            of qr. Pivoting is done by choosing at each step the column with
            the largest sum of squares to be reduced next. These interchanges
            are recorded in the array $pivot[1:n]$. If at any stage the sum of
            squares of the column to be reduced is exactly equal to zero then
            the emergency exit singular is used;

```
begin
      integer i, j, jbar, k;   real beta, sigma, alphak, qrkk;
      array y, sum [1 : n];
      for j := 1 step 1 until n do
      begin comment j-th column sum;
            sum [j] := inner product (i, 1, m, qr [i, j], qr [i, j], 0);
            pivot [j] := j
      end j-th column sum;
      for k := 1 step 1 until n do
      begin comment k-th Householder transformation;
            sigma := sum [k]; jbar := k;
            for j := k + 1 step 1 until n do
                if sigma < sum [j] then
                begin
                      sigma := sum [j]; jbar := j
                end;
      if jbar ≠ k then
      begin comment column interchange;
            i := pivot [k]; pivot [k] := pivot [jbar]; pivot [jbar] := i;
            sum [jbar] := sum [k]; sum [k] := sigma;
            for i := 1 step 1 until m do
            begin
                  sigma := qr [i, k]; qr [i, k] := qr [i, jbar];
                  qr [i, jbar] := sigma
            end i
      end column interchange;
      sigma := inner product (i, k, m, qr [i, k], qr [i, k], 0);
      if sigma = 0 then go to singular;
      qrkk := qr [k, k];
      alphak := alpha [k] := if qrkk < 0 then sqrt (sigma) else − sqrt (sigma);
      beta := 1/(sigma − qrkk × alphak);
      qr [k, k] := qrkk − alphak;
      for j := k + 1 step 1 until n do
          y [j] := beta × inner product (i, k, m, qr [i, k], qr [i, j], 0);
      for j := k + 1 step 1 until n do
      begin
            for i := k step 1 until m do
            qr [i, j] := qr [i, j] − qr [i, k] × y [j];
            sum [j] := sum [j] − qr [k, j] ↑2
      end j
   end k-th Householder transformation
end decompose;


procedure solve (m, n, qr, alpha, pivot, r, y);
   value m, n;
   integer m, n;   array qr, alpha, r, y;   integer array pivot;
   comment nonlocal real procedure inner product;
```

comment Using the vectors u whose nonzero components are stored on and below the main diagonal of $qr[1:m, 1:n]$ *solve* applies the n transformations $(I\text{-}beta\ uu')$ to the right hand side $r[1:m]$. From the reduced matrix given in $alpha[1:n]$ and the upper right triangular part of qr, *solve* then computes by backsubstitution an approximate solution to the linear system. The components of the solution vector are stored in $y[1:n]$ in the order prescribed by $pivot[1:n]$;

begin
    integer i, j;   **real** *gamma*;   **array** $z[1:n]$;
    for $j := 1$ **step** 1 **until** n **do**
    begin comment apply the j-th transformation to the right hand side;
        $gamma := inner\ product(i, j, m, qr[i,j], r[i], 0)/(alpha[j] \times qr[j,j])$;
        for $i := j$ **step** 1 **until** m **do** $r[i] := r[i] + gamma \times qr[i,j]$
    end *j-th transformation*;
    $z[n] := r[n]/alpha[n]$;
    for $i := n-1$ **step** -1 **until** 1 **do**
        $z[i] := -inner\ product(j, i+1, n, qr[i,j], z[j], -r[i])/alpha[i]$;
    for $i := 1$ **step** 1 **until** n **do** $y[pivot[i]] := z[i]$
end *solve*;

integer i, j, k; **real** *normy0, norme0, norme1, eta2*;
array $qr[1:m, 1:n], alpha, e, y[1:n], r[1:m]$; **integer array** $pivot[1:n]$;
for $j := 1$ **step** 1 **until** n **do**
    for $i := 1$ **step** 1 **until** m **do** $qr[i,j] := a[i,j]$;
decompose$(m, n, qr, alpha, pivot, singular)$;
$eta2 := eta \uparrow 2$;
for $k := 1$ **step** 1 **until** p **do**
begin comment solution for the k-th right hand side;
    for $i := 1$ **step** 1 **until** m **do** $r[i] := b[i, k]$;
    solve$(m, n, qr, alpha, pivot, r, y)$;
    for $i := 1$ **step** 1 **until** m **do**
        $r[i] := -inner\ product(j, 1, n, a[i,j], y[j], -b[i, k])$;
    solve$(m, n, qr, alpha, pivot, r, e)$;
    $normy0 := norme1 := 0$;
    for $i := 1$ **step** 1 **until** n **do**
    begin
        $normy0 := normy0 + y[i] \uparrow 2$; $norme1 := norme1 + e[i] \uparrow 2$
    end i;
    if $norme1 > 0.0625 \times normy0$ **then go to** *singular*;
        comment No attempt at obtaining the solution is made unless the norm of the first correction is significantly smaller than the norm of the initial solution;
improve:
    for $i := 1$ **step** 1 **until** n **do** $y[i] := y[i] + e[i]$;
    if $norme1 < eta2 \times normy0$ **then go to** *store*;
        comment Terminate the iteration if the correction was of little significance;

```
    for i:=1 step 1 until m do
        r[i] := − inner product (j, 1, n, a[i,j], y[j], −b[i,k]);
    solve (m, n, qr, alpha, pivot, r, e);
    norme0 := norme1;  norme1 := 0;
    for i:=1 step 1 until n do norme1 := norme1+e[i] ↑2;
    if norme1≦0.0625 ×norme0 then go to improve;
        comment Terminate the iteration also if the norm of the correction
                failed to decrease sufficiently as compared with the norm
                of the previous correction;
store:
    for i:=1 step 1 until n do x[i, k] := y[i]
    end k-th right hand side
end least squares solution;
```

5. Organisational and Notational Details

The array a, containing the original matrix A, is transferred to the array qr which serves as storage for $A^{(k)}$. The non-zero components of the vectors $\boldsymbol{u}^{(k)}$ are stored on and below the leading diagonal of qr. The diagonal elements of $\widetilde{R}$, the reduced matrix, are stored in the array $alpha$.

The column sum of squares, $s_j^{(k)}$, is stored in the array sum. Naturally, the elements of this array are interchanged whenever the columns of $A^{(k+1)}$ are interchanged. The array $pivot$ contains the order in which the columns are selected.

6. Discussion of Numerical Properties

The program uses the iteration scheme described in section 1. Let

$$\boldsymbol{x}^{(q+1)}=\boldsymbol{x}^{(q)}+\boldsymbol{e}^{(q)}, \qquad q=0, 1, \dots$$

with $\boldsymbol{x}^{(0)}=0$ and

$$\boldsymbol{r}^{(q)}=\boldsymbol{b}-A\,\boldsymbol{x}^{(q)}.$$

The residual vector $\boldsymbol{r}^{(q)}$ should be computed using double precision inner products and then rounded to single precision accuracy.

If $\|\boldsymbol{e}^{(1)}\|/\|\boldsymbol{x}^{(1)}\|>0.25$ or if $\sigma_k=0$ at any stage then the singular exit is used. The iteration process is terminated if $\|\boldsymbol{e}^{(k+1)}\|>0.25\|\boldsymbol{e}^{(k)}\|$ or if $\|\boldsymbol{e}^{(k)}\|/\|\boldsymbol{x}^{(0)}\|<eta$.

7. Examples of the Use of the Procedure

In many statistical applications, it is necessary to compute $(A^T A)^{-1}$ or $\det(A^T A)$.

Since

$$(A^T A)=\widetilde{R}^T\widetilde{R}, \qquad (A^T A)^{-1}=\widetilde{R}^{-1}\widetilde{R}^{-T}.$$

In addition,

$$\det(A^T A)=(\det \widetilde{R})^2=r_{11}^2\cdot r_{22}^2\dots\cdot r_{nn}^2.$$

8. Test Results

The procedure was tested on the B 5000 (Stanford University) which has a 39 bit mantissa. All inner products were computed using a double precision inner product routine. We give two examples. The first example consists of

Example I

A

					B
$3.60000_{10}+01$	$-6.30000_{10}+02$	$3.36000_{10}+03$	$-7.56000_{10}+03$	$7.56000_{10}+03$	$4.63000_{10}+02$
$-6.30000_{10}+02$	$1.47000_{10}+04$	$-8.82000_{10}+04$	$2.11680_{10}+05$	$-2.20500_{10}+05$	$-1.38600_{10}+04$
$3.36000_{10}+03$	$-8.82000_{10}+04$	$5.64480_{10}+05$	$-1.41120_{10}+06$	$1.51200_{10}+06$	$9.70200_{10}+04$
$-7.56000_{10}+03$	$2.11680_{10}+05$	$-1.41120_{10}+06$	$3.62880_{10}+06$	$-3.96900_{10}+06$	$-2.58720_{10}+05$
$7.56000_{10}+03$	$-2.20500_{10}+05$	$1.51200_{10}+06$	$-3.96900_{10}+06$	$4.41000_{10}+06$	$2.91060_{10}+05$
$-2.77200_{10}+03$	$8.31600_{10}+04$	$-5.82120_{10}+05$	$1.55232_{10}+06$	$-1.74636_{10}+06$	$-1.16424_{10}+05$

$$x = \begin{bmatrix} 1.0000\,0000\,00_{10}+00 & 5.0000\,0000\,00_{10}-01 & 3.3333\,3333\,33_{10}-01 & 2.5000\,0000\,00_{10}-01 & 2.0000\,0000\,00_{10}-01 \end{bmatrix}$$

Example II

B

$-4.15700_{10}+03$

$-1.78200_{10}+04$

$9.35550_{10}+04$

$-2.61800_{10}+05$

$2.88288_{10}+05$

$-1.18944_{10}+05$

$$x = \begin{bmatrix} 1.0013\,5225\,67_{10}+00 & 5.0045\,0298\,04_{10}-01 & 3.3352\,6028\,36_{10}-01 & 2.5008\,4212\,08_{10}-01 & 2.0002\,9920\,51_{10}-01 \end{bmatrix}$$

the first five columns of the inverse of the 6×6 Hilbert matrix. The vector b was chosen so that $\|b - A x\| = 0$. In the second example, a vector orthogonal to the columns of A was added to b. Thus the solution should be precisely the same as in the first example. Three iterations were required in each case. Note all digits are correct of the computed solution for example I. In example II, however, only the first three digits are correct. Thus, in general not all solutions will be accurate to full working accuracy.

A second test was performed on a Telefunken TR 4 computer at the Kommission für Elektronisches Rechnen der Bayerischen Akademie der Wissenschaften, München.

References

[1] HOUSEHOLDER, A. S.: Unitary Triangularization of a Nonsymmetric Matrix. J. Assoc. Comput. Mach. **5**, 339—342 (1958).
[2] MARTIN, R. S., C. REINSCH, and J. H. WILKINSON: Householder's tridiagonalization of a symmetric matrix. Numer. Math. **11**, 181—195 (1968). Cf. II/2.

Elimination with Weighted Row Combinations
for Solving Linear Equations
and Least Squares Problems*

by F. L. BAUER

1. Theoretical Background

Let A be a matrix of n rows and m columns, $m \leq n$. If and only if the columns are linearly independent, then for any vector b there exists a unique vector x minimizing the Euclidean norm of $b - Ax$, $\|b - Ax\| = \min\limits_{\xi} \|b - A\xi\|$.

It is well known that

(i) the remaining deviation $b - Ax$ is orthogonal to the columns of A, $A^H(b - Ax) = 0$,

(ii) x is determined by $x = A^+ b$, where $A^+ = (A^H A)^{-1} A^H$ is the pseudoinverse of A,

(iii) $b - Ax$ is obtained by a suitable projection on b which annihilates A, $b - Ax = Hb$, where $H = I - A(A^H A)^{-1} A^H$ is a Hermitean idempotent, $H^2 = H$ and $HA = 0$.

The orthodox computational method starts from the normal equations

$$(1) \qquad\qquad A^H A x = A^H b$$

obtained directly from (i). Instead of (1), the equations

$$(A S)^H A x = (A S)^H b,$$

with a nonsingular $m \times m$-matrix S, may be used. For suitable S, $A S = U$ is an orthonormal basis of the columns of A.

The matrix $U^H A$ of the resulting system of linear equations

$$U^H A x = U^H b$$

will have the same columnwise condition** as the original matrix A and in general a better condition than the matrix $A^H A$ in (1).

---

* Prepublished in Numer. Math. 7, 338—352 (1965).

** We define the columnwise condition of a rectangular matrix A to be

$$\operatorname{cond}_c(A) = \max\limits_{\|x\|=1} \|Ax\| \Big/ \min\limits_{\|x\|=1} \|Ax\|.$$

For the Euclidean norm, it is sufficient to show that $(U^H A)^H U^H A = A^H A$. But since S is nonsingular,

$$(U^H A)^H U^H A = A^H U U^H A = A^H U U^H U S^{-1} = A^H U S^{-1} = A^H A.$$

If U is not normalized, the minimum condition number [1] is still unchanged.

Indeed, use of an orthogonalization process on A for obtaining a least squares solution is known in the literature (see [2], p. 73).

The particular computational version used in the ALGOL program to be described here is only slightly different. It is based on a decomposition $A = UDR$, where U consists of m mutually orthogonal, non-zero columns, R is an upper triangular nonsingular $m \times m$-matrix and $D = (U^H U)^{-1}$. Such a decomposition certainly exists if the columns of A are linearly independent.

The condition $U^H(b - Ax) = 0$ yields

$$Rx = U^H b,$$

a nonsingular triangular system which is to be solved by backsubstitution. Notably, $R = U^H A$ and the right side $U^H b$ are obtained by the same operation U^H, applied to A and b respectively.

The computational process of reduction to $Rx = U^H b$ is performed in m steps. In the i-th step, from an intermediate matrix $A^{(i-1)}$ a new intermediate matrix $A^{(i)}$ is formed by orthogonalizing the columns following the i-th column with respect to the i-th column.

Starting with $A^{(0)} = A \,|\, b$ (or rather with $A \,|\, b_1 \, b_2 \ldots b_k$ in the case of several right sides b), $A^{(i-1)}$ has all its columns orthogonal to any of the first $i - 1$ columns; in particular, the first $i - 1$ columns are mutually orthogonal. In deriving $A^{(i)}$ from it, orthogonalization with respect to the i-th column amounts to performing a combination of the rows of $A^{(i-1)}$ with weights given by the corresponding components of the i-th column, and in subtracting a suitable multiple of weighted combination from each row, the multiples being determined only by the requirement that the elements of the i-th column are removed. If we denote the k-th column of $A^{(i)}$ by $a_k^{(i)}$ then we have

$$a_k^{(i)} = a_k^{(i-1)} - a_i^{(i-1)} \frac{[a_i^{(i-1)}]^H a_k^{(i-1)}}{[a_i^{(i-1)}]^H a_i^{(i-1)}} \quad (k > i)$$

$$= a_k^{(i-1)} - \frac{a_i^{(i-1)} [a_i^{(i-1)}]^H}{[a_i^{(i-1)}]^H a_i^{(i-1)}} a_k^{(i-1)}.$$

This is true also if $k < i$ as a consequence of orthogonality of $a_i^{(i-1)}$ and $a_k^{(i-1)}$. The right side is null if $k = i$.

Computationally, of course, only the columns following the i-th will be involved. Thus, the process performed is essentially a sequence of Gauss-eliminations, but with a suitably weighted combination of rows used for elimination instead of a selected row.

It is, however, an orthogonalization process*, and after m steps, the first m columns of $A^{(m)}$ are an orthogonal basis U, $U = AS$, with a nonsingular, preferably a unit upper triangular S.

---

* essentially, a Schmidt orthogonalization process, except that in the latter it is usual to deal once and for all with each column.

The remaining column (or rather, the remaining columns, in the case of several right sides $b_1 \ldots b_k$) being orthogonal to the columns of U, is orthogonal also to the columns of A and therefore is the deviation $b - Ax$ (or rather $b_1 - Ax_1$, $b_2 - Ax_2, \ldots, b_k - Ax_k$).

Moreover, the operation U^H applied to $A \mid b$, is already performed during the elimination process: the i-th row of $U^H A \mid U^H b$ is just the elimination row in the i-th elimination step. To see this, it is to be observed first that this i-th row is the scalar product of the i-th column of U with the columns of $A \mid b$, or the scalar product of the i-th column of $A^{(i)}$ with the columns of $A^{(0)}$, and since corresponding columns of $A^{(0)}$ and $A^{(i)}$ differ only by a vector in the space of the first $i - 1$ columns of $A^{(i)}$, which is orthogonal to the i-th column of $A^{(i)}$, the i-th row of $U^H A \mid U^H b$ is the scalar product of the i-th column of $A^{(i)}$ with the columns of $A^{(i)}$. But this is the elimination row for the i-th step.

Thus, the elimination rows, put aside, form the upper triangular nonsingular system

$$Rx = U^H b = v$$

from which the solution x is obtained. It is to be noted that Gaussian elimination performed with the normal equations leads to the same system, since

$$A^H A = R^H D R.$$

If U were orthonormalized, R would be the Choleski decomposed upper triangle of $A^H A$.

2. Applicability

The algorithm described above may be used for solving least squares problems, systems of linear equations with square matrices and thus also for inverting matrices. Four different procedures were programmed so that an optimum form could be used for each of these problems.

Procedure ORTHOLIN1 yields the solution X of the least squares problem $AX = B$; A is an $n \times m$-matrix, X an $m \times k$-matrix and B an $n \times k$-matrix. If $k = 1$, then B is only a single column vector and procedure ORTHOLIN2 is provided for this special case. For inverting a matrix, the procedure ORTHO1 was developed. Following GOLUB [3], [4] in this respect, these three procedures include iterative improvement of the solution. Omitting this extra section of program a variant ORTHO2 also saves space by using the place of the matrix A first for working storage and then for storing the inverse of A.

3. Formal Parameter List

3.1 *Procedure* ORTHOLIN1 (a, n, m, b, k, eps) *result*: (x) *exit*: $(stop)$.

> a $n \times m$-matrix of the systems of linear equations, to be declared outside with subscripts running from 1 to n, 1 to m,
>
> n number of rows in the matrix a,
>
> m number of columns in the matrix a $(m \le n)$,

b $n \times k$-matrix on the right side of the systems of linear equations, to be declared outside with subscripts running from 1 to n, 1 to k,

k number of columns in the matrix b,

eps   maximal relative rounding error,

x $m \times k$-solution-matrix, to be declared outside with subscripts running from 1 to m, 1 to k,

stop exit used if the iterative improvement is ineffective.

3.2 *Procedure* ORTHOLIN2 *(a, n, m, b, eps) result*:*(x) exit*:*(stop)*.

a $n \times m$-matrix of the system of linear equations, to be declared outside with subscripts running from 1 to n, 1 to m,

n number of rows in the matrix a,

m number of columns in the matrix a $(m \leq n)$,

b n-dimensional vector on the right side of the linear equations, to be declared outside with the subscript running from 1 to n,

eps   maximal relative rounding error,

x n-dimensional solution vector,

stop exit used if the iterative improvement is ineffective.

3.3 *Procedure* ORTHO1 *(a, n, eps) result*:*(inv) exit*:*(stop)*.

a   square matrix,

n degree of the matrix a,

eps   maximal relative rounding error,

inv   inverse of the matrix a,

stop exit used if the iterative improvement is ineffective.

3.4 *Procedure* ORTHO2 *(n) transient*:*(a)*.

n degree of the matrix a,

a   input: square matrix,

a   output: inverse of the input matrix.

4. ALGOL Programs

procedure ORTHOLIN1 *(a, n, m, b, k, eps) result*:*(x) exit*:*(stop)*;
    value n, m, k, eps;
    integer n, m, k; **real** *eps*; **array** a, b, x; **label** *stop*;
    comment ORTHOLIN1 gives least squares solutions for k systems of n linear equations with m unknowns. a is the $n \times m$-matrix of the systems, b the constant $n \times k$-matrix on the right side, *eps* the maximal relative rounding error and x the solution matrix. The exit *stop* is used if the iterative improvement is ineffective;

```
begin
integer g, h, i, j, l, ll, mm; real s, t;
array u[1:n, 1:m], p[1:n], px[1:m], q[1:m×(m+1)/2];
for i:=1 step 1 until n do
for j:=1 step 1 until m do
    u[i,j] := a[i,j];
l:=0;
for i:=1 step 1 until m do
begin s:=0;
        for j:=1 step 1 until n do
        begin p[j]:=t:=u[j,i]; s:=s+t×t end;
        l:=l+1; q[l]:=s;
        comment i-th row of v stored in x;
        for g:=1 step 1 until k do
        begin t:=0;
                for j:=1 step 1 until n do
                    t:=t+p[j]×b[j,g];
                x[i,g]:=t
        end g;
        for g:=i+1 step 1 until m do
        begin t:=0;
                for j:=1 step 1 until n do
                    t:=t+p[j]×u[j,g];
                comment element of the i-th row of r stored in q;
                l:=l+1; q[l]:=t; t:=t/s;
                comment formation of a column of the i-th remaining matrix;
                for j:=1 step 1 until n do
                    u[j,g]:=u[j,g]−p[j]×t
        end g
end formation of the matrices r, v and u;

comment backsubstitution;
ll:=l; mm:=m+2;
for i:=m step −1 until 1 do
begin h:=l−i; t:=q[l];
        for j:=1 step 1 until k do
        begin s:=x[i,j];
                for g:=i+1 step 1 until m do
                    s:=s−q[g+h]×x[g,j];
                x[i,j]:=s:=s/t;
                if j=1 then px[i]:=s
        end j;
        l:=l+i−mm
end backsubstitution;
```

begin comment iterative improvement of the solution;
real *eps2, s0, s1, s2;* **array** *pp* [1:*m*];

real procedure *scalarproduct* (*s, a, i, p, n*);
    value *s, i, n*;
    integer *i, n;* **real** *s;* **array** *a, p*;
    comment The following procedure body should be replaced by a double
             precision scalar product routine;
    begin integer *j;* **real** *t;*
        $t := s;$
        for $j := 1$ **step** 1 **until** *n* **do**
          $t := t + a[i, j] \times p[j];$
        scalarproduct := *t*
    end *scalarproduct*;

    eps2 := *eps* × *eps*;
    for $i := 1$ **step** 1 **until** *k* **do**
    begin *s0* := 0;
        iteration: *l* := *ll;* *s1* := *s2* := 0;
        comment *i*-th column of the residual matrix stored in *p*;
        for $j := 1$ **step** 1 **until** *n* **do**
          p[*j*] := − *scalarproduct* (− *b*[*j, i*], *a, j, px, m*);
        comment *i*-th column of the matrix on the right side stored in *pp*;
        for $j := 1$ **step** 1 **until** *m* **do**
        begin *s* := 0;
            for $g := 1$ **step** 1 **until** *n* **do**
              $s := s + u[g, j] \times p[g];$
            pp[*j*] := *s*
        end *j*;
        comment backsubstitution;
        for $j := m$ **step** −1 **until** 1 **do**
        begin *h* := *l*−*j;* *s* := *pp*[*j*];
            for $g := j + 1$ **step** 1 **until** *m* **do**
              $s := s − q[g + h] \times pp[g];$
            pp[*j*] := *s* := *s/q*[*l*]; *s1* := *s1* + *s* × *s*;
            t := *px*[*j*]; *s2* := *s2* + *t* × *t*;
            px[*j*] := *s* + *t;* *l* := *l* + *j* − *mm*
        end backsubstitution;
        if *s1* ≥ *s2* × 0.25 **then goto** *stop*;
        if *s1* ≥ *s2* × *eps2* ∧
        (*s0* = 0 ∨ *s1* ≤ *s0* × 0.01) **then**
        begin *s0* := *s1;* **goto** *iteration* **end**;
        g := *i* + 1;
        for $j := 1$ **step** 1 **until** *m* **do**

```
      begin x[j, i] := px[j];
            if i ≠ k then px[j] := x[j, g]
         end j
   end i
end iterative improvement
end ORTHOLIN1;
```

procedure ORTHOLIN2 (a, n, m, b, eps) *result*: (x) *exit*: $(stop)$;
   value n, m, eps;
   integer n, m; **real** eps; **array** a, b, x; **label** $stop$;
   comment ORTHOLIN2 gives the least squares solution for a system of n linear
             equations with m unknowns. a is the $n \times m$-matrix of the system,
             b the constant vector on the right side, eps the maximal relative
             rounding error and x the solution vector. The exit $stop$ is used if
             the iterative improvement is ineffective;

```
begin
integer h, i, j, l, ll, mm;  real s, t;
array u[1:n, 1:m], p[1:n], q[1:m × (m+1)/2];
for i := 1 step 1 until n do
for j := 1 step 1 until m do
   u[i, j] := a[i, j];
l := 0;
for i := 1 step 1 until m do
begin s := 0;
      for j := 1 step 1 until n do
      begin p[j] := t := u[j, i];  s := s+t×t end;
      l := l+1;  q[l] := s;  t := 0;
      comment element of v stored in x;
      for j := 1 step 1 until n do
         t := t+p[j]×b[j];
      x[i] := t;
      for h := i+1 step 1 until m do
      begin t := 0;
            for j := 1 step 1 until n do
               t := t+p[j]×u[j, h];
            comment element of the i-th row of r stored in q;
            l := l+1;  q[l] := t;  t := t/s;
            comment formation of a column of the i-th remaining matrix;
            for j := 1 step 1 until n do
               u[j, h] := u[j, h] − p[j]×t
      end h
end formation of the matrices r and u and the vector v;
```

comment backsubstitution;
$ll := l; \; mm := m+2;$
for $i := m$ **step** -1 **until** 1 **do**
begin $h := l-i; \; t := q[l]; \; s := x[i];$
      for $j := i+1$ **step** 1 **until** m **do**
         $s := s - q[j+h] \times x[j];$
      $x[i] := s/t; \; l := l+i-mm$
end backsubstitution;

begin comment iterative improvement of the solution;
real $eps2, s0, s1, s2;$ **array** $pp[1:m];$

real procedure $scalarproduct(s, a, i, p, n);$
      value $s, i, n;$
      integer $i, n;$ **real** $s;$ **array** $a, p;$
      comment The following procedure body should be replaced by a
              double precision scalar product routine;
      begin integer $j;$ **real** $t;$
         $t := s;$
         for $j := 1$ **step** 1 **until** n **do**
            $t := t + a[i,j] \times p[j];$
         $scalarproduct := t$
      end $scalarproduct;$

$eps2 := eps \times eps; \; s0 := 0;$
$iteration: l := ll; \; s1 := s2 := 0;$
comment residual vector stored in $p;$
for $j := 1$ **step** 1 **until** n **do**
   $p[j] := - scalarproduct(-b[j], a, j, x, m);$
comment vector on the right side stored in $pp;$
for $j := 1$ **step** 1 **until** m **do**
begin $s := 0;$
      for $i := 1$ **step** 1 **until** n **do**
         $s := s + u[i,j] \times p[i];$
      $pp[j] := s$
end $j;$
comment backsubstitution;
for $j := m$ **step** -1 **until** 1 **do**
begin $h := l-j; \; s := pp[j];$
      for $i := j+1$ **step** 1 **until** m **do**
         $s := s - q[i+h] \times pp[i];$
      $pp[j] := s := s/q[l]; \; s1 := s1 + s \times s;$

```
        t := x[j]; s2 := s2+t×t;
        x[j] := s+t; l := l+j−mm
    end backsubstitution;
    if s1 ≥ s2×0.25 then goto stop;
    if s1 ≥ s2×eps2∧
        (s0=0 ∨ s1 ≤ s0×0.01) then
        begin s0 :=s1; goto iteration end
end iterative improvement
end ORTHOLIN2;
```

```
procedure ORTHO1 (a, n, eps) result: (inv) exit: (stop);
    value n, eps;
    integer n;
    real eps; array a, inv; label stop;
    comment ORTHO1 inverts a matrix a of order n. eps is the maximal relative
            rounding error. The inverse is called inv. The exit stop is used
            if the iterative improvement is ineffective;
begin
integer g, h, i, j, l, ll, nn;
real s, t;
array u[1:n, 1:n], p[1:n], q[1:n×(n+1)/2];
```

```
for i := 1 step 1 until n do
for j := 1 step 1 until n do
    u[i,j] := a[i,j];
l := 0;
for i := 1 step 1 until n do
begin s := 0;
        for j := 1 step 1 until n do
        begin p[j] :=t := u[j,i]; s := s+t×t end;
        l := l+1; q[l] := s;
        for g := i+1 step 1 until n do
        begin t := 0;
                for j := 1 step 1 until n do
                    t := t+p[j]×u[j,g];
                comment element of the i-th row of r stored in q;
                l := l+1; q[l] := t; t := t/s;
                comment formation of a column of the i-th remaining matrix;
                for j := 1 step 1 until n do
                    u[j,g] := u[j,g] −p[j]×t;
        end g
end formation of the matrices r and u;
```

comment backsubstitution;
$ll := l$; $nn := n+2$;
for $i := n$ **step** -1 **until** 1 **do**
begin $h := l-i$; $t := q[l]$;
     for $j := 1$ **step** 1 **until** n **do**
     begin $s := u[j, i]$;
          for $g := i+1$ **step** 1 **until** n **do**
            $s := s - q[g+h] \times inv[g, j]$;
          $inv[i, j] := s/t$
     end j;
     $l := l+i-nn$
end backsubstitution;

begin comment iterative improvement of the solution;
real $eps2$, $s0$, $s1$, $s2$; **array** pp, $px[1:n]$;

real procedure $scalarproduct(s, a, i, p, n)$;
     value s, i, n;
     integer i, n; **real** s; **array** a, p;
     comment The following procedure body should be replaced by a double
          precision scalar product routine;
     begin integer j; **real** t;
          $t := s$;
          for $j := 1$ **step** 1 **until** n **do**
            $t := t+a[i, j] \times p[j]$;
          $scalarproduct := t$
     end $scalarproduct$;

     $eps2 := eps \times eps$; $s0 := 0$;
$iteration$: $s1 := s2 := 0$;
for $i := 1$ **step** 1 **until** n **do**
begin $l := ll$;
     comment i-th column of the residual matrix stored in p;
     for $j := 1$ **step** 1 **until** n **do**
          $px[j] := inv[j, i]$;
     for $j := 1$ **step** 1 **until** n **do**
          $p[j] := - scalarproduct(\textbf{if } j=i \textbf{ then } -1 \textbf{ else } 0, a, j, px, n)$;
     comment i-th column of the matrix on the right side stored in pp;
     for $j := 1$ **step** 1 **until** n **do**
     begin $s := 0$;
          for $g := 1$ **step** 1 **until** n **do**
            $s := s+u[g, j] \times p[g]$;
          $pp[j] := s$
     end j;

comment backsubstitution;

for $j := n$ **step** -1 **until** 1 **do**

begin $s := pp[j];\ h := l-j;$

    for $g := j+1$ **step** 1 **until** n **do**

        $s := s - q[g+h] \times pp[g];$

    $pp[j] := s := s/q[l];\ s1 := s1 + s \times s;$

    $t := px[j];\ s2 := s2 + t \times t;$

    $inv[j, i] := s+t;\ l := l+j-nn$

end backsubstitution;

end i;

if $s1 \geq s2 \times 0.25$ **then goto** $stop$;

if $s1 \geq s2 \times eps2\ \wedge$

  $(s0 = 0\ \vee\ s1 \leq s0 \times 0.01)$ **then**

begin $s0 := s1$; **goto** $iteration$ **end**

end iterative improvement

end ORTHO1;

procedure ORTHO2 (n) $trans:(a)$;

    value n;

    integer n; **array** a;

    comment ORTHO2 inverts a matrix a of order n. The matrix a is replaced
        by the inverse of a;

begin

integer g, h, i, j, l, nn; **real** s, t;

array $p[1:n], q[1:n \times (n+1)/2]$;

$l := 0$;

for $i := 1$ **step** 1 **until** n **do**

begin $l := l+1;\ s := 0$;

    for $j := 1$ **step** 1 **until** n **do**

    begin $p[j] := t := a[j, i];\ s := s + t \times t$ **end**;

    $q[l] := s$;

    for $g := i+1$ **step** 1 **until** n **do**

    begin $t := 0$;

        for $j := 1$ **step** 1 **until** n **do**

        $t := t + p[j] \times a[j, g]$;

        comment element of the i-th row of r stored in q;

        $l := l+1;\ q[l] := t;\ t := t/s$;

        comment formation of a column of the i-th remaining matrix;

for $j := 1$ **step** 1 **until** n **do**
$$a[j,g] := a[j,g] - p[j] \times t$$
end g
end formation of the matrices r and u;
comment backsubstitution;
$nn := n+2$;
for $i := n$ **step** -1 **until** 1 **do**
begin $h := l - i$; $t := q[l]$;
      for $j := 1$ **step** 1 **until** n **do**
      begin $s := a[j,i]$;
           for $g := i+1$ **step** 1 **until** n **do**
              $s := s - q[g+h] \times a[j,g]$;
           $a[j,i] := s/t$
      end j;
      $l := l + i - nn$
end backsubstitution;

for $i := 1$ **step** 1 **until** n **do**
for $j := i+1$ **step** 1 **until** n **do**
begin $s := a[i,j]$;
      $a[i,j] := a[j,i]$; $a[j,i] := s$
end j
end ORTHO2;

5. Organisational and Notational Details

5.1 *Procedure* ORTHOLIN1. The array a, holding at first the matrix A, serves as storage for the computed columns of the orthogonal matrix U and also for the nontrivial elements of the first m columns of $A^{(i)}$. In order to decrease the computation time the first essential column C of the matrix $A^{(i)}$ will be stored in the array p. In the i-th step, C is the i-th column of the orthogonal matrix U. Therefore the i-th row of the $m \times k$-matrix $V = U^H B$ is computed in this step and stored in the array x. The $m - i + 1$ scalar products, the elements of the i-th row of the upper triangular matrix R, are then formed and stored in the onedimensional array q. After m steps the matrices R, U and V are completely formed. The system of equations $RX = V$ is solved by backsubstitution. The first approximation X of the solution matrix is now improved by iteration. The next approximation, $\bar{X} = X + Z$, is obtained by solving the least squares problem $RZ = U^H (B - AX)$. The residual matrix $B - AX$ has to be formed by using double precision arithmetic. The correction matrix Z is computed column by column. The first iteration step is always executed. Let z_i be the i-th column of Z and x_i the i-th column of X. If $\|z_i\| \geq \|x_i\|/2$ then the iterative improvement is ineffective and the exit *stop* is used. The iteration process at the i-th column is terminated if either $\|z_i\| \leq \|x_i\| \times eps$ or $\|z_i\| \geq \|\tilde{z}_i\|/10$ where eps is the maximal relative rounding error and $\tilde{z}_i$ the i-th column of the correction matrix of the foregoing iteration step.

5.2 *Procedure* ORTHOLIN2. Remarks in 5.1 apply analogously.

5.3 *Procedure* ORTHO1. Remarks in 5.1 apply analogously, moreover the procedure is simplified, because the unit matrix on the right side of the system does not appear explicitly in the program. The inverse INV of A is obtained from $INV^H R^H = U$. The termination criterion for the iteration process is applied to the whole matrix instead of the single columns.

5.4 *Procedure* ORTHO2. Since the original matrix A is no longer needed, at first the orthogonal matrix U is stored in the array containing A and then is replaced by the inverse of A.

6. Discussion of Numerical Properties

Since the process is an elimination process, error analysis of elimination processes applies. Since the intermediate matrix $A^{(i)}$ is obtained by applying an Hermitean projection operator to $A^{(i-1)}$, we have, with the bound norm $lub(.)$ subordinate to the Euclidean norm, $lub(A^{(i)}) \leq lub(A^{(i-1)})$ and therefore $lub(A^{(i)}) \leq lub(A)$.

Accordingly, error analysis results, as in the case of pivot elimination for positive definite matrices, in a relation (for floating point computation)

$$\frac{lub(INV - A^{-1})}{lub(A^{-1})} \leq cond(A) \times h(n) \times eps$$

where INV is the computed inverse, A^{-1} the exact inverse of A, $cond(A) = lub(A) \times lub(A^{-1})$ measures the condition of A, and eps indicates the maximal relative rounding error. $h(n)$ depends on the degree n of the matrix and is $0(n^{\frac{3}{2}})$.

Moreover, equilibration (by rows*) of A will have the effect of decreasing $cond(A)$ and correspondingly improving the accuracy of the resulting columns of INV. The situation is similar for the solution of linear equations and for the least squares problem. In the latter case $cond(A)$ is the columnwise condition, see footnote on page 119.

Scaling of A and $\boldsymbol{b}$ by rows means introducing weights into the problem, the squares of the scale factors being the weights.

The remaining column (or columns) after m elimination steps gives the deviation (which is to be zero in the case $m=n$) only theoretically, and the computed values do not coincide with the deviation (the residual) corresponding to the computed solution. Their order of magnitude, however, is about the same, and if the theoretical deviation is sufficiently larger, the remaining column may indeed be a very good approximation to the residual vector.

---

* Such an equilibration may be done by dividing each row by the sum of its absolute values ([2], p. 80).

7. Examples of the Use of the Procedures

The procedures ORTHOLIN1 and ORTHOLIN2 for the least squares problem produces the right triangular part R of the decomposition $A^H A = R^H D R$. Its diagonal elements are the squares of the Euclidean length of the orthogonalized columns. With a supplementary program, the triangular matrix may be inverted, and $R^{-1} D^{-1} (R^H)^{-1}$ gives the cofactor-matrix.

ORTHO1 and ORTHO2 may be used with advantage in the case of unsymmetric matrices.

8. Test Results

The procedures were tested on a Telefunken TR 4 computer at the Kommission für Elektronisches Rechnen der Bayerischen Akademie der Wissenschaften, München, with a 38 bit mantissa using double precision scalar product routine.

As an example for the application of the procedure ORTHO1 the matrix A of degree six was inverted.

$$A = \begin{pmatrix} -74 & 80 & 18 & -11 & -4 & -8 \\ 14 & -69 & 21 & 28 & 0 & 7 \\ 66 & -72 & -5 & 7 & 1 & 4 \\ -12 & 66 & -30 & -23 & 3 & -3 \\ 3 & 8 & -7 & -4 & 1 & 0 \\ 4 & -12 & 4 & 4 & 0 & 1 \end{pmatrix}.$$

This matrix is extraordinarily ill-conditioned; its condition subordinate to the Euclidean norm is 3.66×10^6. The exact inverse of matrix A is an integer matrix; after three iteration steps the matrix INV given in the table was obtained.

To illustrate the application of the procedure ORTHOLIN1 the least squares problem $\hat{A} X = B$ was solved. $\hat{A}$ is a 6×5-matrix and consists of the first five columns of the matrix A given above. B is the 6×3-matrix

$$B = \begin{pmatrix} 51 & -56 & -5 \\ -61 & 52 & -9 \\ -56 & 764 & 708 \\ 69 & 4096 & 4165 \\ 10 & -13276 & -13266 \\ -12 & 8421 & 8409 \end{pmatrix}.$$

The second column of B, b_2, is orthogonal to the columns of $\hat{A}$ and the third column b_3 is equal to $b_1 + b_2$.

Table

INV

$$= \begin{pmatrix}
1.00000\,00000_{10}0 & -1.99853\,78828_{10}{-15} & -7.00000\,00000_{10}0 & -4.00000\,00000_{10}1 & 1.31000\,00000_{10}2 & -8.40000\,00000_{10}1 \\
2.15205\,56197_{10}{-15} & 1.00000\,00000_{10}0 & 7.00000\,00000_{10}0 & 3.50000\,00000_{10}1 & -1.12000\,00000_{10}2 & 7.00000\,00000_{10}1 \\
-2.00000\,00000_{10}0 & 2.00000\,00000_{10}0 & 2.90000\,00000_{10}1 & 1.55000\,00000_{10}2 & -5.02000\,00000_{10}2 & 3.19000\,00000_{10}2 \\
1.50000\,00000_{10}1 & -1.20000\,00000_{10}1 & -1.92000\,00000_{10}2 & -1.03400\,00000_{10}3 & 3.35400\,00000_{10}3 & -2.13000\,00000_{10}3 \\
4.30000\,00000_{10}1 & -4.20000\,00000_{10}1 & -6.00000\,00000_{10}2 & -3.21100\,00000_{10}3 & 1.04060\,00000_{10}4 & -6.59500\,00000_{10}3 \\
-5.60000\,00000_{10}1 & 5.20000\,00000_{10}1 & 7.64000\,00000_{10}2 & 4.09600\,00000_{10}3 & -1.32760\,00000_{10}4 & 8.42100\,00000_{10}3
\end{pmatrix}$$

The solution matrix X of this example after three iteration steps at each column is

$$X = \begin{pmatrix} 1.00000\,00000_{10}0 & 8.83829\,09477_{10}-7 & 1.00000\,08546_{10}0 \\ 2.00000\,00000_{10}0 & 9.71444\,83846_{10}-7 & 2.00000\,09379_{10}0 \\ -1.00000\,00000_{10}0 & -1.36998\,01699_{10}-7 & -1.00000\,01187_{10}0 \\ 3.00000\,00000_{10}0 & 2.06567\,40920_{10}-6 & 3.00000\,19843_{10}0 \\ -4.00000\,00000_{10}0 & -3.27132\,68907_{10}-6 & -4.00000\,30971_{10}0 \end{pmatrix}.$$

The exact solution is an integer matrix. The first column of X should be equal to the third and the second column should be zero. A second test was performed on the PERM computer of the Technische Hochschule München, a third test on a Siemens S 2002 computer in the Institut für Angewandte Mathematik der Universität Mainz.

References

[1] BAUER, F. L.: Optimally scaled matrices. Numer. Math. 5, 73—87 (1963).
[2] HOUSEHOLDER, A. S.: Principles of numerical analysis. New York 1953.
[3] GOLUB, G. H.: Numerical methods for solving linear least squares problems. Numer. Math. 7, 206—216 (1965).
[4] BUSINGER, P., and G. H. GOLUB: Linear least squares solutions by Householder transformations. Numer. Math. 7, 269—276 (1965). Cf. I/9.

Singular Value Decomposition and Least Squares Solutions*

by G. H. Golub** and C. Reinsch

1. Theoretical Background

1.1. Introduction

Let A be a real $m \times n$ matrix with $m \geq n$. It is well known (cf. [4]) that

$$A = U \Sigma V^T \tag{1}$$

where

$$U^T U = V^T V = V V^T = I_n \quad \text{and} \quad \Sigma = \text{diag}(\sigma_1, \ldots, \sigma_n).$$

The matrix U consists of n orthonormalized eigenvectors associated with the n largest eigenvalues of $A A^T$, and the matrix V consists of the orthonormalized eigenvectors of $A^T A$. The diagonal elements of Σ are the non-negative square roots of the eigenvalues of $A^T A$; they are called *singular values*. We shall assume that

$$\sigma_1 \geq \sigma_2 \geq \cdots \geq \sigma_n \geq 0.$$

Thus if $\text{rank}(A) = r$, $\sigma_{r+1} = \sigma_{r+2} = \cdots = \sigma_n = 0$. The decomposition (1) is called the *singular value decomposition* (SVD).

There are alternative representations to that given by (1). We may write

$$A = U_c \left(\frac{\Sigma}{0} \right) V^T \quad \text{with} \quad U_c^T U_c = I_m$$

or

$$A = U_r \Sigma_r V_r^T \quad \text{with} \quad U_r^T U_r = V_r^T V_r = I_r \quad \text{and} \quad \Sigma_r = \text{diag}(\sigma_1, \ldots, \sigma_r).$$

We use the form (1), however, since it is most useful for computational purposes.

If the matrix U is not needed, it would appear that one could apply the usual diagonalization algorithms to the symmetric matrix $A^T A$ which has to be formed explicitly. However, as in the case of linear least squares problems, the com-

---

* Prepublished in Numer. Math. **14**, 403−420 (1970).

** The work of this author was in part supported by the National Science Foundation and Office of Naval Research.

putation of $A^T A$ involves unnecessary numerical inaccuracy. For example, let

$$A = \begin{bmatrix} 1 & 1 \\ \beta & 0 \\ 0 & \beta \end{bmatrix},$$

then

$$A^T A = \begin{bmatrix} 1+\beta^2 & 1 \\ 1 & 1+\beta^2 \end{bmatrix}$$

so that

$$\sigma_1(A) = (2+\beta^2)^{\frac{1}{2}}, \qquad \sigma_2(A) = |\beta|.$$

If $\beta^2 < \varepsilon_0$, the machine precision, the computed $A^T A$ has the form $\begin{bmatrix} 1 & 1 \\ 1 & 1 \end{bmatrix}$, and the best one may obtain from diagonalization is $\tilde{\sigma}_1 = \sqrt{2}, \tilde{\sigma}_2 = 0$.

To compute the singular value decomposition of a given matrix A, Forsythe and Henrici [2], Hestenes [8], and Kogbetliantz [9] proposed methods based on plane rotations. Kublanovskaya [10] suggested a QR-type method. The program described below first uses Householder transformations to reduce A to bidiagonal form, and then the QR algorithm to find the singular values of the bidiagonal matrix. The two phases properly combined produce the singular value decomposition of A.

1.2. Reduction to Bidiagonal Form

It was shown in [6] how to construct two finite sequences of Householder transformations

$$P^{(k)} = I - 2x^{(k)} x^{(k)T} \quad (k=1, 2, \ldots, n)$$

and

$$Q^{(k)} = I - 2y^{(k)} y^{(k)T} \quad (k=1, 2, \ldots, n-2)$$

(where $x^{(k)T} x^{(k)} = y^{(k)T} y^{(k)} = 1$) such that

$$P^{(n)} \ldots P^{(1)} A Q^{(1)} \ldots Q^{(n-2)} = \begin{bmatrix} q_1 & e_2 & 0 & \cdots & & \cdot & 0 \\ & q_2 & e_3 & & & 0 & \vdots \\ & & \cdot & \cdot & & & 0 \\ 0 & & & \cdot & \cdot & & e_n \\ & & & & & & q_n \\ \hline & & & 0 & & & \end{bmatrix} \begin{array}{l} \\ \\ \\ \\ \\ \end{array} \equiv J^{(0)},$$

$$\left.\phantom{\begin{bmatrix}0\\0\end{bmatrix}}\right\}(m-n)\times n$$

an upper bidiagonal matrix. If we let $A^{(1)} = A$ and define

$$A^{(k+\frac{1}{2})} = P^{(k)} A^{(k)} \quad (k=1, 2, \ldots, n)$$

$$A^{(k+1)} = A^{(k+\frac{1}{2})} Q^{(k)} \quad (k=1, 2, \ldots, n-2)$$

then $P^{(k)}$ is determined such that

$$a_{ik}^{(k+\frac{1}{2})} = 0 \quad (i=k+1, \ldots, m)$$

and $Q^{(k)}$ such that

$$a_{kj}^{(k+1)} = 0 \quad (j=k+2, \ldots, n).$$

The singular values of $J^{(0)}$ are the same as those of A. Thus, if the singular value decomposition of

$$J^{(0)} = G \Sigma H^T$$

then

$$A = PG \Sigma H^T Q^T$$

so that $U = PG$, $V = QH$ with $P \equiv P^{(1)} \dots P^{(n)}$, $Q \equiv Q^{(1)} \dots Q^{(n-2)}$.

1.3. Singular Value Decomposition of the Bidiagonal Matrix

By a variant of the QR algorithm, the matrix $J^{(0)}$ is iteratively diagonalized so that

$$J^{(0)} \to J^{(1)} \to \dots \to \Sigma$$

where

$$J^{(i+1)} = S^{(i)^T} J^{(i)} T^{(i)},$$

and $S^{(i)}$, $T^{(i)}$ are orthogonal. The matrices $T^{(i)}$ are chosen so that the sequence $M^{(i)} = J^{(i)^T} J^{(i)}$ converges to a diagonal matrix while the matrices $S^{(i)}$ are chosen so that all $J^{(i)}$ are of the bidiagonal form. In [7], another technique for deriving $\{S^{(i)}\}$ and $\{T^{(i)}\}$ is given but this is equivalent to the method described below.

For notational convenience, we drop the suffix and use the notation

$$J \equiv J^{(i)}, \quad \bar{J} \equiv J^{(i+1)}, \quad S \equiv S^{(i)}, \quad T \equiv T^{(i)}, \quad M \equiv J^T J, \quad \bar{M} \equiv \bar{J}^T \bar{J}.$$

The transition $J \to \bar{J}$ is achieved by application of Givens rotations to J alternately from the right and the left. Thus

$$\bar{J} = \underbrace{S_n^T S_{(n-1)}^T \dots S_2^T}_{S^T} J \underbrace{T_2 T_3 \dots T_n}_{T} \qquad (2)$$

where

$$S_k = \begin{bmatrix} 1 & 0 & & & & & & & \\ 0 & \ddots & & & & & 0 & & \\ & & 1 & & & & & & \\ & & & \cos\theta_k & -\sin\theta_k & & & & \\ & & & \sin\theta_k & \cos\theta_k & & & & \\ & & & & & 1 & & & \\ & 0 & & & & & \ddots & & 0 \\ & & & & & & & 0 & 1 \end{bmatrix} \begin{matrix} \\ \\ \\ (k-1) \\ (k) \\ \\ \\ \end{matrix}$$

$$\begin{matrix} (k-1) & \quad (k) \end{matrix}$$

and T_k is defined analogously to S_k with φ_k instead of θ_k.

Let the first angle, φ_2, be arbitrary while all the other angles are chosen so that $\bar{J}$ has the same form as J. Thus,

$$\begin{aligned} & T_2 \text{ annihilates nothing,} && \text{generates an entry } \{J\}_{21}, \\ & S_2^T \text{ annihilates } \{J\}_{21}, && \text{generates an entry } \{J\}_{13}, \\ & T_3 \text{ annihilates } \{J\}_{13}, && \text{generates an entry } \{J\}_{32}, \qquad (3) \\ & \quad \vdots \end{aligned}$$

and finally $\quad S_n^T$ annihilates $\{J\}_{n, n-1}$, and generates nothing.

(See Fig. 1.)

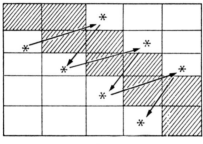

Fig. 1

This process is frequently described as "chasing". Since $\bar{J}=S^T J T$,

$$\bar{M}=\bar{J}^T \bar{J}=T^T M T$$

and $\bar{M}$ is a tri-diagonal matrix just as M is. We show that the first angle, φ_2, which is still undetermined, can be chosen so that the transition $M \rightarrow \bar{M}$ is a QR transformation with a given shift s.

The usual QR algorithm with shifts is described as follows:

$$\begin{aligned} M-sI &= T_s R_s \\ R_s T_s+sI &= \bar{M}_s \end{aligned} \qquad (4)$$

where $T_s^T T_s=I$ and R_s is an upper triangular matrix. Thus $\bar{M}_s=T_s^T M T_s$. It has been shown by Francis [5] that it is not necessary to compute (4) explicitly but it is possible to perform the shift implicitly. Let T be for the moment an arbitrary matrix such that

$$\{T_s\}_{k,1}=\{T\}_{k,1} \quad (k=1, 2, \dots, n),$$

(i.e., the elements of the first column of T_s are equal to the first column of T) and

$$T^T T=I.$$

Then we have the following theorem (Francis): If
   i) $\bar{M}=T^T M T$,
  ii) $\bar{M}$ is a tri-diagonal matrix,
 iii) the sub-diagonal elements of M are non-zero,

it follows that $\bar{M}=D\bar{M}_s D$ where D is a diagonal matrix whose diagonal elements are ± 1.

Thus choosing T_2 in (3) such that its first column is proportional to that of $M-sI$, the same is true for the first column of the product $T=T_2 T_3 \dots T_n$ which therefore is identical to that of T_s. Hence, if the sub-diagonal of M does not contain any non-zero entry the conditions of the Francis theorem are fulfilled and T is therefore identical to T_s (up to a scaling of column ± 1). Thus the transition (2) is equivalent to the QR transformation of $J^T J$ with a given shift s.

The shift parameter s is determined by an eigenvalue of the lower 2×2 minor of M. Wilkinson [13] has shown that for this choice of s, the method converges globally and almost always cubically.

1.4. Test for Convergence

If $|e_n| \leq \delta$, a prescribed tolerance, then $|q_n|$ is accepted as a singular value, and the order of the matrix is dropped by one. If, however, $|e_k| \leq \delta$ for $k \neq n$, the matrix breaks into two, and the singular values of each block may be computed independently.

If $q_k = 0$, then at least one singular value must be equal to zero. In the absence of roundoff error, the matrix will break if a shift of zero is performed. Now, suppose at some stage
$$|q_k| \leq \delta.$$

At this stage an extra sequence of Givens rotations is applied from the left to J involving rows $(k, k+1), (k, k+2), \ldots, (k, n)$ so that

$e_{k+1} \equiv \{J\}_{k,k+1}$ is annihilated, but $\{J\}_{k,k+2}, \{J\}_{k+1,k}$ are generated,

$\quad\quad \{J\}_{k,k+2}$ is annihilated, but $\{J\}_{k,k+3}, \{J\}_{k+2,k}$ are generated,

$\quad\quad \vdots$

and finally $\quad \{J\}_{k,n}$ is annihilated, and $\{J\}_{n,k}$ is generated.

The matrix thus obtained has the form

$$
\bar{J} =
\begin{bmatrix}
q_1 & e_2 & & & & & & & \\
 & \ddots & \ddots & & & & 0 & & \\
 & & \ddots & e_k & & & & & \\
 & & & \bar{q}_k & 0 & & & & \\
 & & & \delta_{k+1} & \bar{q}_{k+1} & \bar{e}_{k+2} & & & \\
 & & & \vdots & & \ddots & \ddots & & \\
 & 0 & & \vdots & & & \ddots & \bar{e}_n & \\
 & & & \delta_n & & & & \bar{q}_n
\end{bmatrix}
\quad (k).
$$

Note that by orthogonality
$$\bar{q}_k^2 + \delta_{k+1}^2 + \cdots + \delta_n^2 = q_k^2 \leq \delta^2.$$

Thus choosing $\delta = \|J^{(0)}\|_\infty \, \varepsilon_0$ (ε_0, the machine precision) ensures that all δ_k are less in magnitude than $\varepsilon_0 \|J^{(0)}\|_\infty$. Elements of $\bar{J}$ not greater than this are neglected. Hence $\bar{J}$ breaks up into two parts which may be treated independently.

2. Applicability

There are a large number of applications of the singular value decomposition; an extensive list is given in [7]. Some of these are as follows:

2.1. Pseudoinverse (Procedure svd)

Let A be a real $m \times n$ matrix. An $n \times m$ matrix X is said to be the pseudo-inverse of A if X satisfies the following four properties:

   i) $\quad AXA = A$,

   ii) $\quad XAX = X$,

   iii) $\quad (AX)^T = AX$,

   iv) $\quad (XA)^T = XA$.

The unique solution is denoted by A^+. It is easy to verify that if $A = U \Sigma V^T$, then $A^+ = V \Sigma^+ U^T$ where $\Sigma^+ = \mathrm{diag}(\sigma_i^+)$ and

$$\sigma_i^+ = \begin{cases} 1/\sigma_i & \text{for} \quad \sigma_i > 0 \\ 0 & \text{for} \quad \sigma_i = 0. \end{cases}$$

Thus the pseudoinverse may easily be computed from the output provided by the procedure *svd*.

2.2. Solution of Homogeneous Equations (Procedure svd or procedure minfit)

Let A be a matrix of rank r, and suppose we wish to solve

$$A x_i = 0 \quad \text{for} \quad i = r+1, \ldots, n$$

where 0 denotes the null vector.

Let

$$U = [u_1, u_2, \ldots, u_n] \quad \text{and} \quad V = [v_1, v_2, \ldots, v_n].$$

Then since $A v_i = \sigma_i u_i$ $(i = 1, 2, \ldots n)$,

$$A v_i = 0 \quad \text{for} \quad i = r+1, \ldots, n$$

and $x_i = v_i$.

Here the procedure *svd* or the procedure *minfit* with $p = 0$ may be used for determining the solution. If the rank of A is known, then a modification of the algorithm of Businger and Golub [1] may be used.

2.3. Solutions of Minimal Length (Procedure minfit)

Let b_1 be a given vector. Suppose we wish to determine a vector x so that

$$\|b_1 - A x\|_2 = \min \tag{5}$$

If the rank of A is less than n then there is no unique solution. Thus we require amongst all x which satisfy (5) that

$$\|\hat{x}\|_2 = \min$$

and this solution is unique. It is easy to verify that

$$\hat{x} = A^+ b_1 = V \Sigma^+ U^T b_1 \equiv V \Sigma^+ c_1.$$

The procedure *minfit* with $p > 0$ will yield $V, \Sigma, c_1, \ldots, c_p$. Thus the user is able to determine which singular values are to be declared as zero.

2.4. A Generalization of the Least Squares Problem (Procedure svd)

Let A be a real $m \times n$ matrix of rank n and let b be a given vector. We wish to construct a vector x such that

$$(A + \Delta A) x = b + \Delta b$$

and

$$\mathrm{trace}(\Delta A^T \Delta A) + K^2 \Delta b^T \Delta b = \min.$$

Here $K > 0$ is a given weight and the standard problem is obtained for $K \to 0$. Introducing the augmented matrices $\bar{A} = (A, Kb)$ and $\Delta\bar{A} = (\Delta A, K\Delta b)$ and the vector

$$\bar{x} = \begin{pmatrix} x \\ -1/K \end{pmatrix},$$

we have to minimize $\mathrm{trace}(\Delta\bar{A}^T \Delta\bar{A})$ under the constraint $(\bar{A} + \Delta\bar{A})\bar{x} = 0$. For fixed $\bar{x}$ the minimum is attained for $\Delta\bar{A} = -\bar{A}\bar{x}\bar{x}^T/\bar{x}^T\bar{x}$ and it has the value $\bar{x}^T\bar{A}^T\bar{A}\bar{x}/\bar{x}^T\bar{x}$. Minimizing with respect to $\bar{x}$ amounts to the computation of the smallest singular value of the matrix $\bar{A}$ and $\bar{x}$ is the corresponding column of the matrix $\bar{V}$ in the decomposition (1) with proper normalization [3].

3. Formal Parameter List

3.1. Input to procedure *svd*

m	number of rows of A, $m \geq n$.
n	number of columns of A.
withu	**true** if U is desired, **false** otherwise.
withv	**true** if V is desired, **false** otherwise.
eps	a constant used in the test for convergence (see Section 5, (iii)); should not be smaller than the machine precision ε_0, i.e., the smallest number for which $1 + \varepsilon_0 > 1$ in computer arithmetic.
tol	a machine dependent constant which should be set equal to β/ε_0 where β is the smallest positive number representable in the computer, see [11].
$a[1:m, 1:n]$	represents the matrix A to be decomposed.

Output of procedure *svd*.

$q[1:n]$	a vector holding the singular values of A; they are non-negative but not necessarily ordered in decreasing sequence.
$u[1:m, 1:n]$	represents the matrix U with orthonormalized columns (if *withu* is **true**, otherwise u is used as a working storage).
$v[1:n, 1:n]$	represents the orthogonal matrix V (if *withv* is **true**, otherwise v is not used).

3.2. Input to procedure *minfit*

m	number of rows of A.
n	number of columns of A.
p	number of columns of B, $p \geq 0$.
eps	same as for procedure *svd*.
tol	same as for procedure *svd*.
$ab[1:max(m,n), 1:n+p]$	$ab[i, j]$ represents $a_{i,j}$, $1 \leq i \leq m$, $1 \leq j \leq n$, $ab[i, n+j]$ represents $b_{i,j}$, $1 \leq i \leq m$, $1 \leq j \leq p$.

Output of procedure *minfit*.

$ab[1:max(m,n), 1:n+p]$ $ab[i, j]$ represents $v_{i,j}, 1 \leq i \leq n, 1 \leq j \leq n,$
                         $ab[i, n+j]$ represents $c_{i,j}, 1 \leq i \leq max(m,n), 1 \leq j \leq p$
                         viz. $C = U_c^T B$.

$q[1:n]$ same as for procedure *svd*.

4. ALGOL Programs

procedure *svd* $(m, n, withu, withv, eps, tol)$ *data*: (a) *result*: (q, u, v);
   value $m, n, withu, withv, eps, tol$;
   integer m, n;
   Boolean *withu, withv*;
   real *eps, tol*;
   array a, q, u, v;

comment Computation of the singular values and complete orthogonal decom-
      position of a real rectangular matrix A,

$$A = U \operatorname{diag}(q) V^T, \quad U^T U = V^T V = I,$$

      where the arrays $a[1:m, 1:n], u[1:m, 1:n], v[1:n, 1:n], q[1:n]$ re-
      present A, U, V, q respectively. The actual parameters corresponding
      to a, u, v may all be identical unless $withu = withv = $ **true**. In this
      case, the actual parameters corresponding to u and v must differ.
      $m \geq n$ is assumed;

```
begin
   integer i, j, k, l, l1;
   real c, f, g, h, s, x, y, z;
   array e[1:n];
   for i := 1 step 1 until m do
   for j := 1 step 1 until n do u[i, j] := a[i, j];
```

comment Householder's reduction to bidiagonal form;

```
   g := x := 0;
   for i := 1 step 1 until n do
   begin
      e[i] := g; s := 0; l := i+1;
      for j := i step 1 until m do s := s + u[j, i]↑2;
      if s < tol then g := 0 else
      begin
         f := u[i, i]; g := if f < 0 then sqrt(s) else −sqrt(s);
         h := f×g−s; u[i, i] := f−g;
         for j := l step 1 until n do
         begin
            s := 0;
            for k := i step 1 until m do s := s + u[k, i]×u[k, j];
            f := s/h;
            for k := i step 1 until m do u[k, j] := u[k, j] + f×u[k, i]
         end j
      end s;
```

```
    q[i] := g; s := 0;
    for j := l step 1 until n do s := s + u[i, j]↑2;
    if s < tol then g := 0 else
    begin
        f := u[i, i+1]; g := if f < 0 then sqrt(s) else −sqrt(s);
        h := f×g−s; u[i, i+1] := f−g;
        for j := l step 1 until n do e[j] := u[i, j]/h;
        for j := l step 1 until m do
        begin
            s := 0;
            for k := l step 1 until n do s := s + u[j, k]×u[i, k];
            for k := l step 1 until n do u[j, k] := u[j, k] + s×e[k]
        end j
    end s;
    y := abs(q[i]) + abs(e[i]); if y > x then x := y
end i;

comment accumulation of right-hand transformations;
    if withv then for i := n step −1 until 1 do
    begin
        if g ≠ 0 then
        begin
            h := u[i, i+1] × g;
            for j := l step 1 until n do v[j, i] := u[i, j]/h;
            for j := l step 1 until n do
            begin
                s := 0;
                for k := l step 1 until n do s := s + u[i, k]×v[k, j];
                for k := l step 1 until n do v[k, j] := v[k, j] + s×v[k, i]
            end j
        end g;
        for j := l step 1 until n do v[i, j] := v[j, i] := 0;
        v[i, i] := 1; g := e[i]; l := i
    end i;

comment accumulation of left-hand transformations;
    if withu then for i := n step −1 until 1 do
    begin
        l := i+1; g := q[i];
        for j := l step 1 until n do u[i, j] := 0;
        if g ≠ 0 then
        begin
            h := u[i, i]×g;
            for j := l step 1 until n do
```

```
    begin
        s := 0;
        for k := l step 1 until m do s := s + u[k, i] × u[k, j];
        f := s/h;
        for k := i step 1 until m do u[k, j] := u[k, j] + f × u[k, i]
    end j;
    for j := i step 1 until m do u[j, i] := u[j, i]/g
  end g
  else for j := i step 1 until m do u[j, i] := 0;
  u[i, i] := u[i, i] + 1
end i;
```

comment diagonalization of the bidiagonal form;

```
eps := eps × x;
for k := n step −1 until 1 do
begin
    test f splitting:
        for l := k step −1 until 1 do
        begin
            if abs (e[l]) ≤ eps then goto test f convergence;
            if abs (q[l − 1]) ≤ eps then goto cancellation
        end l;
```

comment cancellation of $e[l]$ if $l > 1$;

```
cancellation:
    c := 0; s := 1; l1 := l − 1;
    for i := l step 1 until k do
    begin
        f := s × e[i]; e[i] := c × e[i];
        if abs (f) ≤ eps then goto test f convergence;
        g := q[i]; h := q[i] := sqrt (f × f + g × g); c := g/h; s := −f/h;
        if withu then for j := 1 step 1 until m do
        begin
            y := u[j, l1]; z := u[j, i];
            u[j, l1] := y × c + z × s; u[j, i] := −y × s + z × c
        end j
    end i;
test f convergence:
    z := q[k]; if l = k then goto convergence;
```

comment shift from bottom 2 × 2 minor;

```
    x := q[l]; y := q[k − 1]; g := e[k − 1]; h := e[k];
    f := ((y − z) × (y + z) + (g − h) × (g + h))/(2 × h × y); g := sqrt (f × f + 1);
    f := ((x − z) × (x + z) + h × (y/(if f < 0 then f − g else f + g) − h))/x;
```

comment next QR transformation;

```
    c := s := 1;
    for i := l + 1 step 1 until k do
```

```
  begin
    g := e[i]; y := q[i]; h := s×g; g := c×g;
    e[i−1] := z := sqrt(f×f+h×h); c := f/z; s := h/z;
    f := x×c+g×s; g := −x×s+g×c; h := y×s; y := y×c;
    if withv then for j := 1 step 1 until n do
    begin
        x := v[j, i−1]; z := v[j, i];
        v[j, i−1] := x×c+z×s; v[j, i] := −x×s+z×c
    end j;
    q[i−1] := z := sqrt(f×f+h×h); c := f/z; s := h/z;
    f := c×g+s×y; x := −s×g+c×y;
    if withu then for j := 1 step 1 until m do
    begin
        y := u[j, i−1]; z := u[j, i];
        u[j, i−1] := y×c+z×s; u[j, i] := −y×s+z×c
    end j
    end i;
    e[l] := 0; e[k] := f; q[k] := x; goto test f splitting;
convergence:
    if z < 0 then
    begin comment q[k] is made non-negative;
        q[k] := −z;
        if withv then for j := 1 step 1 until n do v[j, k] := −v[j, k]
    end z
  end k
end svd;
```

procedure *minfit* (m, n, p, eps, tol) *trans:* (ab) *result:* (q);
  value m, n, p, eps, tol;
  integer m, n, p;
  real eps, tol;
  array ab, q;

comment Computation of the matrices diag(q), V, and C such that for given
real $m \times n$ matrix A and $m \times p$ matrix B

$$U_c^T A V = \text{diag}(q) \text{ and } U_c^T B = C \text{ with orthogonal matrices } U_c \text{ and } V.$$

The singular values and the matrices V and C may be used to de-
termine $\bar{X}$ minimizing (1) $\|A X - B\|_F$ and (2) $\|X\|_F$ with the solution

$$\bar{X} = V \times \text{Pseudo-inverse of } \text{diag}(q) \times C.$$

The procedure can also be used to determine the complete solution
of an underdetermined linear system, i.e., rank$(A) = m < n$.
  The array $q[1:n]$ represents the matrix diag(q). A and B together
are to be given as the first m rows of the array $ab[1:max(m,n), 1:n+p]$.
V is returned in the first n rows and columns of ab while C is returned
in the last p columns of ab (if $p > 0$);

```
begin
    integer i, j, k, l, l1, n1, np;
    real c, f, g, h, s, x, y, z;
    array e[1:n];

comment Householder's reduction to bidiagonal form;
    g := x := 0; np := n + p;
    for i := 1 step 1 until n do
    begin
        e[i] := g; s := 0; l := i + 1;
        for j := i step 1 until m do s := s + ab[j, i]↑2;
        if s < tol then g := 0 else
        begin
            f := ab[i, i]; g := if f < 0 then sqrt (s) else − sqrt (s);
            h := f × g − s; ab[i, i] := f − g;
            for j := l step 1 until np do
            begin
                s := 0;
                for k := i step 1 until m do s := s + ab[k, i] × ab[k, j];
                f := s/h;
                for k := i step 1 until m do ab[k, j] := ab[k, j] + f × ab[k, i]
            end j
        end s;
        q[i] := g; s := 0;
        if i ≤ m then for j := l step 1 until n do s := s + ab[i, j]↑2;
        if s < tol then g := 0 else
        begin
            f := ab[i, i + 1]; g := if f < 0 then sqrt (s) else − sqrt (s);
            h := f × g − s; ab[i, i + 1] := f − g;
            for j := l step 1 until n do e[j] := ab[i, j]/h;
            for j := l step 1 until m do
            begin
                s := 0;
                for k := l step 1 until n do s := s + ab[j, k] × ab[i, k];
                for k := l step 1 until n do ab[j, k] := ab[j, k] + s × e[k]
            end j
        end s;
        y := abs (q[i]) + abs (e[i]); if y > x then x := y
    end i;

comment accumulation of right-hand transformations;
    for i := n step −1 until 1 do
    begin
        if g ≠ 0 then
        begin
            h := ab[i, i + 1] × g;
            for j := l step 1 until n do ab[j, i] := ab[i, j]/h;
            for j := l step 1 until n do
```

```
      begin
        s := 0;
          for k := l step 1 until n do s := s + ab[i, k] × ab[k, j];
          for k := l step 1 until n do ab[k, j] := ab[k, j] + s × ab[k, i]
        end j
      end g;
      for j := l step 1 until n do ab[i, j] := ab[j, i] := 0;
      ab[i, i] := 1; g := e[i]; l := i
    end i;
  eps := eps × x; n1 := n + 1;
  for i := m + 1 step 1 until n do
  for j := n1 step 1 until np do ab[i, j] := 0;
```

comment diagonalization of the bidiagonal form;

```
  for k := n step −1 until 1 do
  begin
    test f splitting:
      for l := k step −1 until 1 do
      begin
        if abs (e[l]) ≦ eps then goto test f convergence;
        if abs (q[l − 1]) ≦ eps then goto cancellation
      end l;
```

comment cancellation of e[l] if l > 1;

```
    cancellation:
        c := 0; s := 1; l1 := l − 1;
        for i := l step 1 until k do
        begin
          f := s × e[i]; e[i] := c × e[i];
          if abs (f) ≦ eps then goto test f convergence;
          g := q[i]; q[i] := h := sqrt (f × f + g × g); c := g/h; s := − f/h;
          for j := n1 step 1 until np do
          begin
            y := ab[l1, j]; z := ab[i, j];
            ab[l1, j] := c × y + s × z; ab[i, j] := − s × y + c × z
          end j
        end i;
    test f convergence:
        z := q[k]; if l = k then goto convergence;
```

comment shift from bottom 2 × 2 minor;

```
        x := q[l]; y := q[k − 1]; g := e[k − 1]; h := e[k];
        f := ((y − z) × (y + z) + (g − h) × (g + h))/(2 × h × y); g := sqrt (f × f + 1);
        f := ((x − z) × (x + z) + h × (y/(if f < 0 then f − g else f + g) − h))/x;
```

comment next Q R transformation;

```
        c := s := 1;
        for i := l + 1 step 1 until k do
```

```
      begin
          g := e[i]; y := q[i]; h := s×g; g := c×g;
          e[i−1] := z := sqrt(f×f+h×h); c := f/z; s := h/z;
          f := x×c+g×s; g := −x×s+g×c; h := y×s; y := y×c;
          for j := 1 step 1 until n do
          begin
              x := ab[j, i−1]; z := ab[j, i];
              ab[j, i−1] := x×c+z×s; ab[j, i] := −x×s+z×c
          end j;
          q[i−1] := z := sqrt(f×f+h×h); c := f/z; s := h/z;
          f := c×g+s×y; x := −s×g+c×y;
          for j := n1 step 1 until np do
          begin
              y := ab[i−1, j]; z := ab[i, j];
              ab[i−1, j] := c×y+s×z; ab[i, j] := −s×y+c×z
          end j
      end i;
      e[l] := 0; e[k] := f; q[k] := x; goto test f splitting;
  convergence:
      if z < 0 then
      begin comment q[k] is made non-negative;
          q[k] := −z;
          for j := 1 step 1 until n do ab[j, k] := −ab[j, k]
      end z
  end k
end minfit;
```

5. Organisational and Notational Details

(i) The matrix U consists of the first n columns of an orthogonal matrix U_c. The following modification of the procedure *svd* would produce U_c instead of U: After

comment accumulation of left-hand transformations;
insert a statement
if *withu* **then for** $i := n+1$ **step** 1 **until** m **do**
begin
    for $j := n+1$ **step** 1 **until** m **do** $u[i, j] := 0$;
    $u[i, i] := 1$

end i;
Moreover, replace n by m in the fourth and eighth line after that, i.e., write twice **for** $j := l$ **step** 1 **until** m **do**.

(ii) $m \geq n$ is assumed for procedure *svd*. This is no restriction; if $m < n$, store A^T, i.e., use an array $at[1:n, 1:m]$ where $at[i, j]$ represents $a_{j, i}$ and call *svd* $(n, m, withv, withu, eps, tol, at, q, v, u)$ producing the $m \times m$ matrix U and the $n \times m$ matrix V. There is no restriction on the values of m and n for the procedure *minfit*.

(iii) In the iterative part of the procedures an element of $J^{(i)}$ is considered to be negligible and is consequently replaced by zero if it is not larger in magnitude than εx where ε is the given tolerance and

$$x = \max_{1 \le i \le n} (|q_i| + |e_i|).$$

The largest singular value σ_1 is bounded by $x/\sqrt{2} \le \sigma_1 \le x\sqrt{2}$.

(iv) A program organization was chosen which allows us to save storage locations. To this end the actual parameters corresponding to a and u may be identical. In this event the original information stored in a is overwritten by information on the reduction. This, in turn, is overwritten by u if the latter is desired. Likewise, the actual parameters corresponding to a and v may agree. Then v is stored in the upper part of a if it is desired, otherwise a is not changed. Finally, all three parameters a, u, and v may be identical unless $withu = withv = $ **true**.

This special feature, however, increases the number of multiplications needed to form U roughly by a factor m/n.

(v) Shifts are evaluated in a way as to reduce the danger of overflow or underflow of exponents.

(vi) The singular values as delivered in the array q are not necessarily ordered. Any sorting of them should be accompanied by the corresponding sorting of the columns of U and V, and of the rows of C.

(vii) The formal parameter list may be completed by the addition of a limit for the number of iterations to be performed, and by the addition of a failure exit to be taken if no convergence is reached after the specified number of iterations (e.g., 30 per singular value).

6. Discussion of Numerical Properties

The stability of the Householder transformations has been demonstrated by Wilkinson [12]. In addition, he has shown that in the absence of roundoff the QR algorithm has global convergence and asymptotically is almost always cubically convergent.

The numerical experiments indicate that the average number of complete QR iterations on the bidiagonal matrix is usually less than two per singular value. Extra consideration must be given to the implicit shift technique which fails for a split matrix. The difficulties arise when there are small q_k's or e_k's. Using the techniques of Section 1.4, there cannot be numerical instability since stable orthogonal transformations are used but under special circumstances there may be a slowdown in the rate of convergence.

7. Test Results

Tests were carried out on the UNIVAC 1108 Computer of the Andrew R. Jennings Computing Center of Case Western Reserve University. Floating point numbers are represented by a normalized 27 bit mantissa and a 7 bit exponent to the radix 2, whence $eps = 1.5_{10} - 8$, $tol = {}_{10} - 31$. In the following, computed values are marked by a tilde and $m(A)$ denotes $\max |a_{i,j}|$.

First example:

$$
A = \begin{bmatrix}
22 & 10 & 2 & 3 & 7 \\
14 & 7 & 10 & 0 & 8 \\
-1 & 13 & -1 & -11 & 3 \\
-3 & -2 & 13 & -2 & 4 \\
9 & 8 & 1 & -2 & 4 \\
9 & 1 & -7 & 5 & -1 \\
2 & -6 & 6 & 5 & 1 \\
4 & 5 & 0 & -2 & 2
\end{bmatrix},
\quad
B = \begin{bmatrix}
-1 & 1 & 0 \\
2 & -1 & 1 \\
1 & 10 & 11 \\
4 & 0 & 4 \\
0 & -6 & -6 \\
-3 & 6 & 3 \\
1 & 11 & 12 \\
0 & -5 & -5
\end{bmatrix},
$$

$$\sigma_1 = \sqrt{1248}, \quad \sigma_2 = 20, \quad \sigma_3 = \sqrt{384}, \quad \sigma_4 = \sigma_5 = 0.$$

The homogeneous system $A\,x = 0$ has two linearly independent solutions. Six QR transformations were necessary to drop all off-diagonal elements below the internal tolerance $46.4_{10} - 8$. Table 1 gives the singular values in the sequence as computed by procedures *svd* and *minfit*. The accuracy of the achieved decomposition is characterized by

$$m(A - \tilde{U}\tilde{\Sigma}\tilde{V}^T) = 238_{10} - 8, \quad m(\tilde{U}^T\tilde{U} - I) = 8.1_{10} - 8, \quad m(\tilde{V}^T\tilde{V} - I) = 3.3_{10} - 8.$$

Because two singular values are equal to zero, the procedures *svd* and *minfit* may lead to other orderings of the singular values for this matrix when other tolerances are used.

Table 1

$\tilde{\sigma}_k$	$\sigma_k - \tilde{\sigma}_k$	
$0.96_{10} - 7$	-9.6	
$19.595\,916$	191	
$19.999\,999$	143	$\times 10^{-8}$
$1.97_{10} - 7$	-19.7	
$35.327\,038$	518	

The computed solutions of the homogeneous system are given by the first and fourth columns of the matrix $\tilde{V}$ (Table 2).

Table 2

$\tilde{v}_1$	$\tilde{v}_4$	$v_1 - \tilde{v}_1$	$v_4 - \tilde{v}_4$	
$-0.4190\,9545$	0	-1.5	0 (Def.)	
$0.4405\,0912$	$0.4185\,4806$	1.7	0.6	
$-0.0520\,0457$	$0.3487\,9006$	1.2	-1.3	$\times 10^{-8}$
$0.6760\,5915$	$0.2441\,5305$	1.0	0.3	
$0.4129\,7730$	$-0.8022\,1713$	1.3	-0.8	

Procedure *minfit* was used to compute the solutions of the minimization problem of Section 2.3 corresponding to the three right-hand sides as given by the columns of the matrix B. Table 3 lists the exact solutions and the results obtained when the first and fourth values in Table 1 are replaced by zero.

Table 3

x_1	x_2	x_3	$\tilde{x}_1$	$\tilde{x}_2$	$\tilde{x}_3$
$-1/12$	0	$-1/12$	$-0.0833\ 3333$	$0.17_{10}-8$	$-0.0833\ 3333$
0	0	0	$-0.58_{10}-8$	$-1.09_{10}-8$	$-1.11_{10}-8$
$1/4$	0	$1/4$	$0.2500\ 0002$	$1.55_{10}-8$	$0.2500\ 0003$
$-1/12$	0	$-1/12$	$-0.0833\ 3332$	$0.74_{10}-8$	$-0.0833\ 3332$
$1/12$	0	$1/12$	$0.0833\ 3334$	$0.33_{10}-8$	$0.0833\ 3334$
	Residual				
0	$8\sqrt{5}$	$8\sqrt{5}$			

A second example is the 20×21 matrix with entries

$$a_{i,j} = \begin{cases} 0 & \text{if } i > j \\ 21 - i & \text{if } i = j \\ -1 & \text{if } i < j \end{cases} \quad \begin{matrix} 1 \leq i \leq 20 \\ 1 \leq j \leq 21 \end{matrix}$$

which has orthogonal rows and singular values $\sigma_{21-k} = \sqrt{k(k+1)}$, $k = 0, \ldots, 20$. Theoretically, the Householder reduction should produce a matrix $J^{(0)}$ with diagonal $-20, 0, \ldots, 0$ and super-diagonal $-\sqrt{20}, \sigma_2, \ldots, \sigma_{20}$. Under the influence of rounding errors a totally different matrix results. However, within working accuracy its singular values agree with those of the original matrix. Convergence is reached after 32 QR transformations and the $\tilde{\sigma}_k$, $k = 1, \ldots, 20$ are correct within several units in the last digit, $\tilde{\sigma}_{21} = 1.61_{10} - 11$.

A third example is obtained if the diagonal of the foregoing example is changed to

$$a_{i,i} = 1, \quad 1 \leq i \leq 20.$$

This matrix has a cluster of singular values, σ_{10} to σ_{19} lying between 1.5 and 1.6, $\sigma_{20} = \sqrt{2}$, $\sigma_{21} = 0$. Clusters, in general, have a tendency to reduce the number of required iterations; in this example, 26 iterations were necessary for convergence. $\tilde{\sigma}_{21} = 1.49_{10} - 8$ is found in eighteenth position and the corresponding column of $\tilde{V}$ differs from the unique solution of the homogeneous system by less than $3.4_{10} - 8$ in any component.

A second test was made by Dr. Peter Businger on the CDC 6600.

A third test was performed on the IBM 360/67 at Stanford University. The example used was the 30×30 matrix with entries

$$a_{ij} = \begin{cases} 0 & \text{if } i > j \\ 1 & \text{if } i = j \\ -1 & \text{if } i < j. \end{cases}$$

The computed singular values are given in Table 4.

Table 4. *Singular values*

18.2029 0555 7529 2200	6.2231 9652 2604 2340	3.9134 8020 3335 6160	2.9767 9450 2557 7960
2.4904 5062 9660 3570	2.2032 0757 4479 9280	2.0191 8365 4054 5860	1.8943 4154 7685 6890
1.8059 1912 6612 3070	1.7411 3576 7747 9500	1.6923 5654 4395 2610	1.6547 9302 7369 3370
1.6253 2089 2877 9290	1.6018 3335 6666 2670	1.5828 6958 8713 6990	1.5673 9214 4480 0070
1.5546 4889 0109 3720	1.5440 8471 4076 0510	1.5352 8356 5544 9020	1.5279 2951 2160 3040
1.5217 8003 9063 4950	1.5166 4741 2836 7840	1.5123 8547 3899 6950	1.5088 8015 6801 8850
1.5060 4262 0723 9700	1.5038 0424 3812 6520	1.5021 1297 6754 0060	1.5009 3071 1977 0610
1.5002 3143 4775 4370	0.0000 0000 2793 9677		

Note that $\sigma_{30}/\sigma_1 \approx 1.53 \times 10^{-10}$ so that this matrix is very close to being a matrix of rank 29 even though the determinant equals 1.

Acknowledgement. The authors wish to thank Dr. Peter Businger of Bell Telephone Laboratories for his stimulating comments.

References

1. Businger, P., Golub, G.: Linear least squares solutions by Householder transformations. Numer. Math. **7**, 269—276 (1965). Cf. I/8.
2. Forsythe, G. E., Henrici, P.: The cyclic Jacobi method for computing the principal values of a complex matrix. Proc. Amer. Math. Soc. **94**, 1—23 (1960).
3. — Golub, G.: On the stationary values of a second-degree polynomial on the unit sphere. J. Soc. Indust. Appl. Math. **13**, 1050—1068 (1965).
4. — Moler, C. B.: Computer solution of linear algebraic systems. Englewood Cliffs, New Jersey: Prentice-Hall 1967.
5. Francis, J.: The QR transformation. A unitary analogue to the LR transformation. Comput. J. **4**, 265—271 (1961, 1962).
6. Golub, G., Kahan, W.: Calculating the singular values and pseudo-inverse of a matrix. J. SIAM. Numer. Anal., Ser. B **2**, 205—224 (1965).
7. — Least squares, singular values, and matrix approximations. Aplikace Matematiky **13**, 44—51 (1968).
8. Hestenes, M. R.: Inversion of matrices by biorthogonalization and related results. J. Soc. Indust. Appl. Math. **6**, 51—90 (1958).
9. Kogbetliantz, E. G.: Solution of linear equations by diagonalization of coefficients matrix. Quart. Appl. Math. **13**, 123—132 (1955).
10. Kublanovskaja, V. N.: Some algorithms for the solution of the complete problem of eigenvalues. Ž. Vyčisl. Mat. i Mat. Fiz. **1**, 555—570 (1961).
11. Martin, R. S., Reinsch, C., Wilkinson, J. H.: Householder's tridiagonalization of a symmetric matrix. Numer. Math. **11**, 181—195 (1968). Cf. II/2.
12. Wilkinson, J.: Error analysis of transformations based on the use of matrices of the form $I - 2w \, w^H$. Error in digital computation, vol. II, L. B. Rall, ed., p. 77—101. New York: John Wiley & Sons, Inc. 1965.
13. — Global convergence of tridiagonal QR algorithm with origin shifts. Lin. Alg. and its Appl. **1**, 409—420 (1968).

A Realization of the Simplex Method Based on Triangular Decompositions

by R. H. Bartels, J. Stoer, and Ch. Zenger

1. Theoretical Background

1.1. Introduction

Consider the following problem of linear programming

(1.1.1 a)   Minimize $c_0 + c_{-m} x_{-m} + \cdots + c_{-1} x_{-1} + c_1 x_1 + \cdots + c_n x_n$

subject to

(1.1.1 b)
$$x_{-i} + \sum_{k=1}^{n} aa_{ik} x_k = b_i, \qquad i = 1, 2, \ldots, m,$$

(1.1.1 c)
$$x_i \geq 0 \quad \text{for} \quad i \in I^+, \qquad x_i = 0 \quad \text{for} \quad i \in I^0,$$

where I^+, I^0, $I^\pm$ are disjoint index sets with

$$I^+ \cup I^0 \cup I^\pm = N := \{i \mid -m \leq i \leq -1, \ 1 \leq i \leq n\}.$$

The variable x_i is called a nonnegative (zero, free) variable if $i \in I^+$ ($i \in I^0$, $i \in I^\pm$).

Any system of linear relations

$$\sum_{k=1}^{p} \alpha_{ik} z_k \varrho_i \beta_i, \qquad i = 1, 2, \ldots, q,$$

$$\varrho_i = \begin{cases} \leq \\ \geq \\ =, \end{cases}$$

can be written in the form (1.1.1 b, c) by known techniques such as the introduction of slack variables, artificial variables and so on. By introducing the vectors

$$x = (x_{-m}, \ldots, x_{-1}, x_1, \ldots, x_n)^T, \qquad b = (b_1, \ldots, b_m)^T,$$

we may write

(1.1.1 b′)
$$A x = b$$

instead of (1.1.1 b). To each variable x_j corresponds a column A_j of A: If $-m \leq j \leq -1$, then $A_j := e_{-j}$ is the $(-j)$-th axis vector, if $1 \leq j \leq n$ then A_j is the j-th column of AA:

$$A_j = \begin{pmatrix} aa_{1j} \\ \vdots \\ aa_{mj} \end{pmatrix}.$$

Likewise, to each ordered index set

$$J = \{j_1, j_2, \ldots, j_m\} \subseteq N$$

of m indices there correspond the square submatrix

$$A_J := (A_{j_1}, A_{j_2}, \ldots, A_{j_m})$$

of A and the vectors

$$x_J := \begin{pmatrix} x_{j_1} \\ \vdots \\ x_{j_m} \end{pmatrix}, \qquad c_J := (c_{j_1}, \ldots, c_{j_m})^T.$$

If A_J is nonsingular we say that A_J is a *basis* characterized by J and that the variables in x_J are the corresponding basic variables. To every basis A_J belongs a *basic solution* $\bar{x}$ of (1.1.1 b), namely the unique solution $\bar{x}$ of (1.1.1 b) with

(1.1.2)
$$A_J \bar{x}_J = b$$
$$\bar{x}_k = 0 \quad \text{for} \quad k \notin J.$$

There are problems (e.g. in parametric programming) for which it is convenient to consider additional right hand sides $b^{(i)}$, $i = 2, 3, \ldots, rs$, together with "the" right hand side $b =: b^{(1)}$ of (1.1.1 b). Then of course to each of these vectors $b^{(i)}$ and basis A_J belongs a basic solution $\bar{x}^{(i)}$ defined by

(1.1.3)
$$A_J \bar{x}_J^{(i)} = b^{(i)},$$
$$\bar{x}_k^{(i)} = 0 \quad \text{for} \quad k \notin J.$$

For later reference let us assume that B is the $m \times rs$-matrix

$$B = (b^{(1)}, b^{(2)}, \ldots, b^{(rs)})$$

specifying all right hand sides under consideration, in particular, the first column $b^{(1)}$ of which is the vector b of (1.1.1 b). Without additional qualifications, the term "basic solution" will always mean the basic solution corresponding to the first right hand side $b = b^{(1)}$ of (1.1.1 b).

The importance of bases and basic solutions lies in the well-known fact that if the minimization problem (1.1.1) has any optimal solutions at all then there are basic solutions of (1.1.1 b) among them. Moreover, the bases corresponding to optimal (basic) solutions of (1.1.1) have (in general) the additional property that no zero variables, but all free variables are basic. Such a basis will be called a

(1.1.4)   *standard basis*.

The algorithms to solve problems of the kind (1.1.1), in particular the Simplex-method, usually start with an initial basis characterized by an index set $J^{(0)}$ and then proceed in a series of suitable *exchange steps* from one basis $J^{(i)}$ to an adjacent one $J^{(i+1)}$ until the final basis is attained. Here, two bases given by the sets J and J' are called *adjacent*, if J and J' differ by one index: there are exactly two indices l, r with $l \in J$, $l \notin J'$, $r \notin J$, $r \in J'$, and we say that, in the transition $J \to J'$, x_l leaves and x_r enters the basis, or, x_l is exchanged with x_r. Thus a typical step $J \to J'$ of these algorithms can be summarized as follows:

0) Find an initial basis and goto 1).

1) Check by a suitable "*test algorithm*" whether the current basis is the final one; if not proceed to 2), otherwise stop.

(1.1.5) 2) By a suitable "*pivoting algorithm*", find a permissible pair of indices $l \in J$, $r \notin J$. If there is no such pair (l, r) go to *fail*, otherwise to 3).

3) Form J' by exchanging x_l with x_r and return to 1)

fail: stop, the problem has no solution.

Of course, both the "test algorithm" and the "pivoting algorithm" should be efficient in that they lead in finitely many steps either to the final basis or to the decision that there is no solution to the problem.

1.2. The Pivoting Rules

Three subalgorithms (Phases 0, 1, 2) of the type (1.1.5) are important to deal with problem (1.1.1). In Phase 0, a first standard basis (1.1.4) of (1.1.1 b) is found, in Phase 1, a first feasible basic solution, and in Phase 2, an optimal basic solution.

For Phase 0, the initial basis (1.1.5), 0) is given by

$$J := \{-1, -2, \ldots, -m\}.$$

The "test algorithm" (1.1.5), 1) is the following simple check: If no zero variables, but all free variables are basic, i.e. if $I^0 \cap J = \emptyset$ and $I^\pm \subseteq J$, then the current basis J is the final one, that is a standard basis (1.1.4). The "pivoting algorithm" (1.1.5), 2) is only to find a pair of indices l, r such that either

$$l \in J \cap I^0, \qquad r \notin J \cup I^0,$$

or

$$l \in J \setminus I^\pm, \qquad r \in I^\pm \setminus J,$$

for which the matrix $A_{J'}$ is still nonsingular. (Here J' is obtained from J by exchanging l with r.)

For Phase 1, finding a first feasible basic solution to (1.1.1), the initial basis in the sense of (1.1.5), 0) is any standard basis (1.1.4) (e.g. that one obtained in Phase 0). The test algorithm is simply to check whether the current basic solution $\bar{x}$ (1.1.2) corresponding to the current basis A_J is feasible (i.e. satisfies (1.1.1 c)) or not. As A_J is supposed to be a standard basis, there is a $j \in J \cap I^+$ with $\bar{x}_j < 0$, if $\bar{x}$ is not feasible. In this case, one can form the auxiliary objective function

$$\bar{c}^T x = \bar{c}_{-m} x_{-m} + \cdots + \bar{c}_{-1} x_{-1} + \bar{c}_1 x_1 + \cdots + \bar{c}_n x_n$$

with

(1.2.1) $$\bar{c}_j = \begin{cases} -1 & \text{if } j \in J \cap I^+ \text{ and } \bar{x}_j < 0, \\ 0 & \text{otherwise,} \end{cases}$$

($\bar{c}^T x$ is the "sum of the infeasibilities", compare Dantzig [3]). The "pivoting algorithm" consists in finding indices l, r according to the following rules:

1) Find the "Simplex multipliers" $v_1, \ldots, v_m$, that is the solution $v^T = (v_1, \ldots, v_m)$ of the linear equations

(1.2.2) $$v^T A_J = \bar{c}_J^T,$$

and form the vector $u = (u_{-m}, \ldots, u_{-1}, u_1, \ldots, u_n)^T$ of "reduced costs"

$$(1.2.3) \qquad\qquad u := \bar{c} - A^T v.$$

2) Determine the index r of the variable to enter the basis such that

$$r \notin J \cup I^0 \quad \text{and} \quad u_r < 0, \quad \text{e.g.} \quad u_r = \min\{u_k \mid u_k < 0 \,\&\, k \notin J \cup I^0\}.$$

If there is no such r, then (1.1.1) has no feasible solution.

3) Determine the index l of the variable x_l to leave the basis as follows: Let y be the solution of the system of linear equations $A_J y = A_r$ and let $\hat{b}$ be the basic part of the basic solution $\bar{x}$:

$$(1.2.4) \qquad\qquad y := A_J^{-1} A_r, \quad \hat{b} := \bar{x}_J.$$

If there is an index i with $j_i \in I^+ \cap J$, $y_i > 0$, $\hat{b}_i = 0$, find p such that

$$(1.2.5) \qquad\qquad y_p = \max\{y_i \mid j_i \in I^+ \cap J, \ y_i > 0, \ \hat{b}_i = 0\}$$

and set $l := j_p$ (degenerate case). Otherwise, find the quantities ma and p with

$$(1.2.6) \qquad ma := \frac{y_p}{\hat{b}_p} = \max\left\{ \frac{y_i}{\hat{b}_i} \,\middle|\, y_i > 0, \ \hat{b}_i > 0, \ j_i \in I^+ \cap J \right\},$$

where $ma := -\infty$, if the maximum is not defined (in this case, p is left undefined). Likewise,

$$(1.2.7) \qquad mi := \frac{y_q}{\hat{b}_q} := \min\left\{ \frac{y_i}{\hat{b}_i} \,\middle|\, y_i < 0, \ \hat{b}_i < 0, \ j_i \in I^+ \cap J \right\}.$$

The index q and mi are always well defined, if r is determined as in 2). Therefore, the following definition makes sense

$$(1.2.8) \qquad\qquad l := \begin{cases} j_q & \text{if } mi > ma, \\ j_p & \text{otherwise}. \end{cases}$$

These pivoting rules make sure that in each exchange step $J \to J'$ the value of the auxiliary objective function at the basic solution does not increase and that variables which are feasible at the old basic solution corresponding to J stay feasible at the new basic solution with respect to J'.

For Phase 2, any standard basis for which the corresponding basic solution is feasible, e.g. the final basis of Phase 1, can be taken as initial basis (1.1.5), 0). The "test algorithm" in the sense of (1.1.5), 1) is as follows (compare Phase 1): Compute the simplex multipliers $v^T = (v_1, \ldots, v_m)$ by solving

$$(1.2.9) \qquad\qquad v^T A_J = c_J^T,$$

and determine the reduced costs $u = (u_{-m}, \ldots, u_{-1}, u_1, \ldots, u_n)^T$:

$$(1.2.10) \qquad\qquad u := c - A^T v.$$

Find $r \notin I^0$ such that

$$u_r = \min\{u_k \mid k \notin I^0\}.$$

If $u_r \geq 0$, then the algorithm terminates: the current basic solution $\bar{x}$ given by (1.1.2) is also an optimal solution of (1.1.1). If $u_r < 0$, then x_r is the variable to enter the basis. The index l of the variable x_l to leave the current basis, is determined similarly to Phase 1): The vector y is computed as the solution of (1.2.4). The index p and ma is computed as in (1.2.5) or (1.2.6). (The definition of mi and q (1.2.7) does not apply, as $\bar{x}$ is feasible.) If the maximum in (1.2.6) is not defined, then the problem (1.1.1) has no bounded optimal solution. Otherwise, the index l is well defined by $l := j_p$. Again these rules make sure that the objective function does not increase in the exchange step $J \to J'$ and that the basic solutions stay feasible.

1.3. The Exchange Algorithm

As is seen from the previous section $\big($see (1.2.2), (1.2.4), (1.2.9)$\big)$, the computation of the indices l and r of the variables x_l and x_r to be exchanged with each other entails the solution of the following systems of linear equations

$$(1.3.1) \qquad A_J \bar{x}_J^{(i)} = b^{(i)}, \quad i = 1, 2, \ldots, rs; \quad A_J y = A_r; \quad A_J^T v = c_j$$

for each exchange step $J \to J'$. To solve these equations for adjacent bases $J \to J' \to J'' \ldots$ from scratch would involve a prohibitive number of operations; so several schemes have been proposed to reduce the amount of computation. The programs of this article use a device suggested by Golub and Stoer and exploited by Bartels [1]: In his algorithm, each stage is described by a 4-tuple

$$(1.3.2) \qquad\qquad\qquad \{J; \, R, L, \bar{B}\},$$

where J specifies the current basis and has the same meaning as before. L is a nonsingular $m \times m$-matrix, R a nonsingular upper triangular $m \times m$-matrix and $\bar{B}$ is an $m \times rs$-matrix. These matrices are supposed to be related by the equations

$$(1.3.3) \qquad\qquad L A_J = R; \quad \bar{B} := (\bar{b}^{(1)}, \ldots, \bar{b}^{(rs)}) := L \cdot B.$$

Clearly, an initial 4-tuple of this kind for (1.1.1 a), Phase 0, is

$$(1.3.4) \qquad \begin{aligned} J &:= \{-1, -2, \ldots, -m\} \\ L &:= R := I \\ \bar{B} &:= B. \end{aligned}$$

With such a 4-tuple (1.3.2), the systems (1.3.1) are easily solvable. Their solutions $\bar{x}_J^{(i)}$, y, v are obtained from

$$(1.3.5) \qquad \begin{aligned} R\, \bar{x}_J^{(i)} &= \bar{b}^{(i)} &\Rightarrow \bar{x}_J^{(i)} \\ R\, y &= L A_r &\Rightarrow y \\ R^T z &= c_j &\Rightarrow z \\ v &= L^T z &\Rightarrow v. \end{aligned}$$

The 4-tuple $\{J'; \, R', L', \bar{B}'\}$ corresponding to an adjacent basis can be computed as follows: If x_l, $l = j_s \in J$, is leaving the basis and x_r, $r \notin J$ is entering, then

$$(1.3.6) \qquad\qquad J' := \{j_1, \ldots, j_{s-1}, j_{s+1}, \ldots, j_m, r\}.$$

Let R_i be the i-th column of R. Then the matrix

(1.3.7) $$\bar{R} := (R_1, \ldots, R_{s-1}, R_{s+1}, \ldots, R_m, x), \qquad x := LA_r,$$

is almost upper triangular and satisfies

(1.3.8) $$LA_{j'} = \bar{R} =$$

The $m - s$ nonzero subdiagonal elements of $\bar{R}$ can be annihilated in a numerically stable way by premultiplying $\bar{R}$ by suitable $m - s$ permutation matrices P_i and elimination matrices F_i,

(1.3.9) $$R' := F_{m-1} P_{m-1} \ldots F_s P_s \bar{R},$$

where P_i is either the identity matrix or a permutation matrix interchanging row i and row $i+1$:

$$P_i =$$

F_i is a lower triangular Frobenius matrix of the form

$$F_i =$$

These matrices P_i, F_i are chosen such that R' (1.3.9) is again an upper triangular matrix. Then, of course, the matrices

(1.3.10)
$$L' := F_{m-1} P_{m-1} \dots F_s P_s L$$
$$\bar{B}' := F_{m-1} P_{m-1} \dots F_s P_s \bar{B}$$

together with R' (1.3.9) and J' (1.3.6) form the 4-tuple $\{J';\ R', L', \bar{B}'\}$ corresponding to the adjacent basis J'. The computations necessary to form R', L', $\bar{B}'$ from R, L, $\bar{B}$ can be briefly described as follows:

1) Compute $x := L A_r$ and form $\bar{R}$ (1.3.7) and the compound matrix

$$E := (\bar{R}, L, \bar{B}).$$

2) Starting with $E^{(s)} = (\bar{R}^{(s)}, L^{(s)}, \bar{B}^{(s)}) := E = (\bar{R}, L, \bar{B})$ and for $i := s$, $s+1, \dots, m-1$ compute $E^{(i+1)}$ from $E^{(i)}$ in the following way:

(1.3.11)
a) If $|r_{i+1,i}^{(i)}| > |r_{i,i}^{(i)}|$ interchange the rows i and $i+1$ of the matrix $E^{(i)}$ and denote the resulting matrix again by $E^{(i)}$.

b) Set $p := r_{i+1,i}^{(i)}/r_{i,i}^{(i)}$, $|p| \leq 1$, and subtract the p-fold of row i from row $i+1$ of the matrix $E^{(i)}$. The result is the matrix $E^{(i+1)}$.

3) If $E^{(m)} = (\bar{R}^{(m)}, L^{(m)}, \bar{B}^{(m)})$, then the new 4-tuple $\{J';\ R', L', \bar{B}'\}$ is given by (1.3.6) and $R' := \bar{R}^{(m)}$, $L' := L^{(m)}$, $\bar{B}' := \bar{B}^{(m)}$.

Note, that due to the pivot selection (a), the elements of the matrices F_i, P_i are bounded by 1 in modulus. Note also, that one may use $m-s$ Givens-rotations

$$\Omega_{i,i+1} = \begin{bmatrix} 1 \\ & \ddots \\ & & 1 \\ & & & c & s \\ & & & -s & c \\ & & & & & 1 \\ & & & & & & \ddots \\ & & & & & & & 1 \end{bmatrix} \begin{matrix} \\ \\ \\ \leftarrow (i) \\ \leftarrow (i+1), \\ \\ \\ \end{matrix} \quad c^2 + s^2 = 1,$$

to reduce the subdiagonal elements of $\bar{R}$ to zero and thereby to form another triple of matrices satisfying (1.3.3) for J':

$$(R', L', \bar{B}') := \Omega_{m-1,m} \Omega_{m-2,m-1} \dots \Omega_{s,s+1} (\bar{R}, L, \bar{B}).$$

One might think that it is not necessary to compute the solutions $\bar{x}_J^{(i)}$, $v = v(J)$, by actually solving (1.3.5) for successive J, J', $\dots$, inasmuch there are certain simple recursive relations between $\bar{x}_{J'}^{(i)}$, $v(J')$ and $\bar{x}_J^{(i)}$, $v(J)$, which are used for example in the standard realizations of the Simplex method, e.g. the inverse basis method. With the solution y of (1.3.5), the ordered index set $\bar{J}' := \{j_1, \dots, j_{s-1}, r, j_{s+1}, \dots, j_m\}$ (which is a permutation of J' (1.3.6)) and the Frobeniusmatrix

$$(1.3.12) \quad F := \begin{bmatrix} 1 & \alpha_1 & & & 0 \\ & \cdot & & & \\ & \cdot & & & \\ & 1 & \alpha_{s-1} & & \\ & & \alpha_s & & \\ & & \alpha_{s+1} & 1 & \\ & & \cdot & & \cdot \\ & & \cdot & & \\ 0 & & \alpha_n & & 1 \end{bmatrix}, \qquad a_i = \begin{cases} \dfrac{-y_i}{y_s} & \text{for} \quad i \neq s, \\[2mm] \dfrac{1}{y_s} & \text{for} \quad i = s, \end{cases}$$

one has for instance that the solutions $\bar{x}_J^{(i)}$ and $\bar{x}_{J'}^{(i)}$ of

$$A_J \bar{x}_J^{(i)} = b^{(i)}, \qquad A_{J'} \bar{x}_{J'}^{(i)} = b^{(i)}$$

are related by $\bar{x}_{J'}^{(i)} = F \bar{x}_J^{(i)}$. The pivot element y_s, however, may be very small so that, in general, it is numerically unsafe to use these recursions. It is numerically more stable (though more costly) to compute $\bar{x}_{J'}^{(i)}$, v separately for each J by using (1.3.5).

Note also, that the 4-tuple $\{J; R, L, \bar{B}\}$ contains enough information to improve the accuracy of the computed basis solutions and/or simplex multipliers by iteration, if this is wanted. If, for instance, $\bar{x}_J^{(k)}$ (here k denotes an iteration index) is an approximate solution of $A_J \bar{x}_J = b$, if

$$(1.3.13) \qquad r^{(k)} := b - A_J \bar{x}_J^{(k)}$$

is the corresponding residual (which should be computed with double precision arithmetic), then $\bar{x}_J^{(k+1)}$ defined by

$$(1.3.14) \qquad \begin{aligned} s^{(k)} &:= L r^{(k)} \\ R \Delta x^{(k)} &= s^{(k)} \Rightarrow \Delta x^{(k)} \\ \bar{x}_J^{(k+1)} &:= \bar{x}_J^{(k)} + \Delta x^{(k)} \end{aligned}$$

is, in general, a much better approximation to $\bar{x}_J$ than $\bar{x}_J^{(k)}$.

2. Description of the Programs

2.1. General Organization of the Programs

The organization of codes for linear programming problems has to take care of two major difficulties:

a) The complexity of linear optimization problems. The code should be flexible enough to handle all linear programming problems of some importance (e.g. finding feasible, optimal solutions, parametric programming, changes of right hand side, of the objective function etc.).

b) The storage and treatment of large matrices typical for linear programming problems is machine dependent, as it may be necessary to use different storage media. Also, sparse matrices should not be stored and handled in the same way as full matrices.

In the programs of this article, these difficulties are bypassed by the introduction of three procedure parameters *problist, ap, p* of the main procedure *lp*:

a) The problem to be solved is specified by the procedure *problist* which must be supplied by the user in any case.

b) The handling of the matrix AA (1.1.1) is done by means of the procedure ap, and all operations on the matrices $R, L, \bar{B}$ (1.3.2) are performed in the procedure p. For small problems one may use the procedures ap and p of this report which are based on the standard representation of $AA, R, L, \bar{B}$ by rectangular arrays. For big problems with a large and sparse matrix AA, the limitations of his computer may force the user to supply other procedures ap and p which are based on some nonstandard representation of $AA, R, L, \bar{B}$.

2.2. Formal Parameters and Description of the Procedure lp.

m **integer** input parameter, number of rows of the matrix AA (1.1.1 b).

n **integer** input parameter, number of columns of the matrix AA (1.1.1 b).

eps **real** input parameter; eps is the relative machine precision, that is the smallest number such that $1 + eps > 1$ for the floating point arithmetic used.

$problist$ **procedure** parameter specifying the problem to be solved by lp. $problist$ must be declared as follows:

         procedure $problist$ $(problem)$;
         procedure $problem$;
         begin ... **end**;

For the use of $problist$ and $problem$ see Section 2.3.

ap **procedure** parameter specifying the matrix AA (1.1.1 b).
       ap must have a declaration of the form:

         procedure ap (wz, x, k, sk);
         value wz, k;
         integer wz, k;
         array x;
         real sk;
         begin ... **end**;

For the meaning and use of ap see Section 2.4.

p **procedure** parameter for handling the matrices $R, L, \bar{B}$ (1.3.2.).
       p must have a declaration of the form

         procedure p $(case, ind, x)$;
         value $case, ind$;
         integer $case, ind$;
         array x;
         begin ... **end**;

See Section 2.5 for the meaning and use of p.

A call of lp of the form

$$lp\,(m, n, eps, problist, ap, p)$$

has the following consequences:

1) The algorithm is initiated by setting

$$I^0 := I^{\pm} := \emptyset; \quad I^+ := N = \{-m, \ldots, -1, 1, \ldots, n\}$$
$$\bar{B} := 0; \quad c := 0$$
(2.2.1)
$$J := \{-1, -2, \ldots, -m\}$$
$$L := R := I.$$

Note in particular, that all variables are treated as nonnegative variables, if there is no subsequent change of their status (see 2.3.2). Note also that the right hand sides as specified by $\bar{B}$ and the objective function c are zero initially; other values must be provided by *problist* (see Sections 2.3.5, 2.3.11, 2.3.12).

2) After the initialization, a call

$$problist\,(problem)$$

within *lp* of the procedure *problist* using that procedure *problem* as actual para- meter which is declared within the body of *lp*, causes the computations to be executed which are prescribed in the body of *problist*. After calling *problist*, *lp* terminates.

2.3. Formal Parameter Lists and Description of the Procedures problist and problem

The procedure *problist* is to be supplied by the user. Its only formal para- meter is the **procedure**

$$problem.$$

The procedure *problem* which is declared within the body of the procedure *lp* is used as actual procedure parameter in the call of *problist* within *lp*. By properly calling this procedure *problem* within the body of *problist*, the user specifies the linear programming problem to be solved, e.g. specifies or changes right hand sides, objective function, the status of variables (zero, free or nonnegative) etc. Thus, the parameter list of procedure *problem* is the main tool of the user to de- scribe his problem at hand. *problem* has the following procedure heading (see ALGOL programs, Section 3)

procedure *problem (case, u, fail)*;
value *case*;
integer *case*;
array *u*;
label *fail*;

The formal parameters *case, u, fail* of *problem* are input parameters; however, *u* may also be used as output parameter:

case input parameter of type **integer**. *case* may have any value between 1 and 15 inclusive.

u input and output **array** parameter. The meaning and permissible formats of *u* depend on the value of *case*.

fail a **label** parameter used for irregular exits from *problem*. Its meaning also depends on the value of *case*.

As already stated, one may use different calls of *problem* (corresponding to different values of the input parameter *case*) within the body of *problist* in order to describe the problem at hand. With a single exception, the user is completely free (and responsible) in his decision which of the different calls of *problem* he uses. The exception is that the first call of *problem* within *problist* must be a call of the form

$$problem\,(1,\,u,\,fail)$$

with 1 as input value for the parameter *case*. This call (see 2.3.1) causes the matrix AA to be scaled internally, and the reason for the exception is that every matrix AA should be scaled in some way or other before applying the pivoting rules of Section 1.2.

The numbering of the different cases is chosen in a natural way: If *case1*<*case2*, then normally a call *problem* (*case1, u, fail*) precedes a call of the form *problem* (*case2, u, fail*).

Each of the following Sections 2.3.1–2.3.15 contains the description of the effects of calling *problem* with a particular input value of $case = 1, 2, \ldots, 15$.

After a headline indicating the effects of a call with the particular value of *case*, there is a brief description of the format of the array parameter *u*, a statement whether *u* is an input and/or output parameter, and whether the label parameter *fail* is used or not. There follows a detailed description of the meaning and proper use of the particular call.

For a detailed example of a procedure *problist* and a description of further problems to be handled by a proper formulation of *problist*, see Section 7.

2.3.1. $case = 1$

Implicit scaling of AA.

u   array of format $u[-m:n]$ at least.
   $u[0]$ is used as input, $u[i]$, $-m \leq i \leq -1$, $1 \leq i \leq n$ as output parameters.
fail is not used.

If on entry $u[0] = 0$, then scaling factors $u[i]$, $-m \leq i \leq -1$, $1 \leq i \leq n$ are computed for AA (1.1.1 b) such that the maximum norm of each row and each column of the matrix (α_{ik}) with

$$\alpha_{ik} := aa_{ik} \frac{u_k}{u_{-i}}, \quad i = 1, \ldots, m, \quad k = 1, \ldots, n$$

is 1. Simultaneously, the pivot selection rules of the simplex method are modified such that an application of the modified rules to the unscaled matrix AA leads to the same results as the unmodified rules applied the scaled matrix (α_{ik}). (Implicit scaling of AA: AA is not changed.)

If on entry $u[0] > 0$, then a simplified scaling takes place: The scaling factors are now $u[i] := 1$, $i = -m, \ldots, -1$, $u[k] := \dfrac{1}{u[0]}$, $k = 1, \ldots, n$. $u[0]$ should be an upper bound for the maximum norms of all columns of AA. The maximum norm of all columns and rows of the scaled matrix (α_{ik}) then is ≤ 1. AA is only implicitly scaled and is not changed.

A call with $u[0] > 0$ can be used if AA is already a balanced matrix (i.e. the moduli of the elements of AA are not too different).

Important Remark. A call with $case = 1$ must precede any other call of *problem* in *problist*.

2.3.2. $case = 2$

Characterization of nonnegative, zero and free variables; elimination of zero variables from the current basis, introduction of free variables into the current basis (Phase 0).

u array parameter of format $u[-m:n]$ at least.
  $u[k]$, $k \neq 0$, are input parameters;
  $u[0]$, is an output parameter.
fail is used.

If for $k = -m, \ldots, -1, 1, \ldots, n$

$u[k] = 0$, then $k \in I^0 : x_k$ is a zero variable,
$u[k] > 0$, then $k \in I^+ : x_k$ is a nonnegative variable,
$u[k] < 0$, then $k \in I^\pm : x_k$ is a free variable.

Besides the characterization of the variables, the current basis is transformed into a standard basis (1.1.4) by a sequence of exchange steps such that (see Phase 0, Section 1.2)

(a) all free variables become basic,

(b) no zero variable remains basic.

If this is not possible, the exit *fail* is used.

Also, unless the status of the variables is not changed by a further call of *problem* (2, *u*, *fail*), all subsequent exchange steps are such that zero variables will never enter and free variables will never leave the basis.

Upon exit, $u[0]$ contains the total number of exchange steps executed since the last entry of *lp*.

Remark. According to (2.2.1), initially all variables are treated as nonnegative variables and the starting basis is a standard basis (1.1.4) with respect to this status of the variables. Thus a call of *problem* (2, *u*, *fail*) is not necessary, if one does not want to change this status of the variables.

2.3.3. case = 3

Information about the current basic variables.

u output array parameter of format $u[-m:n]$ at least.
fail is not used.

After calling *problem* (3, *u*, *fail*), $u[k]$, $-m \leq k \leq -1$, $1 \leq k \leq n$, contains information about the current status of the variable x_k:

$u[k] \geq 0$ if x_k is not basic and nonnegative;
$u[k] = 0$ if x_k is not basic and a zero variable;
$u[k] = -i < 0$ if $j_i = k$, that is if x_k is basic and the i-th component j_i of J is k.

2.3.4. case = 4

Single exchange step.

u input and output array parameter of format $u[0:1]$ at least.
fail is used as error exit.

11*

If on entry $u[0] = r$, $u[1] = l$ then a single exchange step is performed with the variable x_r entering and x_l leaving the current basis. If

(a) x_r is already basic or a zero variable (that is not permitted to enter a basis), or

(b) x_l is not basic or a free variable (that is not permitted to leave a basis),

then the exit *fail* is used. Upon leaving *problem* by *fail*, the values of $u[0]$ and $u[1]$ are $u[0] = 0$ in case (a) and $u[1] = 0$ in case (b). u is not changed after a normal exit not by *fail*.

Remark. It is not checked whether the exchange step leads to a singular basis matrix $A_{J'}$ (i.e. to a singular upper triangular matrix R'). This has to be prevented, if necessary, by a proper choice of the indices l, r.

2.3.5. case = 5

Input or change of a right hand side.

u input array parameter of format $u[0:m]$ at least; u is not changed.

fail is not used.

If $u[0] = k$, $u[i] = \tilde{b}_i$, $i = 1, 2, \ldots, m$, the vector $\tilde{b}$ replaces the old k-th right hand side (r.h.s.) $b^{(k)}$. More precisely, since the r.h.s. $b^{(j)}$, $j = 1, \ldots, rs$ are not stored for themselves but only in modified form as the matrix $\bar{B} = (\bar{b}^{(1)}, \ldots, \bar{b}^{(rs)}) = L B = L(b^{(1)}, \ldots, b^{(rs)})$, calling *problem* (5, u, *fail*) has only the effect that the k-th column $\bar{b}^{(k)}$ of the matrix $\bar{B}$ belonging to the current 4-tuple (1.3.2) is replaced by the vector $L\tilde{b}$. (Note in this context, that initially (see (2.2.1)) all right hand sides $b^{(i)}$, $i = 1, \ldots, rs$, are defined to be 0.)

In general, *problem* (5, u, *fail*) is called with $u[0] = 1$ for the input of "the" (first) right hand side $b = b^{(1)}$ of (1.1.1). However, $u[0] = 2, 3, \ldots, rs$, may be used to specify additional right hand sides, as for example in parametric programming, where right hand sides of the form $b^{(1)} + \lambda\, b^{(2)}$ with varying λ are considered.

Remarks. 1) Introduction of a r.h.s. by calling *problem* (5, u, *fail*) replaces a column of $\bar{B}$ by another column and thereby increases the number of nonzero columns of $\bar{B}$ (which was 0 initially), in general. The matrix $\bar{B}$ is changed with every exchange step $J \rightarrow J'$ (see 1.3) and the amount of computation per exchange step increases with the number of nonzero columns of $\bar{B}$. Thus a use of *problem* (5, u, *fail*) can be recommended only if the user is interested in the basic solutions for each r.h.s. and for each intermediate basis of the exchange algorithm (for instance, this is the case for "the" first right hand side $b = b^{(1)}$ of (1.1.1) and in parametric programming). If, for a particular r.h.s. the user is only interested in the corresponding basic solution with respect to one particular basis, he should use the call

$$\text{problem}\,(8, u, \text{fail})\quad(\text{see Section 2.3.8}).$$

2) If the r.h.s. to be introduced is an axis vector $e_j = (0, \ldots, 1, \ldots, 0)^T$ with the j-th component 1, one should use the call

$$\text{problem}\,(7, u, \text{fail})\quad(\text{see Section 2.3.7}).$$

3) If the r.h.s. to be introduced is identical with a particular column of AA, one should use the call

$$problem\,(6,\,u,\,fail)\quad\text{(see Section 2.3.6).}$$

2.3.6. case $=6$

Input of a column of AA as right hand side.

u input array parameter of format $u[0:1]$ at least.
fail is not used.

If on entry $u[0]=k$ and $u[1]=i$, $1\leq i\leq n$, k, i integers, then the k-th column $\bar{b}^{(k)}$ of the matrix $\bar{B}$ belonging to the current 4-tuple $\{J;\,R,\,L,\,\bar{B}\}$ (1.3.2) is replaced by the vector LA_i, where A_i is the i-th column of AA. This call can be used if the right hand sides are stored as particular columns of AA (corresponding to zero variables, in general). See also Remark 1) of Section 2.3.5.

2.3.7. case $=7$

Output of a basic solution.

u input and output array parameter of format $u[-m:n]$ at least.
fail is not used.

If on entry $u[0]=k>0$, k an integer, $1\leq k\leq rs$, then upon exit $u[i]$, $-m\leq i\leq-1$, $1\leq i\leq n$, contains the basic solution corresponding to the current basis and to the k-th right hand side.

If on entry, $u[0]=-k$, k an integer with $1\leq k\leq m$, then upon exit $u[i]$, $-m\leq i\leq-1$, $1\leq i\leq n$, contains the basic solution belonging to the current basis and to the right hand side $e_k\,(=k$-th axis vector) (compare 2.3.5, remarks).

In either case, $u[0]$ contains the value of the objective function at the respective basis solution. (See also Section 2.3.8).

2.3.8. case $=8$

Computation of an additional basic solution.

u input and output array parameter of format $u[-m:n1]$ at least, $n1:=\max\,(m,\,n)$.
fail is not used.

If on entry $u[i]=\tilde{b}_i$, $i=1,\,2,\,\ldots,\,m$, then on exit $u[i]$, $-m\leq i\leq-1$, $1\leq i\leq n$ contains the basic solution corresponding to the current basis and to the right hand side $\tilde{b}$. $\tilde{b}$ is not stored, $\bar{B}$ not modified. $u[0]$ contains the value of the objective function at this basic solution.

2.3.9. case $=9$

Computation of a feasible basis (Phase 1).

u input and output array of format $u[0:0]$ at least.
fail is used as error exit.

On entry, $u[0]$ must contain the maximum permissible number of exchange steps to compute a first feasible basis from the current basis.

A call of *problem* $(9, u, fail)$ causes the exchange steps of Phase 1 (see Section 1.2) to be executed: The current basic 4-tuple $\{J; R, L, \bar{B}\}$ is transformed by a suitable number of exchange steps (not exceeding $u[0]$) into a 4-tuple $\{J'; R', L', \bar{B}'\}$ which corresponds to a feasible basis, that is for which the corresponding basic solution is a feasible solution of (1.1.1 b, c).

If a feasible basis is not found then the error exit *fail* is used and the cause is indicated by the output value of $u[0]$:

(a) $u[0] > 0$:   The maximum permissible number of exchange steps was not large enough.

(b) $u[0] = -1$: (1.1.1 b, c) has no (feasible) solution.

In case (a) and if *fail* is not used, $u[0]$ contains the total number of exchange steps executed thus far.

2.3.10. *case* $= 10$

Single step of the (primal) Simplex method.

u input and output array of format $u[0:0]$ at least.

fail is used as error exit.

If on entry $u[0] = r$, $-m \leq r \leq -1$, $1 \leq r \leq n$, r integral, the variable x_r is entered into the current basis. The variable x_l to leave the basis is determined as in Phase 1 (Phase 2) of the Simplex method $\big($See section 1.2, (1.2.4)–(1.2.8)$\big)$.

The error exit *fail* is used if

(a) if the pivoting rules of Phase 1 (Phase 2) do not lead to a variable x_l: There is no variable x_l, $l \in J$, such that exchanging x_l with x_r would lead to a new basic solution $\bar{x}'$ which is such that $\bar{x}'_k$ satisfies (1.1.1c) if $\bar{x}_k$ does ($\bar{x}$ is the old basic solution).

(b) x_r is already basic or is a zero variable (which is not permitted to enter a basis).

Case (a) is indicated by the output value $u[0] = -2$, case (b) by $u[0] = 0$. If *fail* is not used, $u[0]$ is not changed.

2.3.11. *case* $= 11$

Input or change of objective function.

u input array parameter of format $u[-m:n]$ at least.

fail is not used.

If on entry $u[i] = \tilde{c}_i$, $-m \leq i \leq n$, then the current objective function

$$c_0 + c_{-m} x_{-m} + \cdots + c_{-1} x_{-1} + c_1 x_1 + \cdots + c_n x_n$$

is replaced by

$$\tilde{c}_0 + \tilde{c}_{-m} x_{-m} + \cdots + \tilde{c}_{-1} x_{-1} + \tilde{c}_1 x_1 + \cdots + \tilde{c}_n x_n.$$

Note that initially $c_i = 0$, $-m \leq i \leq n$ $\big($see (2.2.1)$\big)$.

u is not changed. (See also Section 2.3.12.)

2.3.12. case = 12

Input of a line of AA as objective function.

u input array parameter of format $u[0:0]$ at least.
fail is not used.

If on entry $u[0] = i$, $1 \leq i \leq m$, i an integer, then the current objective function is replaced by

$$0 \cdot x_{-m} + \cdots + 0 \cdot x_{-1} + aa_{i1} x_1 + aa_{i2} x_2 + \cdots + aa_{in} x_n$$

corresponding to the i-th row of the matrix AA. (See also Section 2.3.11.)

2.3.13. case = 13

Output of reduced costs.

u input and output array parameter of format $u[-m:n]$ at least.
fail is not used.

If on entry $u[0] \neq 0$ then on exit $u[i]$, $-m \leq i \leq -1$, $1 \leq i \leq n$, is equal to the reduced cost u_i corresponding to the variable x_i and the current basis.

    If on entry $u[0] = 0$, then only the reduced costs $u[i]$ for nonnegative and free variables are computed and $u[i] := 0$ for zero variables x_i. (Reduced costs are defined by (1.2.9), (1.2.10).)

2.3.14. case = 14

Computation of an optimal basis (Phases 1 and 2).

u input and output array parameter of format $u[-m:n]$ at least.
fail is used as error exit.

On entry $u[0]$ must contain the maximum permissible number of exchange steps to find the optimal (feasible) basis from the current basis. Calling *problem* (14, u, *fail*) then causes the exchange steps of Phases 1 and 2 (see Section 1.2) to be executed.

    After leaving *problem* by the normal exit, the previous basis has been transformed into an optimal basis, that is a basis for which the basic solution is an optimal solution of (1.1.1). The output values $u[i]$ then have the following meaning:

$u[0]$:   total number of exchange steps executed so far.

$u[i]$, $i \neq 0$: $u[i] = 0$ if x_i is a zero variable
                 $u[i] = $ reduced cost for x_i, if x_i is not a zero variable.

    The error exit *fail* is used in the following cases:

(a)   The maximum permissible number $u[0]$ of exchange steps was not large enough to find an optimal basis.
(b)   The linear program (1.1.1) has no feasible solution.
(c)   The linear program (1.1.1) has feasible solutions, but no bounded optimal solution.

In case (a), the output value $u[0]>0$ gives the total number of exchange steps executed so far. In cases (b) and (c), the output value of $u[0]$ is -1 and -2, respectively.

2.3.15. $case = 15$

Dual Simplex step.

u input and output array parameter of format $u[0:0]$ at least.

fail is used as error exit.

On entry, the current basis must be dually feasible, i.e. all reduced costs corresponding to nonzero variables must be nonnegative. If on entry $u[0]=l$, then a dual simplex step is performed: The variable x_l leaves, and a nonzero variable x_r, $r \notin J$, enters the current vasis J. Here x_r is determined in such a way that the new basis is again dually feasible.

If the normal exit is used then $u[0]$ is not changed. The error exit *fail* is used, if

(a)   a nonzero variable x_r, $r \notin J$, cannot be found such that the exchange step leads again to a dually feasible basis, or if

(b) x_l is not basic, $l \notin J$, or is a free variable (which is not allowed to leave the basis).

In cases (a) and (b) the output value of $u[0]$ is -1 and 0, respectively.

2.4. Formal Parameter List and Description of the Procedure ap.

The procedure *ap* serves to handle the matrix AA of (1.1.1). For small problems, one may use the procedure *ap* of this report; for large and sparse matrices AA however, it may be necessary to provide another procedure *ap* taking care of the special structure of the matrix AA. In these cases the construction of *ap* can be patterned after the procedure *ap* of this report.

The procedure *ap* must have a heading of the following form

```
procedure ap (wz, x, k, sk);
value wz, k;
integer wz, k;
array x;
real sk;
```

parameter list:

wz of type **integer**, input parameter,

x **array** of format $x[1:m]$, input and output parameter,

k of type **integer**, $1 \leq k \leq n$, input parameter,

sk of type **real**, output parameter.

Besides the formal parameters wz, x, k, sk, the procedure *ap* of this report has the following global parameters:

aa **array** of format $aa[1:m, 1:n]$ for storing the matrix AA (1.1.1 b),

m the number of rows of AA, an **integer**.

If $wz \geq 0$, then the procedure computes the scalarproduct

$$sk := \sum_{i=1}^{m} aa_{ik} x_k$$

of the k-th column of AA with the vector x. x is not changed. If $wz < 0$, then the output value of x is the k-th column of AA.

$$x[i] := aa_{ik} \quad i = 1, 2, \ldots, m.$$

2.5. Formal Parameter List and Description of the Procedure p

The procedure p is to handle the matrices L, R and $\bar{B}$ (1.3.2). As with ap, one may use the procedure p of this report for small problems: if the matrices L, R, $\bar{B}$ become too big to store them within the main storage or if one wishes to take advantage of their sparseness, the user has to provide his own procedure p patterned after the procedure p of this report.

The procedure p must have a heading of the form:

procedure p (case, ind, x);
value case, ind;
integer case, ind;
array x;

parameter list:

case input **integer** parameter; permissible values of case: 1, 2, 3, 4, 5, 6, 7.
ind input **integer** parameter with permissible values depending on case.
x   input and output **array** parameter of format $x[1:m]$.

Besides these formal parameters, the procedure p of this report has the following global parameters:

rs an **integer**, $rs \geq 1$, specifying the maximum number of columns of $\bar{B}$ ($=$ the maximum number of right hand sides)
m an **integer**, the number of rows of AA (1.1.1 b)
rl an **array** of format rl $[1:m, 1:2 \times m + rs]$ for storing the compound matrix $(R, L, \bar{B})$ (R is an upper triangular $m \times m$ matrix, L an $m \times m$-matrix, $\bar{B}$ an $m \times rs$-matrix).

The permissible calls and the meaning of p depend on the input parameter case.

2.5.1. case = 1

Initialization of R, L, $\bar{B}$.

ind integer parameter, not used
x   array parameter, not used

A call p (1, ind, x) initializes the matrices $R, L, \bar{B}$: $R := L := I$ ($= m \times m$-identity matrix), $\bar{B} := 0$ (see (2.2.1)).

2.5.2. case = 2

ind input integer parameter with values $0 \leq ind \leq m$.
x   input and output array parameter, format $x[1:m]$.

If $ind = 0$, then p computes $x := Lx$: x is overwritten by the product Lx. If $1 \leq ind \leq m$, then $x := L_{ind}$, that is, x gives column ind of L.

2.5.3. case = 3

ind not used

x input and output array of format $x[1:m]$.

The program computes $x := L^T x$.

2.5.4. case = 4

Computation of $x := R^{-1}g$.

ind input integer parameter with values $-m \leq ind \leq rs$.

x input and output array parameter of format $x[1:m]$.

The program computes the solution $x := R^{-1}g$ of the upper triangular system of linear equations $Rx = g$, where

$$g := \begin{cases} \bar{B}_i (=\text{column } i \text{ of } \bar{B}), & \text{if } ind = i > 0, \\ x \ (=\text{input array}), & \text{if } ind = 0, \\ L_i \ (=\text{column } i \text{ of } L), & \text{if } ind = -i < 0. \end{cases}$$

2.5.5. case = 5

Computation of $x := (R^T)^{-1}g$.

ind input integer parameter with values $1 \leq ind \leq m+1$.

x input and output array parameter of format $x[1:m]$.

The program computes the solution $x := (R^T)^{-1}g$ of the lower triangular system of linear equations $R^T x = g$, where it is supposed that the first ind-1 components $x[i]$, $i = 1, \ldots, ind$-1, of the input array are already the first ind-1 components of the solution x and that the remaining components $x[i]$, $i = ind$, $ind + 1, \ldots, m$ of the input array are those components g_i, $i = ind$, $ind + 1, \ldots, m$ of the right hand side vector g which are still needed to compute the remaining components of the solution x. Thus $x[i]$, $i = 1, \ldots, ind$-1 are not changed, $x[i]$, $i = ind, \ldots, m$ are overwritten.

2.5.6. case = 6

Exchange step.

ind input integer parameter with $1 \leq ind \leq m$.

x input array parameter of format $x[1:m]$.

A call of $p(6, ind, x)$ causes the exchange step as described in Section 1.3 (see in particular (1.3.11), 1), 2), 3)) to be executed: If $ind = s$ then column s is dropped from R, columns $s + 1, s + 2, \ldots, m$ of R are shifted to the left to the positions $s, s + 1, \ldots, m - 1$, respectively, and the vector x given be the input array x is inserted as new column m of R (in applications, x will be the vector $x = LA_r$ (1.3.7)). The resulting almost upper triangular matrix $\bar{R}$ (see (1.3.7)) is then again reduced to an upper triangular matrix R' by $m - s$ permutation and elimination steps and the same transformations are applied to the matrices L and $\bar{B}$ to generate L' and $\bar{B}'$, respectively (the details are described in Section 1.3, (1.3.11), 1), 2), 3)).

2.5.7. case $= 7$

Change of $\bar{B}$.

ind input integer parameter with values $1 \leq ind \leq rs$.

x input array parameter with format $x[1:m]$.

If $ind = s$, then column s of $\bar{B}$ is replaced by x.

3. Applicability

By proper formulation of the procedure *problist* and appropriate use of the proce-dure *problem* as described in Section 2.3 a great many problems concerning systems of linear equations and inequalities of the form (1.1.1 b, c) can be solved. It follows a list of typical applications (which is not complete):

1. Solution of linear equations:

$$A A x = b,$$

where $A A$ is a $m \times m$-matrix and $b \in R^m$. Its solution can be found by scaling the matrix (see (2.3.1)), by setting $n = m$, $I^{\pm} := \{1, 2, \ldots, m\}$, $I^+ = \emptyset$, $I^0 = \{-1, -2, \ldots, -m\}$ (see (2.3.2)), and by computation of a basic solution corresponding to b (see Section 2.3.8).

2. Find a feasible solution of (1.1.1): This can be achieved by proper calls of *problem* $(1, u, fail)$ (scaling), of *problem* $(2, u, fail)$ (specification of the con-straints (1.1.1 c)), of *problem* $(5, u, fail)$ (introduction of right hand side), of *problem* $(9, u, fail)$ (computation of feasible basis) and of *problem* $(7, u, fail)$ (output of feasible solution).

3. Find an optimal solution of (1.1.1) with all variables nonnegative: This is achieved by proper calls of *problem* $(1, u, fail)$ (scaling), *problem* $(5, u, fail)$, *problem* $(11, u, fail)$ (introduction of objective function), *problem* $(14, u, fail)$ (computation of optimal basis, output of optimal reduced costs) and of *problem* $(7, u, fail)$ (output of optimal solution).

4. Iterative refinement of the optimal solution of (1.1.1): Compute (a first approx-imation to) the optimal solution; compute the residual (1.3.13) of the current approximation with double precision, and solve the equations (1.3.14) by proper calls of the form $p(2, 0, x)$ and $p(4, 0, x)$.

See Section 7 for a detailed example of the use of the procedures.

4. ALGOL Programs

4.1. The Procedure lp

```
procedure lp (m, n, eps, problist, ap, p);
    value m, n, eps;
    integer m, n;
    real eps;
```

procedure *problist, ap, p*;
begin
comment Linear programming using Bartels' variant of the simplex method. The
        problem to be solved is specified in **procedure** *problist* which also
        handles input and output of necessary information. The constraint
        matrix is transferred by **procedure** *ap. m* resp. *n* are the numbers of
        rows resp. columns of the constraint matrix. **procedure** *p* performs all
        operations concerning L, R, in the decomposition

$$L \times AB = R$$

        where AB is the matrix of actual basic columns of the constraint
        matrix and R is upper triangular. *eps* is the machine precision;

integer *i, j, k, ip, is, kp, ind, wz, vz, step, stepmax, objrow*;
real *h1, h2, h3, piv, min, bmax, h*;
boolean *con, sol*;
array *b, x, w, v* $[1:m]$, *scale, obj* $[-m:n]$;
integer array *ord* $[1:m]$, *ad* $[-m:n]$;

procedure *problem* (*case, u, fail*);
    value *case*;
    integer *case*;
    array *u*;
    label *fail*;
begin
comment *case* specifies the problem to be solved. The vector *u* has variable
        meaning according to the value of *case*. *fail* is used as an error exit;

    switch *label* := *l1, l2, l3, l4, l5, l6, l7, l8, l9, l10, l11, l12, l13, l9, l15*;

    procedure *rowpiv* (*w, sr*);
        value *sr*; **array** *w*; **boolean** *sr*;
    begin
    comment Selection of pivot row. *w* is the pivot column, *sr* determines the
            pivot selection rule;
        piv := *min* := 0; *ip* := *j* := 0;
        for *i* := 1 **step** 1 **until** *m* **do**
        begin *ind* := *ord* $[i]$; **if** *ind* $\geq -m$ **then**
            begin if *sr* **then** *h1* := *h2* := *abs* (*w* $[i]$) **else**
                begin *h1* := *eps* $\times$ *bmax* $\times$ *scale* $[ind]$ $\times m$;
                   h2 := **if** *abs* (*b* $[i]$) < *h1* **then** *h1* **else** *b* $[i]$;
                   h1 := *w* $[i]$/*h2*;
            end; **if** *h1* > 0 **then**
            begin if *h2* > 0 $\wedge$ *h1* > *piv* **then**
                begin *piv* := *h1*; *ip* := *i* **end else**
                if *h2* $\times$ *scale* $[kp]$ < $-eps \times$ *scale* $[ind]$ $\wedge$
                (*j* = 0 $\vee$ *h1* < *min*) **then**

```
                      begin min := h1; j := i end
                end
             end
          end i;
          if min > piv then
          begin piv := min; ip := j end;
          if ip ≠ 0 then
          begin if abs (w [ip]) × scale [kp] > eps × scale [ord [ip]]
             then goto rp
          end; u [0] := −2; goto fail;
rp:
       end procedure rowpiv;


       procedure colpiv (b1, b2, b3);
          value b2, b3; boolean b1, b2, b3;
       begin
       comment Computation of the reduced costs and selection of the pivot
                column. b1 specifies the selection condition, if b2 = true the
                actual values of the reduced costs are stored in u, if b3 = true
                all nonbasic reduced costs are computed, if b3 = false only
                the reduced costs belonging to nonzero-variables are computed;
          p (5, is, v); for i := 1 step 1 until m do x [i] := v [i];
          p (3, 1, x); h3 := piv := 0; wz := 0;
          for i := 1 step 1 until m do h3 := h3 + abs (x [i]) × scale [−i];
          for k := −m step 1 until −1, 1 step 1 until n do
          if ad [k] > 0 ∨ (ad [k] = 0 ∧ b3) then
          begin if k < 0 then h1 := x [−k] else ap (wz, x, k, h1);
             if b2 then h1 := u [k] := h1 + obj [k]; wz := 0;
             h1 := h1 × scale [k];
             if b1 then begin piv := h1; kp := k; wz := 1 end
          end else if b2 then u [k] := 0
       end procedure colpiv;


       procedure lx;
       begin
       comment Computation of x = L × pivot column;
          if kp > 0 then
             begin ap (−1, x, kp, h1); p (2, 0, x) end else p (2, −kp, x)
       end procedure lx;


       procedure wact;
       begin
       comment computation of transformed pivot column;
          for i := 1 step 1 until m do w [i] := x [i];
          p (4, 0, w)
       end procedure wact;
```

procedure *vt*;
begin
comment computation of simplex multipliers $v[i]$;
   if *objrow* $=0$ **then**
   begin for $i:= is$ **step** 1 **until** m **do**
      begin $k:= ord[i]$; **if** $k<-m$ **then** $k:=k+vz$;
        $v[i]:= -obj[k]$;
     end
   end else
   begin for $i:= 1$ **step** 1 **until** m **do** $v[i]:= 0$; $v[objrow]:= 1$
   end
end *procedure vt*;

procedure *transform*;
begin
comment exchange of basis;
   $p(6, ip, x)$; $ad[ord[ip]]:= 1$; $ad[kp]:= -1$; $sol:=$ **true**;
   for $i:= ip$ **step** 1 **until** $m-1$ **do** $ord[i]:= ord[i+1]$;
   $ord[m]:= kp$; $step:= step+1$
end *procedure transform*;

procedure *bnorm*;
begin
comment Computation of the maximum norm of the scaled actual right hand
        side;
   $bmax:= 0$; **for** $i:= 1$ **step** 1 **until** m **do**
   begin $ind:= ord[i]$; **if** $ind<-m$ **then**
     $ind:= ind+vz$; $h1:= abs(b[i])/scale[ind]$;
     if $h1>bmax$ **then** $bmax:= h1$
   end; **if** $bmax=0$ **then** $bmax:= 1$
end *procedure bnorm*;

goto *label*[*case*];

        comment implicit scaling;
l1:

        if $u[0]>0$ **then**
        begin
        comment The column scaling factors are all identical to $1/u[0]$, the
            row scaling factors are 1;
           for $i:= 1$ **step** 1 **until** m **do** $scale[-i]:= 1$;
           for $i:= 1$ **step** 1 **until** n **do** $scale[i]:= 1/u[0]$
        end else if $u[0]=0$ **then**
        begin
        comment The scaling factors are computed such that if scaling would
            be explictit each row and column of *aa* would have maximum
            norm 1;

```
for i := 1 step 1 until m do scale[−i] := 0;
for k := 1 step 1 until n do
begin ap(−1, x, k, h); h := 0;
    for ip := 1 step 1 until m do
    if abs(x[ip]) > h then h := abs(x[ip]);
    scale[k] := if h = 0 then 1 else 1/h;
    for ip := 1 step 1 until m do
    begin h := abs(x[ip]) × scale[k];
        if h > scale[−ip] then scale[−ip] := h
    end
end;
for i := −1 step −1 until −m do
if scale[i] = 0 then scale[i] := 1
end;
for i := −m step 1 until −1, 1 step 1 until n do u[i] := scale[i];
goto end;


comment Handling of nonnegative, free and zero-variables;
```

l2:

```
for kp := −m step 1 until −1, 1 step 1 until n do
begin if u[kp] > 0 ∧ ad[kp] = 0 then ad[kp] := 1
    else if u[kp] = 0 ∧ ad[kp] > 0 then ad[kp] := 0
end; for ip := 1 step 1 until m do
begin ind := ord[ip]; k := ind + vz; if ind ≥ −m then
    begin if u[ind] < 0 then ord[ip] := ind − vz end
    else if u[k] ≥ 0 then ord[ip] := k
end;
for kp := −m step 1 until −1, 1 step 1 until n do
if u[kp] < 0 ∧ ad[kp] ≥ 0 then
begin lx; wact; rowpiv(w, true);
    is := 1; ind := ord[ip];
    transform; ord[m] := ord[m] − vz;
    if u[ind] = 0 then ad[ind] := 0
end; for ip := 1 step 1 until m do
```

l21:

```
begin ind := ord[ip]; if ind ≥ −m then
begin if u[ind] = 0 then
    begin for i := 1 step 1 until m do v[i] := 0;
        v[ip] := 1;
        is := 1;
        colpiv(abs(h1) > abs(piv), false, false);
        if piv = 0 then goto fail;
        lx; transform; ad[ind] := 0; goto l21;
    end
end
end; u[0] := step; goto end;
```

comment Output of an integer vector indicating the columns of A which
correspond to the columns of R representing the basis;

l3:

for $i := -m$ **step** 1 **until** n **do**
$u[i] := ad[i]$; $u[0] := step$;
for $i := 1$ **step** 1 **until** m **do**
begin $k := ord[i]$; **if** $k < -m$ **then** $k := k + vz$; $u[k] := -i$ **end**;
goto *end*;

comment Exchange step:
variable entering: $x[u[0]]$,
variable leaving: $x[u[1]]$;

l4:

$kp := u[0]$;
for $ip := 1$ **step** 1 **until** m **do**
if $ord[ip] = u[1]$ **then goto** *l41*;
$u[1] := 0$;

l41:

if $ad[kp] \leqq 0$ **then** $u[0] := 0$;
if $u[0] = 0$ ∨ $u[1] = 0$ **then goto** *fail*;
lx; *transform*; $is := 1$;
goto *end*;

comment Input right hand side b,
$b[i] = u[i]$, $i = 1, \ldots, m$;

l5:

for $i := 1$ **step** 1 **until** m **do** $x[i] := u[i]$;
$p(2, 0, x)$; $p(7, u[0], x)$; **if** $u[0] = 1$ **then** $sol := $ **true**;
goto *end*;

comment Input of column no. $u[0]$ of aa as right hand side no. $u[1]$;

l6:

$kp := u[1]$; lx; $p(7, u[0], x)$; $sol := $ **true**;
goto *end*;

comment Output basic solution,
$x[i] = u[i]$, $i = -m, \ldots, n$;

l7:

if sol ∨ $u[0] \neq 1$ **then**
begin $p(4, u[0], x)$; **if** $u[0] = 1$ **then**
begin for $i := 1$ **step** 1 **until** m **do** $b[i] := x[i]$;
$sol := $ **false**
end
end else for $i := 1$ **step** 1 **until** m **do** $x[i] := b[i]$;

l71:

if $case \neq 13$ **then**
begin for $i := -m$ **step** 1 **until** n **do** $u[i] := 0$; **end**;
$u[0] := obj[0]$;
for $i := 1$ **step** 1 **until** m **do**

begin $k := ord[i];$ **if** $k < -m$ **then** $k := k + vz;$
if $case \neq 13$ **then** $u[k] := x[i];$
  $u[0] := u[0] + x[i] \times obj[k]$
end; **goto** *end*;

comment Output of basic solution for the right hand side,
  $u[i], i = 1, \ldots, m, b$ remains unchanged;
l8:

$p(2, 0, u); \ p(4, 0, u);$
for $i := 1$ **step** 1 **until** m **do** $x[i] := u[i];$
goto *l71*;

comment Computation of a feasible basis;
l9:

$stepmax := u[0] + step;$
$is := 1;$
l91:

$con := \textbf{true}; \ u[0] := step;$ **if** sol **then** $p(4, 1, b);$
$sol := \textbf{false}; \ h := 0;$
$bnorm;$
for $i := 1$ **step** 1 **until** m **do**
if $ord[i] \geq -m \land b[i] < -eps \times bmax$ **then**
begin $v[i] := 1/scale[ord[i]]; \ con := \textbf{false end}$
else $v[i] := 0;$
if con **then goto if** $case = 14$ **then** $l14$ **else** *end*;
$colpiv (h1 < piv, \textbf{false}, \textbf{false});$
if $piv > -eps \times h3$ **then**
begin $u[0] := -1;$ **goto** *fail* **end**;
$lx; \ wact; \ rowpiv (w, \textbf{false}); \ transform;$
if $step < stepmax$ **then goto** *l91*;
goto *fail*;

comment Simplex step with variable $x[u[0]]$ entering the basis;
l10:

$kp := u[0];$ **if** $ad[kp] \leq 0$ **then**
begin $u[0] := 0;$ **goto** *fail* **end**;
if sol **then** $p(4, 1, b);$
$lx; \ wact; \ rowpiv (w, \textbf{false}); \ transform; \ is := 1;$
goto *end*;

comment Input objective function.
  $obj[i] = u[i], i = -m, \ldots, n;$
l11:

for $i := -m$ **step** 1 **until** n **do** $obj[i] := u[i];$
$is := 1; \ objrow := 0;$
goto *end*;

comment Input of row $u[0]$ of aa as objective function;
l12:

for $i := -m$ **step** 1 **until** n **do** $obj[i] := 0;$

$objrow := u[0];$
goto *end*;

comment Output of reduced costs
If $u[0]=0$ only the components belonging to nonzero variables
are stored in u;

l13:

vt; *colpiv* (**false, true,** $u[0] \neq 0$);
$u[0] := 1$; **goto** *l7*;

comment Computation of an optimal basis;

l14:

$bnorm$;
vt; *colpiv* $(h1 < piv,$ **true, false**);
$u[0] := step$; **if** $piv > -eps$ **then goto** *end*;
lx; **if** *sol* **then** $p(4, 1, b)$; *wact*; *rowpiv* $(w,$ **false**);
$is := ip$;
transform; **if** $step \leq stepmax$ **then goto** *l14*;
goto *fail*;

comment Dual simplex step with variable $x[u[0]]$
leaving the basis;

l15:

for $ip := 1$ **step** 1 **until** m **do**
if $ord[ip] = u[0]$ **then goto** *l151*;
goto *fail*;

l151:

vt; *colpiv* (**false, true, false**);
for $i := 1$ **step** 1 **until** m **do** $v[i] := 0$;
$v[ip] := 1$; $kp := 1$; $is := ip$;
$piv := 0$;
colpiv $(h1 < 0 \wedge h1 \times u[kp] \times scale[kp] \leq piv \times scale[k] \times u[k],$
false, false); $is := 1$;
if $piv \geq 0$ **then**
begin $u[0] := -1$; **goto** *fail* **end**;
lx; *transform*;
goto *end*;

end:

end *procedure problem*;
for $k := 1$ **step** 1 **until** n **do**
begin $ad[k] := 1$; $obj[k] := 0$ **end**;
for $i := 1$ **step** 1 **until** m **do**
begin $ord[i] := -i$; $ad[-i] := -1$; $obj[-i] := 0$ **end**;
$objrow := step := wz := 0$; $vz := m+n+1$; $ad[0] := 0$; $obj[0] := 0$;
$con :=$ **false**; $sol :=$ **true**; $is := 1$; $bmax := 1$;
$p(1, 0, x)$;
problist (*problem*);
end *procedure lp*;

4.2. The Procedure ap.

procedure $ap\,(wz,\ x,\ k,\ sk)$;
  value $wz,\ k$;
  integer $wz,\ k$;
  array x;
  real sk;
begin
  comment Handling of the matrix aa;
  real h; **integer** i;
  if $wz \geq 0$ **then**
  begin
    comment $sk := x^T \times$ column no. k of aa;
    $h := 0$; **for** $i := 1$ **step** 1 **until** m **do** $h := h + x[i] \times aa[i, k]$;
    $sk := h$;
  end else
  begin
    comment $x :=$ column no. k of aa;
    for $i := 1$ **step** 1 **until** m **do** $x[i] := aa[i, k]$
  end
end procedure ap;

4.3. The Procedure p

procedure $p\,(case,\ ind,\ x)$;
  value $case,\ ind$;
  integer $case,\ ind$;
  array x;
begin
comment Execution of all operations concerning the matrices $L,\ R$ in the de-
        composition

$$L \times AB = R,$$

    where AB is the matrix of all basic columns of the constraint matrix A
    and R is upper triangular, the matrices R, L and the transformed right-
    hand sides BB are in the array $lr = (R, L, BB)$. The parameter $case$ indi-
    cates the operations to be performed, ind and x are parameters of
    variable meaning;
  integer $i,\ j,\ k,\ m2$;
  real $h1,\ piv$; **array** $h[1 : m]$;
  switch $plabel := p1,\ p2,\ p3,\ p4,\ p5,\ p6,\ p7$;
  $m2 := 2 \times m + rs$; **goto** $plabel\ [case]$;

  comment Initial basis;
$p1$:
  for $i := 1$ **step** 1 **until** m **do**
  begin for $k := i + 1$ **step** 1 **until** $m2$ **do**
    $rl[i, k] := 0$;
    $rl[i, i] := rl[i, m + i] := 1$
  end; **goto** $pend$;

    comment $x := L \times x$;
$p2$:
    if $ind = 0$ **then**
    begin
        for $i := 1$ **step** 1 **until** m **do** $h[i] := 0$;
        for $k := 1$ **step** 1 **until** m **do**
        if $x[k] \neq 0$ **then**
        begin for $i := 1$ **step** 1 **until** m **do** $h[i] := h[i] + rl[i, k+m] \times x[k]$ **end**
    end else
    begin $ind := ind + m$;
        for $i := 1$ **step** 1 **until** m **do** $x[i] := rl[i, ind]$;
        goto $pend$
    end;
$p21$:
    for $i := 1$ **step** 1 **until** m **do** $x[i] := h[i]$;
    goto pend;

    comment $x := L^T x$;
$p3$:
    for $i := 1$ **step** 1 **until** m **do** $h[i] := 0$;
    for $k := 1$ **step** 1 **until** m **do**
    if $x[k] \neq 0$ **then**
    for $i := 1$ **step** 1 **until** m **do** $h[i] := h[i] + x[k] \times rl[k, i+m]$;
    goto $p21$;
$p4$:
    if $ind \neq 0$ **then**
    begin comment $x :=$ transformed right hand side;
        $ind := $ **if** $ind \geq 0$ **then** $2 \times m + ind$ **else** $m - ind$;
        for $i := 1$ **step** 1 **until** m **do** $x[i] := rl[i, ind]$
    end;

comment $x := R^{-1} \times x$;
    for $i := m$ **step** -1 **until** 1 **do**
    begin for $k := i + 1$ **step** 1 **until** m **do**
        $x[i] := x[i] - rl[i, k] \times x[k]$;
        $x[i] := x[i]/rl[i, i]$
    end; **goto** $pend$;

comment $x := (R^{-1})^T x$, the components $x[1]$, ..., $x[ind-1]$ are unchanged
          from previous step, only the components $x[ind]$, ..., $x[m]$
          must be computed;
$p5$:
    for $j := 1$ **step** 1 **until** m **do if** $x[j] \neq 0$ **then goto** $p51$;
    goto $pend$;
$p51$:
    if $ind < j$ **then** $ind := j$;
    for $i := ind$ **step** 1 **until** m **do**

begin for $k := i - 1$ **step** -1 **until** j **do**
    $x[i] := x[i] - rl[k, i] \times x[k];$
    $x[i] := x[i]/rl[i, i]$
end; **goto** *pend*;

comment Exchange of basic variables. *ind* is the index of pivot row, x is the
        transformed new basic column, Gaussian elimination is used;
p6:
    for $k := ind$ **step** 1 **until** $m - 1$ **do**
    begin for $i := 1$ **step** 1 **until** k **do** $rl[i, k] := rl[i, k+1];$
        if $rl[k+1, k+1] \neq 0$ **then**
        begin if $abs(rl[k, k]) > abs(rl[k+1, k+1])$ **then**
            begin $piv := rl[k+1, k+1]/rl[k, k];$
                for $i := k+1$ **step** 1 **until** $m2$ **do**
                $rl[k+1, i] := rl[k+1, i] - piv \times rl[k, i];$
                $x[k+1] := x[k+1] - piv \times x[k]$
            end else
            begin $piv := rl[k, k]/rl[k+1, k+1];$ $rl[k, k] := rl[k+1, k+1];$
                for $i := k+1$ **step** 1 **until** $m2$ **do**
                begin $h1 := rl[k, i];$ $rl[k, i] := rl[k+1, i];$
                $rl[k+1, i] := h1 - piv \times rl[k, i]$
                end $i;$ $h1 := x[k];$ $x[k] := x[k+1];$
                $x[k+1] := h1 - piv \times x[k]$
            end
        end
    end $k;$ **for** $i := 1$ **step** 1 **until** m **do** $rl[i, m] := x[i];$
    goto *pend*;
p7:
    for $i := 1$ **step** 1 **until** m **do** $rl[i, 2 \times m + ind] := x[i];$
pend:
end *procedure* $p;$

5. Organisational and Notational Details

There are a few key variables in *lp* which should be explained for the readability of *lp*:

vz $vz := n + m + 1$

step   an integer variable counting the number of exchange steps executed after calling *lp*.

objrow an integer; if *objrow* > 0 then row no. *objrow* of AA gives the coefficients of the current objective function (compare Section 2.3.12);
       if *objrow* $= 0$ then the objective function is stored as the array *obj*.

obj   array of format *obj* $[-m:n]$, contains the coefficients of the objective function if *objrow* $= 0$, otherwise *obj* $= 0$.

v   array of format $v[1:m]$, contains the vector of simplex multipliers.

scale array of format *scale* $[-m:n];$ *scale* $[i]$ is the scaling factor for column i of the matrix A (1.1.1 b') (compare Section 2.3.1).

ord   integer array of format *ord* [1:m], contains information about the index
         vector J belonging to the current basis and about the status of the current
         basic variables: *ord* $[i] \geq -m$: The variable x_k, $k:= obj [i]$, is a non-
         negative variable. *ord* $[i] < -m$: The variable x_k, $k:= obj [i] + n + m + 1$,
         is a free variable.
         Moreover, in both cases, the i-th component j_i of J is $j_i := k$.

sol   a boolean variable; if *sol* = **false** then the array b contains the basic
         solution (i.e. $b[i] := \hat{b}_i = \bar{x}_{j_i}$ (see 1.2.4), where $\bar{x}$ is the basic solution
         belonging to the current basis characterized by the index vector $J = $
         $(j_1, \ldots, j_m)$ and to the first right hand side of (1.1.1)). If *sol* = **true** the
         array b contains a basic solution to a previous basis.

b   an auxiliary array of format $b[1:m]$, contains the current basic solution,
         if *sol* = **false** (see above).

kp   integer specifying the index of the variable to enter the current basis
         (=index of pivot column).

ip   integer specifying the index of the pivot row in each exchange step: If J
         describes the current basis, the variable $x_{j_{ip}}$ is to leave the basis.

The main portion of the body of *lp* is occupied by the declaration of the procedure *problem*. Those parts of *problem* following the labels *l1–l15*, correspond to
the different values of the input parameter *case* of *problem* and have been described
in Sections 2.3.1–2.3.15. There are several subprocedures of *problem*: *rowpiv*,
colpiv, *lx*, *wact*, *vt*, *transform* and *bnorm*. The purpose of these subprocedures will
be clear from the comments associated with them. Only *rowpiv* and *colpiv* need
further explanation. The index *kp* of the variable x_{kp} to enter the current basis
is computed within *colpiv*. The pivot selection rule according to which *kp* is to
be calculated, is specified by the boolean input parameter *b1* of *colpiv* (Note that
b1 is a parameter called by name!). Likewise, the index *ip* of the pivot row and
thereby the variable $x_{j_{ip}}$ to leave the current basis is calculated in the procedure
rowpiv. There, the pivot selection rule is given by the boolean input parameter *sr*
(called by value). The input array w of *rowpiv* gives the pivot column: $w[i] := y_i$,
where $y := A_J^{-1} A_r$, $r (= kp)$ the index of the pivot column (see (1.2.4)). The test
(1.2.5), whether or not $\hat{b}_i = 0$ is replaced by the test whether or not $|\hat{b}_i| < eps$
holds. Similarly, in Phase 1, the test (1.2.1), whether the component $\bar{x}_j$ of the
basic solution is negative, is replaced by checking $\bar{x}_j < -eps$; the tests of Phases 1
and 2 whether the smallest reduced cost u_r (see (1.2.3), (1.2.10)) is nonnegative,
is replaced by checking $u_r > -eps$.

    Note finally that by proper use of the scaling factors *scale* $[i]$ all pivot selection
rules and comparisons with *eps* are realized in a scaling invariant way. Thus, for
example instead of checking $|\hat{b}_i| < eps$, it is checked instead whether or not
$|\hat{b}_i| \leq eps \cdot bmax$, $bmax := \max_j |\hat{b}_j / scale [j]|$.

6. Numerical Properties

    If there were no round-off errors, in each stage, the matrices L, R, A_J satisfy
(1.3.3) exactly. With floating point arithmetic of relative precision ε, however,
one finds matrices L and R which satisfy (1.3.3) only approximately. It has been

shown by Bartels [1] that the matrices L, R found by the computer satisfy a relation of the form

$$(6.1) \qquad L \cdot (A_J + \Delta A_J) = R$$

with $\|\Delta A_J\|$ comparable in size to $\varepsilon \|A_J\|$, provided L becomes not too large, $\|L\| \gg 1$. As L is the product of matrices the components of which are bounded by 1 in modulus, one can prove only $\|L\| \leq 2^k$, if L is the product of k such matrices. In practice, however, one will find $\|L\| \approx 1$, and it is very unlikely that L grows like 2^k with k. Thus, as long as $\|L\| \approx 1$, the solutions $\bar{x}_J^{(i)}$, y, z, v of the linear equations (1.3.5) calculated by the machine can be interpreted as the exact solutions of (1.3.1) with the basis A_J (A_J^T) replaced by a slightly disturbed matrix $A_J + \Delta A_J$ $(A_J^T + \widetilde{\Delta A_J^T})$. This numerical behavior compares favorably with the usual schemes for handling basis exchanges, which are based more or less on the explicit use of the inverse A_J^{-1} of the basis A_J.

For example, with the familiar revised simplex method, the inverse A_J^{-1} of the basis A_J is to be calculated for each stage of the algorithm. If

$$J = \{j_1, \ldots, j_s, \ldots, j_m\}, \; j_s = l,$$

and

$$J' := \{j_1, \ldots, j_{s-1}, r, j_{s+1}, \ldots, j_m\}$$

then the inverse $A_{J'}^{-1}$ of the adjacent basis $A_{J'}$ is computed from A_J^{-1} recursively by premultiplying A_J^{-1} by the Frobenius-matrix F (1.3.12) (with y defined by (1.3.1): $y = A_J^{-1} A_r$).

Since the pair of indices (l, r) and thereby the pivot element y_s of F (1.3.12) is usually chosen only so as to ensure the finiteness of Phases 1,2 (1.2), the pivot y_s may be very small (and F very large) and is not necessarily the numerically best pivot. For this reason, the influence of round off errors may be magnified substantially by a numerically bad pivot choice. Moreover, the bad condition of previous bases A_J limits the accuracy of *all* succeeding inverse bases $A_{J'}^{-1}$: It has been shown by Wilkinson [4, 5] that with floating point arithmetic of relative precision ε one can at best expect to compute an approximate inverse $\tilde{A}_J^{-1}$ of a matrix A_J with $\|\tilde{A}_J^{-1} A_J - I\| \approx \varepsilon \cdot cond(A_J)$. Since $\tilde{A}_J^{-1}$ instead of A_J^{-1} is taken to compute the next inverse $A_{J'}^{-1}$, one gets at best an approximate inverse $\tilde{A}_{J'}^{-1}$ with

$$\|\tilde{A}_{J'}^{-1} A_{J'} - I\| \approx \varepsilon \max \left(cond(A_J), cond(A_{J'}) \right).$$

That is, the precision of $\tilde{A}_{J'}^{-1}$ may be very poor even if $A_{J'}$ is well conditioned. This gradual loss of precision has been observed in practice and one usually tries to overcome it by such devices as multiple precision arithmetic and "sufficiently many" reinversions, that is computation of A_J^{-1} from scratch from time to time.

The following examples are to illustrate the numerical behaviour of the triangularisation method and of the inverse basis method. We set $m = n$ and take as matrix AA (1.1.1) the ill conditioned square matrix given by

$$aa_{ik} = \frac{1}{i+k}, \quad i, k = 1, \ldots, m,$$

and as right hand side

$$b = (b_1, \ldots, b_m)^T, \quad b_i = \sum_{k=1}^{m} \frac{1}{i+k}.$$

Consider the bases given by

$$J_1 := \{-1, -2, \ldots, -m\}, \quad J_2 := \{1, 2, \ldots, m\}.$$

The exact basic solution with respect to J_1 then is given by $\bar{x}_{-i} = b_i$, $i = 1, \ldots, m$, $\bar{x}_i = 0$, $i = 1, 2, \ldots, m$ and the exact basic solution with respect to J_2 is given by $\bar{x}_i = 1$, $i = 1, \ldots, m$, $\bar{x}_i = 0$, $i = -1, \ldots, -m$. We now list the basic solutions obtained by the triangularisation method and the inverse basis method by starting with the basis $A_{J_1} = I$, transforming it by suitable exchange steps into the basis A_{J_2} and then returning by another series of exchange steps to the original basis A_{J_1} (relative machine precision: $\varepsilon \approx 10^{-11}$):

Basis	Exact basic solution		Inverse basis method		Triangularisation method	
Example 1: $m = 5$						
J_2:						
x_1	1		1.000000182_{10}	0	1.0000000786_{10}	0
x_2	1		$9.99999 84079_{10}$	-1	$9.99999 16035_{10}$	-1
x_3	1		1.0000004372_{10}	0	1.0000027212_{10}	0
x_4	1		$9.99999 52118_{10}$	-1	$9.99996 56491_{10}$	-1
x_5	1		1.0000001826_{10}	0	1.0000014837_{10}	0
J_1:						
x_{-1}	1.4500000000_{10}	0	1.4500010511_{10}	0	1.4500000000_{10}	0
x_{-2}	1.0928571428_{10}	0	1.0928579972_{10}	0	1.0928571427_{10}	0
x_{-3}	8.8452380952_{10}	-1	8.8452453057_{10}	-1	8.8452380950_{10}	-1
x_{-4}	7.4563492063_{10}	-1	7.4563554473_{10}	-1	7.4563492060_{10}	-1
x_{-5}	6.4563492063_{10}	-1	6.4563547103_{10}	-1	6.4563492059_{10}	-1
Example 2: $m = 10$						
J_2:						
x_1	1		9.9952516281_{10}	-1	9.9986378918_{10}	-1
x_2	1		1.0185519965_{10}	0	1.0056860944_{10}	0
x_3	1		7.5967519456_{10}	-1	9.1962193606_{10}	-1
x_4	1		2.5020319231_{10}	0	1.5558956482_{10}	0
x_5	1		-4.2655020405_{10}	0	-1.1771963556_{10}	0
x_6	1		1.2085520735_{10}	1	6.1571720228_{10}	0
x_7	1		-1.3310945894_{10}	1	-6.5330762136_{10}	0
x_8	1		1.2081398560_{10}	1	7.6335537514_{10}	0
x_9	1		-3.7178737849_{10}	0	-2.2281412212_{10}	0
x_{10}	1		1.8476072006_{10}	0	1.6666269133_{10}	0
J_1:						
x_1	2.0198773448	0	-1.1999585030_{10}	0	2.0198773448_{10}	0
x_2	1.6032106781	0	-1.1147413211_{10}	0	1.6032106780_{10}	0
x_3	1.3468004217		-9.8764785557_{10}	-1	1.3468004217_{10}	0
x_4	1.1682289932	0	-8.9107523957_{10}	-1	1.1682289933_{10}	0
x_5	1.0348956598	0	-7.9808787783_{10}	-1	1.0348956598_{10}	0
x_6	9.3072899322	-1	-7.1919673857_{10}	-1	9.3072899317_{10}	-1
x_7	8.4669537977_{10}	-1	-6.5234442973_{10}	-1	8.4669537978_{10}	-1
x_8	7.7250093533_{10}	-1	-5.9543089303_{10}	-1	7.7250093530_{10}	-1
x_9	7.1877140316_{10}	-1	-5.4665179294_{10}	-1	7.1877140320_{10}	-1
x_{10}	6.6877140316_{10}	-1	-5.0453744089_{10}	-1	6.6877140328_{10}	-1

One finds that at the ill conditioned basis A_{J_2} the basic solutions given by both methods are inaccurate to the same extent: This inaccuracy reflects the bad condition of A_{J_2} and cannot be avoided except by using higher precision arithmetic.

But after returning to the original well conditioned basis $A_{J_1} = I$ via A_{J_2} there is a tremendous difference: the solution given by the triangularisation method is accurate almost to the last digit whereas the solution given by the inverse basis method is wrong to the same extent as the solution for the basis A_{J_2} (see, in particular the second example): This shows that in the inverse basis method the bad condition of one particular basis limits the accuracy of *all* further bases, whereas this is not the case with the triangularisation method.

7. Examples of the Use of the Procedure *lp*. Test Results

The procedure was tested on the TR 4 (word length: 38 bits in the mantissa) of the Leibnizrechenzentrum der Bayerischen Akademie der Wissenschaften in Munich; the test results have been confirmed at the computing center of the University of Würzburg.

The following relatively complicated example shows the flexibility of the programs. The corresponding procedure *problist* is given explicitly in order to illustrate the typical usage of various calls of *problem* as discussed in Section 2.3.

Let $m = 4$, $n = 3$ and let the matrix AA be given by

$$
AA = \begin{bmatrix} 1 & 1 & 1 \\ -4 & -3 & 2 \\ 2 & 1 & -1 \\ 0 & 0 & 0 \end{bmatrix} \qquad b := \begin{bmatrix} 9 \\ -16 \\ 7 \\ 0 \end{bmatrix}.
$$

Now, consider the following problems (The numbering of the problem is repeated in the corresponding comments of the procedure *problist* given below):

(1) Set $I_+ := \{-2, -1, 1, 2, 3\}$, $I^0 := \{-3\}$, $I^\pm := \{-4\}$.

Then the associated system (1.1.1 b, c) of linear inequalities corresponds to

$$
9 - x_1 - x_2 - x_3 = x_{-1} \geq 0
$$

$$
-16 + 4x_1 + 3x_2 - 2x_3 = x_{-2} \geq 0
$$

(7.1)
$$
7 - 2x_1 - x_2 + x_3 = x_{-3} = 0
$$

$$
0 + 0x_1 + 0x_2 + 0x_3 = x_{-4}
$$

$$
x_1 \geq 0, \qquad x_2 \geq 0, \qquad x_3 \geq 0.
$$

(2) Take as right hand side $b^{(1)} = b$ the vector given above and find a feasible (basic) solution $\bar{x}$ to (7.1) and to this right hand side.

Results.

i	$\bar{x}_i$
-4	0
-3	0
-1	4.5000000000
1	2.5000000000
2	2.0000000000
3	0

value of objective function: 0

(3) Minimize the objective function

$$-x_1 - 2 \cdot x_2$$

subject to the constraints (7.1).

(3 a) Find all reduced costs u_i belonging to the optimal basis.

(3 b) Find an optimal basic solution $\bar{x}$.

Results.

i	u_i	$\bar{x}_i$
-4	0	0
-3	1.0000000000	0
-2	0	6.0000000000
-1	1.0000000000	0
1	2.0000000000	0
2	0	8.0000000000
3	0	1.0000000000

value of objective function $= -1.6000000000_{10}1$

(4) Parametric programming. Consider the right hand side

$$\tilde{b}(p) := b - p\,e_1, \quad e_1 = \begin{pmatrix} 1 \\ 0 \\ 0 \\ 0 \end{pmatrix}, \quad p \geq 0$$

where p is a real nonnegative parameter. Find all optimal solutions (depending on p) of the previous linear program but with the right hand side $\tilde{b}(p)$ for $0 \leq p \leq 6$.

Results.

i	$p_1 = 2.0000000000$ 1st breakpoint $\bar{x}_i^{(1)}$	$p_2 = 4.4999999999$ 2nd breakpoint $\bar{x}_i^{(2)}$
-4	0	0
-3	0	0
-2	5.0000000000	0
-1	0	0
1	0	2.4999999999
2	7.0000000000	2.0000000001
3	0	0

value of objective function $\quad -1.4000000000_{10}1 \qquad -6.5000000001$

There is no feasible solution for $p > p_2$.

Interpretation of the results: If $p_0 := 0$ and $\bar{x}^{(0)}$ is the optimal solution given before under (3) and if $p_i > p > p_{i+1}$, then the optimal solution belonging to the right hand side $\tilde{b}(p)$ is the convex mean

$$\frac{p_{i+1} - p_i}{p_{i+1} - p_i} \bar{x}^{(i)} + \frac{p - p_i}{p_{i+1} - p_i} \bar{x}^{(i+1)}$$

of $\bar{x}^{(i)}$ and $\bar{x}^{(i+1)}$.

(5) Now, let x_1 become a free variable and find the optimal solution $\bar{x}$ of the associated linear program, i.e. the optimal solution of

minimize $\qquad - x_1 - 2 x_2$

subject to:

$$9 - x_1 - x_2 - x_3 = x_{-1} \geq 0$$
$$-16 - 4 x_1 + 3 x_2 - 2 x_3 = x_{-2} \geq 0$$
$$7 - 2 x_1 - x_2 + x_3 = x_{-3} = 0$$
$$0 + 0 x_1 + 0 x_2 + 0 x_3 = x_{-4}$$

(7.2)

$$x_2 \geq 0, \qquad x_3 \geq 0.$$

Results.

i	$\bar{x}_i$
-4	0
-3	0
-2	$9.000000 0002$
-1	0
1	$-2.000000 0000$
2	$1.100000 0000_{10} 1$
3	0

value of objective function $\quad -2.000000 0000_{10} 1$

(6) Replace the right hand side b by the vector $\tilde{b} = (9, -16, 7, 15)^T$, the 4-th constraint

$$0 + 0 x_1 + 0 x_2 + 0 x_3 = x_{-4}$$

by

$$15 - 2 x_1 - 5 x_2 + 0 x_3 = x_{-4} \geq 0$$

and let x_1 again be a nonnegative variable. Find the optimal solution $\bar{x}$ of the resulting linear program, i.e. of

minimize $\qquad - x_1 - 2 x_2$

subject to:

$$9 - x_1 - x_2 - x_3 = x_{-1} \geq 0$$
$$-16 + 4 x_1 + 3 x_2 - 2 x_3 = x_{-2} \geq 0$$
$$7 - 2 x_1 - x_2 + x_3 = x_{-3} = 0$$
$$15 - 2 x_1 - 5 x_2 + 0 x_3 = x_{-4} \geq 0$$
$$x_1 \geq 0, \qquad x_2 \geq 0, \qquad x_3 \geq 0.$$

Results.

i	$\bar{x}_i$
-4	0
-3	0
-2	0
-1	4.5000000005
1	2.4999999997
2	2.0000000001
3	$-4.0745362627_{10} - 10$

value of objective function -6.4999999999

Remark. It should be noted at this point that in addition to the several uses of *problem* as described in Section 2.3, there is a possibility to change a row or column of the constraint matrix AA in certain cases without having to restart *lp* with another matrix AA (that is with another procedure *ap*, in general).

More precisely, any column of the matrix AA which does not correspond to a basic variable can be easily replaced within *problist* if the procedures *ap* and *p* of this report are used: Of course, the following statement within *problist* will cause the replacement of column k of the matrix AA by the vector h, if the variable x_k is not basic:

for $i := 1$ **step** 1 **until** m **do**
$$aa[i, k] := h[i];$$

Likewise, row i of the matrix AA, that is the row vector $(aa_{i1}, \ldots, aa_{in})$, can be easily replaced by another row vector, say $(h_1, \ldots, h_n)$, at every stage of the exchange algorithm within *problist*, provided the variable x_{-i} corresponding to the row to be changed was always a free variable from the beginning of *lp* up to the change of row i. If additionally the procedures *ap* and *p* of this article are used (which are based on the standard representation of the matrices AA, $(R, L, \bar{B})$ by two-dimensional arrays, see 2.4, 2.5, small print), then this replacement together with the necessary changes of $(R, L, \bar{B})$ can be effected by the following piece of program within *problist* (compare part (6) of the preceding example and the corresponding procedure *problist*):

```
for k := step 1 until n do
    aa[i, k] := h[k];
u[0] := 0; problem (1, u, fail);
problem (3, u, fail); j := − u[−i];
for k := 1 step 1 until n do
    if u[k] < 0 then rl [j, − u[k]] := h[k];
```

Here, it is supposed that u is an array of format $u[-m:n]$ at least.

ALGOL Program for the Procedure *problist*

procedure *problist* (*problem*);
    procedure *problem*;

```
begin
    integer i, k, j;
    real min, p;
    array u, v, w [−m:n], b [0:m];

    comment Scaling;
    u [0] := 0; problem (1, u, fail);

    comment (1) Variables no. −2, −1, 1, 2, 3 are nonnegative,
                variable no. −3 is a zero-variable,
                variable no. −4 is a free variable;
    for i := −m step 1 until n do v[i] := 1;
    v[−3] := 0; v[−4] := −1; problem (2, v, fail);

    comment (2) Input right hand side;
        for i := 1 step 1 until m do read (b[i]);
    b[0] := 1; problem (5, b, fail);

    comment (2) Feasible basis;
    u[0] := 10;
    problem (9, u, fail);

    comment (2) Output feasible solution;
    u[0] := 1; problem (7, u, fail);
    for i := −m step 1 until n do print (i, u[i]);

    comment (3) Input objective function;
    for i := −m step 1 until n do read (u[i]);
    problem (11, u, fail);

    comment (3) Computation of optimal basis, output of optimal solution;
    u[0] := 10; problem (14, u, fail);
    u[0] := 1; problem (13, u, fail);
    for i := −m step 1 until n do print (i, u[i]);
    u[0] := 1; problem (7, u, fail);
    for i := −m step 1 until n do print (i, u[i]);

    comment (4) Parametric programming
                right hand side = b − e1 × p, 0 ≦ p ≦ 6,
                e1 = (1, 0, ..., 0);
    p := 0; for j := 1 step 1 until 100 do
    begin
    comment Output solution for right hand side e1;
        w[0] := −1; problem (7, w, fail); min := 6;
        for i := −m step 1 until −1, 1 step 1 until n do
        begin if u[i] < 0 then u[i] := 0;
            if u[i] < min × w[i] then
            begin min := u[i]/w[i]; k := i end;
        end;
        p := p + min; if p > 6 then
        begin min := min + 6 − p; p := 6 end; print (p);
```

```
    for i := −m step 1 until n do
    begin u[i] := u[i] − min × w[i];
        print (i, u[i]);
    end;
    if p = 6 then goto parend;
        w[0] := k; problem (15, w, parend);
    end parametric programming;
```

comment (5) $x[1]$ becomes a free variable;
parend:
    $v[1] := −1$; problem (2, v, fail);

comment (5) Computation of an optimal basis, output solution;
$u[0] := 10$;
problem (14, u, fail); $u[0] := 1$; problem (7, u, fail);
for $i := −m$ **step** 1 **until** n **do** print (i, u[i]);

comment (6) Input new right hand side;
for $i := 1$ **step** 1 **until** m **do** read (b[i]);
$b[0] := 1$; problem (5, b, fail);

comment (6) Change of row 4 of aa;
for $i := 1$ **step** 1 **until** n **do** read (aa[4, i]);
$u[0] := 0$; problem (1, u, fail);
problem (3, u, fail); $j := −u[−4]$;
for $i := 1$ **step** 1 **until** n **do**
if $u[i] < 0$ **then** $rl[j, −u[i]] := aa[4, i]$;
$v[−4] := 1$; $v[1] := 1$;
problem (2, v, fail);

comment (6) Computation of an optimal basis, output solution;
$u[0] := 10$;
problem (14, u, fail); $u[0] := 1$; problem (7, u, fail);
for $i := −m$ **step** 1 **until** n **do** print (i, u[i]);
fail:
end;

References

1. Bartels, R. H.: A numerical investigation of the simplex method. Technical Report No. CS 104, 1968, Computer Science Department, Stanford University, California.
2. — Golub, G. H.: The simplex method of linear programming using LU decomposition. Comm. ACM. **12**, 266−268 (1969).
3. Dantzig, G. B.: Linear programming and extensions. Princeton: Princeton University Press 1963.
4. Wilkinson, J. H.: Rounding errors in algebraic processes. London: Her Majesty's Stationery Office; Englewood Cliffs, N.Y.: Prentice Hall 1963. German edition: Rundungsfehler. Berlin-Heidelberg-New York: Springer 1969.
5. — The algebraic eigenvalue problem. London: Oxford University Press 1965.

Introduction to Part II

The Algebraic Eigenvalue Problem

by J. H. WILKINSON

1. Introduction

The standard algebraic eigenvalue problem, the determination of the non-trivial solutions of $Ax = \lambda x$, is one of the most fascinating of the basic problems of numerical analysis. In spite of the simplicity of its formulation many algorithms are required in order to deal efficiently with the wide range of problems which are encountered in practice.

A variety of factors enter into the determination of the most efficient algorithm for the solution of a given problem; the following are perhaps the most important.

i) Eigenvalues *and* eigenvectors may be required or the eigenvalues only.

ii) The *complete* set of eigenvalues and/or eigenvectors may be required or only a comparatively small number. In the latter case one may require, for example, the k smallest, the k largest, the k nearest to a given value or all those lying within a given range.

iii) The matrix may be symmetric or unsymmetric. In the former case we may include more general eigenproblems which can be reduced to the standard problem with a symmetric matrix.

iv) The matrix may be *sparse*, either with the non-zero elements concentrated on a narrow band centred on the diagonal or alternatively they may be distributed in a less systematic manner. We shall refer to a matrix as *dense* if the percentage of zero elements or its distribution is such as to make it uneconomic to take advantage of their presence.

v) For very large matrices economy of storage may well be the paramount consideration in selecting an algorithm.

vi) The results may be required to only a modest accuracy or high precision may be vital.

We remind the reader that in the algorithms presented in this book no reference is made to the use of a backing store such as a magnetic tape or disc. For this reason the storage requirements in practice may be different from what they appear to be. For example if eigenvalues and eigenvectors of a Hessenberg matrix H are found using procedure *hqr* followed by inverse iteration it is necessary to retain a copy of H while the *hqr* reduction is performed. In practice this copy could be written on a backing store without inconvenience, since it is not required until *hqr* is completed.

2. List of Procedures

To facilitate reference to the procedures they are listed below in alphabetical order, together with the chapters in which they are described. Sets of related procedures are grouped together.

3. Real, Dense, Symmetric Matrices

The computation of the eigensystem of a real symmetric matrix is much simpler than that of a general matrix since the eigenvalues of the former are always well determined.

3.1. Calculation of All the Eigenvalues with or without Eigenvectors

When all the eigenvalues and eigenvectors are required two programs based on the given procedures may be recommended.

a) The Method of Jacobi

The method of Jacobi is the *most elegant* of those developed for solving the complete eigenproblem. The procedure *jacobi* may be used to find all the eigenvalues with or without the eigenvectors. It is an extremely compact procedure and considerable care has been taken to ensure that both eigenvalues and eigenvectors are of the highest precision attainable with the word length that is used. It is not adapted for finding a few selected eigenvalues and eigenvectors and other procedures are generally more efficient for such a purpose.

The computed system of eigenvectors is always orthogonal, almost to working accuracy, and the orthogonality is in no way impaired by the presence of multiple or pathologically close eigenvalues. The accuracy of the orthogonality relation should not be taken to imply a corresponding accuracy in the individual eigenvectors themselves, but there are many applications in which orthogonality is of paramount importance.

b) Householder Tridiagonalization and the QL Algorithm

The *most efficient* program for finding all the eigenvalues alone or all the eigenvalues and eigenvectors is a combination of the Householder tridiagonalization and the QL algorithm. It is usually faster than *jacobi*, in particular if eigenvalues only are required. In this case the optimum combination is *tred 1* and *tql 1*; if economy of storage is paramount then *tred 3* should replace *tred 1* since in the former the original symmetric matrix is stored as a linear array of $\frac{1}{2} n (n+1)$ elements. If both values and vectors are required the optimum combination is *tred 2* and *tql 2*. In each case the appropriate combination is several times faster than *jacobi* except when n is fairly small, say less than ten. *tred 2* and *tql 2* are remarkably effective from the point of view of storage, only $n^2 + O(n)$ locations being required for the complete eigenproblem. The eigenvectors are accurately orthogonal independent of the separations.

The procedures *imtql 1* and *imtql 2* may replace *tql 1* and *tql 2* respectively. The speeds of the two sets of procedures are about the same and when the tridiagonal matrix has elements which diminish rapidly in absolute value as one moves down the diagonal the procedures give the smaller eigenvalues to higher accuracy. When *tql 1* and *2* are used on such matrices they should be orientated so that the smaller elements occur first along the diagonal, if necessary by reflecting in the secondary diagonal. These comments imply that the set *imtql 1* and *2* is preferable to the set *tql 1* and *2* and in practice this appears to be true. However there remains the danger that with *imtql 1* and *2* small subdiagonal elements may inhibit convergence, possibly as a result of underflow, but no such examples have been encountered in practice.

When a procedure is needed to solve the complete eigenproblem of a matrix of modest order as part of a larger main program, *jacobi* might well be preferred because of its compactness. (It is used in *ritzit* (see 5.) for this reason.)

3.2. Calculation of Selected Eigenvalues with or without Eigenvectors

When a comparatively small number of eigenvalues is required the methods of the previous section become inefficient. However it should be emphasized that the combinations *tred 1* and *tql 1* or *imtql 1* and *tred 2* and *tql 2* or *imtql 2* are very effective and when more than, say, 40% of the eigensystem is required the relevant combination will usually provide the optimum solution. Unfortunately, the QL transformation with the standard shift is not well adapted to finding a few specific eigenvalues, and it seems to be difficult to design an aesthetically satisfying shift technique which finds a number of the smallest eigenvalues (a common requirement).

The more effective procedures for finding specified eigenvalues are also based on a preliminary Householder reduction to tridiagonal form using *tred 1* or *tred 3*. Two of the published procedures are then relevant.

a) Givens' Sturm Property Procedure

The procedure *bisect* is based on the Sturm sequence property of the leading principal minors of $T - \lambda I$, where T is the tridiagonal matrix. The published procedure has been designed to find eigenvalues numbered *m1* to *m2*, where they are assumed to be ordered so that $\lambda_1 \leq \lambda_2 \leq \cdots \leq \lambda_n$. Simple modifications to the basic procedure may readily be designed to find, for example, all the eigenvalues in a given interval or the k eigenvalues nearest to a given value. Indeed the virtue of this algorithm is its extreme flexibility. (See also 3.2(c) for a further use of bisection.)

Since the localisation of individual eigenvalues proceeds by repeated bisection the accuracy may be readily varied. In fact the procedure could be speeded up by changing to a method with superlinear convergence after an eigenvalue has been located in an interval. (Modified successive linear interpolation and extrapolation is particularly effective). However, when the matrix has clusters of eigenvalues bisection will, in any case, be pursued until these eigenvalues have been quite closely located. This means that *bisect* is actually faster when there are multiple roots or tight clusters than when they are well separated.

b) Rational QR Transformation with Newton Shift

In the procedure *ratqr*, the normal shift technique of the QR algorithm is modified to a Newton correction such that convergence to the largest or smallest eigenvalues takes place. (With the usual QR shift technique the order in which the eigenvalues are found is largely unpredictable, though there is a tendency to find first the eigenvalue nearest to an eigenvalue of the 2×2 matrix in the bottom right-hand corner.) The shift technique also makes it possible to devise a square-root free version of the QR algorithm. Convergence is not so fast as with *tql 1,2* or *imtql 1,2*, but when a small number of the largest or smallest eigenvalues is required, *ratqr* is considerably more efficient. However, when the required eigenvalues are clustered, *bisect* will be faster since Newton's method has only linear convergence for multiple roots.

c) Calculation of Specified Eigenvectors of a Tridiagonal Matrix

When an accurate eigenvalue of a real symmetric tridiagonal matrix is known the corresponding eigenvector may be computed by inverse iteration and this provides perhaps the most efficient method of determining the eigenvectors corresponding to a relatively small number of selected eigenvalues. When the eigenvalues are well separated inverse iteration provides an elegant and efficient algorithm.

When eigenvectors corresponding to multiple or very close eigenvalues are required, the determination of fully independent eigenvectors (i.e. of eigenvectors giving full digital information) is quite difficult. The procedure *tristurm* is designed to calculate all the eigenvalues of a symmetric tridiagonal matrix lying in a prescribed interval, and to give the corresponding eigenvectors. The orthogonality of the computed vectors is prescribed by a parameter and to assist in the determination of orthogonal vectors, the organisation of the section dealing with the calculation of the eigenvalues themselves, although based on essentially the same principles as *bisect*, has special features which simplify the determination of orthogonal eigenvectors. When a comparatively small percentage of the eigenvalues and the corresponding eigenvectors are required *tristurm* is the most efficient procedure. The individual eigenvectors provided by *tristurm* are fully as accurate as those given by *jacobi*, *tql 2* or *imtql 2* but the orthogonality of the computed eigenvectors will be less satisfactory with *tristurm*. When we are concerned with eigenvectors associated with clusters of eigenvalues *tristurm* is relatively less efficient and less aesthetically satisfying because of the amount of time spent in the reorthogonalization procedures.

4. Symmetric Band Matrices

Matrices in which the non-zero elements are concentrated in a comparatively narrow band arise quite frequently in practice, usually from finite difference approximations to ordinary or partial differential equations. Generally it is inefficient to treat these by the algorithms recommended for dense matrices, and since it is not uncommon for such matrices to be of high order, storage considerations will often preclude this. In addition to the tridiagonal case, which we have already discussed, three procedures are published for dealing with symmetric band matrices in general.

4.1. Calculation of Eigenvalues

The eigenvalue of smallest modulus of a symmetric band matrix A may usually be computed using procedure *bqr*. (The nature of the QR algorithm is such that no absolute guarantee can be given that the computed eigenvalue really is that of minimum modulus but there is a very high probability that this will be true.) This algorithm may be used to find the eigenvalue nearest to an assigned value t by working with $A - tI$. The procedure is designed so that the m eigenvalues nearest to a given value t may be found by m successive calls of *bqr*, the order of the matrix being diminished after each call. If confirmation is needed that the required eigenvalues have been determined the procedure *bandet 2* may be used, since this gives the number of eigenvalues greater than any assigned value k.

An alternative procedure is *bandrd* which reduces the given matrix to symmetric tridiagonal form. The eigenvalues may then be found by *tql 1*, *imtql 1*,

bisect or *ratqr* as described earlier. The use of *bandrd* therefore gives great flexibility, particularly when combined with *bisect*. *bandrd* is more efficient then *bqr*
when any appreciable number of eigenvalues is required.

4.2. Calculation of the Eigenvectors

When eigenvalues of a band matrix have been determined by one of the
preceding algorithms the corresponding eigenvectors may be found by inverse
iteration using the procedure *symray*. This procedure has the additional refinement of computing improved eigenvalues (using the Rayleigh quotient); these
improved values are usually correct to more than single precision.

5. Simultaneous Determination of Dominant Eigenvalues and Eigenvectors of a Symmetric Sparse Matrix

The procedure *ritzit* is designed to find a number of the eigenvalues of maximum modulus and the corresponding eigenvectors of a real symmetric matrix.
The matrix A need not be stored explicitly; all that is needed is a procedure for
computing Ax from a given vector x. This may result in a remarkable economy
of storage. If, for example, A is the matrix arising from a finite difference approximation to a linear partial differential equation with constant coefficients many
of its rows will be the same and will contain very few non-zero elements.

By premultiplying with A^{-1} rather than A the eigenvalues of minimum
modulus may be determined. More generally by iterating with $(A-pI)^{-1}$ the
eigenvalues nearest to p may be determined. In order to take advantage of the
sparseness of A it is essential not to compute $(A-pI)^{-1}$ explicitly: $(A-pI)^{-1}x$
is computed by solving $(A-pI)y=x$. Obviously this algorithm can be used
when A is a narrow band matrix.

Convergence of *ritzit* for the determination of the m absolutely largest eigenvalues is linear and governed by the separation ratio between the smallest of
the considered eigenvalues and the largest remaining one. Therefore, it will
often be advantageous to determine more eigenvalues than are wanted by finding
a complete cluster rather than trying to separate it.

Clearly the relative effectiveness of the procedure depends on the distribution
of the eigenvalues and on the *a priori* knowledge one has of them, since this
enables one to choose m efficiently. In practice there are many situations in which
this is an extremely efficient algorithm. It is the only algorithm in the series
which deals with matrices of an arbitrarily sparse nature.

6. The Generalized Symmetric Eigenvalue Problems
$Ax=\lambda Bx$ and $ABx=\lambda x$

A common problem in practice is the determination of the eigensystems of
$Ax=\lambda Bx$ and $ABx=\lambda x$ where A and B are symmetric and B is positive
definite. Two of the published algorithms are directly relevant in this case.

a) A and B Dense Matrices

The procedures *reduc 1* and *reduc 2* by means of a Cholesky decomposition
of B reduce the problems $Ax=\lambda Bx$ and $ABx=\lambda x$ respectively to the standard

eigenproblem for a symmetric matrix P. (Notice that for the case $Ax = \lambda Bx$, the corresponding matrix P is dense even when A and B are narrow band matrices and in the other case the band widths are added.) The eigenvalue problem for P can then be solved by the methods discussed in Section 3, and accordingly we may find all the eigenvalues or selected eigenvalues with or without the corresponding vectors. An eigenvector of the original problem may be determined from those of the matrix P by the procedure *rebak a* or *rebak b*.

b) A and B Band Matrices

When A and B are narrow band matrices the procedure *ritzit* may be used to compute a number of the eigenvalues of largest modulus. To do this an auxiliary procedure is needed for the Cholesky factorization of B into the product LL^T, where L is a lower triangular band matrix. To find the eigenvalues and eigenvectors of $Ax = \lambda Bx$, *ritzit* is used with $L^{-1}AL^{-T}$ as the iteration matrix. Premultiplication of a vector x with $L^{-1}AL^{-T}$ is performed in the steps

$$L^T y = x, \quad A y = z, \quad L w = z$$

so that full advantage is taken of the band form of A and L. Similarly for the problem $ABx = \lambda x$, *ritzit* is used with $L^T A L$ as the iteration matrix.

7. Hermitian Matrices

No procedures are included to deal specifically with a complex Hermitian matrix $A + iB$, though complex analogues of the published procedures could be designed. Such problems may be dealt with by working with the real symmetric matrix of order $2n$ defined by

$$\begin{bmatrix} A & -B \\ B & A \end{bmatrix}.$$

If the eigenvalues of the Hermitian matrix are $\lambda_1, \lambda_2, \ldots, \lambda_n$ those of the augmented matrix are $\lambda_1, \lambda_1, \lambda_2, \lambda_2, \ldots, \lambda_n, \lambda_n$. The presence of the double eigenvalues does not lead to a loss of accuracy. If w is one eigenvector of the augmented matrix corresponding to some (double) eigenvalue λ and w^T is partitioned in the form (u^T, v^T) then $u + iv$ is the eigenvector of $A + iB$ corresponding to λ. The other (orthogonal) eigenvector of the augmented matrix turns out to be $(-v^T, u^T)$ and gives up to a factor i the same eigenvector of $A + iB$. If (u^T, v^T) is normalized, so is $u + iv$. The augmented matrix requires twice as much storage as the complex matrix from which it is derived and with efficient coding the analogous complex procedure could usually be about twice as fast as the real procedure applied to the augmented matrix. However, experience suggests that many "complex packages" are so inefficient that it is not uncommon for the real procedure to be the faster.

8. Real Dense Unsymmetric Matrices

It is impossible to design procedures for solving the general unsymmetric eigenproblem which are as satisfactory numerically as those for symmetric (or Hermitian) matrices, since the eigenvalues themselves may be very sensitive to small changes in the matrix elements. Moreover the matrix may be defective in which case there is no complete set of eigenvectors.

In practice it is virtually impossible to take this into account unless one deals with matrices having integer elements and rational eigenvalues and eigenvectors. With procedures involving rounding errors one cannot demonstrate that a matrix is defective and the published procedures always attempt to find the same number of eigenvectors as eigenvalues. It will often be evident from the computed eigenvectors that the matrix is defective; the corresponding eigenvectors will be almost parallel. It should be emphasized that the computed eigenvalues corresponding to a non-linear elementary divisor will often be by no means close.

When a matrix is defective or close to a defective matrix, it would be more satisfactory to determine invariant subspaces associated with groups of eigenvalues. The most difficult practical problem in designing such a procedure is to determine which eigenvalues to group together. The larger one takes the groups the easier it becomes to determine accurate invariant subspaces, but if the groups are made unnecessarily large then less information is made available. Algorithms of this type are still under development.

The determination of an accurate eigensystem may present practical difficulties if the matrix is badly balanced, that is, if corresponding rows and columns have very different norms. Before using any of the algorithms described in this section it is advisable to balance the matrix using the procedure *balance*. This will usually give a substantial improvement in the accuracy of the computed eigenvalues when the original matrix is badly balanced. If the original matrix is fairly well balanced the use of this procedure will have little effect but since the time taken is quite small compared with that required to find the eigensystem, it is a good discipline to use *balance* in all cases. The procedure *balance* also recognises 'isolated' eigenvalues, that is eigenvalues which are available by inspection without any computation and its use ensures that such eigenvalues are determined exactly *however ill conditioned they may be*.

8.1. Calculation of All the Eigenvalues with or without Eigenvectors

a) The Generalized Jacobi Method

The procedure *eigen* is based on a generalization of the Jacobi method for real symmetric matrices. It seems unlikely that there is a generalization which retains much of the elegance of the Jacobi method but *eigen* is very compact and gives results which, in general, are almost as accurate as can be expected from the sensitivity of the original problem. It is designed to give all the eigenvalues, and to give either the left-hand system of eigenvectors or the right-hand system or neither, as required. The use of eigen is recommended only for matrices expected to have real eigenvalues (see 10.1 (c)).

b) The QR Algorithm

The alternative algorithm for finding all eigenvalues and eigenvectors of a general matrix is based on a combination of a reduction to Hessenberg form using one of the pairs of procedures *elmhes* and *elmtrans*, *dirhes* and *dirtrans*, or *orthes* and *ortrans* followed by the QR algorithm *hqr 2*. If the procedure *balance* has been used, the eigenvectors of the original matrix are recovered from those of the balanced matrix using the procedure *balbak*. If the eigenvalues only are required then *hqr* is used in place of *hqr 2*.

The choice between *elmhes*, *dirhes* and *orthes* depends on the following considerations. *orthes* has the advantage of leaving the condition numbers of the individual eigenvalues unaltered. It is therefore valuable when one is interested in analysing a matrix rather carefully. It requires twice as much computation as either *dirhes* or *elmhes* and for this reason cannot be regarded as a general purpose procedure.

In *elmhes* considerable advantage can be taken of the presence of zero elements in A, and if A is very sparse *elmhes* is strongly recommended. In *dirhes* inner-products are accumulated whenever possible during the reduction and hence the effect of rounding errors is minimized. The advantage can be appreciable with matrices of high order but on some computers the accumulation of inner-products is so inefficient that the time taken by *dirhes* is prohibitive.

Each eigenvalue λ_i and its corresponding eigenvector v_i is always exact for some matrix $A + E_i$ where $\|E_i\|/\|A\|$ is of the order of magnitude of the machine precision. The residual vector $A v_i - \lambda v_i$ is therefore always small relative to $\|A\|$, independent of its 'accuracy', regarding A as exact. When A has multiple eigenvalues corresponding to linear divisors *hqr 2* determines fully independent vectors. *hqr 2* is very economical on storage.

8.2. The Computation of Selected Eigenvectors

There are no procedures in the series for finding selected eigenvalues, but when all eigenvalues have been computed using *hqr*, eigenvectors corresponding to a small number of selected eigenvalues may be found using the procedure *invit*. As is usual with inverse iteration the speed and accuracy of *invit* are very satisfactory for well separated eigenvalues and the procedure provides an elegant solution to the problem. For coincident and pathologically close eigenvalues the procedure is aesthetically less satisfying. If morethan 25% of the eigenvectors are wanted it is almost invariably more satisfactory to use *hqr 2*.

9. Unsymmetric Band Matrices

The only published procedure dealing specifically with band matrices is *unsray*. This is designed to find accurate eigenvalues and eigenvectors corresponding to a number of approximate eigenvalues. It can accept matrices with a different number of non-zero super-diagonal and sub-diagonal lines. Both left-hand and right-hand eigenvectors are found and the generalised Rayleigh values are computed from them, thereby providing eigenvalues which are usually accurate to more than single precision. It is valuable in connexion with finite difference approximations to partial differential equation eigenvalue problems. The complete problem may be solved with a coarse mesh (ignoring band form) and then selected eigenvalues and eigenvectors determined accurately using a fine mesh.

The method may, of course, be used even if the matrix involved happens to be symmetric; symmetry is in any case ignored in the execution of the algorithm.

10. Dense Unsymmetric Matrices with Complex Elements
10.1. Calculation of All the Eigenvalues with or without Eigenvectors
a) Reduction to Real Case

A complex matrix $A + iB$ may be treated by working with the augmented matrix

$$\begin{bmatrix} A & -B \\ B & A \end{bmatrix}.$$

Corresponding to each *complex* eigenvalue λ_k and eigenvector x_k of the original matrix, the augmented matrix has eigenvalues λ_k and $\bar{\lambda}_k$ and (with exact computation) the corresponding eigenvectors are of the forms

$$\begin{bmatrix} x_k \\ -i\,x_k \end{bmatrix} \quad \text{and} \quad \begin{bmatrix} \bar{x}_k \\ i\bar{x}_k \end{bmatrix}.$$

Corresponding to a *real* eigenvalue λ_k (with eigenvector $x_k = u_k + iv_k$) of the original matrix, the augmented matrix has a double eigenvalue λ_k and independent vectors are given by

$$\begin{bmatrix} u_k \\ v_k \end{bmatrix} \quad \text{and} \quad \begin{bmatrix} -v_k \\ u_k \end{bmatrix}.$$

If the eigenvalues only are computed there is no simple way of telling which of a complex conjugate pair of eigenvalues 'belongs' to the original matrix. The computation of the eigenvectors resolves this difficulty. Partitioning a typical eigenvector of the augmented matrix into u and v we see that corresponding to a complex conjugate pair one of the vectors will be such that $u + iv$ is null (to working accuracy); the associated eigenvalue is the one to be discarded and the vector $u + iv$ derived from the other will be the required vector. Corresponding to a double real eigenvalue the eigenvector $u + iv$ derived from either of the computed eigenvectors gives an eigenvector of the original.

b) The Complex LR Algorithm

There are two procedures *comlr* and *comlr 2* based on the LR algorithm which deal directly with complex matrices. Both depend on a preliminary reduction of the matrix to Hessenberg form using *comhes*. If eigenvalues only are required *comlr* should be used while if both values and vectors are required *comlr 2* is the relevant procedure. For multiple eigenvalues corresponding to linear elementary divisors *comlr 2* gives fully independent eigenvectors. No attempt to detect non-linear elementary divisors is made, but the near parallel nature of the computed eigenvectors usually gives a strong indication. Although one cannot as with *hqr 2* guarantee that *comlr 2* will give eigenvalues and eigenvectors with small residuals this is almost certain to be true.

The LR algorithm has been preferred to the complex QR algorithm because of its simplicity; in the real case the QR algorithm was preferred because it deals satisfactorily with complex conjugate eigenvalues.

c) The Generalized Jacobi Method

The procedure *comeig* is based on a generalization of the Jacobi method analogous to that used in *eigen*. Its convergence properties are more satisfactory than those of *eigen* and as regards speed it is to be preferred to the latter for a real matrix with some complex conjugate eigenvalues. It provides both eigenvalues and right-hand eigenvectors. It can be modified in obvious ways to give left-hand eigenvectors or both left-hand and right-hand eigenvectors. For a real matrix *hqr 2* is superior to *comeig* both in speed and storage since *comeig* takes no advantage of the matrix being real. For complex matrices *comeig* is fully comparable with *comlr 2*.

10.2. Calculation of Selected Eigenvectors of a Complex Matrix

There are no procedures for computing selected eigenvalues of a complex matrix. However, when all the eigenvalues have been computed using *comlr*, selected eigenvectors may be found using *cxinvit* and if the number required is small this combination is much more effective. As is usually the case with procedures based on inverse iteration, the algorithm is less pleasing when there are coincident or pathologically close eigenvalues.

Contribution II/1

The Jacobi Method for Real Symmetric Matrices*

by H. Rutishauser †

1. Theoretical Background

As is well known, a real symmetric matrix can be transformed iteratively into diagonal form through a sequence of appropriately chosen *elementary orthogonal transformations* (in the following called *Jacobi rotations*):

$$A_k \rightarrow A_{k+1} = U_k^T A_k U_k \qquad (A_0 = \text{given matrix}),$$

where $U_k = U_k(p, q, \varphi)$ is an orthogonal matrix which deviates from the unit matrix only in the elements

$$u_{pp} = u_{qq} = \cos(\varphi) \quad \text{and} \quad u_{pq} = -u_{qp} = \sin(\varphi).$$

On the whole the Jacobi method performs (approximately) the operation

$$A \rightarrow D = V^T A V,$$

where D is diagonal and V is orthogonal, D being the limit of A_k for $k \rightarrow \infty$, while V is the product of all Jacobi rotations which were used for achieving the diagonalisation:

$$V = U_0 U_1 U_2 U_3 \dots .$$

The main virtues of the Jacobi method stem from the fact that the Jacobi rotations produce a systematic decrease of the sum of the squares of the off-diagonal elements; the convergence of this sum to zero with increasing k guarantees the convergence of the diagonalisation process. Various schemes have been developed:

Jacobi [4] inspected the matrix A_k for the largest off-diagonal element a_{pq}** and then chose the rotation angle φ such that in the matrix A_{k+1} the p, q-element vanished. After the rediscovery the method in 1952 [2], when for the first time it was used in automatic computing, it was applied in such a way that p, q ran row-wise through all superdiagonal positions of the matrix, and again the rotation angle φ was chosen every time so as to annihilate the p, q-element in the matrix A_{k+1}. Later Pope and Tompkins [5] suggested a strategy which tended to avoid inefficient rotations and thus achieved diagonalisation with less effort.

---

* Prepublished in Numer. Math. 9, 1–10 (1966).

** Because of the symmetry of the matrices involved, only the matrix elements in and above the diagonal need be taken into consideration.

Proofs as well as estimates for the convergence properties of the Jacobi method have been given by HENRICI [3], SCHOENHAGE [7] and WILKINSON [8]. As these papers show, both the original Jacobi method as well as the more computer-oriented method with row-wise scanning of the upper triangle have *quadratic convergence*, and this is what makes the Jacobi method interesting.

2. Applicability

The present algorithm — which essentially uses row-wise scanning of the upper triangle — can be used and is absolutely foolproof for all real symmetric matrices. It is designed such that it requires no tolerance limit for termination but proceeds just as long as it is both necessary and meaningful on the computer which performs the calculation*. Usually this will take of the order of 6 to 10 sweeps, i.e. from $3n^2$ to $5n^2$ Jacobi rotations.

On the other hand it should be recognized that while the Jacobi process is more compact and elegant and also produces eigenvectors without difficulties, it is also more time-consuming than the combination of the Householder-transformation with a method for computing eigenvalues of tridiagonal matrices (e.g. the combination of procedure *tred2* [10] and *tql2* [11] or *imtql2* [12], or of *tred1* [10], *tristurm* [13] and *trbak1* [10] as published in this Handbook). For large-order matrices, therefore, the use of the latter procedures in place of procedure *jacobi* may be more appropriate. Moreover, the Jacobi method is comparatively uneconomical for symmetric bandmatrices whose bandwidth is small compared to n; for matrices of this kind the LR-transformation [6] will be more efficient.

3. Formal Parameter List

a) quantities to be given:

n the order of the matrix A.

eivec **true**, if eigenvectors are desired, otherwise **false**.

a **array** $a[1:n, 1:n]$ containing the elements of the matrix A. To be precise, only the diagonal and superdiagonal elements (the $a[i, k]$ with $k \geq i$) are actually used.

b) results produced by the procedure:

a The superdiagonal elements of the array a are destroyed by the process, but the diagonal and subdiagonal elements are unchanged (therefore full information on the given matrix A is still contained in array a, if the subdiagonal elements had also been filled in).

d The diagonal elements $d[i]$ of the matrix D, i.e. the approximate eigenvalues of A.

---

* Well-timed termination of the present algorithm requires that the Boolean expression $a + eps = a$ with a and eps non-negative have the value **true** provided eps is small compared to a within the computer accuracy. For computers in which this condition is not fulfilled the termination of the algorithm may be be delayed. Furthermore, procedure *jacobi* may produce erroneous results with computers which upon underflow produce values other than zero.

v **array** $v[1:n, 1:n]$ containing (provided $eivec =$ **true**) the elements of the matrix V, the k.th column of V being the normalized eigenvector to the eigenvalue $d[k]$. ★

rot The number of Jacobi rotations which have been needed to achieve the diagonalisation.

c) An example of application:

Since procedure *jacobi* does not order the computed eigenvalues, such ordering, if desired, must be done by the user of the procedure, e.g.

```
comment n, a[i, k] are assumed as being given;
begin
        integer array r[1:n];   integer aux;
        jacobi (n, true, a, d, v, rot);
        for k := 1 step 1 until n do r[k] := k;
        for k := 1 step 1 until n − 1 do
            for l := k+1 step 1 until n do
                if d[r[k]] < d[r[l]] then
                    begin aux := r[k];   r[k] := r[l];   r[l] := aux end;
comment : now d[r[k]] is the k.th eigenvalue in descending order and v[j, r[k]]
            is the j.th component of the corresponding eigenvector;
```

4. ALGOL Program

```
procedure jacobi (n, eivec) trans : (a) res : (d, v, rot);
    value n, eivec;
    integer n, rot;   boolean eivec;   array a, d, v;
    begin
            real sm, c, s, t, h, g, tau, theta, tresh;
            integer p, q, i, j;
            array b, z[1:n];
program:
            if eivec then
                for p := 1 step 1 until n do
                    for q := 1 step 1 until n do
                        v[p, q] := if p=q then 1.0 else 0.0;
                for p := 1 step 1 until n do
                    begin b[p] := d[p] := a[p, p];   z[p] := 0 end;
                rot := 0;
                for i := 1 step 1 until 50 do
swp:
            begin
                sm := 0;
```

★ Even if no eigenvectors are desired (*eivec* = **false**) the rules of ALGOL require that *jacobi* be called with some 2-dimensional array as actual counterpart of v, e.g., *jacobi* (16, **false**, a, d, a, rot).

```
for p := 1 step 1 until n − 1 do
    for q := p + 1 step 1 until n do
    sm := sm + abs (a[p, q]);
if sm = 0 then goto out;
tresh := if i < 4 then 0.2 × sm/n ↑ 2 else 0.0;
for p := 1 step 1 until n − 1 do
    for q := p + 1 step 1 until n do
    begin
        g := 100 × abs (a[p, q]);
        if i > 4 ∧ abs (d[p]) + g = abs (d[p]) ∧
        abs (d[q]) + g = abs (d[q]) then a[p, q] := 0
        else
          if abs (a[p, q]) > tresh then
            begin
                h := d[q] − d[p];
                if abs (h) + g = abs (h) then t := a[p, q]/h
                else
                    begin
                        theta := 0.5 × h/a[p, q];
                        t := 1/(abs (theta) + sqrt (1 + theta ↑ 2));
                        if theta < 0 then t := − t
                    end computing tan of rotation angle;
                c := 1/sqrt (1 + t ↑ 2);
                s := t × c;
                tau := s/(1 + c);
                h := t × a[p, q];
                z[p] := z[p] − h;
                z[q] := z[q] + h;
                d[p] := d[p] − h;
                d[q] := d[q] + h;
                a[p, q] := 0;
                for j := 1 step 1 until p − 1 do
                begin
                    g := a[j, p];  h := a[j, q];
                    a[j, p] := g − s × (h + g × tau);
                    a[j, q] := h + s × (g − h × tau)
                end of case 1 ≤ j < p;
                for j := p + 1 step 1 until q − 1 do
                begin
                    g := a[p, j];  h := a[j, q];
                    a[p, j] := g − s × (h + g × tau);
                    a[j, q] := h + s × (g − h × tau)
                end of case p < j < q;
                for j := q + 1 step 1 until n do
                begin
                    g := a[p, j];  h := a[q, j];
                    a[p, j] := g − s × (h + g × tau);
```

rotate:

$$a[q, j] := h + s \times (g - h \times tau)$$
end of case $q < j \leq n$;
if *eivec* **then**
    for $j := 1$ **step** 1 **until** n **do**
    begin
        $g := v[j, p];$ $h := v[j, q];$
        $v[j, p] := g - s \times (h + g \times tau);$
        $v[j, q] := h + s \times (g - h \times tau)$
    end of case v;
    $rot := rot + 1;$
  end *rotate*;
end;
for $p := 1$ **step** 1 **until** n **do**
begin
    $d[p] := b[p] := b[p] + z[p];$
    $z[p] := 0$
end p
end *swp*;
out:
end *jacobi*;

5. Organisational and Notational Details

The pivots of the Jacobi rotations, i.e. the elements $a[p, q]$ which are annihilated by the rotations, are chosen in a row-wise pattern, namely

for $p := 1$ **step** 1 **until** $n - 1$ **do**
  for $q := i + 1$ **step** 1 **until** n **do**
    rotate in rows and columns p and q;

Such a sequence of $\binom{n}{2}$ Jacobi rotations is called one *sweep*; immediately after completion of one sweep, the next sweep is begun, etc., until finally the diagonalisation is completed.

However, the present program contains the following additional features:

a) During the first three sweeps it performs only those rotations for which

$$abs(a[p, q]) > tresh \quad (= 0.2 \times sm/n \uparrow 2),$$

where *sm* is the sum of the moduli of all superdiagonal elements. In the later sweeps *tresh* is set to zero.

b) If before the p, q-rotation the element $a[p, q]$ is small compared to $a[p, p]$ *and* small compared to $a[q, q]$, then $a[p, q]$ is set to zero and the p, q-rotation is skipped.

This is certainly meaningful since it produces no larger error than would be produced anyhow if the rotation had been performed; however, in order that the procedure can be used for computing eigenvectors of perturbed diagonal★ matrices, this device is suppressed during the first four sweeps.

★ It should be recognized that this feature does not work for perturbed non-diagonal matrices.

c) In order to annihilate the p, q-element, the rotation parameters c, s are computed as follows: Compute first the quantity

$$theta = cot(2\varphi) = \frac{a[q, q] - a[p, p]}{2a[p, q]};$$

then $tan(\varphi)$ as the smaller root (in modulus) of the equation

$$t^2 + 2t \times theta = 1 \quad \text{(or } t = 0.5/theta, \text{ if } theta \text{ is large)},$$

and thereof $c = cos(\varphi)$, $s = sin(\varphi)$ and $tau = tan(\varphi/2)$, these latter entering into the rotation formulae.

d) With these features, the procedure will sooner or later make all super-diagonal elements zero (machine representation of zero), whereupon the process is discontinued.

e) *jacobi* does not actually operate on the diagonal elements of the array a but transfers them at the beginning into an array $d[1:n]$ and then performs all operations on $d[x]$ instead of upon $a[x, x]$.

6. Discussion of Numerical Properties

The present program attempts to diminish the accumulation of roundoff-errors as produced by the many Jacobi rotations needed for diagonalisation. This is achieved by the following measures:

a) It does not use the usual formulae

$$a_{pp}^{(new)} = c^2 \times a_{pp} - 2cs \times a_{pq} + s^2 \times a_{qq},$$
$$a_{qq}^{(new)} = s^2 \times a_{pp} + 2cs \times a_{pq} + c^2 \times a_{qq},$$
$$a_{pq}^{(new)} = (c^2 - s^2) \times a_{pq} + cs \times (a_{pp} - a_{qq})$$

for computing the p, p-, q, q- and p, q-elements of the matrix A_{k+1}, but the equivalent formulae

$$a_{pp}^{(new)} = a_{pp} - t \times a_{pq},$$
$$a_{qq}^{(new)} = a_{qq} + t \times a_{pq},$$
$$a_{pq}^{(new)} = 0.$$

b) Likewise also the formulae for the off-diagonal elements (and similarly for the components of the matrix V) are modified, e.g.

$$a_{pj}^{(new)} = c \times a_{pj} - s \times a_{qj},$$
$$a_{qj}^{(new)} = s \times a_{pj} + c \times a_{qj}$$

into the equivalent formulae

$$a_{pj}^{(new)} = a_{pj} - s \times (a_{qj} + tau \times a_{pj}),$$
$$a_{qj}^{(new)} = a_{qj} + s \times (a_{pj} - tau \times a_{qj}),$$

where $tau = tan(\varphi/2) = s/(1+c)$.

c) The terms $t \times a_{pq}$, by which the diagonal elements are changed, are accumulated separately (in an array $z[1:n]$), and only at the end of every sweep are the accumulated increments of all diagonal elements used to compute new (and better) values of the latter.

Now, according to WILKINSON [9], the total error incurred in the diagonalisation of a symmetric matrix is for every eigenvalue at most $E = 18.2 n^{1.5} r \|A_0\| \Theta$ ($\| \; \| =$ Schur Norm), where r is the number of sweeps required to perform the diagonalisation, and Θ is the smallest positive number such that in the computer $1 + \Theta \neq 1$.

This error estimate is influenced by the measures described above only insofar as the contributions of the roundoff-errors become insignificant as soon as the off-diagonal elements of the iterated matrix A_k become small as compared to the diagonal elements. Thus with the present algorithm we can in fact substitute a smaller r, 3 say, in the above formula for E.

7. Test Results

The following tests have been performed with procedure *jacobi* on the CDC 1604-A computer★ of the Swiss Federal Institute of Technology, Zurich:

a) Matrix A with $a[i, k] = max(i, k)$. The following table gives some of the eigenvalues found for $n = 30$:

1) with the present procedure *jacobi*, requiring a total of 2339 Jacobi rotations.

2) With another procedure for the Jacobi method, not containing any special measures against the accumulation of roundoff-errors.

3) With the qd-algorithm, using the fact that $-A^{-1}$ is a tridiagonal matrix corresponding to the qd-line

$$q_k = e_k = 1 \qquad (k = 1, 2, \ldots, n-1),$$

$$q_n = -1/n, \qquad e_n = 0.$$

The eigenvalues corresponding to this qd-line have been computed with the ALCOR multiple precision procedures and rounded correctly to the number of digits given:

	λ_1	λ_2	λ_3
1)	639.629 434 44	−0.250 687 020 21	−0.252 763 251 41
2)	639.629 435 87	−0.250 687 021 37	−0.252 763 252 02
3)	639.629 434 44	−0.250 687 020 23	−0.252 763 251 51

	λ_{16}	λ_{29}	λ_{30}
1)	−0.500 273 498 39	−24.077 530 173	−114.511 176 46
2)	−0.500 273 500 09	−24.077 530 237	−114.511 176 74
3)	−0.500 273 498 45	−24.077 530 172	−114.511 176 46.

Thus the present procedure *jacobi* is for all eigenvalues superior to the older Jacobi-program, and the actual errors are far below the estimate given in the

---

★ The 1604-A has a 36-bit mantissa and a binary exponent ranging from −1024 to 1023.

foregoing section (numerical properties), which would be $E=9_{10}-5$ in this example.

b) The matrix $B=8J-5J^2+J^3$ of order 44, where J denotes the tridiagonal matrix with $j[i, i]=2, j[i, i+1]=1$, i.e.

$$B=\begin{pmatrix} 5 & 2 & 1 & 1 & & & & \\ 2 & 6 & 3 & 1 & 1 & & & \\ 1 & 3 & 6 & 3 & 1 & 1 & & \\ 1 & 1 & 3 & 6 & 3 & 1 & 1 & \\ & 1 & 1 & 3 & 6 & 3 & 1 & 1 \\ & & & & & & & \\ & & & & 1 & 1 & 3 & 6 & 2 \\ & & & & & 1 & 1 & 2 & 5 \end{pmatrix}$$

Eleven of the eigenvalues of this matrix are in the interval $4\leq\lambda\leq4.163$, the length of which is 1% of the total length of the spectrum:

$$4.162\,503\,824\,429\,765$$
$$4.145\,898\,033\,750\,315$$
$$4.140\,771\,754\,112\,427$$
$$4.120\,834\,653\,401\,856$$
$$4.094\,745\,370\,081\,830$$
$$4.078\,725\,692\,426\,601$$
$$4.052\,841\,589\,388\,415$$
$$4.034\,568\,007\,676\,363$$
$$4.005\,211\,953\,160\,050$$
$$4.004\,531\,845\,800\,653$$
$$4$$

The following table contains for 6 eigenvalues and the first five components of the corresponding eigenvectors:

1) The values computed by procedure *jacobi*, requiring a total of 6260 Jacobi rotations.

2) The values computed by another procedure for the Jacobi method, as in the previous example.

3) The precise values, correctly rounded to the given number of digits.

Eigenvalues:

	λ_1	λ_{15}	λ_{28}
1)	15.922\,215\,640	6.000\,000\,0001	4.005\,211\,9528
2)	15.922\,215\,715	6.000\,000\,0270	4.005\,211\,9740
3)	15.922\,215\,641	6	4.005\,211\,9532

	λ_{29}	λ_{30}	λ_{44}
1)	4.004\,531\,8456	4.000\,000\,0000	0.038\,856\,634\,495
2)	4.004\,531\,8652	4.000\,000\,0193	0.038\,856\,634\,615
3)	4.004\,531\,8458	4	0.038\,856\,634\,457

Eigenvectors (components 1 through 5):

	v_1	v_{15}	v_{28}
1)	.014 705 95592	$-$.182 574 18584	$-$.210 690 17490
	.029 340 26590	$-$.182 574 18584	$-$.014 705 96163
	.043 831 63305	.000 000 00000	.209 663 71574
	.058 109 45688	.182 574 18579	.029 340 25522
	.072 104 17727	.182 574 18580	$-$.207 615 79602
2)	.014 705 95557	$-$.182 574 18615	$-$.210 690 00855
	.029 340 26517	$-$.182 574 18611	$-$.014 705 94683
	.043 831 63200	.000 000 00004	.209 663 53291
	.058 109 45555	.182 574 18622	.029 340 26251
	.072 104 17569	.182 574 18617	$-$.207 615 59730

	v_{29}	v_{30}	v_{44}
1)	.210 689 99816	$-$.182 574 18407	.014 705 95593
	$-$.014 705 96619	.182 574 18523	$-$.029 340 26589
	$-$.209 663 53377	.000 000 00164	.043 831 63301
	.029 340 28251	$-$.182 574 18485	$-$.058 109 45685
	.207 615 60979	.182 574 18745	.072 104 17724
2)	.210 690 16797	$-$.182 574 18989	.014 705 95596
	$-$.014 705 93265	.182 574 18996	$-$.029 340 26597
	$-$.209 663 72355	.000 000 00282	.043 831 63316
	.029 340 23626	$-$.182 574 19290	$-$.058 109 45703
	.207 615 81820	.182 574 18454	.072 104 17751

3) The precise values of the moduli of the numbers occurring among these components are (correctly rounded to the number of digits given)

.014 705 95590	.182 574 18584
.029 340 26587	.207 615 70379
.043 831 63301	.209 663 62482
.058 109 45684	.210 690 08574
.072 104 17724	

As these values indicate, procedure *jacobi* also produces better eigenvectors; only the vectors v_{28} and v_{29} corresponding to a very close pair of eigenvalues are slightly worse than with the older Jacobi process. However, the two vectors v_{28} and v_{29} computed by *jacobi* define the plane better than the corresponding vectors as produced by the earlier Jacobi procedure.

c) In order to check the performance of procedure *jacobi* when applied to perturbed diagonal matrices, the following matrix C of order 10 was tested, and satisfactory results were obtained:

$$c_{kk} = 1 - 10 \uparrow (1 - k),$$

$$c_{ik} = 10^{-12}, \quad \text{if} \quad i \neq k, \quad i - k = \text{even},$$

$$c_{ik} = 10^{-15}, \quad \text{if} \quad i - k = \text{odd}.$$

A second test of procedure *jacobi* was performed by Dr. W. BARTH, Rechenzentrum der Technischen Hochschule, Darmstadt. The present author wishes to thank Dr. BARTH for this service.

References

[1] BARTH, W., R. S. MARTIN, and J. H. WILKINSON: Calculation of the eigenvectors of a symmetric tridiagonal matrix by the method of bisection. Numer. Math. **9**, 386—393 (1967). Cf. II/5.

[2] GREGORY, R. T.: Computing eigenvalues and eigenvectors of a symmetric matrix on the ILLIAC. Math. Tab. and other Aids to Comp. **7**, 215—220 (1953).

[3] HENRICI, P.: On the speed of convergence of cyclic and quasicyclic Jacobi methods for computing eigenvalues of Hermitian matrices. J. Soc. Indust. Appl. Math. **6**, 144—162 (1958).

[4] JACOBI, C. G. J.: Über ein leichtes Verfahren, die in der Theorie der Säkularstörungen vorkommenden Gleichungen numerisch aufzulösen. Crelle's Journal **30**, 51—94 (1846).

[5] POPE, D. A., and C. TOMPKINS: Maximizing functions of rotations-experiments concerning speed of diagonalisation of symmetric matrices using Jacobi's method. J. Assoc. Comput. Mach. **4**, 459—466 (1957).

[6] RUTISHAUSER, H., and H. R. SCHWARZ: The LR-transformation method for symmetric matrices. Numer. Math. **5**, 273—289 (1963).

[7] SCHOENHAGE, A.: Zur Konvergenz des Jacobi-Verfahrens. Numer. Math. **3**, 374—380 (1961).

[8] WILKINSON, J. H.: Note on the quadratic convergence of the cyclic Jacobi process. Numer. Math. **4**, 296—300 (1962).

[9] — The algebraic eigenvalue problem, 662 p. Oxford: Clarendon Press 1965.

[10] MARTIN, R. S., C. REINSCH, and J. H. WILKINSON: Householder's tridiagonalization of a symmetric matrix. Numer. Math. **11**, 181—195 (1968). Cf. II/2.

[11] BOWDLER, H., R. S. MARTIN, C. REINSCH, and J. H. WILKINSON: The QR and QL algorithms for symmetric matrices. Numer. Math. **11**, 293—306 (1968). Cf. II/3.

[12] DUBRULLE, A., R. S. MARTIN, and J. H. WILKINSON: The implicit QL algorithm. Numer. Math. **12**, 377—383 (1968). Cf. II/4.

[13] PETERS, G., and J. H. WILKINSON: The calculation of specified eigenvectors by inverse iteration. Cf. II/18.

Contribution II/2

Householder's Tridiagonalization of a Symmetric Matrix[*]

by R. S. Martin, C. Reinsch, and J. H. Wilkinson

1. Theoretical Background

In an early paper in this series [4] Householder's algorithm for the tri-diagonalization of a real symmetric matrix was discussed. In the light of experience gained since its publication and in view of its importance it seems worthwhile to issue improved versions of the procedure given there. More than one variant is given since the most efficient form of the procedure depends on the method used to solve the eigenproblem of the derived tridiagonal matrix.

The symmetric matrix $A = A_1$ is reduced to the symmetric tridiagonal matrix A_{n-1} by $n-2$ orthogonal transformations. The notation is essentially the same as in the earlier paper [4] and the method has been very fully expounded in [6]. We therefore content ourselves with a statement of the following relations defining the algorithm.

$$A_{i+1} = P_i A_i P_i, \quad i = 1, 2, \ldots, n-2, \tag{1}$$

where

$$P_i = I - u_i u_i^T / H_i, \quad H_i = \tfrac{1}{2} u_i^T u_i, \tag{2}$$

$$u_i^T = [a_{l,1}^{(i)}; a_{l,2}^{(i)}; \ldots; a_{l,l-2}^{(i)}; a_{l,l-1}^{(i)} \pm \sigma_i^{\frac{1}{2}}; 0; \ldots; 0], \tag{3}$$

where $l = n - i + 1$ and

$$\sigma_i = (a_{l,1}^{(i)})^2 + (a_{l,2}^{(i)})^2 + \cdots + (a_{l,l-1}^{(i)})^2, \tag{4}$$

$$p_i = A_i u_i / H_i, \quad K_i = u_i^T p_i / 2 H_i, \tag{5}$$

$$q_i = p_i - K_i u_i, \tag{6}$$

$$A_{i+1} = (I - u_i u_i^T / H_i) A_i (I - u_i u_i^T / H_i)$$
$$= A_i - u_i q_i^T - q_i u_i^T. \tag{7}$$

The matrix A_i is of tridiagonal form in its last $i-1$ rows and columns. From the form of A_i and u_i it is evident that p_i is of the form

$$p_i^T = [p_{i,1}; p_{i,2}; \ldots; p_{i,l-1}; p_{i,l}; 0; \ldots; 0] \tag{8}$$

so that q_i^T is given by

$$q_i^T = [p_{i,1} - K_i u_{i,1}; p_{i,2} - K_i u_{i,2}; \ldots; p_{i,l-1} - K_i u_{i,l-1}; p_{i,l}; 0; \ldots; 0]. \tag{9}$$

If z is an eigenvector of the tridiagonal matrix A_{n-1} then $P_1 P_2 \ldots P_{n-2} z$ is an eigenvector of A_1.

---

[*] Prepublished in Numer. Math. **11**, 181–195 (1968).

2. Applicability

The procedures *tred 1*, *tred 2*, *tred 3* all give the Householder reduction of a real symmetric matrix A_1 to a symmetric tridiagonal matrix A_{n-1}. They can be used for arbitrary real symmetric matrices, possibly with multiple or pathologically close eigenvalues. The selection of the appropriate procedure depends on the following considerations.

tred 1 stores A_1 in a square $n \times n$ array and can preserve full information on A_1 if required for use in computing residuals and or Rayleigh quotients. It should be used if eigenvalues only are required or if certain selected eigenvalues and the corresponding eigenvectors are to be determined via the procedures *bisect* [1] and *tridiinverse iteration* [5] (an improved version is published here [7]). *tred 1* can be used with any procedure for finding the eigenvalues or eigenvectors of a symmetric tridiagonal matrix (for example [2, 8]).

tred 2 also stores the input matrix A_1 in a square $n \times n$ array and can preserve full information on A_1. Its use is mandatory when all eigenvalues and eigenvectors are to be found via those of A_{n-1} using procedure *tql 2* [2], which is a variant of the QR algorithm. The combination *tred 2* and *tql 2* is probably the most efficient of known methods when the complete eigensystem of A_1 is required. It produces eigenvectors always orthogonal to working accuracy and provides the possibility of "overwriting" the eigenvectors on the columns of A_1. *tred 2* should never be used if eigenvectors are not required.

tred 3 differs from *tred 1* with respect to the storage arrangement. The lower triangle of A_1 has to be stored as a linear array of $\frac{1}{2}n(n+1)$ elements. Its main advantage is in the economy of storage, but it is marginally slower than *tred 1* and the original information is lost. *tred 3* may be used as *tred 1*.

A procedure *tred 4* acting as *tred 2* but with the storage arrangement of *tred 3* may be obtained if the first statement in the body of procedure *tred 2* is replaced by

$k := 0;$
for $i := 1$ **step** 1 **until** n **do**
for $j := 1$ **step** 1 **until** i **do**
begin $k := k+1;$ $x[i, j] := a[k]$ **end** $ij;$

When *tred 2* is used in conjunction with *tql 2* [2], the eigenvectors of the original matrix, A_1, are automatically provided. *tred 1* and *tred 3* on the other hand are used with procedures which compute the eigenvectors of A_{n-1} and these must be transformed into the corresponding eigenvectors of A_1. The relevant procedures for doing this are *trbak 1* and *trbak 3* respectively; there is no need for a procedure *trbak 2*.

trbak 1 derives eigenvectors *m1* to *m2* of A_1 from eigenvectors *m1* to *m2* of A_{n-1} making use of the details of the transformation from A_1 to A_{n-1} provided as output from *tred 1*. *trbak 3* derives eigenvectors *m1* to *m2* of A_1 from eigenvectors *m1* to *m2* of A_{n-1} making use of the details of the transformation from A_1 to A_{n-1} provided as output from *tred 3*.

3. Formal Parameter List

3.1 (a) Input to procedure *tred 1*

n order of the real symmetric matrix $A = A_1$.

tol   a machine dependent constant. Should be set equal to *eta/macheps* where *eta* is the smallest positive number representable in the computer and *macheps* is the smallest number for which

$$1 + macheps > 1.$$

a   an $n \times n$ array giving the elements of the symmetric matrix $A = A_1$. Only the lower-triangle of elements is actually used but the upper triangle of elements is preserved. This makes it possible to compute residuals subsequently if the complete A_1 array is given.

Output from procedure *tred 1*

a   an $n \times n$ array; the strict lower triangle provides information on the Householder transformation; the full upper triangle is the original upper triangle of the input matrix *a*.

d   an $n \times 1$ array giving the diagonal elements of the tridiagonal matrix A_{n-1}.

e   an $n \times 1$ array of which $e[2]$ to $e[n]$ give the $n-1$ off-diagonal elements of the tridiagonal matrix A_{n-1}. The element $e[1]$ is set to zero.

e2   an $n \times 1$ array such that $e2[i] = (e[i])^2$ for use in procedure *bisect* [1], a call with $e = e2$ is admissible producing no squares.

3.1 (b) Input to procedure *trbak 1*

n   order of the real symmetric matrix A_1.

m1, m2 eigenvectors *m1* to *m2* of the tridiagonal matrix A_{n-1} have been found and normalized according to the Euclidean norm.

a   the strict lower triangle of this $n \times n$ array contains details of the Householder reduction of A_1 as produced by *tred 1*.

e   an $n \times 1$ array giving the $n-1$ off-diagonal elements of the tridiagonal matrix A_{n-1} with $e[1] = 0$, as produced by *tred 1*.

z   an $n \times (m2 - m1 + 1)$ array giving eigenvectors *m1* to *m2* of A_{n-1} normalized according to the Euclidean norm.

Output from procedure *trbak 1*

z   an $n \times (m2 - m1 + 1)$ array giving eigenvectors *m1* to *m2* of A_1 normalized according to the Euclidean norm.

3.2. Input to procedure *tred 2*

n   order of the real symmetric matrix $A = A_1$.

tol   machine dependent constant. The same as in *tred 1*.

a   an $n \times n$ array giving the elements of the symmetric matrix $A = A_1$. Only the lower-triangle is actually used in *tred 2*. The initial array *a* is copied into the array *z*. If the actual parameters corresponding to *z* and *a* are chosen to be different then the original matrix A_1 is completely preserved, if they are chosen to be identical then A_1 is completely lost.

Output from procedure *tred 2*

a   an $n \times n$ array which is identical with the input array *a* unless the actual parameter *z* is chosen to be *a*. In this case the output *a* is the output array *z* described below.

d an $n \times 1$ array giving the diagonal elements of the tridiagonal matrix A_{n-1}.

e an $n \times 1$ array of which $e[2]$ to $e[n]$ give the $n-1$ off-diagonal elements of the tridiagonal matrix A_{n-1}. The element $e[1]$ is set to zero.

z an $n \times n$ array whose elements are those of the orthogonal matrix Q such that $Q^T A_1 Q = A_{n-1}$ (i.e. the product of the Householder transformation matrices.

3.3 (a) Input to procedure *tred 3*

n order of the real symmetric matrix $A = A_1$.

tol   machine dependent constant. The same as in *tred 1*.

a a linear array of dimension $\frac{1}{2}n(n+1)$ giving the elements of A_1 in the order $a_{11}; a_{21}, a_{22}; \dots; a_{n1}, a_{n2}, \dots, a_{nn}$.

Output from procedure *tred 3*

a a linear array of dimension $\frac{1}{2}n(n+1)$ giving details of the Householder reduction of A_1.

d an $n \times 1$ array giving the diagonal elements of the tridiagonal matrix A_{n-1}.

e an $n \times 1$ array of which $e[2]$ to $e[n]$ give the $n-1$ off-diagonal elements of the tridiagonal matrix A_{n-1}. The element $e[1]$ is set to zero.

$e2$ an $n \times 1$ array such that $e2[i] = (e[i])^2$ for use in procedure *bisect* [1], a call with $e = e2$ is admissible producing no squares.

3.3 (b) Input to procedure *trbak 3*

n order of the real symmetric matrix.

m1, m2 eigenvectors *m1* to *m2* of the tridiagonal matrix A_{n-1} have been found and normalized according to the Euclidean norm.

a a linear array of order $\frac{1}{2}n(n+1)$ giving details of the Householder reduction of A_1 as produced by *tred 3*.

z an $n \times (m2 - m1 + 1)$ array giving eigenvectors *m1* to *m2* of A_{n-1} normalized according to the Euclidean norm.

Output from procedure *trbak 3*

z an $n \times (m2 - m1 + 1)$ array giving eigenvectors *m1* to *m2* of A_1 normalized according to the Euclidean norm.

4. ALGOL Programs

procedure *tred1 (n, tol) trans: (a) result: (d, e, e2)*;
value n, *tol*; **integer** n; **real** *tol*; **array** $a, d, e, e2$;
comment This procedure reduces the given lower triangle of a symmetric matrix, A, stored in the array $a[1:n, 1:n]$, to tridiagonal form using Householder's reduction. The diagonal of the result is stored in the array $d[1:n]$ and the sub-diagonal in the last $n-1$ stores of the array $e[1:n]$

(with the additional element $e[1]=0$). $e2[i]$ is set to equal $e[i]\uparrow 2$. The strictly lower triangle of the array a, together with the array e, is used to store sufficient information for the details of the transformation to be recoverable in the procedure *trbak 1*. The upper triangle of the array a is left unaltered;

```
begin      integer i, j, k, l;
           real f, g, h;
           for i := 1 step 1 until n do
           d[i] := a[i, i];
           for i := n step −1 until 1 do
           begin l := i −1;   h := 0;
                    for k := 1 step 1 until l do
                    h := h+a[i, k]×a[i, k];
                    comment if h is too small for orthogonality to be guaranteed,
                                the transformation is skipped;
                    if h ≤ tol then
                    begin e[i] := e2[i] := 0;   go to skip
                    end;
                    e2[i] := h;   f := a[i, i −1];
                    e[i] := g := if f ≥ 0 then − sqrt (h) else sqrt (h);
                    h := h−f×g;   a[i, i −1] := f −g;   f := 0;
                    for j := 1 step 1 until l do
                    begin g := 0;
                             comment form element of A ×u;
                             for k := 1 step 1 until j do
                             g := g+a[j, k]×a[i, k];
                             for k := j +1 step 1 until l do
                             g := g+a[k, j]×a[i, k];
                             comment form element of p;
                             g := e[j] := g/h;   f := f +g×a[i, j]
                    end j;
                    comment form K;
                    h := f/(h+h);
                    comment form reduced A;
                    for j := 1 step 1 until l do
                    begin f := a[i, j];   g := e[j] := e[j] −h×f;
                             for k := 1 step 1 until j do
                             a[j, k] := a[j, k] −f×e[k] −g×a[i, k]
                    end j;
           skip:  h := d[i];   d[i] := a[i, i];   a[i, i] := h
           end i
end tred 1;
```

procedure *trbak 1* (n) *data*: $(m1, m2, a, e)$ *trans*: (z);
value $n, m1, m2$;   **integer** $n, m1, m2$;   **array** a, e, z;
comment This procedure performs, on the matrix of eigenvectors, Z, stored in the array $z[1:n, m1:m2]$, a backtransformation to form the eigen-

vectors of the original symmetric matrix from the eigenvectors of the tridiagonal matrix. The new vectors are overwritten on the old ones.

The sub-diagonal of the tridiagonal matrix must be stored in the array $e[1:n]$, and the details of the Householder reduction must be stored in the lower triangle of the array $a[1:n, 1:n]$, as left by the procedure *tred 1*.

If z denotes any column of the resultant matrix Z, then z satisfies $z^T \times z = z \text{ (input)}^T \times z \text{ (input)}$;

```
begin      integer i, j, k, l;
           real h, s;
           for i := 2 step 1 until n do
           if e[i] ≠ 0 then
           begin l := i − 1;   h := e[i] × a[i, i − 1];
                      for j := m1 step 1 until m2 do
                      begin s := 0;
                                 for k := 1 step 1 until l do
                                 s := s + a[i, k] × z[k, j];
                                 s := s/h;
                                 for k := 1 step 1 until l do
                                 z[k, j] := z[k, j] + s × a[i, k]
                      end j
           end i
end trbak 1;
```

```
procedure tred 2 (n, tol) data: (a) result: (d, e, z);
value n, tol;  integer n;  real tol;  array a, d, e, z;
comment This procedure reduces the given lower triangle of a symmetric matrix,
           A, stored in the array a[1:n, 1:n], to tridiagonal form using House-
           holder's reduction. The diagonal of the result is stored in the array
           d[1:n] and the sub-diagonal in the last n − 1 stores of the array e[1:n]
           (with the additional element e[1] = 0). The transformation matrices
           are accumulated in the array z[1:n, 1:n]. The array a is left unaltered
           unless the actual parameters corresponding to a and z are identical;
begin      integer i, j, k, l;
           real f, g, h, hh;
           for i := 1 step 1 until n do
           for j := 1 step 1 until i do
           z[i, j] := a[i, j];
           for i := n step −1 until 2 do
           begin l := i − 2;   f := z[i, i − 1];   g := 0;
                      for k := 1 step 1 until l do
                      g := g + z[i, k] × z[i, k];
                      h := g + f × f;
                      comment if g is too small for orthogonality to be guaranteed,
                                      the transformation is skipped;
                      if g ≤ tol then
```

```
                begin e[i] := f;   h := 0;   go to skip
                end;
                l := l + 1;
                g := e[i] := if f ≧ 0 then − sqrt (h) else sqrt (h);
                h := h − f × g;   z[i, i − 1] := f − g;   f := 0;
                for j := 1 step 1 until l do
                begin z[j, i] := z[i, j]/h;   g := 0;
                        comment form element of A × u;
                        for k := 1 step 1 until j do
                        g := g + z[j, k] × z[i, k];
                        for k := j + 1 step 1 until l do
                        g := g + z[k, j] × z[i, k];
                        comment form element of p;
                        e[j] := g/h;   f := f + g × z[j, i]
                end j;
                comment form K;
                hh := f/(h + h);
                comment form reduced A;
                for j := 1 step 1 until l do
                begin f := z[i, j];   g := e[j] := e[j] − hh × f;
                        for k := 1 step 1 until j do
                        z[j, k] := z[j, k] − f × e[k] − g × z[i, k]
                end j;
        skip:   d[i] := h
        end i;

        d[1] := e[1] := 0;
        comment accumulation of transformation matrices;
        for i := 1 step 1 until n do
        begin l := i − 1;
                if d[i] ≠ 0 then
                for j := 1 step 1 until l do
                begin g := 0;
                        for k := 1 step 1 until l do
                        g := g + z[i, k] × z[k, j];
                        for k := 1 step 1 until l do
                        z[k, j] := z[k, j] − g × z[k, i]
                end j;
                d[i] := z[i, i];   z[i, i] := 1;
                for j := 1 step 1 until l do
                z[i, j] := z[j, i] := 0
        end i
end tred2;
```

procedure *tred3* (*n, tol*) *trans*: (*a*) *result*: (*d, e, e2*);
value *n, tol*; **integer** *n*; **real** *tol*; **array** *a, d, e, e2*;
comment This procedure reduces the given lower triangle of a symmetric matrix, *A*, stored row by row in the array $a[1 : n(n + 1)/2]$, to tridiagonal form

using Householder's reduction. The diagonal of the result is stored in the array $d[1:n]$ and the sub-diagonal in the last $n-1$ stores of the array $e[1:n]$ (with the additional element $e[1]=0$). $e2[i]$ is set to equal $e[i]\uparrow 2$. The array a is used to store sufficient information for the details of the transformation to be recoverable in the procedure $trbak\,3$;

```
begin       integer i, iz, j, jk, k, l;
            real f, g, h, hh;
            for i := n step −1 until 1 do
            begin l := i − 1;   iz := i × l/2;   h := 0;
                  for k := 1 step 1 until l do
                  begin f := d[k] := a[iz + k];   h := h + f × f
                  end k;
                  comment if h is too small for orthogonality to be guaranteed,
                             the transformation is skipped;
                  if h ≤ tol then
                  begin e[i] := e2[i] := h := 0;   go to skip
                  end;
                  e2[i] := h;
                  e[i] := g := if f ≥ 0 then − sqrt (h) else sqrt (h);
                  h := h − f × g;   d[l] := a[iz + l] := f − g;   f := 0;
                  for j := 1 step 1 until l do
                  begin g := 0;   jk := j × (j − 1)/2 + 1;
                        comment form element of A × u;
                        for k := 1 step 1 until l do
                        begin g := g + a[jk] × d[k];
                              jk := jk + (if k < j then 1 else k)
                        end k;
                        comment form element of p;
                        g := e[j] := g/h;   f := f + g × d[j]
                  end j;
                  comment form K;
                  hh := f/(h + h);   jk := 0;
                  comment form reduced A;
                  for j := 1 step 1 until l do
                  begin f := d[j];   g := e[j] := e[j] − hh × f;
                        for k := 1 step 1 until j do
                        begin jk := jk + 1;   a[jk] := a[jk] − f × e[k] − g × d[k]
                        end k
                  end j;
            skip:  d[i] := a[iz + i];   a[iz + i] := h
            end i
end tred 3;

procedure trbak 3 (n) data: (m1, m2, a) trans: (z);
value n, m1, m2;  integer n, m1, m2;  array a, z;
comment This procedure performs, on the matrix of eigenvectors, Z, stored in
            the array z[1:n, m1:m2], a backtransformation to form the eigen-
```

vectors of the original symmetric matrix from the eigenvectors of the
tridiagonal matrix. The new vectors are overwritten on the old ones.

The details of the Householder reduction must be stored in the
array $a[1:n(n+1)/2]$, as left by the procedure *tred 3*.

If z denotes any column of the resultant matrix Z, then z satisfies
$z^T \times z = z \, (\text{input})^T \times z \, (\text{input})$;

begin   **integer** i, iz, j, k, l;
   real h, s;
   for $i := 2$ **step** 1 **until** n **do**
   begin $l := i - 1$; $iz := i \times l/2$; $h := a[iz + i]$;
    if $h \neq 0$ **then**
    for $j := m1$ **step** 1 **until** $m2$ **do**
    begin $s := 0$;
      for $k := 1$ **step** 1 **until** l **do**
      $s := s + a[iz + k] \times z[k, j]$;
      $s := s/h$;
      for $k := 1$ **step** 1 **until** l **do**
      $z[k, j] := z[k, j] - s \times a[iz + k]$
    end j
   end i
end *trbak 3*;

5. Organisational and Notational Details

In general the notation used in the present procedures is the same as that
used in the earlier procedure [4]. The main differences in the new versions are
outlined in the following discussion.

In the original treatment, after forming σ_i defined by

$$\sigma_i = a_{l1}^2 + \cdots + a_{l, l-1}^2 \quad (l = n - i + 1) \tag{10}$$

the current transformation was skipped if $\sigma_i = 0$. This saved time and also avoided
indeterminacy in the elements of the transformation matrix. A more careful
treatment here is important.

If A_1 has any multiple roots, then with exact computation at least one σ_i must
be zero, since $\sigma_i^{\frac{1}{2}} = e_i$, an off-diagonal element of the tridiagonal matrix. This
implies that for some l we must have $a_{lj} = 0 \; (j = 1, \ldots, l-1)$. Hence in practice
it is quite common for a complete set of a_{lj} to be very small. Suppose for example
on a computer having numbers in the range 2^{-128} to 2^{128} we have at the stage $l = 4$

$$a_{41} = 2^{-63}, \quad a_{42} = 2^{-65}, \quad a_{43} = 2^{-63}. \tag{11}$$

Then $a_{42}^2 = 2^{-130}$ and will probably be replaced by zero. As a result the transformation which is dependent on the ratios of the a_{4j} and σ_{n-3} will differ substantially from an orthogonal matrix.

In the published procedures we have replaced "negligible" σ_i by zero and
skipped the corresponding transformation. A computed σ_i is regarded as negligible if

$$\sigma_i \leq eta/macheps = tol \, (\text{say}) \tag{12}$$

where *eta* is the smallest positive number representable in the computer and *macheps* is the relative machine precision. This ensures that the transformation is skipped whenever there is a danger that it might not be accurately orthogonal.

The maximum error in any eigenvalue as a result of the use of this criterion is $2 \times tol^{\frac{1}{2}}$. On a computer with a range 2^{-256} to 2^{256} and a 40 digit mantissa we have

$$2 \times tol^{\frac{1}{2}} = 2^{-107}.$$

The resulting error would therefore be of little consequence in general; indeed larger errors than this would usually be inevitable in any case as a result of rounding errors.

However, there are some matrices which have eigenvalues which are determined with a very small error compared with $\|A\|_2$. Consider for example the matrix

$$A = \begin{bmatrix} "10^{-12}" & "10^{-12}" & "10^{-12}" \\ "10^{-12}" & "10^{-6}" & "10^{-6}" \\ "10^{-12}" & "10^{-6}" & "1" \end{bmatrix} \tag{13}$$

where " 10^{-k} " denotes a number of the order of magnitude 10^{-k}. Such a matrix has an eigenvalue λ of order 10^{-12}. If we assume that the elements of A are prescribed to within one part in 10^t then in general, the eigenvalue λ will itself be determined to within a few parts in 10^t, that is with an absolute error of order 10^{-12-t}.

Clearly in order to deal with such matrices satisfactorily one would wish to make the criterion for skipping transformations as strict as possible. The range of permissable σ_i could be extended without losing orthogonality by proceeding as follows. Let

$$m_i = \max_{j=1}^{l-1} |a_{lj}|, \quad r_{lj} = a_{lj}/m_i \quad (j = 1, \ldots, l-1). \tag{14}$$

Then the transformation may be determined using the r_{lj} instead of the a_{lj} and we skip the transformation only if $m_i = 0$. We have decided in favour of the simpler technique described earlier. It should be appreciated that if the eigenvalues of the tridiagonal matrix are to be found by the procedure *bisect* [1] it is the squares of the subdiagonal elements which are used, i.e. the σ_i themselves. Hence when $\sigma_i < eta$ the corresponding e_i^2 is necessarily replaced by zero and we cannot reduce the bound for the errors introduced in the eigenvalues below $2(eta)^{\frac{1}{2}}$. For the computer specified above even the extreme precautions merely reduce the error bound from 2^{-107} to 2^{-127}.

When dealing with matrices having elements of widely varying orders of magnitude it is important that the smaller elements be in the top left-hand corner. The matrix should *not* be presented in the form

$$\begin{bmatrix} "1" & "10^{-6}" & "10^{-12}" \\ "10^{-6}" & "10^{-6}" & "10^{-12}" \\ "10^{-12}" & "10^{-12}" & "10^{-12}" \end{bmatrix}. \tag{15}$$

This is because the Householder reduction is performed starting from the last row and the consequences this has for rounding errors. To give maximum accuracy when dealing with matrices of type (13) all inner products are formed in direction of ascending subscripts.

The procedure *tred 2* is designed specifically for use with the procedure *tql 2* [2] when all eigenvalues and eigenvectors are required. With this combination it is more economical to have the orthogonal transformation Q reducing A_1 to tridiagonal form before solving the eigenproblem of A_{n-1}. We have

$$Q^T A_1 Q = A_{n-1} \qquad (16)$$

and

$$Q = P_1 P_2 \ldots P_{n-2}. \qquad (17)$$

The product Q is computed after completing the Householder reduction and it is derived in the steps

$$\begin{aligned} Q_{n-2} &= P_{n-2} \\ Q_i &= P_i Q_{i+1} \quad i = n-3, \ldots, 1 \\ Q &= Q_1. \end{aligned} \qquad (18)$$

Because of the nature of the P_i the matrix Q_i is equal to the identity matrix in rows and columns n, $n-1, \ldots, n-i+1$. Hence it is possible to overwrite the relevant parts of successive Q_i on the same $n \times n$ array which initially stores details of the transformation $P_1, \ldots, P_{n-2}$. The details of u_i are stored in row l of the $n \times n$ array; it proves convenient to store also u_i/H_i in column l of the $n \times n$ array. If information on the original matrix A_1 is required for subsequent calculation of residuals, A_1 must be copied before starting the Householder reduction and multiplying the transformation matrices. The working copy is the array z. If A_1 is not required for subsequent use, the $n \times n$ extra storage locations may be saved by taking $z = a$, in which case A_1 is effectively "overwritten on itself".

In all three procedures *tred 1, tred 2, tred 3* the non-zero elements of each p_i are stored in the storage locations required for the vector e. At each stage the number of storage locations not yet occupied by elements of e is adequate to store the non-zero elements of p_i. The vector p_i is subsequently overwritten with q_i.

tred 2 contains one feature which is not included in the other two procedures. If, at any stage in the reduction, the row which is about to be made tridiagonal is already of this form, the current transformation can be skipped. This is done in *tred 2* since the economy achieved is more worthwhile. If *tred 2* is provided with a matrix which is already entirely tridiagonal then virtually no computation is performed.

The procedure *trbak 1* and *trbak 3* call for little comment. The details of the transformations from A_1 to A_{n-1} are provided in the outputs a and e from *tred 1* and *tred 3* respectively.

6. Discussion of Numerical Properties

The procedures *tred 1, tred 2, tred 3* are essentially the same apart from the organization and give a very stable reduction. It has been shown [3, 6] that the computed tridiagonal matrix is orthogonally similar to $A + E$ where

$$\|E\|_F \le k_1 n^2 2^{-t} \|A\|_F \qquad (19)$$

where k_1 is a constant depending on the rounding procedure, t is the number of binary digits in mantissa and $\|\cdot\|_F$ denotes the Frobenius norm. If an inner product procedure which accumulates without rounding is used when computing σ_i, p_i and K_i then

$$\|E\|_F \le k_2\, n\, 2^{-t}\|A\|_F \tag{20}$$

where k_2 is a second constant. Both these bounds, satisfactory though they are, prove to be substantial overestimates in practice. The bounds ignore the problem of underflow, but the additional error arising from this cause when the criterion (12) is used is usually negligible compared with these bounds.

The error involved in using *trbak 1* and *trbak 3* to recover an eigenvector of A_1 from a computed eigenvector of computed A_{n-1} may be described as follows. Suppose that by any method we have obtained a computed normalized eigenvector z of A_{n-1} and can show that it is an exact eigenvector of $A_{n-1}+F$ with a known bound for $\|F\|_F$. If $\bar{x}$ is the corresponding computed eigenvector of A_1 then

$$\bar{x} = x + f$$

where x is an exact eigenvector of A_1+G with some G satisfying

$$\|G\|_F \le \|E\|_F + \|F\|_F,$$

(E being as defined above) and f satisfies

$$\|f\|_2 \le k_3\, n\, 2^{-t}.$$

In practice the bound for $\|F\|_F$ is usually smaller than that for $\|E\|_F$ and easily the most important source of error in x is that due to the inherent sensitivity of the relevant eigenvector to the perturbation E in A_1.

Again the computed transformation $\bar{Q}$ given by *tred 2* satisfies the relations

$$\bar{Q} = Q + R, \quad \|R\|_F \le k_4 n^2 2^{-t}$$

where Q is the exact orthogonal matrix such that

$$Q^T(A_1+E)\,Q = A_{n-1}$$

and again the most important consideration is the effect of the perturbation E.

7. Test Results

As a formal test of the procedures described here the matrices

$$A = \begin{bmatrix} 10 & 1 & 2 & 3 & 4 \\ 1 & 9 & -1 & 2 & -3 \\ 2 & -1 & 7 & 3 & -5 \\ 3 & 2 & 3 & 12 & -1 \\ 4 & -3 & -5 & -1 & 15 \end{bmatrix} \quad \text{and} \quad B = \begin{bmatrix} 5 & 1 & -2 & 0 & -2 & 5 \\ 1 & 6 & -3 & 2 & 0 & 6 \\ -2 & -3 & 8 & -5 & -6 & 0 \\ 0 & 2 & -5 & 5 & 1 & -2 \\ -2 & 0 & -6 & 1 & 6 & -3 \\ 5 & 6 & 0 & -2 & -3 & 8 \end{bmatrix}$$

were used. The second matrix was chosen so as to illustrate the case of multiple eigenvalues. The use of the procedures on more sophisticated examples will be illustrated in later papers describing the solution of the eigenproblem for tridiagonal matrices [2, 7]. The results obtained on KDF 9, a computer using a 39 binary digit mantissa, are given in Tables 1—3.

In Table 1 we give the subdiagonal and diagonal of the tridiagonal matrices determined using *tred 1*, *tred 2* and *tred 3* and the squares of the subdiagonal elements determined using *tred 1* and *tred 3* (*tred 2* does not output this vector). Procedures *tred 1* and *tred 3* always give identical results. The results obtained with *tred 2* may differ for three reasons. First the signs of sub-diagonal elements may differ because *tred 2* skips any step in which all elements to be annihilated are zero. Secondly if a σ_i is negligible it is replaced by zero in *tred 1* and *tred 3* but not in *tred 2*. Thirdly the arithmetic operations are performed in a slightly different order in *tred 2* and because the digital operations are, in general, non-commutative the rounding errors may be different.

As far as matrix A is concerned results obtained on a computer with a different word length and different rounding procedures should agree with those given almost to full working accuracy. For matrix B, however, different tridiagonal matrices may be obtained on different computers since the determination of the elements of the tridiagonal matrix itself is not stable when there are multiple roots. Nevertheless the *eigenvalues* should be preserved almost to full working accuracy. Note that with matrix B the element e_4 is of order 10^{-11}, that is of the order $macheps \times \|B\|_2$. With exact computation it should be zero. On KDF 9, e_4 was not small enough for the transformation to be skipped.

Table 1

Matrix A, tred 1 and tred 3

Sub-diagonal	Diagonal	Sub-diagonal squared
$+ 0.00000000000;$	$+ 9.2952021775_{10} + 0$	$+ 0.0000000000$
$+ 7.4948467774_{10} - 1$	$+ 1.1626711556_{10} + 1$	$+ 5.6172728216_{10} - 1$
$- 4.4962682012_{10} + 0$	$+ 1.0960439207_{10} + 1$	$+ 2.0216427737_{10} + 1$
$- 2.1570409908_{10} + 0$	$+ 6.1176470588_{10} + 0$	$+ 4.6528258362_{10} + 0$
$+ 7.1414284285_{10} + 0$	$+ 1.5000000000_{10} + 1$	$+ 5.1000000000_{10} + 1$

Matrix A, tred 2

Sub-diagonal	Diagonal
$+ 0.0000000000$	$+ 9.2952021775_{10} + 0$
$- 7.4948467774_{10} - 1$	$+ 1.1626711556_{10} + 1$
$- 4.4962682012_{10} + 0$	$+ 1.0960439207_{10} + 1$
$- 2.1570409908_{10} + 0$	$+ 6.1176470588_{10} + 0$
$+ 7.1414284285_{10} + 0$	$+ 1.5000000000_{10} + 1$

Matrix B, tred 1 and tred 3

Sub-diagonal	Diagonal	Sub-diagonal squared
$+ 0.0000000000$	$+ 4.4002127539_{10} + 0$	$+ 0.000000000$
$- 7.8376880058_{10} - 1$	$+ 3.8034768161_{10} + 0$	$+ 6.1429353276_{10} - 1$
$- 8.1050226977_{10} + 0$	$+ 1.0796310430_{10} + 1$	$+ 6.5691392931_{10} + 1$
$- 2.9798586700_{10} - 11$	$+ 4.2567567567_{10} + 0$	$+ 8.8795576935_{10} - 22$
$- 1.9246678253_{10} + 0$	$+ 6.7432432432_{10} + 0$	$+ 3.7043462381_{10} + 0$
$+ 8.6023252670_{10} + 0$	$+ 8.0000000000_{10} + 0$	$+ 7.4000000000_{10} + 1$

Matrix B, tred 2

Sub-diagonal	Diagonal
$+0.0000\,0000\,000$	$+4.4002\,1275\,387_{10}+0$
$+7.8376\,8800\,584_{10}-1$	$+3.8034\,7681\,618_{10}+0$
$-8.1050\,2269\,777_{10}+0$	$+1.0796\,3104\,303_{10}+1$
$-2.9798\,5867\,006_{10}-11$	$+4.2567\,5675\,677_{10}+0$
$-1.9246\,6782\,539_{10}+0$	$+6.7432\,4324\,323_{10}+0$
$+8.6023\,2526\,704_{10}+0$	$+8.0000\,0000\,000_{10}+0$

In Table 2 we give the eigenvalues of matrices A and B. The values presented are those obtained via *tred 1* and procedure *bisect* [1]. The values obtained with *tred 3* were, of course, identical with these, while those obtained using *tred 2* and *tql 2* differed by a maximum of 2 *macheps* $\|\cdot\|_2$.

Table 2. Eigenvalues

Matrix A	Matrix B
$+1.6552\,6620\,775_{10}+0$	$-1.5987\,3429\,358_{10}+0$
$+6.9948\,3783\,064_{10}+0$	$-1.5987\,3429\,346_{10}+0$
$+9.3655\,5492\,016_{10}+0$	$+4.4559\,8963\,847_{10}+0$
$+1.5808\,9207\,645_{10}+1$	$+4.4559\,8963\,855_{10}+0$
$+1.9175\,4202\,773_{10}+1$	$+1.6142\,7446\,551_{10}+1$
	$+1.6142\,7446\,553_{10}+1$

Finally in Table 3 we give the transformation matrices derived using *tred 2*. It will be observed that the last row and column are e_n^T and e_n respectively.

Table 3. Transformation matrix for matrix A

$-3.8454\,1619\,403_{10}-2$	$-8.2695\,9819\,664_{10}-1$	$+3.0549\,0408\,455_{10}-2$
$-8.6856\,5336\,976_{10}-1$	$-2.4019\,0402\,438_{10}-1$	$+1.0692\,1643\,009_{10}-1$
$+4.4924\,4356\,071_{10}-1$	$-5.0371\,0280\,447_{10}-1$	$-2.3293\,6436\,565_{10}-1$
$+2.0565\,7582\,815_{10}-1$	$-6.8716\,6691\,116_{10}-2$	$+9.6611\,3417\,195_{10}-1$
$+0.0000\,0000\,000$	$+0.0000\,0000\,000$	$+0.0000\,0000\,000$
$+5.6011\,2033\,610_{10}-1$	$+0.0000\,0000\,000$	
$-4.2008\,4025\,208_{10}-1$	$+0.0000\,0000\,000$	
$-7.0014\,0042\,012_{10}-1$	$+0.0000\,0000\,000$	
$-1.4002\,8008\,398_{10}-1$	$+0.0000\,0000\,000$	
$+0.0000\,0000\,000$	$+1.0000\,0000\,000_{10}+0$	

Transformation matrix for matrix B

$+7.3489\,9893\,569_{10}-1$	$-2.8079\,3152\,869_{10}-1$	$-6.2466\,7838\,579_{10}-2$
$-5.9928\,5559\,556_{10}-1$	$-3.0290\,2839\,915_{10}-1$	$+1.3769\,9891\,160_{10}-1$
$-1.4389\,1033\,659_{10}-1$	$-1.5931\,7058\,391_{10}-1$	$-9.7668\,4926\,340_{10}-1$
$-2.2262\,4697\,547_{10}-1$	$-5.5879\,3489\,748_{10}-1$	$+1.2394\,8910\,869_{10}-1$
$+1.7467\,8501\,867_{10}-1$	$-7.0126\,5274\,773_{10}-1$	$+8.8655\,8686\,443_{10}-2$
$+0.0000\,0000\,000$	$+0.0000\,0000\,000$	$+0.0000\,0000\,000$
$-1.9833\,6619\,947_{10}-1$	$+5.8123\,8193\,719_{10}-1$	$+0.0000\,0000\,000$
$+2.0894\,7221\,014_{10}-1$	$+6.9748\,5832\,461_{10}-1$	$+0.0000\,0000\,000$
$+0.0000\,0000\,000$	$+0.0000\,0000\,000$	$+0.0000\,0000\,000$
$-7.5416\,8875\,847_{10}-1$	$-2.3249\,5277\,487_{10}-1$	$+0.0000\,0000\,000$
$+5.9011\,2659\,345_{10}-1$	$-3.4874\,2916\,231_{10}-1$	$+0.0000\,0000\,000$
$+0.0000\,0000\,000$	$+0.0000\,0000\,000$	$+1.0000\,0000\,000_{10}+0$

The tests were confirmed at the Technische Hochschule München.

Acknowledgements. The work of R. S. Martin and J. H. Wilkinson has been carried out at the National Physical Laboratory and that of C. Reinsch at the Technische Hochschule München.

References

1. Barth, W., R. S. Martin, and J. H. Wilkinson: Calculation of the eigenvalues of a symmetric tridiagonal matrix by the method of bisection. Numer. Math. **9**, 386—393 (1967). Cf. II/5.
2. Bowdler, H., R. S. Martin, C. Reinsch, and J. H. Wilkinson: The QR and QL algorithms for symmetric matrices. Numer. Math. **11**, 293—306 (1968). Cf. II/3.
3. Ortega, J. M.: An error analysis of Householder's method for the symmetric eigenvalue problem. Numer. Math. **5**, 211—225 (1963).
4. Wilkinson, J. H.: Householder's method for symmetric matrices. Numer. Math. **4**, 354—361 (1962).
5. — Calculation of the eigenvectors of a symmetric tridiagonal matrix by inverse iteration. Numer. Math. **4**, 368—376 (1962).
6. — The algebraic eigenvalue problem. London: Oxford University Press 1965.
7. Peters, G., and J. H. Wilkinson: The calculation of specified eigenvectors by inverse iteration. Cf. II/18.
8. Reinsch, C., and F. L. Bauer: Rational QR transformation with Newton shift for symmetric tridiagonal matrices. Numer. Math. **11**, 264—272 (1968). Cf. II/6.

Contribution II/3

The QR and QL Algorithms for Symmetric Matrices *

by H. Bowdler, R. S. Martin, C. Reinsch, and J. H. Wilkinson

1. Theoretical Background

The QR algorithm as developed by Francis [2] and Kublanovskaya [4] is conceptually related to the LR algorithm of Rutishauser [7]. It is based on the observation that if

$$A = QR \quad \text{and} \quad B = RQ, \tag{1}$$

where Q is unitary and R is upper-triangular then

$$B = RQ = Q^H A Q, \tag{2}$$

that is, B is unitarily similar to A. By repeated application of the above result a sequence of matrices which are unitarily similar to a given matrix A_1 may be derived from the relations

$$A_s = Q_s R_s, \quad A_{s+1} = R_s Q_s = Q_s^H A_s Q_s \tag{3}$$

and, in general, A_s tends to upper-triangular form.

In this paper, for reasons which we explain later, we use an algorithm which is derived by a trivial modification of the above. It is defined by the relations

$$A_s = Q_s L_s, \quad A_{s+1} = L_s Q_s = Q_s^H A_s Q_s, \tag{4}$$

where Q_s is unitary and L_s is lower-triangular. We restate the results for the QR algorithm in the modified form appropriate to this "QL algorithm".

In general when A_1 has eigenvalues of distinct modulus the matrix A_s tends to lower-triangular form, the limiting matrix A_∞ having its diagonal elements equal to the eigenvalues of A_1 arranged in order of increasing absolute magnitude. When A_1 has a number of eigenvalues of equal modulus the limiting matrix is not triangular but, in general, corresponding to a $|\lambda_i|$ of multiplicity p, A_s ultimately has an associated diagonal block matrix of order p of which the eigenvalues tend to the corresponding p eigenvalues, though the block-diagonal matrix does not itself tend to a limit. (See, for example [2, 4, 8, 9].)

In general the super-diagonal element $(A_s)_{ij}$ behaves asymptotically like $k_{ij}(\lambda_i/\lambda_j)^s$ where k_{ij} is constant. The speed of convergence of the algorithm in

---

* Prepublished in Numer. Math. **11**, 293−306 (1968).

15*

this form is generally inadequate. The rate may be improved by working with $A_s - k_s I$ (for suitably chosen k_s) at each stage rather than with A_s itself. The process then becomes

$$A_s - k_s I = Q_s L_s, \qquad A_{s+1} = L_s Q_s + k_s I \tag{5}$$

giving

$$A_{s+1} = Q_s^H A_s Q_s. \tag{6}$$

The A_s obtained in this way are still unitarily similar to A_1 but the contribution of the s-th step to the convergence of the (i, j) element is determined by $(\lambda_i - k_s)/(\lambda_j - k_s)$ rather than λ_i/λ_j. If k_s could be chosen close to λ_1 (the eigenvalue of smallest modulus) then ultimately the off-diagonal elements in the 1-st row would decrease rapidly. When they are negligible to working accuracy $a_{11}^{(s)}$ can be accepted as an eigenvalue and the other eigenvalues are those of the remaining principal matrix of order $n-1$. Now if the QL algorithm is performed without acceleration then, in general, $a_{11}^{(s)} \to \lambda_1$ and hence the algorithm ultimately gives automatically a suitable value for k_s.

When A_1 is hermitian then obviously all A_s are hermitian; further if A_1 is real and symmetric all Q_s are orthogonal and all A_s are real and symmetric. Finally if A_1 is of band form it is easy to show that all A_s are of this form (see for example [8], pp. 557—558).

In the case when A_1 is symmetric and tridiagonal the process is particularly simple. Any real symmetric matrix may be reduced to tridiagonal form by HOUSEHOLDER's method [6] and it is more economical to do this first. It is convenient to recast the algorithm in the form

$$Q_s(A_s - k_s I) = L_s, \qquad A_{s+1} = L_s Q_s^T. \tag{7}$$

(Note that we do not add back $k_s I$, so that A_{s+1} is similar to $A_1 - \sum k_i I$ rather than A_1.) When A_s is tridiagonal Q_s is obtained in the factorized form

$$Q_s = P_1^{(s)} P_2^{(s)} \dots P_{n-1}^{(s)}. \tag{8}$$

Here the $P_i^{(s)}$ are determined in the order $P_{n-1}^{(s)}, \dots, P_1^{(s)}$, $P_i^{(s)}$ being a rotation in the $(i, i+1)$ plane designed to annihilate the $(i, i+1)$ element. If we write

$$P_i^{(s)} = \begin{bmatrix} 1 & & & & & & & \\ & 1 & & & & & & \\ & & \ddots & & & & & \\ & & & c_i & -s_i & & & \\ & & & s_i & c_i & & & \\ & & & & & \ddots & & \\ & & & & & & 1 & \\ & & & & & & & 1 \end{bmatrix} \begin{matrix} \\ \\ \\ \leftarrow \text{row } i \\ \leftarrow \text{row } i+1 \\ \\ \\ \end{matrix} \tag{9}$$

then omitting the upper suffix in the $P_i^{(s)}$

$$P_1 P_2 \dots P_{n-1}(A_s - k_s I) = L_s, \qquad A_{s+1} = L_s P_{n-1}^T \dots P_2^T P_1^T. \tag{10}$$

The matrix L_s is of the form

$$
\begin{bmatrix}
r_1 \\
w_2 & r_2 & & & & & \text{\Large 0} \\
z_3 & w_3 & r_3 \\
 & z_4 & w_4 & r_4 \\
 & & & & \cdot \\
\text{\Large 0} & & & z_{n-1} & w_{n-1} & r_{n-1} \\
 & & & & z_n & w_n & r_n
\end{bmatrix}
\tag{11}
$$

but it is evident that post-multiplication by $P_{n-1}^T \ldots P_1^T$ introduces only one super-diagonal line of non-zero elements, and since symmetry is restored, the second sub-diagonal line of elements must vanish. Hence there is no need to determine the z_i.

The reduction to L_s and the determination of A_{s+1} can proceed *pari passu* and a little algebraic manipulation shows that a single iteration is defined in the following way. Let $d_i^{(s)}$ $(i = 1, \ldots, n)$ be the diagonal elements and $e_i^{(s)}$ $(i = 1, \ldots, n-1)$* be the super-diagonal elements. Then we have, omitting the upper suffix s in everything except d and e,

$$
p_n = d_n^{(s)} - k_s, \quad c_n = 1, \quad s_n = 0
$$

$$
\left.
\begin{aligned}
r_{i+1} &= (p_{i+1}^2 + (e_i^{(s)})^2)^{\frac{1}{2}} \\
g_{i+1} &= c_{i+1} e_i^{(s)} \\
h_{i+1} &= c_{i+1} p_{i+1} \\
e_{i+1}^{(s+1)} &= s_{i+1} r_{i+1} \\
c_i &= p_{i+1}/r_{i+1} \\
s_i &= e_i^{(s)}/r_{i+1} \\
p_i &= c_i(d_i^{(s)} - k_s) - s_i g_{i+1} \\
d_{i+1}^{(s+1)} &= h_{i+1} + s_i(c_i g_{i+1} + s_i(d_i^{(s)} - k_s))
\end{aligned}
\right\}
\quad i = n-1, \ldots 1
\tag{12}
$$

$$
e_1^{(s+1)} = s_1 p_1, \quad d_1^{(s+1)} = c_1 p_1.
$$

Since the eigenvalues are real, those of equal modulus are either coincident or merely of opposite signs. Further if a tridiagonal matrix has an eigenvalue of multiplicity r then it must have at least $r-1$ zero off-diagonal elements. The diagonal blocks referred to in connexion with multiple $|\lambda_i|$ in the general case are therefore of little significance for matrices of tridiagonal form.

If at any stage an intermediate $e_i^{(s)}$ is "negligible" the matrix splits into the direct sum of smaller tridiagonal matrices and the eigenvalues of each submatrix may be found without reference to the others. At each stage therefore the $e_i^{(s)}$ are examined to see if any of them is negligible. If A_1 has been derived from a matrix with coincident or pathologically close eigenvalues by HOUSEHOLDER's tridiagonalization [6] then it will usually split in this way at an early stage, and the resulting economy is substantial. (If the original matrix has exact multiple eigen-

---

* It is more convenient in the QL algorithm to number the off-diagonal elements $e_1, e_2, \ldots, e_{n-1}$.

values then with exact computation the tridiagonal matrix would split *ab initio*
but in practice this is not necessarily true.)

The determination of the shift k_s at each stage presents some problems. In
practice the following has proved very effective. Suppose at the current stage
we are working with a submatrix beginning at row r. Then k_s is taken to be
the eigenvalue of the 2×2 matrix

$$\begin{bmatrix} d_r^{(s)} & e_r^{(s)} \\ e_r^{(s)} & d_{r+1}^{(s)} \end{bmatrix} \tag{13}$$

which is nearer to $d_r^{(s)}$. This choice of k_s has been used throughout in the algorithm
in this paper; there is no initial stage at which k_s is taken equal to zero. It can
be shown that this choice always gives convergence [11] (in the absence of
rounding errors) and usually gives cubic convergence [8, 11]. In practice the
rate of convergence is very impressive and the average number of iterations per
root is often less than two. To guard against the remote possibility of excessively
slow convergence a limit is set to the number of iterations taken to isolate any
one eigenvalue.

The order in which the eigenvalues are found is to some extent arbitrary.
Since the shift used in the first iteration is k_1 there is a marked tendency for
the eigenvalue nearest to this quantity to be the first to be isolated. The algorithm
is therefore a little unsatisfactory if we require, say, a few of the smallest eigenvalues of a very large matrix. In such cases we have found it practical to include
a small number of initial iterations with a zero shift, but on the whole the method
of bisection [1] is aesthetically more pleasing since this is specifically designed to
isolate individual eigenvalues. (However, see procedure *bqr* in II/7.)

It is perhaps worth mentioning that during the factorization of $(A_s - k_s I)$ the
information obtained is adequate for the determination of the signs of the principal
minors of orders $1, \ldots, n$ in the bottom right-hand corner of $A_s - k_s I$ and hence,
from the Sturm sequence property, the number of eigenvalues greater than k_s is
readily available. Some form of combination of the QL algorithm with the Sturm
sequence property is therefore possible.

When the norms of successive rows of the matrix A_1 vary widely in their
orders of magnitude the use of the shift k_s in the way we have described may be
undesirable. Consider for example a matrix A with elements of the orders of
magnitude given by

$$A = \begin{bmatrix} 10^{12} & 10^4 & & \\ 10^4 & 10^8 & 10^2 & \\ & 10^2 & 10^4 & 1 \\ & & 1 & 1 \end{bmatrix} \tag{14}$$

which has eigenvalues of orders 10^{12}, 10^8, 10^4 and 1. The first shift determined
from the leading 2×2 matrix is then obviously of order 10^{12}. Suppose we are
using a 10 digit decimal computer. When the shift is subtracted from a_{44} we get
the same results for all values of a_{44} from -50 to $+50$. The small eigenvalue,
which is well determined by the data, is inevitably lost. In this example the
difficulty could be avoided by reflecting the matrix in the secondary diagonal.
The first shift is then of order unity and no loss of accuracy results. However,

in general, the danger exists when our version is used as a subprocedure within a large program.

The danger may be avoided by the use of an alternative version of the algorithm due to FRANCIS [2, 8]. He described a method of performing the QR transformation without actually subtracting the shift from the diagonal elements. This is sometimes known as the "implicit shift technique". Francis originally described the implicit version in connexion with performing two steps of the QR algorithms in the unsymmetric case; a single step of the QL algorithm may be performed in an analogous way. This alternative algorithm will be described in a later paper in this series.

2. Applicability

The procedure *tql1* is designed to find all the eigenvalues of a symmetric tridiagonal matrix.

The procedure *tql2* is designed to find all the eigenvalues *and eigenvectors* of a symmetric tridiagonal matrix. If the tridiagonal matrix T has arisen from the Householder reduction of a full symmetric matrix A using *tred2* [6] then *tql2* may be used to find the eigenvectors of A directly without first finding the eigenvectors of T. In all cases computed vectors are orthonormal almost to *working accuracy*.

3. Formal Parameter List

3.1. Input to procedure *tql1*.

n order of tridiagonal matrix T.

macheps the smallest number representable on the computer for which $1 + macheps > 1$.

d the $n \times 1$ array of diagonal elements of T.

e the $n \times 1$ array of subdiagonal elements of T. $e[1]$ is not used and may be arbitrary.

Output from procedure *tql1*.

d the $n \times 1$ array giving the eigenvalues of T in ascending order.

e is used as working storage, and the original information stored in this vector is lost.

fail   the exit used if more than 30 iterations are required to isolate any one eigenvalue.

3.2. Input to procedure *tql2*.

n order of tridiagonal matrix T.

macheps the smallest number representable on the computer for which $1 + macheps > 1$.

d the $n \times 1$ array of diagonal elements of T.

e the $n \times 1$ array of subdiagonal elements of T. $e[1]$ is not used and may be arbitrary.

z an $n \times n$ array which is equal to the identity matrix if we require the eigenvectors of T itself. If T was derived from a full matrix using procedure *tred2* then z should be the matrix of the Householder transformation provided as output by that procedure.

Output from procedure *tql2*.

d the $n \times 1$ array giving the eigenvalues of T in ascending order.

e is used as working storage, and the original information stored in this vector is lost.

z the $n \times n$ array giving the normalized eigenvectors column by column of T if the input z was the identity matrix, but giving the eigenvectors of a full matrix if T has been derived using *tred2*.

4. ALGOL Programs

procedure *tql1* $(n, macheps)$ *trans*: (d, e) *exit*: $(fail)$;
value n, *macheps*; **integer** n; **real** *macheps*; **array** d, e; **label** *fail*;
comment This procedure finds the eigenvalues of a tridiagonal matrix, T, given with its diagonal elements in the array $d[1:n]$ and its subdiagonal elements in the last $n-1$ stores of the array $e[1:n]$, using QL transformations. The eigenvalues are overwritten on the diagonal elements in the array d in ascending order. The procedure will fail if any one eigenvalue takes more than 30 iterations;

```
begin    integer i, j, l, m;
         real b, c, f, g, h, p, r, s;
         for i := 2 step 1 until n do e[i-1] := e[i];
         e[n] := b := f := 0;
         for l := 1 step 1 until n do
         begin j := 0;  h := macheps × (abs (d[l]) + abs (e[l]));
                   if b < h then b := h;
                   comment look for small sub-diagonal element;
                   for m := l step 1 until n do
                   if abs (e[m]) ≤ b then go to cont1;
         cont1: if m = l then go to root;
         nextit: if j = 30 then go to fail;
                   j := j + 1;

                   comment form shift;
                   g := d[l];  p := (d[l+1] - g)/(2×e[l]);  r := sqrt (p↑2+1);
                   d[l] := e[l]/(if p < 0 then p - r else p + r);  h := g - d[l];
                   for i := l+1 step 1 until n do d[i] := d[i] - h;
                   f := f + h;

                   comment QL transformation;
                   p := d[m];  c := 1;  s := 0;
                   for i := m-1 step -1 until l do
                   begin g := c × e[i];  h := c × p;
                             if abs (p) ≥ abs (e[i]) then
```

```
          begin c := e[i]/p; r := sqrt(c↑2+1);
                e[i+1] := s×p×r; s := c/r; c := 1/r
          end
          else
          begin c := p/e[i]; r := sqrt(c↑2+1);
                e[i+1] := s×e[i]×r; s := 1/r; c := c/r
          end;
          p := c×d[i] − s×g;
          d[i+1] := h + s×(c×g + s×d[i])
      end i;
      e[l] := s×p; d[l] := c×p;
      if abs(e[l]) > b then go to nextit;

root:  p := d[l]+f;
      comment order eigenvalue;
      for i := l step −1 until 2 do
      if p < d[i−1] then d[i] := d[i−1] else go to cont2;
      i := 1;
cont2: d[i] := p
      end l
end tql1;
```

```
procedure tql2 (n, macheps) trans: (d, e, z) exit: (fail);
value n, macheps; integer n; real macheps; array d, e, z; label fail;
comment Finds the eigenvalues and eigenvectors of $ZTZ^T$, the tridiagonal
        matrix $T$ given with its diagonal elements in the array $d[1:n]$ and
        its sub-diagonal elements in the last $n-1$ stores of the array $e[1:n]$,
        using $QL$ transformations. The eigenvalues are overwritten on the
        diagonal elements in the array $d$ in ascending order. The eigenvectors
        are formed in the array $z[1:n, 1:n]$, overwriting the supplied
        orthogonal transformation matrix $Z$. The procedure will fail if any
        one eigenvalue takes more than 30 iterations;
begin   integer i, j, k, l, m;
        real b, c, f, g, h, p, r, s;
        for i := 2 step 1 until n do e[i−1] := e[i];
        e[n] := b := f := 0;
        for l := 1 step 1 until n do
        begin j := 0; h := macheps × (abs(d[l]) + abs(e[l]));
              if b < h then b := h;
              comment look for small sub-diagonal element;
              for m := l step 1 until n do
              if abs(e[m]) ≤ b then go to cont;
cont:   if m = l then go to root;
nextit: if j = 30 then go to fail;
              j := j+1;
```

```
                  comment form shift;
                  g := d[l]; p := (d[l+1] − g)/(2×e[l]); r := sqrt(p↑2+1);
                  d[l] := e[l]/(if p < 0 then p − r else p + r); h := g − d[l];
                  for i := l+1 step 1 until n do d[i] := d[i] − h;
                  f := f + h;

                  comment QL transformation;
                  p := d[m]; c := 1; s := 0;
                  for i := m − 1 step − 1 until l do
                  begin g := c×e[i]; h := c×p;
                        if abs(p) ≥ abs(e[i]) then
                        begin c := e[i]/p; r := sqrt(c↑2+1);
                              e[i+1] := s×p×r; s := c/r; c := 1/r
                        end
                        else
                        begin c := p/e[i]; r := sqrt(c↑2+1);
                              e[i+1] := s×e[i]×r; s := 1/r; c := c/r
                        end;
                        p := c×d[i] − s×g;
                        d[i+1] := h + s×(c×g + s×d[i]);
                        comment form vector;
                        for k := 1 step 1 until n do
                        begin h := z[k, i+1];
                              z[k, i+1] := s×z[k, i] + c×h;
                              z[k, i] := c×z[k, i] − s×h
                        end k
                  end i;
                  e[l] := s×p; d[l] := c×p;
                  if abs(e[l]) > b then go to nextit;

         root:    d[l] := d[l] + f
                  end l;
         comment order eigenvalues and eigenvectors;
         for i := 1 step 1 until n do
         begin k := i; p := d[i];
               for j := i+1 step 1 until n do
               if d[j] < p then
               begin k := j; p := d[j]
               end;
               if k ≠ i then
               begin d[k] := d[i]; d[i] := p;
                     for j := 1 step 1 until n do
                     begin p := z[j, i]; z[j, i] := z[j, k]; z[j, k] := p
                     end j
               end
         end i
end tql2;
```

5. Organisational and Notational Details

In both *tql1* and *tql2* the vector e of subdiagonal elements of the tridiagonal matrix T is given with e_1 arbitrary, $e_2, \ldots, e_n$ being the true subdiagonal elements. This is for consistency with other procedures in the series for which this ordering is convenient. Within the procedures *tql1* and *tql2* elements $e_2, \ldots, e_n$ are immediately transferred to positions $e_1, \ldots, e_{n-1}$.

The procedures will most frequently be used in combination with *tred1*, *tred2*, and *tred3* [6] for solving the standard eigenproblem for a full matrix and also in combination with *reduc1*, *reduc2*, *rebak a*, *rebak b* [5] for solving the generalized eigenproblems $A x = \lambda B x$, $A B x = \lambda x$ etc. In *tred2* the Householder reduction is performed starting with the last row and column and hence it is important that if the rows of A vary widely in norm, A should be presented so that the last row and column of A have the largest norm. This means that the tridiagonal matrices produced by *tred2* will have the largest elements in the bottom right-hand corner and this in turn makes it necessary to start the QR algorithm working from the bottom upwards. Hence the use of the QL algorithm rather than the more conventional QR algorithm. If the tridiagonal matrix is the primary data then this too must be given with the largest elements in the bottom right hand corner if they vary considerably in order of magnitude.

The volume of computation may be reduced by detecting "negligible" subdiagonal elements. The determination of a satisfactory criterion for deciding when an element is negligible presents some difficulty. A criterion is required which is not unnecessarily elaborate but makes it possible to determine small eigenvalues accurately when this is meaningful. A reliable criterion in practice is to regard $e_i^{(s)}$ as negligible when

$$|e_i^{(s)}| \leq macheps \, (|d_{i+1}^{(s)}| + |d_i^{(s)}|),\qquad(15)$$

(we assume here that the $e_i^{(s)}$ are numbered $e_1^{(s)}$ to $e_{n-1}^{(s)}$) but this was regarded as unnecessarily elaborate. The following simpler criterion was adopted. Immediately before iterating for the r-th eigenvalue $h_r = macheps \, (|d_r^{(s)}| + |e_r^{(s)}|)$ is computed (where $d_r^{(s)}$ and $e_r^{(s)}$ are the current values) and before starting the iteration any $|e_i^{(s)}|$ smaller than $b_r = \max\limits_{i=1}^{r} h_i$ is regarded as negligible. The effectiveness of this criterion is illustrated in the last section.

If *tql2* is used to find the eigenvectors of the tridiagonal matrix itself the input matrix z must be the identity matrix of order n. If it is being used to find the eigenvectors of a general matrix A which has previously been reduced to tridiagonal form using *tred2* the input z must be the output transformation matrix given by *tred2*.

The number of iterations required to find any one root has been restricted (somewhat arbitrarily to 30). In no case has six iterations been exceeded.

6. Discussion of Numerical Properties

The eigenvalues given by *tql1* and *tql2* *always have errors which are small compared with* $\|T\|_2$. These algorithms are covered by the rather general error analysis given in [8]. If s iterations are needed altogether the error in any one eigenvalue is certainly bounded by $2^{-t}ks\|T\|_2$ where k is of order unity and computation with a t digit binary mantissa is assumed. For a well-balanced matrix this is a very satisfactory bound, but for matrices having elements which vary widely in order of magnitude such a bound may be useless as far as the smaller eigenvalues are concerned and they may be well-determined by the data. In general provided such matrices are presented with elements in order of increasing absolute magnitude eigenvalues are usually determined with errors which correspond to changes of one or two units in the last significant figure of each element.

The eigenvectors given by *tql2* are *always* very accurately orthogonal even if T (or the original full matrix) has roots of high multiplicity or many very close roots. Every vector is also an exact eigenvector of some matrix which is close to the original. The accuracy of individual eigenvectors is of course dependent on their inherent sensitivity to changes in the original data. (See for example the discussion in [8].)

7. Test Results

The procedures *tql1* and *tql2* were tested very extensively on the computer KDF 9. We give first the results obtained with the matrices A and B used earlier [6] to test procedures *tred1*, *tred2* and *tred3*. The eigenvalues obtained with *tql1* and *tql2* and the total number of iterations are given in Table 1. (Note that the two procedures give identical eigenvalues.)

The eigenvectors produced by *tql2* for these matrices are given in Table 2. Matrix B has three pairs of double roots; it is immediately apparent that the two vectors corresponding to each pair are orthogonal to working accuracy since each pair is almost exactly of the form

$$\begin{bmatrix} u \\ \cdots \\ v \end{bmatrix} \quad \text{and} \quad \begin{bmatrix} -v \\ \cdots \\ u \end{bmatrix}$$

where u and v are vectors of order three.

It should be appreciated that any pair of orthogonal vectors in the 2 space is equally good, and vectors parallel to

$$\begin{bmatrix} u - \alpha v \\ \cdots \\ v + \alpha u \end{bmatrix} \quad \text{and} \quad \begin{bmatrix} \alpha u + v \\ \cdots \\ \alpha v - u \end{bmatrix}$$

may be obtained on a computer with a different rounding procedure.

Table 1

Eigenvalues of matrix A	Eigenvalues of matrix B
$+1.6552\,6620\,792_{10}+0$	$-1.5987\,3429\,360_{10}+0$
$+6.9948\,3783\,061_{10}+0$	$-1.5987\,3429\,346_{10}+0$
$+9.3655\,5492\,014_{10}+0$	$+4.4559\,8963\,849_{10}+0$
$+1.5808\,9207\,644_{10}+1$	$+4.4559\,8963\,849_{10}+0$
$+1.9175\,4202\,773_{10}+1$	$+1.6142\,7446\,551_{10}+1$
8 iterations	$+1.6142\,7446\,553_{10}+1$
	6 iterations

Table 2

Eigenvectors of matrix A

$+\,3.8729\,6874\,886_{10}-1$	$+\,6.5408\,2984\,085_{10}-1$	$+\,5.2151\,1178\,463_{10}-2$
$-\,3.6622\,1021\,131_{10}-1$	$+\,1.9968\,1268\,959_{10}-1$	$-\,8.5996\,3866\,689_{10}-1$
$-\,7.0437\,7266\,220_{10}-1$	$+\,2.5651\,0456\,336_{10}-1$	$+\,5.0557\,5072\,575_{10}-1$
$+\,1.1892\,6222\,076_{10}-1$	$-\,6.6040\,2722\,389_{10}-1$	$+\,2.0116\,6650\,650_{10}-4$
$-\,4.5342\,3108\,037_{10}-1$	$-\,1.7427\,9863\,500_{10}-1$	$-\,4.6219\,1996\,239_{10}-2$

$-\,6.2370\,2499\,852_{10}-1$	$+\,1.7450\,5109\,459_{10}-1$
$-\,1.5910\,1120\,870_{10}-1$	$-\,2.4730\,2518\,851_{10}-1$
$-\,2.2729\,7494\,237_{10}-1$	$-\,3.6164\,1739\,446_{10}-1$
$-\,6.9268\,4385\,756_{10}-1$	$-\,2.6441\,0853\,099_{10}-1$
$-\,2.3282\,2283\,880_{10}-1$	$+\,8.4124\,4069\,212_{10}-1$

Eigenvectors of matrix B

$+\,3.6056\,0828\,403_{10}-1$	$+\,3.6066\,6817\,179_{10}-1$	$-\,4.2485\,2821\,114_{10}-2$
$+\,5.4982\,5485\,630_{10}-1$	$+\,1.3178\,6716\,209_{10}-1$	$-\,4.2288\,5282\,424_{10}-2$
$-\,2.3279\,7640\,369_{10}-2$	$+\,6.4784\,0709\,878_{10}-1$	$-\,6.7803\,4796\,717_{10}-2$
$-\,3.6066\,6817\,169_{10}-1$	$+\,3.6056\,0828\,404_{10}-1$	$-\,7.4620\,0546\,442_{10}-1$
$-\,1.3178\,6716\,212_{10}-1$	$+\,5.4982\,5485\,636_{10}-1$	$+\,6.1709\,0441\,108_{10}-1$
$-\,6.4784\,0709\,888_{10}-1$	$-\,2.3279\,7640\,382_{10}-2$	$+\,2.3279\,4976\,959_{10}-1$

$+\,7.4620\,0546\,439_{10}-1$	$+\,4.1905\,6528\,968_{10}-1$	$-\,7.5410\,2568\,144_{10}-2$
$-\,6.1709\,0441\,116_{10}-1$	$+\,4.5342\,0630\,029_{10}-1$	$-\,3.0355\,3943\,991_{10}-1$
$-\,2.3279\,4976\,968_{10}-1$	$-\,5.9886\,6735\,976_{10}-13$	$+\,7.2178\,2276\,070_{10}-1$
$-\,4.2485\,2821\,170_{10}-2$	$-\,7.5410\,2568\,151_{10}-2$	$-\,4.1905\,6528\,965_{10}-1$
$-\,4.2288\,5282\,501_{10}-2$	$-\,3.0355\,3943\,992_{10}-1$	$-\,4.5342\,0630\,026_{10}-1$
$-\,6.7803\,4796\,695_{10}-2$	$+\,7.2178\,2276\,069_{10}-1$	$+\,9.0949\,4701\,773_{10}-13$

The next example is associated with matrices F and G used earlier in [5] to test procedure *reduc1*, *reduc2*, *rebak a* and *rebak b*. Table 3 gives all eigenvalues, the total iteration count and eigenvectors of $Fx=\lambda Gx$ derived using *reduc1* (to reduce the problem to the standard problem), *tred2* (to reduce the standard problem to the tridiagonal problem), *tql2* (to find the eigenvalues and eigenvectors of the standard problem) and finally *rebak a* to find the eigenvectors of $Fx=\lambda Gx$.

Table 3

Eigenvalues of $Fx=\lambda Gx$

$$+\,4.3278\,7211\,030_{10}-1$$
$$+\,6.6366\,2748\,393_{10}-1$$
$$+\,9.4385\,9004\,679_{10}-1$$
$$+\,1.1092\,8454\,002_{10}+0$$
$$+\,1.4923\,5323\,254_{10}+0$$

8 iterations

Table 3 (Continued)

	Eigenvectors of $Fx = \lambda Gx$	
$+1.3459\,0573\,963_{10} - 1$	$+8.2919\,8064\,839_{10} - 2$	$-1.9171\,0031\,578_{10} - 1$
$-6.1294\,7224\,713_{10} - 2$	$+1.5314\,8395\,666_{10} - 1$	$+1.5899\,1211\,510_{10} - 1$
$-1.5790\,2562\,212_{10} - 1$	$-1.1860\,3667\,910_{10} - 1$	$-7.4839\,0709\,418_{10} - 2$
$+1.0946\,5787\,726_{10} - 1$	$-1.8281\,3041\,785_{10} - 1$	$+1.3746\,8929\,465_{10} - 1$
$-4.1473\,0117\,972_{10} - 2$	$+3.5617\,2036\,811_{10} - 3$	$-8.8977\,8923\,529_{10} - 2$
	$+1.4201\,1959\,880_{10} - 1$	$-7.6386\,7178\,789_{10} - 2$
	$+1.4241\,9950\,553_{10} - 1$	$+1.7098\,0018\,729_{10} - 2$
	$+1.2099\,7623\,003_{10} - 1$	$-6.6664\,5336\,719_{10} - 2$
	$+1.2553\,1015\,192_{10} - 1$	$+8.6048\,0093\,077_{10} - 2$
	$+7.6922\,0728\,043_{10} - 3$	$+2.8943\,3414\,167_{10} - 1$

As the final tests we have taken two matrices with elements varying from 1 to 10^{12}. Matrix X is defined by

$$d_1, \ldots, d_7 = 1,\ 10^2,\ 10^4,\ 10^6,\ 10^8,\ 10^{10},\ 10^{12}$$
$$e_2, \ldots, e_7 = \quad 10,\ 10^3,\ 10^5,\ 10^7,\ 10^9,\ 10^{11}$$

while matrix Y is defined by

$$d_1, \ldots, d_7 = 1.4,\ 10^2,\ 10^4,\ 10^6,\ 10^8,\ 10^{10},\ 10^{12}$$
$$e_2, \ldots, e_7 = \quad 10,\ 10^3,\ 10^5,\ 10^7,\ 10^9,\ 10^{11}$$

and to illustrate the necessity for having the elements of d and e in order of increasing absolute magnitude results are also given for matrices $\overline{X}$ and $\overline{Y}$ defined respectively by

$$d_1, \ldots, d_7 = 10^{12},\ 10^{10},\ 10^8,\ 10^6,\ 10^4,\ 10^2,\ 1$$
$$e_2, \ldots, e_7 = \quad 10^{11},\ 10^9,\ 10^7,\ 10^5,\ 10^3,\ 10$$
$$d_1, \ldots, d_7 = 10^{12},\ 10^{10},\ 10^8,\ 10^6,\ 10^4,\ 10^2,\ 1.4$$
$$e_2, \ldots, e_7 = \quad 10^{11},\ 10^9,\ 10^7,\ 10^5,\ 10^3,\ 10$$

i.e. $\overline{X}$ and $\overline{Y}$ are obtained by reflecting X and Y in their secondary diagonals.

Table 4 gives the eigenvalues obtained and the iteration counts for matrices X and $\overline{X}$, and matrices Y and $\overline{Y}$. Matrices X and $\overline{X}$ should have identical eigenvalues. In fact the values obtained for X are all correct to within a few digits in the last significant figures, so that even the small eigenvalues have very low relative errors. Those obtained for $\overline{X}$ have errors which are all, of course, negligible with respect to $\|\overline{X}\|_2$ but the smaller eigenvalues have high relative errors. This illustrates the importance of presenting heavily graded matrices with elements in increasing order.

Table 4

Computed eigenvalues of X	Computed eigenvalues of $\overline{X}$
$-9.4634\,7415\,650_{10} + 8$	$-9.4634\,7415\,100_{10} + 8$
$-9.4634\,6919\,714_{10} + 2$	$-9.4632\,7834\,737_{10} + 2$
$+9.9989\,9020\,191_{10} - 1$	$+1.9998\,0118\,871_{10} + 0$
$+1.0463\,3721\,479_{10} + 3$	$+1.0463\,5443\,889_{10} + 3$
$+1.0098\,9903\,020_{10} + 6$	$+1.0098\,9897\,555_{10} + 6$
$+1.0463\,3771\,269_{10} + 9$	$+1.0463\,3771\,409_{10} + 9$
$+1.0100\,0000\,981_{10} + 12$	$+1.0100\,0000\,980_{10} + 12$
6 iterations	8 iterations

Table 5 gives the computed eigenvalues and iteration counts for Y and $\overline{Y}$ (again these should have identical eigenvalues). All computed eigenvalues of Y have low relative errors while those of $\overline{Y}$ merely have errors which are small relative to $\|\overline{Y}\|_2$. Notice that the computed eigenvalues for $\overline{X}$ and $\overline{Y}$ are identical! This is because the first shift is of order 10^{12} and as a result the element 1.4 is treated as though it were 1.0 when the shift is subtracted from it.

In none of these examples does the average number of iterations exceed 1.6 per eigenvalue. Where the matrix is heavily graded or has multiple eigenvalues the count is particularly low.

Table 5

Computed eigenvalues of Y	Computed eigenvalues of $\overline{Y}$
$-9.4634\,7415\,646_{10}+8$	$-9.4634\,7415\,100_{10}+8$
$-9.4634\,6898\,550_{10}+2$	$-9.4632\,7834\,737_{10}+2$
$+1.3998\,5863\,385_{10}+0$	$+1.9998\,0118\,871_{10}+0$
$+1.0463\,3723\,402_{10}+3$	$+1.0463\,5443\,889_{10}+3$
$+1.0098\,9903\,020_{10}+6$	$+1.0098\,9897\,555_{10}+6$
$+1.0463\,3771\,268_{10}+9$	$+1.0463\,3771\,409_{10}+9$
$+1.0100\,0000\,980_{10}+12$	$+1.0100\,0000\,980_{10}+12$
6 iterations	8 iterations

Finally in Table 6 we give the computed eigenvectors of X and Y corresponding to the eigenvalue λ_3 of smallest modulus in each case.

Table 6

Eigenvectors of X corresponding to λ_3	Eigenvectors of Y corresponding to λ_3
$+9.9994\,9513\,816_{10}-1$	$+9.9994\,9517\,747_{10}-1$
$-1.0097\,4708\,972_{10}-5$	$-1.4135\,9021\,336_{10}-5$
$-9.9984\,9548\,751_{10}-3$	$-9.9981\,0137\,554_{10}-3$
$+9.9985\,0548\,606_{10}-4$	$+9.9981\,1537\,287_{10}-4$
$-1.0097\,4707\,962_{10}-14$	$-1.4218\,8114\,076_{10}-14$
$-9.9985\,0548\,503_{10}-6$	$-9.9981\,1537\,142_{10}-6$
$+9.9985\,0548\,501_{10}-7$	$+9.9981\,1537\,145_{10}-7$

Acknowledgements. The work of H. BOWDLER, R. S. MARTIN and J. H. WILKINSON has been carried out at the National Physical Laboratory and that of C. REINSCH at the Technische Hochschule München.

References

1. BARTH, W., R. S. MARTIN, and J. H. WILKINSON: Calculation of the Eigenvalues of a symmetric tridiagonal matrix by the method of bisection. Numer. Math. 9, 386—393 (1967). Cf. II/5.

2. FRANCIS, J. G. F.: The QR transformation, Parts I and II. Comput. J. 4, 265—271, 332—345 (1961, 1962).

3. GIVENS, J. W.: Numerical computation of the characteristic values of a real symmetric matrix. Oak Ridge National Laboratory, ORNL-1574 (1954).

4. KUBLANOVSKAYA, V. N.: On some algorithms for the solution of the complete eigenvalue problem. Ž. Vyčisl. Mat. i Mat. Fiz. 1, 555—570 (1961).

5. MARTIN, R. S., and J. H. WILKINSON: Reduction of the symmetric eigenproblem $Ax=\lambda Bx$ and related problem to standard form. Numer. Math. **11**, 99—110 (1968). Cf.II/10.

6. — C. REINSCH, and J. H. WILKINSON: HOUSEHOLDER's tridiagonalization of a symmetric matrix. Numer. Math. **11**, 181—195 (1968). Cf.II/2.

7. RUTISHAUSER, H.: Solution of eigenvalue problems with the LR-transformation. Nat. Bur. Standards Appl. Math. Ser. **49**, 47—81 (1958).

8. WILKINSON, J. H.: The algebraic eigenvalue problem. London: Oxford University Press 1965.

9. — Convergence of the LR, QR and related algorithms. Comput. J. **8**, 77—84 (1965).

10. — The QR algorithm for real symmetric matrices with multiple eigenvalues. Comput. J. **8**, 85—87 (1965).

11. — Global convergence of tridiagonal QR algorithm with origin shifts. Lin. Alg. and its Appl. **1**, 409—420 (1968).

Contribution II/4

The Implicit QL Algorithm*

by A. Dubrulle, R. S. Martin and J. H. Wilkinson

1. Theoretical Background

In [1] an algorithm was described for carrying out the QL algorithm for a real symmetric matrix using shifts of origin. This algorithm is described by the relations

$$Q_s(A_s - k_s I) = L_s, \quad A_{s+1} = L_s Q_s^T + k_s I, \quad \text{giving} \quad A_{s+1} = Q_s A_s Q_s^T, \quad (1)$$

where Q_s is orthogonal, L_s is lower triangular and k_s is the shift of origin determined from the leading 2×2 matrix of A_s.

When the elements of A_s vary widely in their orders of magnitude the subtraction of k_s from the diagonal elements may result in a severe loss of accuracy in the eigenvalues of smaller magnitude. This loss of accuracy can be avoided to some extent by an alternative algorithm due to Francis [2] in which A_{s+1} defined by Eq. (1) is determined without subtracting $k_s I$ from A_s. With exact computation the A_{s+1} determined by the two algorithms would be identical, but the rounding errors are entirely different. The algorithm is based on the lemma that if A is symmetric and non-singular, and

$$BQ = QA \qquad (2)$$

with Q orthogonal and B tridiagonal with positive off-diagonal elements, then Q and B are fully determined when the last row of Q is specified. This can be verified by equating successively rows n, $n-1$, ..., 1 in Eq. (2). Hence if by any method we determine a tridiagonal matrix $\bar{A}_{s+1}$ such that

$$\bar{A}_{s+1} = \bar{Q}_s A_s \bar{Q}_s^T, \qquad (3)$$

where $\bar{Q}_s$ is orthogonal, then provided $\bar{Q}_s$ has the same last row as the Q_s given by (1), $Q_s = \bar{Q}_s$ and $\bar{A}_{s+1} = A_{s+1}$. In the original algorithm Q_s is given as the product of $n-1$ plane rotations

$$Q_s = P_1^{(s)} P_2^{(s)} \dots P_{n-1}^{(s)}, \qquad (4)$$

where $P_i^{(s)}$ is a rotation in the plane $(i, i+1)$. It is clear that the last row of Q_s is the last row of $P_{n-1}^{(s)}$. In the alternative algorithm a $\bar{Q}_s$ is determined satisfying (3), again as the product of rotations $\bar{P}_1^{(s)} \bar{P}_2^{(s)} \dots \bar{P}_{n-2}^{(s)} P_{n-1}^{(s)}$, the rotation $P_{n-1}^{(s)}$ being

---

* Prepublished in Numer. Math. **12**, 377–383 (1968).

that determined by the original algorithm. This ensures that $\overline{Q}_s = Q_s$. Writing

$$
P_{n-1}^{(s)} = \begin{bmatrix} 1 & & & & \\ & 1 & & & \\ & & \ddots & & \\ & & & 1 & \\ & & & c_{n-1}^{(s)} & -s_{n-1}^{(s)} \\ & & & s_{n-1}^{(s)} & c_{n-1}^{(s)} \end{bmatrix}, \qquad A^{(s)} = \begin{bmatrix} d_1^{(s)} & e_1^{(s)} & & & \\ e_1^{(s)} & d_2^{(s)} & e_2^{(s)} & & \\ & & \ddots & & \\ & & & e_{n-1}^{(s)} & d_n^{(s)} \end{bmatrix}, \qquad (5)
$$

where $P_{n-1}^{(s)}$ is defined as in the original algorithm, we have

$$
c_{n-1}^{(s)} = \frac{d_n^{(s)} - k_s}{p_{n-1}^{(s)}}, \qquad s_{n-1}^{(s)} = \frac{e_{n-1}^{(s)}}{p_{n-1}^{(s)}}, \qquad p_{n-1}^{(s)} = [(d_n^{(s)} - k_s)^2 + (e_{n-1}^{(s)})^2]^{\frac{1}{2}}. \qquad (6)
$$

The matrix $P_{n-1}^{(s)} A_s (P_{n-1}^{(s)})^T$ is of the form illustrated by

$$
\begin{bmatrix} x & x & & & & \\ x & x & x & & & \\ & x & x & x & x & \\ & & x & x & x & \\ & & x & x & x & \end{bmatrix}, \qquad (7)
$$

that is, it is equal to a tridiagonal matrix with additional elements in positions $(n, n-2)$ and $(n-2, n)$. This matrix can be reduced to tridiagonal form by $n-2$ orthogonal similarity transformations with matrices $\overline{P}_{n-2}^{(s)}, \ldots, \overline{P}_1^{(s)}$ respectively. (This is essentially Given's algorithm [3] for the reduction of a real symmetric matrix to tridiagonal form, performed starting with row n rather than row 1 and taking advantage of the special form of the original matrix.) Immediately before the transformation with matrix $P_r^{(s)}$ the current matrix is of the form illustrated when $n=6$, $r=3$ by

$$
\begin{bmatrix} x & x & & & & \\ x & x & x & & & \\ & x & x & x & x & \\ & & x & x & x & \\ & & x & x & x & x \\ & & & & x & x \end{bmatrix}. \qquad (8)
$$

Notice that the transformations are performed on matrix A_s itself and not on $A_s - k_s I$.

As in the original algorithm economy of computation is achieved by examining the $e_i^{(s)}$ in order to determine whether the A_s has effectively split into the direct sum of smaller tridiagonal matrices.

2. Applicability

The procedure *imtql1* is designed to find all the eigenvalues of a symmetric tridiagonal matrix.

The procedure *imtql2* is designed to find all the eigenvalues *and eigenvectors* of a symmetric tridiagonal matrix. If the tridiagonal matrix T has arisen from the Householder reduction of a full symmetric matrix A using *tred2* [4], then *imtql2* may be used to find the eigenvectors of A directly without first finding the eigenvectors of T.

3. Formal Parameter List

3.1. Input to procedure *imtql1*

n	order of tridiagonal matrix T.
macheps	the smallest number representable on the computer for which $1 + macheps > 1$.
d	the $n \times 1$ array of diagonal elements of T.
e	the $n \times 1$ array of subdiagonal elements of T, $e[1]$ may be arbitrary.

Output from procedure *imtql1*

d	the $n \times 1$ array giving the eigenvalues of T in ascending order.
e	is used as working storage and the original information is lost.
fail	the exit used if more than 30 iterations are used to isolate any one eigenvalue.

3.2. Input to procedure *imtql2*

n	order of tridiagonal matrix T.
macheps	the smallest number representable on the computer for which $1 + macheps > 1$.
d	the $n \times 1$ array of diagonal elements of T.
e	the $n \times 1$ array of subdiagonal elements of T, $e[1]$ may be arbitrary.
z	an $n \times n$ array which is equal to the identity matrix if we require the eigenvalues of T itself. If T was derived from a full matrix using procedure *tred2* then z should be the matrix of the Householder transformation provided as output by that procedure.

Output from procedure *imtql2*

d	the $n \times 1$ array giving the eigenvalues of T in ascending order.
e	is used as working storage and the original information is lost.
z	the $n \times n$ array giving the eigenvectors column by column of T if the input z was the identity matrix but giving the eigenvectors of a full matrix if T has been derived using *tred2*.

4. ALGOL Procedures

procedure *imtql1* $(n, macheps)$ *trans*: (d, e) *exit*: $(fail)$;
value n, *macheps*; **integer** n; **real** *macheps*; **array** d, e; **label** *fail*;
begin
    integer i, j, k, m, its; **real** h, c, p, q, s, t, u;

    for $i := 2$ **step** 1 **until** n **do** $e[i-1] := e[i]$;
    $e[n] := 0.0$; $k := n-1$;
    for $j := 1$ **step** 1 **until** n **do**
    begin
        $its := 0$;
        comment look for single small sub-diagonal element;

test:

    for $m := j$ **step** 1 **until** k **do**
       if $abs\,(e\,[m]) \leqq macheps \times (abs\,(d\,[m]) + abs\,(d\,[m+1]))$ **then goto** *cont1*;
    $m := n$;
cont1:

    $u := d\,[j]$;
    if $m \neq j$ **then**
    begin
      if $its = 30$ **then goto** *fail*;
      $its := its + 1$;
      comment form shift;
      $q := (d\,[j+1] - u)/(2 \times e\,[j])$; $t := sqrt\,(1 + q{\uparrow}2)$;
      $q := d\,[m] - u + e\,[j]/(\textbf{if } q < 0.0 \textbf{ then } q - t \textbf{ else } q + t)$;
      $u := 0.0$; $s := c := 1$;
      for $i := m - 1$ **step** -1 **until** j **do**
      begin
        $p := s \times e\,[i]$; $h := c \times e\,[i]$;
        if $abs\,(p) \geqq abs\,(q)$ **then**
        begin
          $c := q/p$; $t := sqrt\,(c{\uparrow}2 + 1.0)$;
          $e\,[i+1] := p \times t$; $s := 1.0/t$; $c := c \times s$
        end
        else
        begin
          $s := p/q$; $t := sqrt\,(s{\uparrow}2 + 1.0)$;
          $e\,[i+1] := q \times t$; $c := 1.0/t$; $s := s \times c$
        end;
        $q := d\,[i+1] - u$; $t := (d\,[i] - q) \times s + 2.0 \times c \times h$;
        $u := s \times t$; $d\,[i+1] := q + u$; $q := c \times t - h$
      end i;
      $d\,[j] := d\,[j] - u$; $e\,[j] := q$; $e\,[m] := 0.0$;   **goto** *test*
    end $m \neq j$;

    comment order eigenvalue;
    for $i := j$ **step** -1 **until** 2 **do**
      if $u < d\,[i-1]$ **then** $d\,[i] := d\,[i-1]$ **else goto** *cont2*;
    $i := 1$;
cont2:

      $d\,[i] := u$
   end j
end *imtql1*;

procedure *imtql2* $(n, macheps)$ *trans*: (d, e, z) *exit*: $(fail)$;
value n, *macheps*;   **integer** n;   **real** *macheps*;   **array** d, e, z;   **label** *fail*;
begin
   integer i, ia, j, k, m, its;   **real** h, c, p, q, s, t, u;
   for $i := 2$ **step** 1 **until** n **do** $e\,[i-1] := e\,[i]$;
   $e\,[n] := 0.0$; $k := n - 1$;

```
for j := 1 step 1 until n do
begin
    its := 0;
    comment look for single small sub-diagonal element;
test:
    for m := j step 1 until k do
        if abs(e[m]) ≤ macheps × (abs(d[m]) + abs(d[m+1])) then goto cont1;
    m := n;
cont1:
    u := d[j];
    if m ≠ j then
    begin
        if its = 30 then goto fail;
        its := its + 1;
        comment form shift;
        q := (d[j+1] − u)/(2×e[j]); t := sqrt(1 + q↑2);
        q := d[m] − u + e[j]/(if q < 0.0 then q − t else q + t);
        u := 0.0; s := c := 1;
        for i := m − 1 step − 1 until j do
        begin
            p := s×e[i]; h := c×e[i];
            if abs(p) ≥ abs(q) then
            begin
                c := q/p; t := sqrt(c↑2 + 1.0);
                e[i+1] := p×t; s := 1.0/t; c := c×s
            end
            else
            begin
                s := p/q; t := sqrt(s↑2 + 1.0);
                e[i+1] := q×t; c := 1.0/t; s := s×c
            end;
            q := d[i+1] − u; t := (d[i] − q)×s + 2.0×c×h;
            u := s×t; d[i+1] := q + u; q := c×t − h;
            comment form vector;
            for ia := 1 step 1 until n do
            begin
                p := z[ia, i+1];
                z[ia, i+1] := s×z[ia, i] + c×p;
                z[ia, i] := c×z[ia, i] − s×p
            end ia
        end i;
        d[j] := d[j] − u; e[j] := q; e[m] := 0.0; goto test
    end m ≠ j
end j;
comment order eigenvalues and eigenvectors;
for i := 1 step 1 until n do
begin
```

```
    k := i;  p := d[i];
    for j := i + 1 step 1 until n do
    if d[j] < p then
    begin
        k := j;  p := d[j]
    end;
    if k ≠ i then
    begin
        d[k] := d[i];  d[i] := p;
        for j := 1 step 1 until n do
        begin
            p := z[j, i];  z[j, i] := z[j, k];  z[j, k] := p
        end
    end
  end i
end imtql2;
```

5. Organisational and Notational Details

In general the comments given in [1] on the explicit QL algorithm apply equally to the implicit algorithm. However, with the explicit algorithm it was vital that tridiagonal matrices having elements which change rapidly in order of magnitude with progress along the diagonal should be presented with the largest elements in the bottom right-hand corner. With the implicit algorithm this is no longer so important. The implicit algorithm will be used in practice mainly when the tridiagonal matrix is the primary data. If the original matrix is full, it is already essential in the Householder reduction that it be orientated with the largest elements in the bottom right-hand corner and if this has been done the tridiagonal matrix will be in the appropriate form for the explicit QL algorithm. The speeds of the two algorithms are almost exactly the same.

After each iteration the subdiagonal elements are examined to determine whether A_s can be split. The element $e_i^{(s)}$ is regarded as negligible when

$$|e_i^{(s)}| \leq macheps\,(|d_{i+1}^{(s)}| + |d_i^{(s)}|). \tag{9}$$

6. Discussion of Numerical Properties

The comments given in [1] again apply in general except that for matrices of the special type described in Section 5 the small eigenvalues are given more accurately than with the explicit algorithm when the tridiagonal matrix is presented with its largest element in the top left-hand corner.

7. Test Results

The procedure *imtql2* was used on the matrices X and $\overline{X}$ defined respectively by

$$d_1, \ldots, d_7 = 1, 10^2, 10^4, 10^6, 10^8, 10^{10}, 10^{12},$$
$$e_2, \ldots, e_7 = 10, 10^3, 10^5, 10^7, 10^9, 10^{11},$$
$$d_1, \ldots, d_7 = 10^{12}, 10^{10}, 10^8, 10^6, 10^4, 10^2, 1,$$
$$e_2, \ldots, e_7 = 10^{11}, 10^9, 10^7, 10^5, 10^3, 10.$$

The eigenvalues given on the computer KDF9 are given in Table 1.

Table 1

Computed eigenvalues of X	Computed eigenvalues of $\overline{X}$
$-9.4634\,7415\,648_{10}+8;$	$-9.4634\,7415\,707_{10}+8;$
$-9.4634\,6919\,727_{10}+2;$	$-9.4634\,6919\,876_{10}+2;$
$+9.9989\,9020\,189_{10}-1;$	$+9.9989\,9020\,1\,57_{10}-1;$
$+1.0463\,3721\,478_{10}+3;$	$+1.0463\,3721\,466_{10}+3;$
$+1.0098\,9903\,020_{10}+6;$	$+1.0098\,9903\,020_{10}+6;$
$+1.0463\,3771\,269_{10}+9;$	$+1.0463\,3771\,265_{10}+9;$
$+1.0100\,0000\,980_{10}+12;$	$+1.0100\,0000\,980_{10}+12;$

The computed eigenvalues of X are of very high accuracy, even those which are small with respect to $\|X\|_2$; the computed eigenvalues of $\overline{X}$ ($\overline{X}$ has the same eigenvalues as X) are not in general quite as accurate, but all of them have at least ten correct decimals. This is in spite of the fact that in $\overline{X}$ the largest elements are in the top left-hand corner. The results obtained with $\overline{X}$ compare very favourably with those obtained using the explicit algorithm [1]; with that algorithm the computed eigenvalue of smallest modulus was incorrect even in its most significant figure.

In Table 2 we give the eigenvector corresponding to the eigenvalue of smallest modulus for each matrix. (The exact vectors should have the same components but in opposite orders.) It will be seen that the eigenvectors are in quite good agreement (except for the component of order 10^{-14} which one can hardly expect to be accurate). With the explicit algorithm the corresponding computed eigenvector of $\overline{X}$ is extremely inaccurate.

Table 2

Computed 3rd eigenvector of X	Computed 3rd eigenvector of $\overline{X}$
$+9.9994\,9513\,816_{10}-1;$	$+9.9985\,0548\,432_{10}-7;$
$-1.0097\,4711\,993_{10}-5;$	$-9.9985\,0548\,417_{10}-6;$
$-9.9984\,9548\,745_{10}-3;$	$-1.1252\,6974\,437_{10}-14;$
$+9.9985\,0548\,602_{10}-4;$	$+9.9985\,0548\,546_{10}-4;$
$-1.0097\,4710\,982_{10}-14;$	$-9.9984\,9548\,703_{10}-3;$
$-9.9985\,0548\,500_{10}-6;$	$-1.0097\,4740\,462_{10}-5;$
$+9.9985\,0548\,498_{10}-7;$	$+9.9994\,9513\,813_{10}-1;$

These results have been confirmed at the Technische Hochschule München.

Acknowledgements. The work described here has been carried out at the National Physical Laboratory. The authors wish to thank Dr. C. REINSCH of the Technische Hochschule München for a number of helpful suggestions and for carrying out the test runs.

References

1. BOWDLER, HILARY, R. S. MARTIN, C. REINSCH, and J. H. WILKINSON: The QR and QL algorithms for symmetric matrices. Numer. Math. **11**, 293—306 (1968). Cf. II/3.

2. FRANCIS, J. G. F.: The QR transformation, Part I and II. Comput. J. **4**, 265—271, 332—345 (1961, 1962).

3. GIVENS, J. W.: A method for computing eigenvalues and eigenvectors suggested by classical results on symmetric matrices. Nat. Bur. Standards Appl. Math. Ser. **29**, 117—122 (1953).

4. MARTIN, R. S., C. REINSCH, and J. H. WILKINSON: Householder's tridiagonalization of a symmetric matrix. Numer. Math. **11**, 181—195 (1968). Cf. II/2.

Calculation of the Eigenvalues of a Symmetric Tridiagonal Matrix by the Method of Bisection*

by W. Barth, R. S. Martin, and J. H. Wilkinson

1. Theoretical Background

The procedure *bisect* is designed to replace the procedures *tridibi 1* and *2* given in [5]. All three procedures are based essentially on the following theorem.

Given a symmetric tridiagonal matrix with diagonal elements c_1 to c_n and off-diagonal elements b_2 to b_n (with b_1 set equal to zero for completeness) and $b_i \neq 0$, then for any number λ let the sequence $p_0(\lambda), \ldots, p_n(\lambda)$ be defined by

$$p_0(\lambda) = 1, \qquad p_1(\lambda) = c_1 - \lambda, \tag{1}$$

$$p_i(\lambda) = (c_i - \lambda) p_{i-1}(\lambda) - b_i^2 p_{i-2}(\lambda) \qquad (i = 2, \ldots, n). \tag{2}$$

Then, in general, the number, $a(\lambda)$, of disagreements in sign between consecutive numbers of the sequence is equal to the number of eigenvalues smaller than λ.

When carried out in floating-point arithmetic, this method is extremely stable but unfortunately, even for matrices of quite modest order, it is common for the later $p_i(\lambda)$ to pass outside the range of permissible numbers. In practice both underflow and overflow occur and it is almost impossible to scale the matrix to avoid this. The difficulty is particularly acute when there are a number of very close eigenvalues since $p_n(\lambda) = \Pi(\lambda - \lambda_i)$ is then very small for any λ in their neighbourhood. The zeros of each $p_r(\lambda)$ separate those of $p_{r+1}(\lambda)$ and accordingly quite a number of the $p_i(\lambda)$ other than $p_n(\lambda)$ may also be very small and give rise to underflow.

The difficulty is avoided by replacing the sequence of $p_i(\lambda)$ by a sequence of $q_i(\lambda)$ defined by

$$q_i(\lambda) = p_i(\lambda)/p_{i-1}(\lambda) \qquad (i = 1, \ldots, n) \tag{3}$$

$a(\lambda)$ is now given by the number of negative $q_i(\lambda)$. The $q_i(\lambda)$ satisfy the relations

$$q_1(\lambda) = c_1 - \lambda, \tag{4}$$

$$q_i(\lambda) = (c_i - \lambda) - b_i^2/q_{i-1}(\lambda) \qquad (i = 2, \ldots, n). \tag{5}$$

---

* Prepublished in Numer. Math. 9, 386—393 (1967).

At first sight these relations look very dangerous, since it is possible for $q_{i-1}(\lambda)$ to be zero for some i; but in such cases it is merely necessary to replace the zero $q_{i-1}(\lambda)$ by a suitably small quantity and the error analysis for the $p_i(\lambda)$ sequence [4, 6] applies almost unaltered to the $q_i(\lambda)$. The $q_i(\lambda)$ do not suffer from the problems of underflow and overflow associated with the $p_i(\lambda)$. Comparing the computation of $q_i(\lambda)$ with that of $p_i(\lambda)$ we observe that two multiplications have been replaced by one division; further we have only to detect the sign of each $q_i(\lambda)$ instead of comparing the signs of $p_{i-1}(\lambda)$ and $p_i(\lambda)$. It is easy to see that when working with the $q_i(\lambda)$ the condition $b_i \neq 0$ can be omitted.

In the earlier paper [5] each eigenvalue was found independently by bisection. This is inefficient since, as observed by GIVENS in his original paper [2], some of the information obtained when determining one eigenvalue is, in general, of significance in the determination of other eigenvalues. In *bisect* full advantage is taken of all relevant information and this results in a very substantial saving of time when the matrix has a number of close or coincident eigenvalues.

2. Applicability

bisect may be used to find the eigenvalues $\lambda_{m1}, \lambda_{m1+1}, \ldots, \lambda_{m2}$ $(\lambda_{i+1} \geq \lambda_i)$ of a symmetric tridiagonal matrix of order n. Unsymmetric tridiagonal matrices A with $a_{i,i+1} = f_i$ and $a_{i+1,i} = g_i$ may be treated provided $f_i g_i \geq 0$ $(i = 1, \ldots, n-1)$ by taking $b_{i+1}^2 = f_i g_i$.

3. Formal Parameter List

Input to procedure *bisect*

c an $n \times 1$ array giving the diagonal elements of a tridiagonal matrix.

b an $n \times 1$ array giving the sub-diagonal elements. $b[1]$ may be arbitrary but is replaced by zero in the procedure.

beta   an $n \times 1$ array giving the squares of the sub-diagonal elements. *beta*[1] may be arbitrary but is replaced by zero in the procedure. Both $b[i]$ and *beta*[i] are given since the squares may be the primary data. If storage economy is important then one of these can be dispensed with in an obvious way.

n the order of the tridiagonal matrix.

m1, m2 the eigenvalues $\lambda_{m1}, \lambda_{m1+1}, \ldots, \lambda_{m2}$ are calculated (λ_1 is the smallest eigenvalue). $m1 \leq m2$ must hold otherwise no eigenvalues are computed.

eps1   a quantity affecting the precision to which the eigenvalues are computed (see "Discussion of numerical properties").

relfeh the smallest number for which $1 + relfeh > 1$ on the computer.

Output of procedure *bisect*

eps2   gives information concerning the accuracy of the results (see "Discussion of numerical properties").

z total number of bisections to find all required eigenvalues.

x array $x[m1:m2]$ contains the computed eigenvalues.

4. ALGOL Program

procedure *bisect* $(c, b, beta, n, m1, m2, eps1, relfeh)$ **res**: $(eps2, z, x)$;
value $n, m1, m2, eps1, relfeh$;
real $eps1, eps2, relfeh$; **integer** $n, m1, m2, z$; **array** $c, b, x, beta$;
comment c is the diagonal, b the sub-diagonal and *beta* the squared subdiagonal
of a symmetric tridiagonal matrix of order n. The eigenvalues
$lambda[m1], \ldots, lambda[m2]$, where $m2$ is not less than $m1$ and
$lambda[i+1]$ is not less than $lambda[i]$, are calculated by the method
of bisection and stored in the vector x. Bisection is continued until the
upper and lower bounds for an eigenvalue differ by less than $eps1$,
unless at some earlier stage, the upper and lower bounds differ only
in the least significant digits. $eps2$ gives an extreme upper bound for
the error in any eigenvalue, but for certain types of matrices the small
eigenvalues are determined to a very much higher accuracy. In this
case, $eps1$ should be set equal to the error to be tolerated in the smallest
eigenvalue. It must not be set equal to zero;
begin   **real** $h, xmin, xmax$; **integer** i;
comment Calculation of $xmin, xmax$;
$beta[1] := b[1] := 0$;
$xmin := c[n] - abs(b[n])$;
$xmax := c[n] + abs(b[n])$;
for $i := n-1$ **step** -1 **until** 1 **do**
begin $h := abs(b[i]) + abs(b[i+1])$;
$\qquad$ **if** $c[i]+h > xmax$ **then** $xmax := c[i]+h$;
$\qquad$ **if** $c[i]-h < xmin$ **then** $xmin := c[i]-h$;
end i;
$eps2 := relfeh \times ($**if** $xmin+xmax > 0$ **then** $xmax$ **else** $-xmin)$;
if $eps1 \leq 0$ **then** $eps1 := eps2$;
$eps2 := 0.5 \times eps1 + 7 \times eps2$;
comment Inner block;
begin$\qquad$ **integer** a, k; **real** $q, x1, xu, x0$; **array** $wu[m1:m2]$;
$\qquad\qquad$ $x0 := xmax$;
$\qquad\qquad$ **for** $i := m1$ **step** 1 **until** $m2$ **do**
$\qquad\qquad$ **begin** $x[i] := xmax$; $wu[i] := xmin$
$\qquad\qquad$ **end** i;
$\qquad\qquad$ $z := 0$;
$\qquad\qquad$ **comment** Loop for the k-th eigenvalue;
$\qquad\qquad$ **for** $k := m2$ **step** -1 **until** $m1$ **do**
$\qquad\qquad$ **begin** $xu := xmin$;
$\qquad\qquad\qquad$ **for** $i := k$ **step** -1 **until** $m1$ **do**
$\qquad\qquad\qquad$ **begin if** $xu < wu[i]$ **then**
$\qquad\qquad\qquad\qquad$ **begin** $xu := wu[i]$; **go to** *contin*
$\qquad\qquad\qquad\qquad$ **end**
$\qquad\qquad\qquad$ **end** i;
$\qquad\qquad$ *contin*:$\quad$ **if** $x0 > x[k]$ **then** $x0 := x[k]$;
$\qquad\qquad\qquad$ **for** $x1 := (xu+x0)/2$ **while** $x0 - xu > 2 \times relfeh \times$
$\qquad\qquad\qquad\qquad\qquad\qquad$ $(abs(xu) + abs(x0)) + eps1$ **do**

begin $z := z+1$;
  comment Sturms sequence;
  $a := 0$; $q := 1$;
  for $i := 1$ **step** 1 **until** n **do**
  begin
    $q := c[i] - x1 - ($**if** $q \neq 0$ **then** $beta[i]/q$
            else $abs\,(b\,[i])/relfeh)$;
    if $q < 0$ **then** $a := a+1$
  end i;
  if $a < k$ **then**
  begin if $a < m1$ **then**
    $xu := wu\,[m1] := x1$
    else
    begin $xu := wu\,[a+1] := x1$;
      if $x\,[a] > x1$ **then** $x\,[a] := x1$
    end
  end
  else $x0 := x1$
  end $x1$;
  $x\,[k] := (x0 + xu)/2$
end k
end *inner block*
end *bisect*;

5. Organisational and Notational Details

The value of *relfeh* is machine dependent and is directly related to the precision of the arithmetic which is used. On a computer with a t digit binary mantissa *relfeh* is of the order of 2^{-t}. Errors in c_i and b_i of up to $|c_i|$ *relfeh* and $|b_i|$ *relfeh* in absolute magnitude may be introduced by the digital representation.

From GERSCHGORINS' theorem the eigenvalues are all contained in the union of the n intervals $c_i \pm (|b_i| + |b_{i+1}|)$ with $b_1 = b_{n+1} = 0$. Hence *xmax* and *x min* defined by

$$\frac{xmax}{xmin} = \frac{max}{min}\,\{c_i \pm (|b_i| + |b_{i+1}|)\} \tag{6}$$

$$i = 1, 2, \ldots, n$$

can be used as initial upper and lower bounds for the eigenvalues.

When computing a typical eigenvalue λ_k two quantities $x0$ and xu which are current upper and lower bounds for λ_k are stored. Bisection is continued as long as

$$x0 - xu > 2\,relfeh\,(|xu| + |x0|) + eps1 \tag{7}$$

where *eps 1* is a preassigned tolerance. The significance of this criterion is discussed in the next section. When (7) is no longer satisfied $\frac{1}{2}(x0 + xu)$ gives the current eigenvalue λ_k.

If at any stage in the computation of a Sturm sequence $q_{i-1}(\lambda)$ is zero then (5) is replaced by

$$q_i(\lambda) = (c_i - \lambda) - |b_i|/relfeh. \tag{8}$$

This means in effect that $q_{i-1}(\lambda)$ is replaced by $|b_i|$ *relfeh* (note that this is positive; this is essential because the zero $q_{i-1}(\lambda)$ is treated as positive) and this is equivalent to replacing c_{i-1} by $c_{i-1}+|b_i|$ *relfeh*.

The eigenvalues are found in the order $\lambda_{m2}, \lambda_{m2-1}, \ldots, \lambda_{m1}$. Two arrays $wu[m1:m2]$ and $x[m1:m2]$ are used to store relevant information on the lower and upper bounds of the required eigenvalues. Initially we have $xu = wu[i] = xmin$, $x0 = x[i] = xmax$ $(i = m1, \ldots, m2)$. In practice it is not necessary to keep all the $wu[i]$ and $x[i]$ fully updated provided it is possible to determine the current best upper and lower bounds for any eigenvalue when required. This is achieved as follows.

If, during the calculation of λ_k corresponding to an argument $x1$, we have $a(x1) \geq k$, the only useful deduction is $x0 := x1$. If, however, $a(x1) < k$, then $x1$ is an upper bound for $\lambda_{m1}, \lambda_{m1+1}, \ldots, \lambda_a$ and a lower bound for $\lambda_{a+1}, \lambda_{a+2}, \ldots, \lambda_k$. Hence in any case $xu := x1$. If $a < m1$ then $wu[m1] := x1$; otherwise $wu[a+1] := x1$ and $x[a] := x1$ if this is an improved upper bound.

To find the initial values of $x0$ and xu when computing λ_k we take xu to be the largest of $xmin$ and $wu[i]$ $(i = m1, \ldots, k)$, and $x0$ to be the smaller of $x0$ and $x[k]$. No superfluous information is stored and the total number of bisections required when there are multiple or close eigenvalues is significantly less than when the eigenvalues are well separated.

6. Discussion of Numerical Properties

To understand the criterion for the termination of the bisection process it is necessary to consider the limitations on the accuracy obtainable by the bisection method. *In general*, using floating point computation with a t digit mantissa, errors of the order of magnitude $2^{-t} max[|xmax|, |xmin|]$ are inevitable in all eigenvalues and they cannot be reduced by increasing the number of bisection steps. This means that, in general, the relative error in the smaller eigenvalues is higher than in larger ones.

However, there are certain types of matrices for which it is possible to determine the lower eigenvalues with the same low relative error as the higher eigenvalues. An example of such a matrix is given by

$$c_i = i^4, \qquad b_i = i - 1 \qquad (i = 1, \ldots, 30). \tag{9}$$

This matrix has eigenvalues which are roughly of the order of magnitude i^4 $(i = 1, \ldots, 30)$ and variations of the elements by up to one part in 2^t change all the eigenvalues by roughly one part in 2^t. The smaller eigenvalues are therefore determined by the data to the same relative accuracy as the larger ones. Moreover by continuing the bisection process long enough this accuracy can actually be attained in practice though it may be more efficient to use the QD-algorithm [3] or the symmetric QR algorithm [1]. Matrices of this type are common in theoretical physics where they arise by truncating infinite tridiagonal matrices. Here it is the smallest eigenvalues of the infinite matrix which are required and by taking a truncated matrix of sufficiently high order these can be obtained to any prescribed accuracy.

The criterion (7) enables us to deal with matrices of the non-special or special types with considerable flexibility. If we have a non-special matrix, then a value of *eps1* approximately equal to $2^{-t} max[|xmax|, |xmin|]$ will give the maximum attainable accuracy in each eigenvalues without any superfluous bisection steps. If a lower accuracy is adequate then this is achieved by taking a larger value of *eps 1*; for example if eigenvalues are required up to the third decimal place (*not the third significant figure*), *eps 1* may be taken equal to $\frac{1}{2} 10^{-3}$ and if the norm of the matrix is of order unity the term $2 \, relfeh \times (|xu| + |x0|)$ on the right hand side of (7) will be of negligible significance.

With the special type of matrix on the other hand, when the small eigenvalues are required with low relative error it is the term $2 \, relfeh (|xu| + |x0|)$ which plays the dominant role. If the term *eps1* is omitted, then, in general, bisection would continue until all but the last one or two digits of $x0$ and xu agree. However, if one of the eigenvalues were zero, this condition might never be attained and accordingly *eps1* should be set equal to the error which is acceptable in the smallest eigenvalue. To guard against the accidental assignment of a non-positive value of *eps1*, such values are replaced by $relfeh \times max[|xmax|, |xmin|]$, thus avoiding the danger of indefinite cycling.

Bounds for the errors may be obtained by the method described in [4, 6]. The overall bound for the error in any eigenvalue is given by

$$\tfrac{1}{2} \, eps1 + 7 \, relfeh \times max[|x \, max|, |x \, min|] \tag{10}$$

and this quantity is computed and output as *eps 2*. It should be appreciated that for the matrices of special type the errors in the smaller eigenvalues will be much smaller than this.

7. Test Results

The procedure *bisect* has been tested on a number of matrices using the IBM 7040 at Darmstadt and the KDF 9 at the National Physical Laboratory. Tests of an earlier version of the procedure were also carried out on the CDC 1604 computer of the Swiss Federal Institute of Technology, Zurich, Switzerland. The following illustrate the points of interest discussed earlier.

The first tridiagonal matrix was obtained by the Householder reduction of a matrix A of order 50 with $a_{ij}=1$ (all i, j). This matrix has the elements

$$c_1=1, \qquad c_2=49, \qquad c_i=0 \qquad (i=3,\ldots,50),$$
$$b_2=7, \qquad b_i=0 \qquad\qquad\;\; (i=3,\ldots,50),$$

and has one eigenvalue equal to 50 and the other 49 equal to zero. This matrix gives rise to underflow with the original procedures *tridibi 1* and *2* [5]. The computed eigenvalues on KDF 9 ($relfeh=2^{-39}$) and with $eps1=10^{-10}$ were

$$\lambda_1, \lambda_2, \ldots, \lambda_{49} = 2.27373\,675443_{10} - 11, \qquad \lambda_{50} = 5.00000\,000001_{10} + 1$$

and the output value of *eps2* was $7.6304_{10}-10$. A total of 73 bisections were needed, none being required for the eigenvalues $\lambda_1, \ldots, \lambda_{48}$. The number of iterations needed with the earlier procedures would have been 50×39.

The second matrix is of the special type and is of order 30 with the elements

$$c_i=i^4, \qquad b_i=i-1 \qquad (i=1, 2, \ldots, 30).$$

This was solved on KDF 9 using $eps1=10^{-8}$, 10^{-10}, and 10^{-12} respectively. The results obtained are given in Table 1.

Table 1

$eps1=10^{-8}$ eigenvalues		$eps1=10^{-10}$ eigenvalues	$eps1=10^{-12}$ eigenvalues
$9.3340\,7085\,678_{10}-1$	$6.5536\,0008\,557_{10}+4$	$9.3340\,7084\,825_{10}-1$	$9.3340\,7084\,869_{10}-1$
$1.6005\,0653\,722_{10}+1$	$8.3521\,0007\,641_{10}+4$	$1.6005\,0653\,704_{10}+1$	$1.6005\,0653\,703_{10}+1$
$8.1010\,1005\,468_{10}+1$	$1.0497\,6000\,687_{10}+5$	$8.1010\,1005\,456_{10}+1$	The remaining
$2.5600\,8066\,892_{10}+2$	$1.3032\,1000\,620_{10}+5$	$2.5600\,8066\,893_{10}+2$	values as for
$6.2500\,6102\,376_{10}+2$	$1.6000\,0000\,563_{10}+5$	$6.2500\,6102\,372_{10}+2$	$eps1=10^{-10}$
$1.2960\,0467\,858_{10}+3$	$1.9448\,1000\,514_{10}+5$	$1.2960\,0467\,858_{10}+3$	
$2.4010\,0367\,061_{10}+3$	$2.3425\,6000\,470_{10}+5$	$2.4010\,0367\,062_{10}+3$	
$4.0960\,0294\,507_{10}+3$	$2.7984\,1000\,431_{10}+5$	The remaining values	
$6.5610\,0241\,013_{10}+3$	$3.3177\,6000\,399_{10}+5$	as for $eps1=10^{-8}$	
$1.0000\,0020\,063_{10}+4$	$3.9062\,5000\,369_{10}+5$		
$1.4641\,0016\,947_{10}+4$	$4.5697\,6000\,342_{10}+5$		
$2.0736\,0014\,498_{10}+4$	$5.3144\,1000\,319_{10}+5$		
$2.8561\,0012\,540_{10}+4$	$6.1465\,6000\,296_{10}+5$		
$3.8416\,0010\,949_{10}+4$	$7.0728\,1000\,278_{10}+5$		
$5.0625\,0009\,644_{10}+4$	$8.1000\,0008\,188_{10}+5$		
$eps2=1.0319_{10}-5$		$eps2=1.0314_{10}-5$	$eps2=1.0314_{10}-5$

With a value of $eps1$ of 10^{-12} all eigenvalues were given correctly to within five units in the last significant figure; this represents a far smaller error than $eps2$ in the smaller eigenvalues.

The final matrix is defined by

$$c_i=110-10i \quad (i=1,\ldots,11), \qquad c_i=10i-110 \quad (i=12,\ldots,21),$$
$$b_i=1 \quad (i=2,\ldots,21).$$

This matrix has a number of extremely close (but not coincident) eigenvalues. It was solved on KDF 9 using $eps1=10^{-7}$ and the results are given in Table 2.

Table 2

$-1.9709\,2929\,389_{10}-1$	$5.9999\,9999\,539_{10}+1$
$9.9004\,9425\,559_{10}+0$	$5.9999\,9999\,539_{10}+1$
$1.0096\,5954\,703_{10}+1$	$7.0000\,0000\,470_{10}+1$
$1.9999\,5066\,165_{10}+1$	$7.0000\,0000\,470_{10}+1$
$2.0000\,4966\,711_{10}+1$	$8.0000\,0008\,116_{10}+1$
$2.9999\,9991\,949_{10}+1$	$8.0000\,0008\,116_{10}+1$
$3.0000\,0008\,256_{10}+1$	$9.0000\,4933\,900_{10}+1$
$3.9999\,9999\,595_{10}+1$	$9.0000\,4933\,900_{10}+1$
$3.9999\,9999\,595_{10}+1$	$1.0009\,9505\,751_{10}+2$
$4.9999\,9999\,567_{10}+1$	$1.0009\,9505\,751_{10}+2$
$4.9999\,9999\,567_{10}+1$	$eps2=1.0128_{10}-7$

The error in each computed values is a few units in the underlined figure and is in each case less than $\frac{1}{2}10^{-7}$ in absolute magnitude. The total number of iteration was 345 an average of 16.4 per eigenvalue. It should be appreciated that by using a smaller value of $eps1$ the results can be obtained correct to working accuracy.

Acknowledgements. The authors wish to thank Professor Dr. H. Rutishauser for important suggestions, and Dipl.-Math. L. Krauss for help in writing and testing a preliminary version of the procedure made at Darmstadt. The contribution by R. S. Martin and J. H. Wilkinson has been carried out at the National Physical Laboratory.

References

[1] Bowdler, Hilary, R. S. Martin, C. Reinsch and J. H. Wilkinson: The QR and QL algorithms for symmetric matrices. Numer. Math. **11**, 293—306 (1968). Cf. II/3.

[2] Givens, J. W.: Numerical computation of the characteristic values of a real symmetric matrix. Oak Ridge National Laboratory, ORNL-1574 (1954).

[3] Rutishauser, H.: Stabile Sonderfälle des Quotienten-Differenzen-Algorithmus. Numer. Math. **5**, 95—11 (1963).

[4] Wilkinson, J. H.: Error analysis of floating-point computation. Numer. Math. **2**, 319—340 (1960).

[5] — Calculation of the eigenvalue of a symmetric tridiagonal matrix by the method of bisection. Numer. Math. **4**, 362—367 (1962).

[6] — The algebraic eigenvalue problem. London: Oxford University Press 1965.

Contribution II/6

Rational QR Transformation with Newton Shift for Symmetric Tridiagonal Matrices*

by C. REINSCH and F. L. BAUER

1. Theoretical Background

1.1. Introduction

If some of the smallest or some of the largest eigenvalues of a symmetric (tridiagonal) matrix are wanted, it suggests itself to use monotonic Newton corrections in combination with QR steps. If an initial shift has rendered the matrix positive or negative definite, then this property is preserved throughout the iteration. Thus, the QR step may be achieved by two successive Cholesky LR steps or equivalently, since the matrix is tridiagonal, by two QD steps which are numerically stable [4] and avoid square roots. The rational QR step used here needs slightly fewer additions than the Ortega-Kaiser step [3].

Moreover, it turns out that the next Newton correction can be easily calculated from the q's of the QD decompositions [8]. The combination of Newton corrections and diagonalization by QR transformations guarantees global convergence and stable deflation.

1.2. The Rational QR Algorithm

Given are the entries of a tridiagonal matrix,

$$
(1) \qquad A_0 = \begin{bmatrix} d_1 & 1 & & & 0 \\ b_2^2 & d_2 & 1 & & \\ & b_3^2 & \cdot & \cdot & \\ & & \cdot & \cdot & \cdot & 1 \\ 0 & & & b_n^2 & d_n \end{bmatrix}
$$

either as primary data or as obtained by some reduction of an Hermitian matrix to tridiagonal form. From GERSHGORIN's theorem a lower bound for the spectrum is obtained, and a corresponding shift produces a starting matrix, A, with a positive spectrum (an upward shift being skipped if the original matrix is known to be positive definite).

* Prepublished in Numer. Math. **11**, 264—272 (1968).

The rational QR step starts with a shift by s, $A - sI$ is decomposed in the QD way

$$(2) \qquad A - sI = LR = \begin{bmatrix} 1 & & & & 0 \\ e_2 & 1 & & & \\ & e_3 & 1 & & \\ & & & \ddots & \\ 0 & & & e_n & 1 \end{bmatrix} \begin{bmatrix} q_1 & 1 & & & 0 \\ & q_2 & 1 & & \\ & & q_3 & & \\ & & & \ddots & 1 \\ 0 & & & & q_n \end{bmatrix}.$$

Then $RL = A'$ is formed and is again decomposed, $A' = L'R'$. Finally, $A'' = R'L'$ is formed, which is A in the next step. This gives the algorithm ($b_{n+1} = 0$):

$$(3) \qquad \begin{aligned} q_1 &= d_1 - s, \\ e_2 &= b_2^2/q_1, & q_1' &= q_1 + e_2, \\ q_i &= d_i - s - e_i, & e_i' &= e_i(q_i/q_{i-1}'), & d_{i-1}'' &= q_{i-1}' + e_i', \\ e_{i+1} &= b_{i+1}^2/q_i, & q_i' &= q_i - e_i' + e_{i+1}, & b_i''^2 &= e_i' q_i', \\ & & & d_n'' = q_n'. \end{aligned} \left.\begin{aligned} & \\ & \\ & \\ & \end{aligned}\right\} i = 2(1)\,n$$

We have ($e_1' = 0$)

$$(4) \qquad e_i/q_{i-1}' = e_i'/q_i \quad \text{and} \quad q_i' - e_{i+1} = q_i - e_i'.$$

Therefore

$$q_i' - e_{i+1} = \frac{q_i}{q_{i-1}'}(q_{i-1}' - e_i) = \frac{q_i \cdots q_2}{q_{i-1}' \cdots q_1'}(q_1' - e_2)$$

and

$$(5) \qquad q_i - e_i' = \frac{q_i \cdots q_1}{q_{i-1}' \cdots q_1'}.$$

1.3. Calculation of the Newton Shift

According to (2),

$$f(s) := \det(A - sI) = q_1 \cdots q_n = q_1' \cdots q_n'$$

is the value of the characteristic polynomial of A at the point s. The next shift is the Newton correction $-f(s)/\dot{f}(s)$, and the calculation of the derivative $\dot{f}(s) := df/ds$ at the point s is the crucial step. From the recursion $q_i = d_i - s - b_i^2/q_{i-1}$ it follows by differentiation

$$\dot{q}_i = -1 + b_i^2\,\dot{q}_{i-1}/q_{i-1}^2 = (e_i\,\dot{q}_{i-1} - q_{i-1})/q_{i-1}.$$

By induction we find with the help of (5) and (4)

$$\dot{q}_i = -\frac{q_1' \cdots q_{i-1}'}{q_1 \cdots q_{i-1}}.$$

Introducing

$$p_i := \frac{d}{ds}(-q_1 \cdots q_i)/q_1' \cdots q_{i-1}',$$

we have

$$p_1 = 1$$

and

$$p_i = \left[\frac{d}{ds} (-q_1 \cdots q_{i-1}) \cdot q_i + (-q_1 \cdots q_{i-1}) \cdot \frac{dq_i}{ds} \right] / q_1' \cdots q_{i-1}',$$

therefore

$$p_i = p_{i-1}(q_i/q_{i-1}') + 1.$$

But q_n'/p_n is the wanted Newton correction. Since q_i/q_{i-1}' occurs already in the rational QR algorithm (3), only n additional multiplications and additions are needed to form the sequence of p's and thereby the next Newton correction. Note that this correction can be expressed in the quantities $q_i - e_i'$, in fact we have

$$-f(s)/\dot{f}(s) = 1 \bigg/ \sum_{i=1}^{n} (q_i - e_i')^{-1}.$$

1.4. Test for Convergence and Deflation

We tolerate an error not greater than $k\delta$ in the k-th computed eigenvalue, where δ is a given non-negative number. To this end iteration is terminated if any of the quantities $q_i - e_i'$ is not greater than δ.

Proof. The matrices L and R' can be factorized,

$$L = L_1 \ldots L_n, \qquad R' = R_n' \ldots R_1',$$

where L_i is the unit matrix with an additional entry e_{i+1} at position $(i+1, i)$, while R_i' differs from the unit matrix in positions (i, i) and $(i, i+1)$ where it has entries q_i' and 1 respectively. In particular, $L_n = I$ and $R_n' = \mathrm{diag}(1, \ldots, 1, q_n')$. R_i' commutes with $L_{i+1}, \ldots, L_n$, hence

$$A'' = R_n' L_n^{-1} \ldots R_1' L_1^{-1} (A - sI) L_1 R_1'^{-1} \ldots L_n R_n'^{-1}.$$

The similarity transformations $X \to R_i' L_i^{-1} X L_i R_n'^{-1}$ may be considered as minor steps in the iteration, the $i-1^{\text{st}}$ minor step producing a matrix of form

(6)

$$\text{column } i$$

Due to (5) all elements are positive. Hence it is easily seen that the matrix (6) can be rendered symmetric by real scaling. The diagonal element $q_i - e_i'$ is an upper bound for the smallest eigenvalue. Moreover, if we replace all elements of the i-th row and i-th column by zero, this changes no eigenvalue by more than $q_i - e_i'$, according to the following theorem of WILKINSON [7], p. 553 (see also [5]):

17*

Let M be a positive definite matrix with eigenvalues $\lambda_1 \leq \cdots \leq \lambda_n$, and $P = \mathrm{diag}(1, \ldots, 1, 0, 1, \ldots, 1)$ a projector with zero in its i-th column. Let $M' = P M P$ have eigenvalues $0 = \lambda'_1 \leq \cdots \leq \lambda'_n$. Then

$$0 \leq \lambda_k - \lambda'_k \leq m_{ii}, \qquad k = 1, \ldots, n.$$

The left-hand inequality follows from the well known separation theorem for symmetric matrices. The right-hand inequality comes from the fact that the sum of all positive differences $\lambda_k - \lambda'_k$ equals m_{ii}.

Hence, if $q_i - e'_i \leq \delta$ we accept zero as lowest eigenvalue of the current matrix and deflate by deleting the i-th row and column. The deflated matrix is again tridiagonal of form (1), its subdiagonal entries being

$$b''^2_2, \ldots, b''^2_{i-1}, e'_i e_{i+1}, b^2_{i+2}, \ldots, b^2_n.$$

For the extra element, we have

$$e'_i e_{i+1} = e_i (q_i/q'_{i-1}) e_{i+1} = (e_i/q'_{i-1}) b^2_{i+1},$$

where the factor e_i/q'_{i-1} is less than unity.

2. Applicability

The algorithm as written computes the m *lowest* eigenvalues of a symmetric tridiagonal matrix T, given by its diagonal elements, d_i, $(i = 1, \ldots, n)$ and the squares of its off-diagonal elements, b^2_i, $(i = 2, \ldots, n)$. The *upper* eigenvalues are found, if the signs of the diagonal entries are reversed before and after the call of the procedure. A non-symmetric tridiagonal matrix, T, may be handled, provided $t_{i-1,i} t_{i,i-1} \geq 0$, $(i = 2, \ldots, n)$, by taking $b^2_i = t_{i-1,i} t_{i,i-1}$.

Due to its quadratic rate of convergence the algorithm is well suited if high precision results are wanted. However, the method of bisection [1] may be preferable, if clusters of eigenvalues are to be expected.

3. Formal Parameter List

Input to procedure *ratqr*

n the order of the matrix T,

m the number of smallest eigenvalues wanted,

posdef **true**, if the matrix is known to be positive definite, **false** otherwise,

dlam   specifies a tolerance for the theoretical error of the computed eigenvalues: the theoretical error of the k-th eigenvalue is usually not greater than $k \times dlam$, $(dlam = 0$ is admissible),

eps   the machine precision, i.e. the smallest number x for which $1 + x > 1$ on the computer,

d a vector with n components, giving the diagonal of T,

b2   a vector with n components, giving the squares of the off-diagonal elements of T, $b2[1]$ may be arbitrary.

Output of procedure *ratqr*

d   the computed eigenvalues, stored as the lower components of the vector
        d in non-decreasing sequence, while the remaining components are lost,

b2 $b2[k]$ is a bound for the theoretical error of the k-th computed eigen-
        value $d[k]$, $k = 1, \ldots, m$. This bound does not comprise the effect of
        rounding errors. The remaining components are lost.

4. ALGOL Program

procedure *ratqr* $(n, m, posdef, dlam, eps)$ *trans*: $(d, b2)$;
   value $n, m, posdef, dlam, eps$;
   integer n, m;
   Boolean *posdef*;
   real *dlam, eps*;
   array $d, b2$;
comment QR algorithm for the computation of the lowest eigenvalues of a
       symmetric tridiagonal matrix. A rational variant of the QR trans-
       formation is used, consisting of two successive QD steps per iteration.
       A shift of the spectrum after each iteration gives an accelerated rate
       of convergence. A Newton correction, derived from the characteristic
       polynomial, is used as shift.
       Formats: $d, b2\ [1:n]$;
begin
     integer i, j, k;   **real** *delta, e, ep, err, p, q, qp, r, s, tot*;

comment Lower bound for eigenvalues from Gershgorin, initial shift;
       $b2[1] := err := q := s := 0$; $tot := d[1]$;
       for $i := n$ **step** -1 **until** 1 **do**
       begin
           $p := q$; $q := sqrt(b2[i])$; $e := d[i] - p - q$;
           if $e < tot$ **then** $tot := e$
       end i;
       if *posdef* $\wedge$ $tot < 0$ **then** $tot := 0$ **else**
       for $i := 1$ **step** 1 **until** n **do** $d[i] := d[i] - tot$;

       for $k := 1$ **step** 1 **until** m **do**
       begin
next QR transformation:
           $tot := tot + s$; $delta := d[n] - s$; $i := n$;
           $e := abs(eps \times tot)$;   **if** $dlam < e$ **then** $dlam := e$;
           if $delta \leq dlam$ **then goto** *convergence*;
           $e := b2[n]/delta$; $qp := delta + e$; $p := 1$;
           for $i := n - 1$ **step** -1 **until** k **do**
           begin
               $q := d[i] - s - e$; $r := q/qp$; $p := p \times r + 1$;
               $ep := e \times r$; $d[i+1] := qp + ep$; $delta := q - ep$;

if $delta \leq dlam$ **then goto** convergence;
$\qquad e := b2[i]/q; \quad qp := delta + e; \quad b2[i+1] := qp \times ep$
end i;
$d[k] := qp; \quad s := qp/p;$
if $tot + s > tot$ **then goto** next QR transformation;

comment Irregular end of iteration,
$\qquad$ deflate minimum diagonal element;
$\qquad s := 0; \quad i := k; \quad delta := qp;$
$\qquad$ **for** $j := k+1$ **step** 1 **until** n **do**
$\qquad\qquad$ **if** $d[j] < delta$ **then**
$\qquad\qquad\qquad$ **begin** $i := j; \quad delta := d[j]$ **end**;
convergence:
$\qquad$ **if** $i < n$ **then** $b2[i+1] := b2[i] \times e/qp;$
$\qquad$ **for** $j := i-1$ **step** -1 **until** k **do**
$\qquad\qquad$ **begin** $d[j+1] := d[j] - s; \quad b2[j+1] := b2[j]$ **end** j;
$\qquad d[k] := tot; \quad b2[k] := err := err + abs(delta)$
end k
end ratqr;

5. Organisational and Notational Details

(i) The initial shift is computed from GERSHGORIN'S theorem applied to the symmetric matrix with sub-diagonal entries $|b_2|, ..., |b_n|$. These quantities are computed from the given b_i^2 which are often primary data. If, however, the entries b_i are available, one may add them to the formal parameter list ($b_1 = 0$) and replace the expression $sqrt(b2[i])$ in line 7 of the procedure body by $abs(b[i])$.

(ii) As the first shift in the sequence of iterations for the determination of an eigenvalue we use the last shift of the foregoing sequence. It can be shown that this shift never exceeds the current eigenvalue.

(iii) The QD step is done in backward direction.

(iv) The iteration is repeated until the computed value of some difference $q_i - e_i'$ is not larger than the prescribed tolerance δ. Thus, all divisors are larger than δ *.

(v) Under the influence of rounding errors the computed values of $q_i - e_i'$ may never reach the specified tolerance. In fact, we can scarcely expect the relative error of the computed eigenvalue λ to be smaller than the machine precision ε. Hence, $\varepsilon |\lambda|$ is a lower limit for a reasonable tolerance and a smaller δ is replaced by this limit.

(vi) Iteration is also terminated if the sequence of total shifts does no longer increase. If such an irregular end of iteration occurs, the minimum diagonal element is selected and the standard deflation process is applied to the corresponding row and column. This situation, however, was never encountered and the appropriate piece of program was artificially tested by setting $\varepsilon = 0$.

---

* We assume a machine rounding such that the computed value of $q_i - (q_i/q_{i-1}') e_i$ is not positive if $q_i < 0$ and $0 < e_i \leq q_{i-1}'$. Here, all quantities denote computed values.

(vii) The theoretical error of the k-th computed eigenvalue is bounded by the sum of k deflated diagonal elements. In general this sum is not larger than $k\delta$, where δ is the prescribed tolerance. Of course, this sum which is also output does not comprise the effect of rounding errors.

6. Discussion of Numerical Properties

The accuracy of the computed eigenvalues is determined by the effect of rounding errors made in the QD steps. The error analysis of WILKINSON [6] gives for the first QD step of each full QR step

$$A-sI+F=LR, \qquad RL+F'=A'.$$

A, L, R, and A' denote computed matrices, F and F' are zero except for the diagonal and sub-diagonal:

$$|f_{ii}| \leq 2\varepsilon |d_i - s|, \qquad |f'_{ii}| \leq \varepsilon |d'_i|,$$
$$|f_{i+1,i}| \leq \varepsilon b^2_{i+1}, \qquad |f'_{i+1,i}| \leq \varepsilon b'^2_{i+1},$$
$$(\varepsilon = \text{machine precision}).$$

A similar result holds for the second QD step. For the symmetrized form of the matrices A, A', and A'' we may describe the effect of rounding errors of one QR step as the successive superposition of three symmetric and tridiagonal perturbation matrices E, E', E'' which have diagonal elements bounded by

$$2\varepsilon|d_i - s|, \qquad 2\varepsilon|d'_i|, \qquad \varepsilon|d''_i|,$$

and, neglecting ε^2, off-diagonal elements bounded by

$$\frac{\varepsilon}{2}|b_{i+1}|, \qquad 2\frac{\varepsilon}{2}|b'_{i+1}|, \qquad \frac{\varepsilon}{2}|b''_{i+1}|$$

respectively. For the corresponding perturbation of the i-th eigenvalue follows

$$|\varDelta \lambda_i| \leq \text{lub}_2(E) + \text{lub}_2(E') + \text{lub}_2(E''),$$

and a result of KAHAN [2] gives

$$|\varDelta \lambda_i| \leq \left(4\varepsilon^2 + \frac{\varepsilon^2}{4}\right)^{\frac{1}{2}} \varrho(A - sI) + (4\varepsilon^2 + \varepsilon^2)^{\frac{1}{2}} \varrho(A') + \left(\varepsilon^2 + \frac{\varepsilon^2}{4}\right)^{\frac{1}{2}} \varrho(A'').$$

After k iterations we have

(7) $$|\varDelta \lambda_i| < 6k\,\varepsilon\,\varrho(A_1).$$

Here, $\varrho(A_1)$ denotes the spectral radius of the matrix produced by the initial shift.

Most commonly, the actual errors are much smaller than this bound. This holds in particular, if the elements of the input matrix A_0, roughly speaking, strongly increase along the diagonal. Usually, this property is preserved throughout the iteration. Then, the perturbations describing the effect of rounding errors are relatively small, not only with respect to the current matrix A, but also with respect to the initial matrix A_0. Hence, the computed eigenvalues may be interpreted as the exact results of an input matrix $A_0 + F_0$, where $|F_0| \leq \varepsilon\,c|A_0|$ and c is a constant of order unity.

It is recommended to store the elements of the input matrix in reversed order if these elements are strongly decreasing along the diagonal.

7. Test Results

The following results were obtained using the TR 4 computer of the *Leibniz-Rechenzentrum der Bayerischen Akademie der Wissenschaften, München*. The machine precision is $\varepsilon = {}_2{-}35 \approx 2.9_{10}{-}11$. We list truncated values.

(a) $n = 5$, $d_i = 0$, $b_i^2 = 2i - 3$, $i = 1, 2, 3, 4, 5$

Prescribed tolerance $_{10}-3$ number of iterations 25 computed eigenvalue	actual error	computed bounds f. theoret. error
$-3.68344\,59391$	$0.00000\,13063$	$0.00000\,13064$
$-1.55956\,40495$	$0.00000\,14066$	$0.00000\,27129$
$-0.00097\,19575$	$0.00097\,19575$	$0.00097\,46705$
$1.55956\,07676$	$0.00000\,18753$	$0.00097\,65459$
$3.68344\,46326$	$0.00000\,00001$	$0.00097\,65459$

Prescribed tolerance 0 number of iterations 30 computed eigenvalue	actual error	computed bounds f. theoret. error
$-3.68344\,46327$	$-4.89_{10}-11$	$1.82_{10}-12$
$-1.55956\,26428$	$-8.72_{10}-11$	$4.14_{10}-12$
$-1.98_{10}-12$	$1.98_{10}-12$	$4.83_{10}-12$
$1.55956\,26429$	$-2.91_{10}-11$	$7.25_{10}-12$
$3.68344\,46327$	$4.89_{10}-11$	$7.25_{10}-12$

Evaluation of (7) yields the bound $4.5_{10}-8$ for the errors due to rounding.

(b) Calculation of the ten smallest eigenvalues of a matrix given by WILKINSON [1]:

$$n = 1000, \quad d_i = i^4, \quad b_i^2 = 1, \quad i - 1, \ldots, 1000$$

Prescribed tolerance $_{10}-10$ number of iterations 69 computed eigenvalue	actual error	computed bounds f. theoret. error
$9.33572\,19691_{10}-1$	$-8.22_{10}-13$	$1.27_{10}-17$
$1.60510\,42885_{10}+1$	$-8.32_{10}-10$	$7.58_{10}-17$
$8.10096\,70624_{10}+1$	$-1.31_{10}-9$	$1.54_{10}-10$
$2.56003\,00426_{10}+2$	$-3.67_{10}-10$	$1.54_{10}-10$
$6.25001\,21966_{10}+2$	$5.46_{10}-8$	$1.57_{10}-10$
$1.29600\,05853_{10}+3$	$3.14_{10}-9$	$1.13_{10}-9$
$2.40100\,03149_{10}+3$	$4.10_{10}-8$	$6.59_{10}-8$
$4.09600\,01843_{10}+3$	$-6.57_{10}-9$	$6.59_{10}-8$
$6.56100\,01161_{10}+3$	$-1.21_{10}-6$	$6.59_{10}-8$
$1.00000\,00075_{10}+4$	$-5.05_{10}-7$	$6.59_{10}-8$

The ninth computed eigenvalue has the largest relative error, which is less than seven times the machine precision. If the diagonal is stored in reversed order we obtain less satisfactory results. Then the actual errors range from $-9.02_{10}-8$ for the smallest eigenvalue to $4.50_{10}-6$ for the tenth.

A second test was carried out on the CDC 1604 computer of Eidgenössische Technische Hochschule, Zürich, Switzerland.

Acknowledgement. The authors wish to thank Professor Dr. H. RUTISHAUSER for valuable comments.

References

1. BARTH, W., R. S. MARTIN, and J. H. WILKINSON: Handbook series linear algebra. Calculation of the eigenvalues of a symmetric tridiagonal matrix by the method of bisection. Numer. Math. **9**, 386—393 (1967). Cf. II/5.
2. KAHAN, W.: Accurate eigenvalues of a symmetric tri-diagonal matrix. Technical Report No. CS 41, Computer Science Department, Stanford University 1966.
3. ORTEGA, J. M., and H. F. KAISER: The LL^T and QR methods for symmetric tridiagonal matrices. Comput. J. **6**, 99—101 (1963).
4. RUTISHAUSER, H.: Stabile Sonderfälle des Quotienten-Differenzen-Algorithmus. Numer. Math. **5**, 95—112 (1963).
5. — and H. R. SCHWARZ: The LR transformation method for symmetric matrices. Numer. Math. **5**, 273—289 (1963).
6. WILKINSON, J. H.: Rounding errors in algebraic processes. Notes on applied science No. 32. London: Her Majesty's Stationary Office 1963. German edition: Rundungsfehler. Berlin-Göttingen-Heidelberg: Springer 1969.
7. — The algebraic eigenvalue problem. Oxford: Clarendon Press 1965.
8. BAUER, F. L.: QD-method with Newton shift. Techn. Rep. 56. Computer Science Department, Stanford University 1967.

Contribution II/7

The QR Algorithm for Band Symmetric Matrices*

by R. S. MARTIN, C. REINSCH, and J. H. WILKINSON

1. Theoretical Background

The QR algorithm with shifts of origin may be used to determine the eigen-values of a band symmetric matrix A. The algorithm is described by the relations

$$A_s - k_s I = Q_s R_s, \qquad R_s Q_s = A_{s+1} \quad (s = 1, 2, \ldots) \tag{1}$$

where the Q_s are orthogonal and the R_s are upper triangular. It has been described in some detail by Wilkinson [5, pp. 557–561]. An essential feature is that all the A_s remain of band symmetric form and R_s is also a band matrix, so that when the width of the band is small compared with the order of the A_s, the volume of computation in each step is quite modest. If the shifts k_s are appropriately chosen the off-diagonal elements in the last row and column tend rapidly to zero, thereby giving an eigenvalue.

2. Applicability

Procedure *bqr* may be used to find eigenvalues of a symmetric band matrix. It is designed specifically to find *one* eigenvalue, and to find p eigenvalues, p con-secutive calls are required. If *all* eigenvalues are required the procedure *bandrd* [4] is much to be preferred.

Denoting the original n-th order band matrix by B, if the eigenvalue nearest to t is required, the procedure is provided with the input matrix $A = B - tI$, and the input parameters t and n. *Usually* the procedure will find the eigenvalue λ of A which is of smallest modulus (see discussion in Section 5), and hence the eigenvalue of B nearest to t. The computed eigenvalue is given as the output parameter t and the output $A + tI$ is a band matrix similar to the input $A + tI$. In order to calculate a further eigenvalue nearest to that already calculated, the procedure must be entered again using as input t and A the output t and A obtained in the previous application, and with n replaced by $n - 1$.

A common application is to find the k smallest eigenvalues of a positive definite band matrix. This is done using $t = 0$ in the first call and the output t as input t in the subsequent calls.

The inability to guarantee the correct ordinal of the computed eigenvalue is a weakness of the method, but in practice one does in fact usually find the required eigenvalue. If it is essential to guarantee the ordinal of the computed eigenvalues this can be done by using *bandet 2* [2] as is illustrated in Section 7. The eigen-vectors may be found by inverse iteration [2].

* Prepublished in Numer. Math. **16**, 85—92 (1970).

3. Formal Parameter List

Input to procedure bqr

n order of the symmetric band matrix.

m the width of the full band matrix is $2m+1$.

tol the smallest positive floating point number representable within the computer divided by the machine precision.

a an $n \times (m+1)$ array, $a[1:n, 0:m]$ defining the band symmetric matrix by giving the elements on and above the diagonal. The (i, k) element of the band matrix A is stored as element $a[i, k-i]$, $i=1(1)n$ and $k=1(1)$ $\min(i+m, n)$.

t the procedure determines an eigenvalue of $A+tI$, usually it is $\lambda(A)+t$ where $\lambda(A)$ is the eigenvalue of smallest modulus of the input matrix A.

r for the first call $r=0$. For subsequent calls r is the output value from the previous call. It is used to determine when the last row and column of the current transformed band matrix can be regarded as negligible.

$fail$ the exit used when 30 QR transformations fail to isolate an eigenvalue. A second call could then be used (if it was felt this was advisable) using the same value of n. A further set of 30 iterations would then be available.

Output from procedure bqr

a the transformed band matrix with the same storage arrangements as the input matrix. The matrix $(A+tI)$ derived from the output parameters is similar to the input $(A+tI)$ to within rounding errors. Its last row and column are null unless the failure exit has been used.

t is the computed eigenvalue unless the failure exit has been used. It is the input t for the next call, if the next required eigenvalue is the one nearest to the currently computed eigenvalue.

r $max(s, \text{input } r)$ where $s=\text{sum-norm}$ of the last column of the input matrix A.

4. ALGOL Program

procedure bqr (n, m, tol) $trans:(a, t, r)$ $exit:(fail)$;
value n, m, tol;
integer n, m;
real tol, t, r;
array a;
label $fail$;

comment QR transformation with explicit shift for the computation of one eigenvalue of a symmetric band matrix $(A+t \times I)$, where t is a given shift. Usually, the computed eigenvalue is $\lambda(A)+t$, where $\lambda(A)$ is the eigenvalue of A with smallest modulus;

begin
    integer i, j, k, l, its;
    real f, g, q, s;
    array $b[-m:2 \times m], u[0:m, -m:m], h[-m:m]$;

```
if m ≧ n then m := n − 1;
for i := n + 1 − m step 1 until n do
for k := n + 1 − i  step 1 until m do a [i, k] := 0;
its := 0; s := 0;
for k := 1 step 1 until m do s := s + abs (a [n − k, k]);
if s > r then r := s;
```
test for convergence:
```
    g := a [n, 0]; f := 0;
    for k := 1 step 1 until m do f := f + abs (a [n − k, k]);
    if r + f ≦ r then goto convergence;
    if its = 30 then goto fail;
    its := its + 1;
```
comment form shift from last 2×2 block;
```
    if f ≦ 0.25 × r ∨ its ≧ 5 then
    begin
        f := a [n − 1, 1];
        if f ≠ 0 then
        begin
            q := (a [n − 1, 0] − g)/(2 × f);
            s := sqrt (q × q + 1);
            g := g − f/(if q < 0 then q − s else q + s)
        end;
        t := t + g;
        for i := 1 step 1 until n do a [i, 0] := a [i, 0] − g
    end shift;
    for k := − m step 1 until m do h [k] := 0;
    for i := 1 − m step 1 until n do
    begin
        if i + m > n then goto contin;
```
comment $l = max (0, 1 − i)$, *form* $(i + m)$-th *column of* $A − g × I$;
```
        l := if i > 0 then 0 else 1 − i;
        for k := − m step 1 until − 1 do b [k] := 0;
        for k := l step 1 until m do b [k] := a [i + k, m − k];
        for k := 1 step 1 until m do b [k + m] := a [i + m, k];
```
comment pre-multiply with Householder reflections $i − m$ through $i + m − 1$;
```
    for j := − m step 1 until m − 1 do
    if h [j] ≠ 0 then
    begin
        f := 0;
        for k := 0 step 1 until m do f := f + u [k, j] × b [k + j];
        f := f/h [j];
        for k := 0 step 1 until m do b [k + j] := b [k + j] − u [k, j] × f
    end j;
```

comment $(i+m)$-th Householder reflection;

    $f:=b[m];\ s:=0;$

    for $k:=1$ **step** 1 **until** m **do** $s:=s+b[k+m]\uparrow2;$

    if $s\le tol$ **then** $h[m]:=0$ **else**

    begin

        $s:=s+f\times f;$

        $b[m]:=g:=$ **if** $f<0$ **then** $sqrt(s)$ **else** $-sqrt(s);$

        $h[m]:=s-f\times g;\ u[0,m]:=f-g;$

        for $k:=1$ **step** 1 **until** m **do** $u[k,m]:=b[k+m]$

    end $s>tol;$

comment $save\,(i+m)$-th column of R, get i-th row of R;

    for $k:=l$ **step** 1 **until** m **do** $a[i+k,m-k]:=b[k];$

    if $i\le0$ **then goto** $update;$

contin:

    for $k:=0$ **step** 1 **until** m **do**

    begin $b[-k]:=0;\ b[k]:=a[i,k]$ **end** $k;$

comment post-multiply with Householder reflections $i-m$ through $i;$

    for $j:=-m$ **step** 1 **until** 0 **do**

    if $h[j]\neq0$ **then**

    begin

        $f:=0;$

        for $k:=0$ **step** 1 **until** m **do** $f:=f+u[k,j]\times b[k+j];$

        $f:=f/h[j];$

        for $k:=0$ **step** 1 **until** m **do** $b[k+j]:=b[k+j]-u[k,j]\times f$

    end $j;$

comment $l=min(m,i-1)$, store i-th column of the new A, up-date u and $h;$

    $l:=$ **if** $m<i$ **then** m **else** $i-1;$

    for $k:=0$ **step** 1 **until** l **do** $a[i-k,k]:=b[-k];$

update:

    for $j:=1-m$ **step** 1 **until** m **do**

    begin

        $h[j-1]:=h[j];$

        for $k:=0$ **step** 1 **until** m **do** $u[k,j-1]:=u[k,j]$

    end j

    end i and QR transformation;

    goto $test\ for\ convergence;$

convergence:

    $t:=t+g;$

    for $i:=1$ **step** 1 **until** n **do** $a[i,0]:=a[i,0]-g;$

    for $k:=0$ **step** 1 **until** m **do** $a[n-k,k]:=0$

end $bqr;$

5. Organisational and Notational Details

The upper half of the symmetric band matrix of width $2m+1$ and order n is stored as an $n \times (m+1)$ array, the diagonal elements forming the first column of the array. The steps described by Eq. (1) are carried out in such a way as to economise on the storage required. Denoting the shifted matrix by $A_s - k_s I$ the reduction to upper triangular form is effected by premultiplication with elementary Hermitians, $P_1, P_2, \ldots, P_{n-1}$ so that we have

$$P_{n-1} \ldots P_2 P_1 [A_s - k_s I] = R_s \quad \text{and} \quad A_{s+1} = R_s P_1 P_2 \ldots P_{n-1}, \tag{2}$$

where

$$P_i = I - u_i u_i^T / H_i. \tag{3}$$

An iteration is carried out in $m+n$ minor steps numbered $1-m$ to n; in the i-th minor step the $(i+m)$-th column of R_s and the left-hand part of the i-th row of A_{s+1} are derived. The latter, by symmetry gives the upper half of the i-th column of A_{s+1}. The i-th minor step consists of the following operations.

If $i+m \leq n$ perform the three steps

(i) Form b, the $(i+m)$-th column of $A_s - k_s I$ extended by an initial m zeros to give a vector of $3m+1$ components, $b[-m:2m]$.

(ii) Form $P_{i+m} \ldots P_{i-m} b$ to give the upper part of the $(i+m)$-th column of R_s.

(iii) Store the elements $r_{i,i+m}, \ldots, r_{i+m,i+m}$ as the corresponding entries of the array a.

If $i \geq 1$ perform the additional steps

(iv) Extract the vector $r^T = (0, \ldots, 0, r_{i,i}, \ldots, r_{i,i+m})$ and store as first $2m+1$ elements of b.

(v) Compute $r^T P_{i-m} \ldots P_i$ to give the left-hand half of the i-th row of A_{s+1} (notice that elements $r_{i,i+m+1}, \ldots, r_{i,i+2m}$ are not required in this computation.)

(vi) Store these elements as the upper half of the i-th column of A_{s+1}.

In addition to the auxiliary vector b used for the two purposes described above, an $m \times (2m+1)$ array is needed to store the non zero elements of u_{i+j} ($j = -m, \ldots, m$) at each stage and a $(2m+1) \times 1$ array to store H_{i+j} ($j = -m, \ldots, m$). The total number of multiplications and additions in each minor step is of the order of $3(m+1)(2m+1)$.

The choice of the origin shift at each step is based on the following consideration. The aim is to obtain the eigenvalue of A of smallest modulus, i.e. the eigenvalue of $B = A + tI$ nearest to t. In general, this would be achieved if shifts were not used at all (and if the matrix A is not block-diagonal). Unfortunately, the rate of convergence is too small without shifts. Thus, no shifts are only used in the "initial" stages while in the "final" stages the shift k_s is taken to be the eigenvalue of the 2×2 matrix in the bottom right-hand corner of A_s closer to $a_{nn}^{(s)}$. This usually gives cubic convergence. The problem is when to change from the "initial" strategy to the "final" strategy. In order to be virtually certain of getting the required eigenvalue the change over would need to be delayed until the off diagonal elements of the last row and column of A_s were "small". Usually this would still be too slow. Instead, the changeover takes place when

$$\sum_{i=1}^{m} |a^{(s)}_{n-i,\,n}| \leq 0.25\,r \qquad (4)$$

where the tolerance quantity $r = max \left(\text{input } r,\ \sum_{i=1}^{m} |a^{(0)}_{n-i,\,n}|\right)$ is also output. For the first call the input $r = 0$ and for subsequent calls the input r is the output r from the previous call. Hence when the p-th call is made, r is the greatest value assumed by the sum norm of the last off diagonal column of the p input band matrices used up to that stage. This gives a relevant criterion when the row-norms of the initial A are of the same order of magnitude or rapidly decreasing. (Notice that when A has rows with norms of widely different orders of magnitude it should be input with the larger rows first, if necessary by reflecting the matrix in its secondary diagonal.) This avoids using the second shift technique until the last off diagonal column is significantly smaller than on entry, but this strategy cannot guarantee convergence to the required eigenvalue. If it is essential to know the ordinal to be associated with a computed eigenvalue this can be obtained by using *bandet 2* [2]. Its determination takes much less time than one QR iteration. In practice this shortcoming is not usually serious. The algorithm has proved very satisfactory, for example, for finding several of the eigenvalues of smallest modulus of large band matrices arising from partial differential equations.

The procedure will accept a full matrix or a diagonal matrix but the input n must be positive.

6. Discussion of Numerical Properties

As is usual with the QR algorithm applied to real symmetric matrices convergence is very rapid and if several eigenvalues are required the average number of iterations per eigenvalue is usually less than two (cf. [1, 3]). An *a priori* error analysis [5, pp. 561–562] shows that the computed eigenvalues must have errors which are small compared with $\|A\|_2$, i.e. the eigenvalue of A of largest modulus, but in general one cannot guarantee that a small eigenvalue will have a low relative error. However when A has rows of rapidly decreasing norms and the small eigenvalues are insensitive to small relative changes in the elements these eigenvalues usually will be determined with low relative errors.

7. Test Results

The procedure *bqr* was tested using the computer KDF 9 (*39 digit binary mantissa*) at the National Physical Laboratory.

The first test matrix was of order 7 with $a_{i,\,i} = 10^{7-i}$, $a_{i,\,i+1} = 10$, $a_{i,\,i+2} = 1$ so that $n = 7$ and $m = 2$. Taking $t = 0$ the eigenvalues were found in increasing order. The computed values and the number of iterations are given in Table 1. As is usual with such good eigenvalue separations, the convergence was very rapid.

The second test matrix was of order 30 with non-zero elements (taking account of symmetry) defined by

$$b_{3i+1,\,3i+1} = b_{3i+2,\,3i+2} = b_{3i+3,\,3i+3} = 10 - i \quad (i = 0, \ldots, 9),$$
$$b_{1,\,2} = b_{1,\,3} = 1, \quad b_{i,\,i+3} = 1 \quad (i = 1, \ldots, 27).$$

Table 1

Computed eigenvalues	Number of iterations
$-\,5.6722\,8961\,580_{10}0$	4
$1.5530\,6282\,221_{10}1$	1
$1.0102\,9236\,852_{10}2$	1
$1.0001\,0120\,029_{10}3$	1
$1.0000\,0101\,019_{10}4$	1
$1.0000\,0001\,010_{10}5$	1
$1.0000\,0000\,011_{10}6$	0

This matrix was used by Schwarz [4], it has three eigenvalues close to 5 and all three were found in a total of 4 iterations. To this end 5 was subtracted along the diagonal and three calls of the procedure were made, the first one with starting values $r = 0$ and $t = 5$ for the transit parameters. The results are given in Table 2.

Table 2

	Computed eigenvalue t	Number of iterations
First call:	$4.9997\,8247\,775_{10}0$	3
Second call:	$4.9998\,3258\,576_{10}0$	1
Third call:	$4.9996\,8956\,628_{10}0$	0

Notice the first eigenvalue to be found was not the closest to the initial t. (This reversal of order is fairly common with close eigenvalues.) However, the second eigenvalue was the closest to the first. The procedure *bandet 2* showed that there are indeed three eigenvalues between 4.9 and 5.0 and that the computed eigenvalues are the 17-th, 16-th and 18-th respectively. (Numbered in decreasing order.) These results were confirmed at the Technische Universität München. A considerable number of high order matrices have been solved at the National Physical Laboratory using an earlier version of the procedure, the performance of which might be expected to be broadly similar.

Acknowledgements. The work of R. S. Martin and J. H. Wilkinson has been carried out at the National Physical Laboratory and that of C. Reinsch at the Technische Universität München. The authors wish to thank Mr. R. Pitfield for testing the procedure.

References

1. Bowdler, Hilary, Martin, R. S., Reinsch, C., Wilkinson, J. H.: The QR and QL algorithms for symmetric matrices. Numer. Math. **11**, 293–306 (1968). Cf. II/3.
2. Martin, R. S., Wilkinson, J. H.: Solution of symmetric and unsymmetric band equations and the calculation of eigenvectors of band matrices. Numer. Math. **9**, 279–301 (1967). Cf. I/6.
3. — — The implicit QL algorithm. Numer. Math. **12**, 377–383 (1968). Cf. II/4.
4. Schwarz, H. R.: Tridiagonalization of a symmetric band matrix. Numer. Math. **12**, 231–241 (1968). Cf. II/8.
5. Wilkinson, J. H.: The algebraic eigenvalue problem, 662 p. Oxford: Clarendon Press 1965.

Contribution II/8

Tridiagonalization of a Symmetric Band Matrix*

by H. R. Schwarz

1. Theoretical Background

The well known method proposed by Givens [1] reduces a full symmetric matrix $A = (a_{ik})$ of order n by a sequence of appropriately chosen elementary orthogonal transformations (in the following called Jacobi rotations) to tridiagonal form. This is achieved by $(n-1)(n-2)/2$ Jacobi rotations, each of which annihilates one of the elements a_{ik} with $|i-k| > 1$. If this process is applied in one of its usual ways to a symmetric band matrix $A = (a_{ik})$ of order n and with the band width $m > 1$, i.e. with

(1) $$a_{ik} = 0 \quad \text{for all } i \text{ and } k \text{ with } |i-k| > m,$$

it would of course produce a tridiagonal matrix, too. But the rotations generate immediately nonvanishing elements outside the original band that show the tendency to fill out the matrix. Thus it seems that little profit with respect to computational and storage requirements may be taken from the property of the given matrix A to be of band type.

However, by a generalization of an idea of Rutishauser [2], it is indeed possible to reduce a symmetric band matrix to tridiagonal form by a suitable sequence of Jacobi rotations while preserving the band property of the given matrix throughout the whole process (see [3]). A single step consists of the well known similarity transformation

(2) $$A_{k+1} = U_k^T A_k U_k \quad (A_0 = \text{given matrix } A),$$

where $U_k = U_k(p, q, \varphi)$ is an orthogonal matrix which deviates from the unit matrix only in the elements

(3) $$u_{pp} = u_{qq} = cos(\varphi), \quad u_{pq} = -u_{qp} = sin(\varphi) \quad \text{with} \quad p < q.$$

Herein φ denotes the real angle of rotation in the (p, q)-coordinate plane. The effect of the transformation (2) on A_k is a linear combination of the rows and columns with indices p and q, while all other elements of A_k remain unchanged.

---

* Prepublished in Numer. Math. **12**, 231—241 (1968).

The basic idea of the transformation is the following: The elimination of an element within the band of the matrix by an appropriate Jacobi rotation generates just two identical, in general nonzero elements g in symmetric positions outside the band. To maintain the band type of the matrix these elements are eliminated by a sequence of additional rotations with the effect of shifting them downwards over m rows and m columns and finally beyond the border of the matrix.

For $n = 10$, $m = 3$ the complete process for the elimination of the element a_{14} is illustrated in (4), where for the reason of symmetry only the elements in and above the main diagonal are considered.

(4)

A first rotation with $U_1(3, 4, \varphi_1)$ and appropriate chosen angle φ_1 eliminates a_{14}, but generates an element g in the third row and seventh column. This element g is annihilated by a second rotation with $U_2(6, 7, \varphi_2)$ which on its turn produces another element g in the sixth row and the tenth column. A third rotation with $U_3(9, 10, \varphi_3)$ completes the elimination of a_{14}.

By analogy the element a_{13} is eliminated next by starting with a rotation defined by $U_4(2, 3, \varphi_4)$. Two additional rotations will sweep the appearing element g in the second row and sixth column downwards and over the border of the matrix.

So far the first row (and column) has been brought to the desired form. It is almost obvious that continuation of the process with the following rows will never destroy the already generated zero elements within the band and will terminate with the transformed tridiagonal matrix.

The sequence of Jacobi rotations for the complete transformation of a band matrix A of order n and band width $m > 1$ is fully described in Table 1 *.

---

* A different strategy for the transformation of a band matrix to tridiagonal form consists of a systematic reduction of the band width of the matrix (see [4, 5]). However, the present strategy is less time-consuming, since it requires a smaller number of Jacobi rotations and a marginally smaller number of multiplications. A similar reduction to that given here is achieved by the Householder transformation [2]. On the whole the extra complexity of the organization of HOUSEHOLDER in this case means that it does not have its usual superiority over GIVENS.

Table 1. *Sequence of rotations*

Elimination of the elements in row j	First rotation	Position of the first element g (if any)	Additional rotations (if necessary)
$a_{j,j+k}$ ($k=m$, $m-1$, ..., 2) only if $j+k\leq n$ $j=1, 2, ..., n-2$	$U(j+k-1, j+k, \varphi)$	$(j+k-1, j+k+m)$ only if $j+k+m\leq n$	$U(j+k+\mu m-1, j+k+\mu m, \varphi)$ $\left(\mu=1, 2, ..., \left[\dfrac{n-k-j}{m}\right]\right)$

It should be noted that for all rotations the two indices p and q differ by one, that is the corresponding transformations change in all cases two adjacent rows and columns. Furthermore the first rotation is just an extrapolated case of the additional rotations. This simplifies the procedure. There exists only a slight difference in the determination of the angles φ. For the first rotation the angle φ is determined from the equations

$$(5) \qquad a_{j,j+k}\cos \varphi + a_{j,j+k-1}\sin \varphi = 0, \qquad \cos^2 \varphi + \sin^2 \varphi = 1,$$

whereas for the additional rotations $(\mu=1, 2, ..., [(n-k-j)/m])$ the equations

$$(6) \qquad g \cdot \cos \varphi + a_{j+k+(\mu-1)m-1, j+k+\mu m-1}\sin \varphi = 0, \qquad \cos^2 \varphi + \sin^2 \varphi = 1$$

are to be used.

The complete method as described performs the operation

$$(7) \qquad A \rightarrow J = V^T A V$$

in a finite number of steps N_{rot}, where J is tridiagonal and V is orthogonal. V is the product of all Jacobi rotations which were used for achieving the transformation (7), i.e.

$$(8) \qquad V = U_1 U_2 U_3 \ldots U_{N_{\text{rot}}}.$$

2. Applicability

The present algorithm may be applied to any symmetric band matrix A of order n and band width $m>1$ to reduce it to a tridiagonal matrix by a finite sequence of Jacobi rotations. The method is especially advantageous if $m \ll n$.

The application of the process is to be recommended, if *all* the eigenvalues of a symmetric band matrix are to be computed, since the combination of this bandreduction with the QD-algorithm shows a remarkable saving of computing time in comparison with the determination of all eigenvalues by the LR-transformation. The relation of the corresponding amount of computing time is approximately $1:m$.

In addition, the present procedure gives the connection to the save method of bisection for computing the eigenvalues in a given interval or the eigenvalues with a prescribed index [6], or for computing the eigenvalues by the rational QR-transformation [7].

3. Formal Parameter List

Input to procedure *bandrd*

n order of the matrix A (and of the matrix V).

m band width of the matrix A (= number of diagonals above the main diagonal).

matv boolean variable to decide whether the transformation matrix V (8) is to be computed (*matv* = **true**) or not (*matv* = **false**).

a **array** $a\,[1:n,\,0:m]$ are the elements of the given band matrix A arranged in a rectangular array as described in "Organizational and Notational Details". The elements $a\,[i,\,k]$ with $i+k>n$ are irrelevant.

Output of procedure *bandrd*

a the elements of the resulting tridiagonal matrix J (7). The diagonal elements of J are given by $a\,[i,\,0]$ ($i=1,\,2,\,\ldots,\,n$), the elements of the superdiagonal are given by $a\,[i,\,1]$ ($i=1,\,2,\,\ldots,\,n-1$). The other elements $a\,[i,\,2],\,a\,[i,\,3],\,\ldots,\,a\,[i,\,m]$ are meaningless.

v if *matv* = **true**, the **array** $v\,[1:n,\,1:n]$ contains the elements of the orthogonal transformation matrix V (8). This is a full square matrix. If *matv* = **false**, no values are assigned to $v\,[i,\,k]$.

4. ALGOL Program

```
procedure bandrd (n, m, matv) trans: (a) result: (v);
      value n, m, matv;
      integer n, m; Boolean matv; array a, v;
      comment Transforms a real and symmetric band matrix A of order n
              and band width m (i.e. with a[i, k]=0 for all i and k with
              abs(i − k)>m) to tridiagonal form by an appropriate finite
              sequence of Jacobi rotations. The elements of A in and above
              the main diagonal are supposed as a rectangular array (array
              a[1:n, 0:m]), where a[i, j] denotes the element in the i-th
              row and (i+j)-th column of the usual representation of a
              matrix. During the transformation the property of the band
              matrix is maintained. The method yields the tridiagonal
              matrix J = (Vᵀ)AV (where Vᵀ is the transpose of V), the ele-
              ments of which are a[i, 0] (i=1 (1) n) and a[i, 1] (i=1 (1) n−1).
              If the parameter matv is given the value true, the orthogonal
              matrix V is delivered as array v[1:n, 1:n], otherwise not;
      begin
         integer j, k, l, r, maxr, maxl, ugl;
         real b, c, s, c2, s2, cs, u, u1, g;
         if matv then
             for j := 1 step 1 until n do
                 for k := 1 step 1 until n do
                     v[j, k] := if j=k then 1 else 0;
         for k := 1 step 1 until n − 2 do
```

```
begin
    maxr := if n − k < m then n − k else m;
    for r := maxr step −1 until 2 do
    begin
        for j := k + r step m until n do
        begin
            if j = k + r then
            begin
                if a [k, r] = 0 then goto endr;
                b := − a [k, r − 1]/a [k, r];
                ugl := k
            end
            else
                begin
                    if g = 0 then goto endr;
                    b := − a [j − m − 1, m]/g;
                    ugl := j − m
                end;
            s := 1/sqrt (1 + b↑2); c := b × s;
            c2 := c↑2; s2 := s↑2; cs := c × s;
            u := c2 × a [j − 1, 0] − 2 × cs × a [j − 1, 1] + s2 × a [j, 0];
            u1 := s2 × a [j − 1, 0] + 2 × cs × a [j − 1, 1] + c2 × a [j, 0];
            a [j − 1, 1] := cs × (a [j − 1, 0] − a [j, 0]) + (c2 − s2) × a [j − 1, 1];
            a [j − 1, 0] := u;
            a [j, 0] := u1;
            for l := ugl step 1 until j − 2 do
            begin
                u := c × a [l, j − l − 1] − s × a [l, j − l];
                a [l, j − l] := s × a [l, j − l − 1] + c × a [l, j − l];
                a [l, j − l − 1] := u
            end l;
            if j ≠ k + r then
                a [j − m − 1, m] := c × a [j − m − 1, m] − s × g;
            maxl := if n − j < m − 1 then n − j else m − 1;
            for l := 1 step 1 until maxl do
            begin
                u        := c × a [j − 1, l + 1] − s × a [j, l];
                a [j, l] := s × a [j − 1, l + 1] + c × a [j, l];
                a [j − 1, l + 1] := u
            end l;
            if j + m ≤ n then
            begin
                g := − s × a [j, m];
                a [j, m] := c × a [j, m]
            end;
            if matv then
                for l := 1 step 1 until n do
```

```
              begin
                 u      := c×v[l, j−1]−s×v[l, j];
                 v[l, j]:= s×v[l, j−1]+c×v[l, j];
                 v[l, j−1]:= u
              end l;
           end j;
endr:    end r
       end k;
       a[n, 1]:= 0
     end bandrd;
```

5. Organisational and Notational Details

Since symmetry of the given matrix A is preserved throughout the process, it is sufficient to know and to work with the elements of A in and above the main diagonal. The band matrix A is stored as an **array** $a[1:n, 0:m]$, where $a[i, j]$ $(i=1, 2, \ldots, n; j=0, 1, \ldots, m)$ denotes the element in the i-th row and $(i+j)$-th column of A in its usual arrangement. Thus $a[i, 0]$ is the i-th element in the diagonal, $a[i, 1]$ the element in the first superdiagonal of the i-th row, and so on. For example, the matrix

$$(9) \qquad A = \begin{bmatrix} 5 & -4 & 1 & & & & \\ -4 & 6 & -4 & 1 & & & \\ 1 & -4 & 6 & -4 & 1 & & \\ & 1 & -4 & 6 & -4 & 1 & \\ & & 1 & -4 & 6 & -4 & 1 \\ & & & 1 & -4 & 6 & -4 \\ & & & & 1 & -4 & 5 \end{bmatrix}$$

of order $n=7$ with $m=2$ is stored as

$$(10) \qquad \begin{array}{ccc} 5 & -1 & 1 \\ 6 & -4 & 1 \\ 6 & -4 & 1 \\ 6 & -4 & 1 \\ 6 & -4 & 1 \\ 6 & -4 & \times \\ 5 & \times & \times \end{array}$$

The elements in the lower right triangle marked by an $\times$ are irrelevant, since they are not used in the procedure *bandrd*.

The resulting tridiagonal matrix $J=V^T A V$ is contained in the same array a, i.e. the diagonal elements of J are given by $a[i, 0]$, the superdiagonal elements by $a[i, 1]$. For convenience the value of $a[n, 1]$ (that lies actually outside the matrix) is set equal to zero.

If required (*matv* = **true**), the transformation matrix V is delivered as **array** $v[1:n, 1:n]$. If on the other hand V is not of interest (*matv* = **false**), the actual counterpart of v can be any declared twodimensional array, thus e.g. a call like *bandrd* (70, 5, **false**, a, a) is meaningful.

The Jacobi rotations with $j-1$ and j as rotation indices and $c = \cos \varphi$, $s = \sin \varphi$ as determined from (5) or (6) are performed in three parts:

a) The new elements that are transformed by both the row and the column combinations are computed according to the usual formulas.

b) The new elements above the diagonal, undergoing only the column combination, are formed. Distinction has to be made whether the rotation annihilates an element within or the element g outside the band of the matrix.

c) The new elements to the right of the diagonal, transformed only by the row combination, are computed. Two cases have to be distinguished whether the rotation produces a new element g outside the band or not.

6. Numerical Properties

For the total number of rotations N_{rot} of the process an upper bound may be derived by neglecting the effects at the lower end of the band:

(11) $$N_{rot} \leq n^2 \cdot (m-1)/(2m).$$

Each rotation involves a square root and in the normal case $8m + 13$ multiplications. The total number of multiplications needed for the complete transformation is therefore bounded by the expression

(12) $$N_{mult} \leq n^2 (m-1)(4+6.5/m).$$

For $m \ll n$ the method is indeed an n^2-process.

According to WILKINSON [5] we have the approximate error bound

(13) $$\left[\frac{\Sigma(\mu_i - \lambda_i)^2}{\Sigma \lambda_i^2} \right]^{\frac{1}{2}} \leq 12 \cdot \Theta \cdot n^{1.5} (1+6\Theta)^{4n-7} \cdot \frac{m-1}{m},$$

where λ_i are the eigenvalues of the given matrix A, μ_i are the eigenvalues of the computed tridiagonal matrix J and Θ is the smallest positive number such that in the computer $1+\Theta \neq 1$.

7. Test Results

The procedure *bandrd* has been tested on the computer CDC 1604-A★ of the Swiss Federal Institute of Technology, Zurich for the following examples:

a) The matrix (9) has been transformed to tridiagonal form J, and the eigenvalues of J have been computed by the method of bisection. The elements of J as well as the eigenvalues μ_i of J together with the eigenvalues λ_i of A as determined by the method of JACOBI [8] are listed in Table 2.

★ The CDC 1604-A has a 36-bit mantissa and a binary exponent ranging from −1024 to +1023.

Table 2

i	$j\,[i, i]$	$j\,[i, i+1]$	$\mu\,[i]$	$\lambda\,[i]$
1	5.0000000000	−4.1231056257	0.0231 7730229	0.0231 7730233
2	7.8823529418	−4.0348825039	0.3431 4575048	0.3431 4575054
3	7.9535662945	−4.0166055335	1.5243 189787	1.5243 189787
4	7.9748041817	−3.9975334001	4.0000000001	3.9999999999
5	7.6058064229	−2.9758282822	7.6472538968	7.6472538965
6	3.3461613147	−0.4538408623	11.6568 54250	11.6568 54250
7	0.2373088458	0	14.8052498 23	14.8052498 23

b) The matrix $B = 8C - 5C^2 + C^3$ of order 44, where C denotes the tridiagonal matrix with elements $c\,[i, i] = 2$, $c\,[i, i+1] = 1$, i.e.

$$(14) \qquad B = \begin{bmatrix}
5 & 2 & 1 & 1 & & & & & & & \\
2 & 6 & 3 & 1 & 1 & & & & & & \\
1 & 3 & 6 & 3 & 1 & 1 & & & & & \\
1 & 1 & 3 & 6 & 3 & 1 & 1 & & & & \\
& 1 & 1 & 3 & 6 & 3 & 1 & 1 & & & \\
& & & \cdot & \cdot & \cdot & & \cdot & \cdot & \cdot & \\
& & & & & & 1 & 1 & 3 & 6 & 2 \\
& & & & & & & 1 & 1 & 2 & 5
\end{bmatrix}$$

has been used to determine its eigenvalues. The elements of the transformed tridiagonal matrix J are listed in Table 3 together with the computed eigenvalues μ_i of J by the method of bisection as well as the exact eigenvalues λ_i of B with

$$(15) \qquad \lambda_i = s_i^3 - 5 s_i^2 + 8 s_i, \qquad s_i = 4 \cdot \sin^2(2 i^0) \qquad (i = 1, 2, \ldots, 44).$$

c) The matrix A of order $n = 30$ with $m = 3$

$$(16) \qquad A = \begin{bmatrix}
10 & 1 & 1 & 1 & & & & & & & & \\
1 & 10 & 0 & 0 & 1 & & & & & & & \\
1 & 0 & 10 & 0 & 0 & 1 & & & & & & \\
1 & 0 & 0 & 9 & 0 & 0 & 1 & & & & & \\
& 1 & 0 & 0 & 9 & 0 & 0 & 1 & & & & \\
& & 1 & 0 & 0 & 9 & 0 & 0 & 1 & & & \\
& & & \cdot & \cdot & \cdot & & \cdot & \cdot & \cdot & \cdot & \\
& & & & & & 1 & 0 & 0 & 2 & 0 & 0 & 1 \\
& & & & & & & 1 & 0 & 0 & 1 & 0 & 0 \\
& & & & & & & & 1 & 0 & 0 & 1 & 0 \\
& & & & & & & & & 1 & 0 & 0 & 1
\end{bmatrix}$$

has been used to compute all its eigenvalues. The method of the band reduction yields the matrix J of Table 4. The subsequent application of the method of bisection delivers the eigenvalues μ_i of J. For comparison the eigenvalues λ_i of A have been determined by the Jacobi algorithm [8]. Although some of the

Table 3

i	$j\,[i, i]$	$j\,[i, i+1]$	$\mu\,[i]$	$\lambda\,[i]$
1	5.0000000000	2.4494897427	0.0388 5663462	0.0388 5663446
2	9.6666666667	3.7043517953	0.1538 2406432	0.1538 2406414
3	7.5114709853	4.4182460949	0.3401 7132218	0.3401 7132213
4	7.6884525213	3.8121582980	0.5902 6480414	0.5902 6480397
5	8.5830360828	3.8635865610	0.8939 3316153	0.8939 3316152
6	7.7597731312	4.2264403956	1.2389 544781	1.2389 544781
7	8.1073 58423	3.8888572252	1.6116433233	1.6116433232
8	8.3723450440	3.9142790950	1.9975106432	1.9975106432
9	7.8258325770	4.1582582958	2.3819660114	2.3819660113
10	7.8711218853	3.9164482712	2.7510296968	2.7510296966
11	8.2749744193	3.9408159364	3.0920214042	3.0920214041
12	7.8578585475	4.1212019076	3.3941934243	3.3941934241
13	7.9082005882	3.9312475121	3.6492782723	3.6492782721
14	8.2159441281	3.9577102096	3.8519245939	3.8519245938
15	7.8781074404	4.0968242675	4.0000000001	4.0000000000
16	7.9337598720	−3.9410934000	4.0045318454	4.0045318458
17	8.1746472896	3.9695920067	4.0052119528	4.0052119532
18	7.8932128858	4.0789034129	4.0345680073	4.0345680077
19	7.9523352541	3.9480048752	4.0528415894	4.0528415894
20	8.1365859494	3.9651689592	4.0787256923	4.0787256924
21	7.8178365964	3.9479925008	4.0947453703	4.0947453700
22	7.4628371074	3.4781440573	4.1208346530	4.1208346534
23	6.6175117962	2.9637963604	4.1407717538	4.1407717541
24	5.2534192105	−2.4517781222	4.1458980334	4.1458980338
25	4.5422812540	2.0208165905	4.1625038242	4.1625038244
26	3.7256107620	1.7905673660	4.3473749752	4.3473749752
27	3.4574191781	1.5646029996	4.6180339887	4.6180339887
28	3.0729391709	−1.3383315621	4.9818690444	4.9818690445
29	3.1365536812	1.1419380160	5.4426087284	5.4426087286
30	3.2563743813	0.8506159836	6.0000000000	6.0000000000
31	3.4230726357	0.6344781230	6.6496490325	6.6496490325
32	3.6359837740	−0.4394343961	7.3830341453	7.3830341455
33	3.7783048010	0.3049897050	8.1876927225	8.1876927229
34	3.9169434055	0.1636998047	9.0475765823	9.0475765827
35	3.9304903009	0.0759018266	9.9435630140	9.9435630140
36	4.0638930104	−0.0428400764	10.854101966	10.854101966
37	4.0911967582	−0.0405738637	11.755973944	11.755973944
38	4.0900741534	−0.0336363562	12.625128291	12.625128291
39	4.0790565672	−0.0435345313	13.437567947	13.437567947
40	4.0615724516	0.0363513025	14.170244610	14.170244611
41	4.0592745087	−0.0318706517	14.801927580	14.801927581
42	4.0946552432	−0.0312974483	15.314010508	15.314010508
43	4.1121615408	−0.0084580378	15.691222719	15.691222719
44	4.0055665978	0	15.922215640	15.922215641

eigenvalues of A are almost triple, the transformed tridiagonal matrix J does not even nearly decompose as one might expect.

Table 4

i	$j\,[i,\,i]$	$j\,[i,\,i+1]$	$\mu\,[i]$	$\lambda\,[i]$
1	10.000000000	1.7320508076	0.2538058171	0.2538058170
2	9.6666666667	1.1055415967	0.2538058171	0.2538058171
3	8.7878787883	1.2694763675	0.2538058173	0.2538058171
4	8.0993006993	1.3870918026	1.7893213523	1.7893213522
5	7.3649630481	1.4295019112	1.7893213527	1.7893213527
6	6.6161666284	1.6030841144	1.7893213531	1.7893213529
7	6.5076783627	2.0653254573	2.9610588361	2.9610588358
8	6.2709028289	1.8863971727	2.9610588841	2.9610588842
9	4.6403533769	1.5370064289	2.9610589165	2.9610589161
10	3.5244488962	−1.7523840047	3.9960456418	3.9960456418
11	4.3171991227	2.9080849479	3.9960482020	3.9960482014
12	4.6866497255	1.8471929353	3.9960497561	3.9960497565
13	2.5184783050	0.2786808032	4.9996895667	4.9996895662
14	6.9959884597	−0.0512104592	4.9997824781	4.9997824777
15	6.0011669284	−0.0463833001	4.9998325860	4.9998325857
16	5.0562247743	−0.5697872379	5.9978378497	5.9978378494
17	10.640540849	0.5706708476	6.0002175230	6.0002175223
18	4.0431585993	0.0701584281	6.0012633582	6.0012633582
19	3.4977933009	1.7549123478	6.9637264189	6.9637264186
20	8.6523751242	−0.3854162757	7.0039517991	7.0039517985
21	1.2007407688	0.8906616138	7.0163502621	7.0163502620
22	2.8056476826	−3.1493000913	7.7615665168	7.7615665166
23	6.1014455162	0.0083686212	8.0389411172	8.0389411158
24	7.0033308959	0.0258639754 9	8.1138105888	8.1138105886
25	5.9994477055	0.03794630954	8.6878418569	8.6878418564
26	4.9947088899	−0.0811264 8663	9.2106786487	9.2106786475
27	3.9754502629	0.17699886640	9.4734106881	9.4734106874
28	2.5233531950	1.0249844551	10.174892584	10.174892583
29	0.7188981895	0.0265661 5397	10.746194184	10.746194183
30	1.7890424196	0	11.8093 10237	11.8093 10238

A second test has been performed at the National Physical Laboratory, Teddington. The results have been confirmed, but J. H. WILKINSON points out that although the algorithm is stable, the elements of the tridiagonal matrix are not always determined in a stable way. In the last example the later elements disagreed entirely (i.e. in the most significant figure) from those given in Table 4. But the eigenvalues, of course, agreed almost exactly as is proved by backward error analysis. This disagreement usually occurs with multiple or very close eigenvalues.

References

1. GIVENS, W.: A method for computing eigenvalues and eigenvectors suggested by classical results on symmetric matrices. Nat. Bur. Standards Appl. Math. Ser. 29, 117—122 (1953).
2. RUTISHAUSER, H.: On Jacobi rotation patterns. Proceedings of Symposia in Applied Mathematics, Vol. 15 Experimental Arithmetic, High Speed Computing and Mathematics, 1963, 219—239.
3. SCHWARZ, H. R.: Die Reduktion einer symmetrischen Bandmatrix auf tridiagonale Form. Z. Angew. Math. Mech. (Sonderheft) 45, T 76—T 77 (1965).

4. SCHWARZ, H. R.: Reduction of a symmetric bandmatrix to triple diagonal form. Comm. ACM **6**, 315—316 (1963).

5. WILKINSON, J. H.: The algebraic eigenvalue problem, 662 p. Oxford: Clarendon Press 1965.

6. BARTH, W., R. S. MARTIN, and J. H. WILKINSON: Calculation of the eigenvalues of a symmetric tridiagonal matrix by the method of bisection. Numer. Math. **9**, 386—393 (1967). Cf. II/5.

7. REINSCH, C., and F. L. BAUER: Rational QR transformation with Newton shift for symmetric tridiagonal matrices. Numer. Math. **11**, 264—272 (1968). Cf. II/6.

8. RUTISHAUSER, H.: The Jacobi method for real symmetric matrices. Numer. Math. **9**, 1—10 (1966). Cf. II/1.

Contribution II/9

Simultaneous Iteration Method for Symmetric Matrices*

by H. Rutishauser †

1. Theoretical Background

The "ordinary" iteration method with one single iteration vector (sometimes called v. Mises-Geiringer iteration) can often yield an eigenvector and its eigenvalue in very short time. But since this cannot be guaranteed, not even with improvements such as shifts of origin, Aitken-Wynn acceleration or Richardson's purification, the method cannot be recommended for general use. In order to prevent possible poor convergence, the computation is carried in parallel with several iteration vectors, between which an orthogonality relation is maintained.

While the principle of this new method is already described by Bauer [2] for general matrices, in the meantime certain refinements for the symmetric case have been proposed [5, 6, 12, 13]. In [11] the present author gave a survey of facts which are relevant for the actual numerical computation, however with a restriction to positive definite matrices. The present paper makes use of these facts, but is also valid for the non-definite case.

2. Applicability

The present algorithm—procedure *ritzit*—can be used whenever some of the absolutely greatest eigenvalues and corresponding eigenvectors of a symmetric matrix are to be computed. It is essential that the matrix A need not be given as an array of $n \times n$ numbers, but rather as a rule for computing the vector $A \times v$ for an arbitrary given vector v. This rule must be described by a procedure to be used as actual counterpart of the formal parameter *op* in a call of *ritzit*.

It should be observed however, that in order to compute the *em* absolutely largest eigenvalues $(\lambda_1, \lambda_2, \ldots, \lambda_{em})$ of A, the iteration must be performed with $p > em$ simultaneous iteration vectors, where p is chosen in accordance with the convergence properties. Indeed, the convergence quotient is between $|\lambda_p/\lambda_{em}|$ and $exp(-ArCosh(|\lambda_{em}/\lambda_p|))$; it is nearer to the first value, if $|\lambda_1/\lambda_{em}|$ is big, and nearer to the second, if this quotient is near to one. Accordingly *ritzit* may be comparatively fast for matrices whose spectrum is dense at its upper end, while situations like that exhibited by testexample b) allow no acceleration unless λ_1 is removed by deflation of A.

---

* Prepublished in Numer. Math. **16**, 205–223 (1970).

Comparing *ritzit* with other methods, it must be said that the Jacobi method [9] and even more the Householder-bisection method [1, 3, 7] are more efficient if the above considerations require p to be chosen bigger than $n/10$, say. This rule practically confines the use of *ritzit* to matrices with $n > 20$.

Warning: Non-symmetry of the operator described by procedure *op* may cause serious malfunction of procedure *ritzit*. Furthermore, *ritzit* cannot be used for matrices A for which λ_1^2 is in the overflow range. Underflow, on the other hand will not have serious consequences.

3. Formal Parameter List

a) Quantities to be given:

n the order of the matrix A.

p the number of simultaneous iteration vectors.

km maximum number of iteration steps to be performed.

eps tolerance for accepting eigenvectors (see §6 of [11]).

op **procedure** $op(n, v, w)$; **integer** n; **array** v, w; **code** defining the matrix A: the body of this procedure must be such that it computes $w = A v$, but must not change v;

inf **procedure** $inf(ks, g, h, f)$; **integer** ks, g, h; **array** f; **code** which may be used for obtaining information or to exert control during execution of *ritzit*; *inf* is activated only after every Jacobi-step. Its parameters have the following meaning:

    $ks =$ number of the next iteration step.

    $g\ =$ number of already accepted eigenvectors.

    $h\ =$ number of already accepted eigenvalues.

    $f\ :$ **array** $f[1:p] =$ error quantities f_k (in the sense of [11], § 6) for the vectors $x_1 \ldots x_p$*.

em number of eigenvalues to be computed. em must be $< p$.

b) Quantities produced by *ritzit*:

em If km steps were not sufficient to compute em eigenvectors, the terminal value of em indicates the number of accepted eigenvectors.

x **array** $x[1:n, 1:p]$, containing in its em first columns normalized eigenvectors to the eigenvalues $d[1]$ through $d[em]$ of A. The later columns are more or less approximations to the eigenvectors corresponding to $d[em+1]$, etc.

d **array** $d[1:p]$, of which $d[1] \ldots d[em]$ are the em absolutely biggest eigenvalues of A, while $d[em+1]$ through $d[p-1]$ are more or less approximations to smaller eigenvalues. $d[p]$ on the other hand is the most recently computed value of e, $(-e, e)$ being the Chebyshev interval.

---

    * The error quantity $f[j]$ has the value 4, until the corresponding eigenvalue $d[j]$ has been accepted.

c) Special case.

If km is given as a negative number, then it is expected that starting values to the p iteration vectors $x_1, \ldots, x_p$ are given, i.e. the user must provide an initial matrix $x[1:n, 1:p]$, with which the iteration then begins. In this way one may make use of approximately known eigenvectors. It is understood that in this case $-km$ is used as maximum number of iteration steps.

4. Examples of Application

a) Computation of the 5 lowest eigenvalues (and corresponding vectors) of a positive definite symmetric matrix A of order n, with a given upper bound t for the spectral norm. In this case iteration is performed with $t \times I - A$, though we know that convergence may be quite slow and must be compensated for by a larger value of p.

```
begin
    comment insert here procedure ritzit;
    real s, t;
    integer n, p, em, j;
    comment n, p, t on input medium 1;
    inreal (1, t);  n := t;
    inreal (1, t);  p := t;
    inreal (1, t);
    em := 5;
    begin
        array x[1:n, 1:p], d[1:p];
        procedure op (n, v, w);  comment globals (t);
            value n;
            integer n;  array v, w;
            begin
                integer k;
                comment insert here piece of program which computes the vector
                        w = A × v without changing v;
                for k := 1 step 1 until n do w[k] := t × v[k] − w[k];
            end op;
        procedure dummy (a, b, c, d);  integer a, b, c;  array d;  ;
        ritzit (n, p, 1 000, 10 − 7, op, dummy, em, x, d);
        for j := 1 step 1 until p do d[j] := t − d[j];
        for j := 1, 2, 3, 4 do outsymbol (2, ' em = ⊔', j);
        outreal (2, em);
        for j := 1, 2, 3 do outsymbol (2, ' d = ⊔', j);
        outarray (2, d);
        for j := 1, 2, 3 do outsymbol (2, ' x = ⊔', j);
        outarray (2, x);
    end;
end;
```

b) Computation of the 5 lowest eigenvalues and corresponding vectors of the eigenvalue problem $Bx = \lambda A x$, where A, B both are symmetric and positive definite bandmatrices of bandwidth m. The solution of this problem can be found by iteration with $R^{-T} A R^{-1}$, where $R^T R$ is the Cholesky decomposition of B. We use procedure *cholbd* as described in [8] and iterate with 8 parallel vectors:

```
begin
    comment insert here procedures ritzit, cholbd;
    real s, t;
    integer n, m, em, k, l, j;
    comment n, m, a[1:n, 0:m], b[1:n, 0:m] on input medium 1;
    inreal (1, t);  n := t;
    inreal (1, t);  m := t;
    em := 5;
    begin
        array a[1:n, −m:m], b[1:n, 0:m], x[1:n, 1:8], e[1:5], d[1:8],
              y[1:n, 1:5], z[1−m:n+m];
        procedure opp (n, v, w);  comment globals (m, a, b, z, cholbd);
            value n;
            integer n;  array v, w;
            begin
                integer k, l;
                real s;
                for k := 1 step 1 until n do z[k] := v[k];
                cholbd (n, m, 3, b, z, j);
                for k := 1 step 1 until n do
                begin
                    s := 0;
                    for l := −m step 1 until m do s := s + a[k, l] × z[k + l];
                    w[k] := s;
                end for k;
                cholbd (n, m, 5, b, w, j);
            end opp;
        procedure ifo (ks, g, h, j);  integer ks, g, h;  array j;
            begin
                comment channel 3 is intended for informational output;
                outreal (3, ks);
                outreal (3, g);
                outreal (3, h);
                outarray (3, d);
                comment array d is available here because it is actual name para-
                        meter corresponding to the formal parameter d of pro-
                        cedure ritzit;
                outarray (3, j);
            end ifo;
        inarray (1, b);
        for k := 1 step 1 until n do
```

```
    begin
        for l := 0 step 1 until m do a[k, l] := b[k, l];
        for l := −m step 1 until −1 do
                a[k, l] := if k + l > 0 then a[k + l, −l] else 0.0;
    end for k;
    inarray (1, b);
    cholbd (n, m, 6, b, z, f);
    for k := 1 step 1 until m do z[1 − k] := z[n + k] := 0;
    ritzit (n, 8, 1 000, 10 − 6, opp, ifo, em, x, d);
    for j := 1 step 1 until 5 do
    begin
        e[j] := 1/d[j];
        for l := 1 step 1 until n do z[l] := x[l, j];
        cholbd (n, m, 3, b, z, f);
        for l := 1 step 1 until n do y[l, j] := z[l];
    end for j;
    for j := 1 step 1 until 11 do outsymbol (2, 'eigenvalues', j);
    outarray (2, e);
    for j := 1 step 1 until 12 do outsymbol (2, 'eigenvectors', j);
    outarray (2, y);
f:  end block;
end of program;
```

5. ALGOL Program

Since the following procedure makes use of a prepublished procedure *jacobi*
(cf. [9]), the latter is indicated in the body of *ritzit* merely by a pseudodeclaration
(cf. § 47.2 of [10]).

Procedure *inprod* which is declared in full inside *ritzit*, contributes a great
proportion of the total computing time. It would therefore increase the efficiency
of *ritzit*, if only the body of this procedure—and possibly that of *op*—were written
in machine code.

Furthermore, it should be recognized, that *ritzit* is a storage-saving version
of a simultaneous iteration program. In fact, despite the Chebyshev iteration,
the total storage requirement including the formal parameters (but excluding
possible storage space for the matrix A) is

$$(p + 3) \times n + 2 \times p^2 + 5 \times p + \text{const.}$$

This is possible because only one $n \times p$ matrix (the formal array x) is used.
With three such matrices a more efficient program could have been written,
but it seems that storage space is more often a bottleneck than computing time.

```
procedure ritzit (n, p, km, eps, op, inf) trans:(em) res:(x, d);
    value n, p, km, eps;
    integer n, p, km, em; real eps; procedure op, inf; array x, d;
    begin
        integer g, h, i, j, k, l, l1, m, m1, ks, z, z1, z2, z3;
        real e, e1, e2, ee2, mc, s, t;
```

array $u, v, w[1:n], b, q[1:p, 1:p], rq, f, cx[1:p]$;
integer array $r[1:p]$;
procedure *jacobi* $(n, eivec)$ *trans*: (a) *res*: (d, v, rot);
   value $n, eivec$;
   integer n, rot; **Boolean** *eivec*; **array** a, d, v;
   code see Contribution II/1;
real procedure *inprod* (n, k, l, x); **comment** *no globals*;
   value n, k, l;
   integer n, k, l; **array** x;
   comment computes inner product of columns k and l of matrix
           $x[1:n, 1:p]$;
   begin
     real s;
     integer i;
     $s := 0$;
     for $i := 1$ **step** 1 **until** n **do** $s := s + x[i, k] \times x[i, l]$;
     $inprod := s$;
   end *inprod*;
procedure *random* (l); **comment** *globals* (n, z, x);
   value l;
   integer l;
   comment fills random elements into column l of matrix x;
   begin
     integer k, f, g, h, y, m;
     $m := entier (z/4096 + 0.0001)$;
     $y := z - 4096 \times m$;
     for $k := 1$ **step** 1 **until** n **do**
     begin
       $f := 1001 \times y + 3721$;
       $g := entier (f/4096 + 0.0001)$;
       $f := f - 4096 \times g$;
       $h := 1001 \times (m + y) + g + 173$;
       $g := entier (h/4096 + 0.0001)$;
       $h := h - 4096 \times g$;
       $m := h$;
       $y := f$;
       $x[k, l] := (m + y/4096)/2048 - 1$;
     end *for* k;
     $z := 4096 \times m + y$;
   end *random*;
procedure *orthog* (f, p, b); **comment** *globals* $(n, x, random, inprod, mc)$;
   value f, p;
   integer f, p; **array** b;
   comment performs orthonormalisation of columns 1 through p of
         array $x[1:n, 1:p]$, assuming that columns 1 through f are
         already orthonormal;

```
            begin
                real s, t;
                integer i, j, k;
                Boolean orig;
                for k := f + 1 step 1 until p do
                begin
                    orig := true;
repeat:             t := 0;
                    for i := 1 step 1 until k − 1 do
                    begin
                        s := inprod (n, i, k, x);
                        if orig then b [p − k + 1, i] := s;
                        t := t + s × s;
                        for j := 1 step 1 until n do
                                x [j, k] := x [j, k] − s × x [j, i];
                    end for i;
                    s := inprod (n, k, k, x);
                    t := s + t;
                    if s ≤ t/100 ∨ t × mc = 0 then
                    begin
                        orig := false;
                        if s × mc = 0 then s := 0 else goto repeat;
                    end if s;
                    s := b [p − k + 1, k] := sqrt (s);
                    if s ≠ 0 then s := 1/s;
                    for j := 1 step 1 until n do x [j, k] := s × x [j, k];
                end for k;
            end orthog;
start:
        ee2 := 1 + eps/10;
        e := 0;
        for s := 10 − 6, s/2 while 1 + s ≠ 1 do mc := s;
        g := h := z := z1 := z2 := ks := 0;
        m := 1;
        for l := 1 step 1 until p do
        begin
            f [l] := 4;
            cx [l] := rq [l] := 0;
        end for l;
        if km > 0 then
            for l := 1 step 1 until p do random (l);
        orthog (0, p, b);
        km := abs (km);
loop:   ; comment jacobi-step modified;
        for k := g + 1 step 1 until p do
```

```
begin
    for j := 1 step 1 until n do v[j] := x[j, k];
    op(n, v, w);
    for j := 1 step 1 until n do x[j, k] := w[j];
end for k;
orthog(g, p, b);
if ks = 0 then
begin
    comment measures against unhappy choice of initial vectors;
    for k := 1 step 1 until p do
        if b[p − k + 1, k] = 0 then
        begin
            random(k);
            ks := 1;
        end if b, for k;
    if ks = 1 then
    begin
        orthog(0, p, b);
        goto loop;
    end if ks = 1;
end if ks = 0;
for k := g + 1 step 1 until p do
begin
    j := k − g;
    for l := k step 1 until p do q[j, l − g] := inprod(p − l + 1, k, l, b);
end for k;
jacobi(j, true, q, w, b, k);
comment reordering eigenvalues and -vectors according to size of the
        former;
for k := g + 1 step 1 until p do r[k] := k − g;
for k := g + 1 step 1 until p − 1 do
    for l := k + 1 step 1 until p do
        if w[r[k]] < w[r[l]] then
            begin i := r[k]; r[k] := r[l]; r[l] := i end;
for k := g + 1 step 1 until p do
begin
    i := r[k];
    d[k] := if w[i] ≤ 0 then 0.0 else sqrt(w[i]);
    for l := g + 1 step 1 until p do q[l, k] := b[l − g, i];
end for k;
for j := 1 step 1 until n do
begin
    for k := g + 1 step 1 until p do
```

```
begin
    s := 0;
    for l := g +1 step 1 until p do s := s + x[j, l] ×q[l, k];
    v[k] := s;
end for k;
    for k := g +1 step 1 until p do x[j, k] := v[k];
end for j;
ks := ks +1;
if d[p] > e then e := d[p];
comment randomisation;
if z1 < 3 then
begin
    random (p);
    orthog (p −1, p, b);
end if z1;
comment compute control quantities cx[i];
for k := g +1 step 1 until p −1 do
begin
    s := (d[k] − e) ×(d[k] + e);
    cx[k] := if s ≦ 0 then 0.0
             else if e = 0 then 1 000 + ln (d[k])
             else ln ((d[k] + sqrt (s))/e);
end for k;
comment acceptance test for eigenvalues including adjustment of em
        and h such that d[em] > e, d[h] > e and d[em] does not oscillate
        strongly;
for k := g +1 step 1 until em do
    if d[k] − e ≦ 0 ∨ (z1 > 1 ∧ d[k] ≦ 0.999 ×rq[k]) then
            begin em := k −1; goto ex4 end;
```

ex4:   `if em = 0 then goto ex;`

```
for k := h +1 while d[k] ≠ 0 ∧ d[k] ≦ ee2 ×rq[k] do h := k;
for k := h +1, k −1 while k > em do if d[k] − e ≦ 0 then h := k −1;
comment acceptance test for eigenvectors;
l := g;
e2 := 0;
for k := g +1 step 1 until p −1 do
begin
    if k = l +1 then
    begin
        comment check for nested eigenvalues;
        l := l1 := k;
        for j := k +1 step 1 until p −1 do
            if cx[j] ×(cx[j] + 0.5/ks) +1/(ks ×m) > cx[j −1]↑2 then l := j
            else goto ex5;
```

```
ex5:          if l > h then begin l := l1 − 1; goto ex6 end;
              end if k = l + 1;
              for i := 1 step 1 until n do v[i] := x[i, k];
              op (n, v, w);
              s := 0;
              for j := 1 step 1 until l do
                  if abs (d[j] − d[k]) < 0.01 × d[k] then
                  begin
                      t := 0;
                      for i := 1 step 1 until n do t := t + w[i] × x[i, j];
                      for i := 1 step 1 until n do w[i] := w[i] − t × x[i, j];
                      s := s + t↑2;
                  end if abs, for j;
              t := 0;
              for i := 1 step 1 until n do t := t + w[i]↑2;
              t := if s = 0 then 1 else sqrt (t/(s + t));
              if t > e2 then e2 := t;
              if k = l then
              begin
                  comment test for acceptance of group of eigenvectors;
                  if l ≥ em ∧ d[em] × f[em] < eps × (d[em] − e) then g := em;
                  if e2 < f[l] then
                          for j := l1 step 1 until l do f[j] := e2;
                  if l ≤ em ∧ d[l] × f[l] < eps × (d[l] − e) then g := l;
              end if k = l;
          end for k;
ex6:      ; comment adjust m;
          if e ≤ 0.04 × d[1] then m := k := 1
          else
              begin
                  e2 := 2/e;
                  e1 := 0.51 × e2;
                  k := 2 × entier (4/cx[1]);
                  if m > k then m := k;
              end if else;
          comment reduce em if convergence would be too slow;
          if f[em] ≠ 0 ∧ ks < 0.9 × km then
          begin
              t := if k × cx[em] < 0.05 then 0.5 × k × cx[em]↑2
                      else cx[em] + ln ((1 + exp (− 2 × k × cx[em]))/2)/k;
              s := ln (d[em] × f[em]/(eps × (d[em] − e)))/t;
              if (km − ks) × km < s × ks then em := em − 1;
          end if f[em];
ex2:      for k := g + 1 step 1 until p − 1 do rq[k] := d[k];
          inf (ks, g, h, f);
          if g ≥ em ∨ ks ≥ km then goto ex;
```

ex1:   **if** $ks + m > km$ **then**
    begin
       $z2 := -1$;
       if $m > 1$ **then** $m := 2 \times entier((km - ks)/2 + 0.6)$;
    end *if ks*;
    $m1 := m$;
    comment shortcut last intermediate block if all $f[i]$ are sufficiently small;
    if $l \geqq em$ **then**
    begin
       $s := d[em] \times f[em]/(eps \times (d[em] - e))$;
       $t := s\uparrow 2 - 1$;
       if $t \leq 0$ **then goto** *loop*;
       $s := ln(s + sqrt(t))/(cx[em] - cx[h + 1])$;
       $m1 := 2 \times entier(s/2 + 1.01)$;
       if $m1 > m$ **then** $m1 := m$ **else** $z2 := -1$;
    end *if h*;
    comment chebyshev iteration;
    if $m = 1$ **then**
      for $k := g + 1$ **step** 1 **until** p **do**
      begin
         for $i := 1$ **step** 1 **until** n **do** $v[i] := x[i, k]$;
         $op(n, v, u)$;
         for $i := 1$ **step** 1 **until** n **do** $x[i, k] := u[i]$;
      end *for k, if m*;
    if $m > 1$ **then**
      for $k := g + 1$ **step** 1 **until** p **do**
      begin
         for $i := 1$ **step** 1 **until** n **do** $v[i] := x[i, k]$;
         $op(n, v, u)$;
         for $i := 1$ **step** 1 **until** n **do** $w[i] := e1 \times u[i]$;
         $op(n, w, u)$;
         for $i := 1$ **step** 1 **until** n **do** $v[i] := e2 \times u[i] - v[i]$;
         for $j := 4$ **step** 2 **until** $m1$ **do**
         begin
            $op(n, v, u)$;
            for $i := 1$ **step** 1 **until** n **do** $w[i] := e2 \times u[i] - w[i]$;
            $op(n, w, u)$;
            for $i := 1$ **step** 1 **until** n **do** $v[i] := e2 \times u[i] - v[i]$;
         end *for j*;
         for $i := 1$ **step** 1 **until** n **do** $x[i, k] := v[i]$;
      end *for k, if m*;
   orthog (g, p, q);
    comment discounting the error quantities *f*;
    $t := exp(-m1 \times cx[h + 1])$;
    for $k := g + 1$ **step** 1 **until** h **do**
      if $m = 1$ **then** $f[k] := f[k] \times (d[h + 1]/d[k])$
      else

```
              begin
                 s := exp(−m1×(cx[k]−cx[h+1]));
                 f[k] := s×f[k]×(1+t↑2)/(1+(s×t)↑2);
              end for k, if else;
        ks := ks+m1;
        z2 := z2−m1;
        comment possible repetition of intermediate steps;
        if z2≥0 then goto ex1;
        z1 := z1+1;
        z2 := 2×z1;
        m := 2×m;
        goto loop;
ex:     em := g;
        comment solve eigenvalue problem of projection of matrix a;
        orthog(0, p−1, b);
        for k := 1 step 1 until p−1 do
        begin
           for j := 1 step 1 until n do v[j] := x[j, k];
           op(n, v, w);
           for j := 1 step 1 until n do x[j, p] := w[j];
           for i := 1 step 1 until k do q[i, k] := inprod(n, i, p, x);
        end for k;
        jacobi(p−1, true, q, d, b, k);
        for j := 1 step 1 until n do
        begin
           for i := 1 step 1 until p−1 do
           begin
              s := 0;
              for k := 1 step 1 until p−1 do s := s+x[j, k]×b[k, i];
              v[i] := s;
           end for i;
           for i := 1 step 1 until p−1 do x[j, i] := v[i];
        end for j;
        d[p] := e;
end ritzit;
```

6. Organisational and Notational Details

ritzit follows closely the recommendations given in [11]:

1. X_0 is chosen as a random $n \times p$ matrix whose columns are subsequently orthonormalized.

2. The iteration process includes 4 devices:

a) Jacobi-steps. In *ritzit* these are not performed as shown in [11], § 3, but—according to a proposal of Reinsch, Technische Hochschule München—as follows:

(1) $Y := A X_k$.

(2) $Y =: U R$ (Gram-Schmidt orthogonalisation).

(3) $Q := R R^T$.

(4) $Q \to B^T Q B = \mathrm{diag}\,(d_1^2, d_2^2, \ldots, d_p^2)$ (Jacobi method).

(5) $X_{k+1} := U B$.

b) Intermediate steps ([11], § 4).

c) Randomisation ([11], § 5).

d) Acceptance tests ([11], § 6).

3. In order that the process can be applied also to indefinite matrices, the intermediate steps of [11] have been modified as follows:

a) The Chebyshev interval is now $(-e, e)$ with $e = \bar{d}_{pp}$.

b) Set out with X_h.

c) If $m = 1$, set $X_{h+1} := A X_h$.

If $m > 1$, however, use d), e) below:

d) $X_{h+1} := \dfrac{1.02}{e}\, A X_h{}^{\star}, \qquad X_{h+2} := \dfrac{2}{e}\, A X_{h+1} - X_h.$

e) $X_{h+j} := \dfrac{2}{e}\, A X_{h+j-1} - X_{h+j-2} \quad (j = 3, 4, \ldots, m).$

f) Orthonormalize the columns of X_{h+m}.

g) Increase step counter by m.

In the following, 3 b) through 3 g) are called an *intermediate block*.

4. The method to delimit m is essentially the same as in [11], but in the present program m is either 1 or an even number. Formulae 3 d), 3 e) imply ($\| \; \| = $ 2-norm for the columns of X):

$$\|X_{h+k}\| \approx Cosh\,(c \times k),$$

where $c = ArCosh\,(\lambda_1/e)$, the latter being approximately the $cx\,[1]$ occurring in *ritzit*. Desiring no amplification more than $Cosh\,(8) = 1490.479$ in any one column of X during an intermediate block, m is defined as

1 ,   if $e < 0.04 \times d\,[1]$ (this being equivalent to $cx\,[1] > 3.91$),

$2 \times entier\,(4/cx\,[1])$ otherwise.

5. Formulae 3 d), 3 e) are realized by performing them serially column by column of X. For column index k, the k-th column of X_{h+j} is held in vector v, if j is even, in w if j is odd, while $A X_{h+j}$ is always stored in u. This accounts for the 3 extra vectors $u, v, w\,[1:n]$ declared within *ritzit*. The case $m = 1$ is treated as a special case.

6. Extensive tests revealed that convergence speed can depend critically upon the relative frequencies of the devices 2 a) through 2 d). Indeed,

a) if Jacobi steps are too frequent, computing time is increased; not so much because of the Jacobi method applied to the $p \times p$ matrix Q, but because of the operations involving the columns of X. As a consequence, Jacobi steps are made more and more infrequent, either by increasing m continually or—if its upper

---

★ With this factor 1.02 iteration occurs actually with a mixture of Chebyshev T- and U-polynomials. This improves the stability of the process.

limit as stated in 4. above is reached—by performing several intermediate blocks between Jacobi steps.

b) If randomisation is too frequent, then the Rayleigh quotient $d[p]\star$ for the last column of X, which after every randomisation is reduced considerably, may possibly never come near to the eigenvalue λ_p. Since this would cause a too small interval $(-e, e)$ being taken as basic interval for the Chebyshev acceleration, the latter would be far from being as powerful as it might be. As a consequence, randomisation is performed *only after the first three* Jacobi steps.

c) Acceptance tests are also rather time-consuming and are therefore performed only after every Jacobi-step.

7. In view of these observations the following method for governing the various devices has been adopted**:

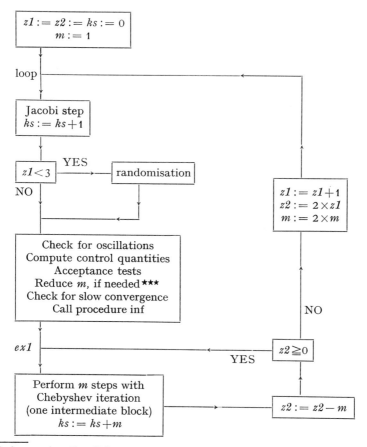

---

    * To be precise, the $d[j]$ are square roots of Rayleigh quotients for A^2.

    ** Note that ks counts the iteration steps, m is the number of steps combined into one intermediate block, $z1$ counts the Jacobi steps, while $z2$ governs the number of intermediate blocks executed between two Jacobi steps.

    *** I.e. either sets m to 1 or makes that $T_m(d[1]/e)$ should not exceed 1 500.—Note that m may be reduced also at the end of the iteration process in order to save unneeded iteration steps.

From this diagram it follows that if $|\lambda_p|$ is very near to $|\lambda_1|$, then m is steadily increasing such that Jacobi steps have numbers 0, 2, 7, 16, 25, 42, 75, 140, 269, ... If on the other hand $|\lambda_p|$ is appreciably smaller than $|\lambda_1|$, then m will never become big, which means that—because $z1$ is steadily increasing—a growing number of intermediate blocks are performed between Jacobi steps. If $|\lambda_p/\lambda_1| < 0.04$, we have always $m = 1$; then the Jacobi steps have step numbers 0, 2, 6, 12, 20, 30, 42, ..., while in the general case the behaviour of *ritzit* is between these extremes.

7. Discussion of Numerical Properties

As the iteration proceeds, the Rayleigh quotients $d[j]$ approach the respective eigenvalues. As soon as the relative increase of $d[h+1]$ is below $eps/10$, this eigenvalue is accepted and h is increased by one.

Somewhat slower than the $d[j]$ (theoretically half as fast) also the columns of X approach the respective eigenvectors of A. However, since *ritzit* allows for negative eigenvalues, but in the Jacobi steps Rayleigh quotients are actually computed with respect to A^2, eigenvalues of equal or nearly equal moduli and opposite sign are badly separated. This fact requires the acceptance test for eigenvectors as given in [11] to be modified as follows:

a) If eigenvalues occur in groups of equal or nearly equal moduli, the corresponding group of eigenvectors is only accepted as a whole.

b) *ritzit* computes—after every Jacobi step—the quantities

$$g_k = \|A\,x_k - y_l\|/\|A\,x_k\|,$$

where y_l is the projection of $A\,x_k$ onto the space spanned by $x_1, \ldots, x_l$, l being defined by the fact that d_l is the last of a group of nearly equal d's among which d_k occurs. The biggest g_j for this group—implicitly multiplied by $d_k/(d_k - e)$—is then used as error quantity f_k in the same sense as the quantity on the right side of (27) in [11].

c) Discounting—performed after every intermediate block—is done in accordance with the properties of the Chebyshev polynomials and somewhat more carefully than proposed in [11]. Indeed, it cannot be taken for granted that $\bar{d}_{pp}$ is a sufficiently safe upper bound for $|\lambda_{p+1}|$. Since a too small value of $\bar{d}_{pp}$ might cause a too optimistic discounting and thus a premature acceptance of an eigenvector, $d[h+1]$—where h is the number of already accepted eigenvalues—is taken as an upper bound for $|\lambda_{p+1}|$. This yields the following discounting rule:

$$f_j := f_j \times Cosh\left(m \times ArCosh\,(d_{h+1}/e)\right)/Cosh\left(m \times ArCosh\,(d_j/e)\right).$$

d) If it can be foreseen that the next intermediate block would reduce the error quantity f_{em} (for the last vector to be accepted) to far less than eps, this intermediate block will be performed with a smaller value of m (called $m1$).

e) In order to separate eigenvectors to eigenvalues of equal moduli but opposite in sign, *ritzit* solves—immediately before termination—the eigenvalue problem for the projection of A onto the space spanned by $x_1, \ldots, x_{p-1}$.

Concerning the effects of roundoff errors, it must be admitted that *ritzit* may compute certain eigenvectors—mainly those corresponding to small or close eigenvalues—less precisely than prescribed by the formal parameter *eps*.

Sometimes it is meaningful to compute such eigenvectors all the same; as an example for the eigenvectors corresponding to a close pair of eigenvalues *ritzit* certainly does its best, but also small eigenvalues and corresponding vectors may be computed correctly in certain cases, e.g. for a matrix like

$$
\begin{array}{ccc}
10^8 & 10^4 & 1 \\
10^4 & 10^4 & 1 \\
1 & 1 & 1.
\end{array}
$$

Therefore we are faced with the problem to decide under what circumstances it will be useful to compute λ_k and v_k. Of course in procedure *ritzit* we can prescribe the number *em* of eigenvectors to be computed, but in certain situations we may spend a lot of computing time only to see that parts of the results are useless. A strategy has to be developed therefore which allows to decrease *em* during execution of *ritzit* in order to avoid unneeded calculations. This is done as follows:

First, if an accepted Rayleigh quotient $d[k]$ decreases more than 0.1 percent between consecutive Jacobi steps, *em* is decreased to $k-1$.

Second, if one of the Rayleigh quotients $d[k]$ $(k=1, \ldots, em)$ falls below the quantity $\bar{d}_{pp}$ (in the program called e), then *em* is also decreased to $k-1$.

If $d[k]$ begins to oscillate strongly (which indicates severe cancellation for the k-th iteration vector) sooner or later one of these two devices comes into action (cf. the statements following the comment "acceptance test for eigenvalues including adjustment of *em* and h ...").

Third, if the convergence quotients derived from the $d[k]$'s indicate that even with an optimistic judgement *km* steps are insufficient for computing *em* eigenvectors, then again *em* is reduced (see comment "reduce *em* if convergence would be too slow").

Special care has been given to the orthonormalisation problem: Not only is the orthogonalisation repeated whenever a column of X is reduced to less than 1/10-th of its length by the process, but in addition a column is set to zero if underflow threatens the validity of the orthonormalisation process. This null vector—and any other null vector possibly produced if the rank of A is less than p—will later automatically be shifted into the last column of X where it can do no harm, and the corresponding eigenvalue is set to zero. If several such null vectors are produced, it is taken care of that *em* is reduced such that $x_1, \ldots, x_{em}$ are non-null vectors.

8. Test Results *

a) Computation of the 4 lowest eigenvalues and corresponding-vectors of the plate matrix (of order 70) which is defined by the difference operators given on

---

* All calculations were performed with the CDC-1604A computer of the Swiss Federal Institute of Technology. This computer has a 48 bit wordlength, of which 36 are available for the mantissa.

p. 20/22 of [4]. The 9 lowest eigenvalues are

0.0415   0.0570   0.1881   0.3216   0.3332   0.6248   0.8123   0.8810   0.8879,

while the highest is 61.307. The simultaneous iteration method was applied to A^{-1} with $p=5$ and to $32 \times I - A$ with $p=8$. The higher value of p is suggested by the poorer convergence of the shifted iteration as compared to the inverse iteration. The following number of iteration steps and computing time were required:

		Iteration steps	Time (sec)
A^{-1},	$eps = {}_{10}-5$	63	141
A^{-1},	$eps = {}_{10}-8$	99	200
$32I - A$,	$eps = {}_{10}-5$	111	204
$32I - A$,	$eps = {}_{10}-8$	146	280

It had been expected that despite the low convergence rate the iteration with $32I - A$ could, by virtue of the faster execution of procedure op in that case, complete with the inverse iteration. That this is not so, is due to the fact that inverse iteration, performed via Cholesky-decomposition, is favored not only by the bandform of the matrix A, but also by the smaller number of parallel iteration vectors.

Concerning the precision of the results, these of the inverse iteration were always better than those obtained by shifted iteration. The following table shows the computed 2-norms $\|A x_k - d_k x_k\|$ for $k = 1, 2, 3, 4$:

		$k = 1$	$k = 2$	$k = 3$	$k = 4$
A^{-1},	$eps = {}_{10}-5$	$6.4_{10}-10$	$6.8_{10}-10$	$1.1_{10}-9$	$2.2_{10}-7$
A^{-1},	$eps = {}_{10}-8$	$6.4_{10}-10$	$7.4_{10}-10$	$1.1_{10}-9$	$5.0_{10}-10$
$32I - A$,	$eps = {}_{10}-5$	$3.7_{10}-6$	$6.3_{10}-6$	$1.5_{10}-7$	$5.4_{10}-8$
$32I - A$,	$eps = {}_{10}-8$	$1.6_{10}-9$	$3.0_{10}-9$	$1.5_{10}-9$	$3.0_{10}-9$

It may be seen that a smaller value of eps does not always yield also a smaller $\|A x - dx\|$, but this is due to the fact that overshooting a prescribed precision is rather commonplace in quickly converging iteration processes.

Furthermore it occurred that for the shifted iteration and $eps = {}_{10}-8$ the computed eigenvectors x corresponding to $\lambda_1, \lambda_2, \lambda_4$, deviate by more than eps from the true eigenvectors v:

$$\|x_1 - v_1\| = 5.9_{10} - 8, \qquad \|x_2 - v_2\| = 5.8_{10} - 8, \qquad \|x_4 - v_4\| = 1.2_{10} - 7.$$

In all other cases, $\|v - x\|$ was below eps. The exception is explained by the relative closeness of the pairs $(32 - \lambda_1, 32 - \lambda_2)$ and $(32 - \lambda_4, 32 - \lambda_5)$ which makes a more precise computation with a 36-bit mantissa impossible. On the other hand, the space spanned by x_1, x_2 is well within 8-digit precision.

b) Symmetric matrix with elements $a_{ij} = a_{ji} = M + d_k$, where M is a big constant, $d_0, d_1 d_2 d_3 \ldots$ is the decimal fraction of π, and $k = (n - i/2) \times (i - 1) + j$.

E.g. for $M = 1\,000$, $n = 4$:

$$A = \begin{pmatrix} 1\,001 & 1\,004 & 1\,001 & 1\,005 \\ & 1\,009 & 1\,002 & 1\,006 \\ & & 1\,005 & 1\,003 \\ \text{symm.} & & & 1\,005 \end{pmatrix}.$$

For $n = 40$, $M = 1\,000$, the dominant eigenvalues are

$$\begin{aligned} & 40\,179.804\,75 \\ & 32.135\,3323 \\ & -32.128\,2489 \\ & -30.792\,3596 \\ & 28.896\,6912 \\ & -28.540\,8701 \\ & 27.713\,1029 \end{aligned}$$

Since already λ_1 and λ_2 differ to such an extent, that the Chebyshev iteration occurs permanently with $m = 1$, the convergence is good only for the first iteration vector. Indeed, with $p = 5$, $em = 1$, $eps = 10-6$, 7 steps were sufficient to produce the eigenvector v_1 with 10-digit precision.

With $p = 3$, $em = 2$, $eps = 10-8$, however, the convergence quotient for the second eigenvector is 0.99978 which leaves no hope for computing v_2 within e.g. 500 steps. As a consequence, if this case is tried with $km = 500$, then *ritzit* itself reduces em after 43 steps from 2 to 1, whereupon the process is terminated since v_1 has been accepted long before.

With $p = 8$, $em = 5$, $eps = 10-8$, 164 steps were sufficient for computing 5 eigenvalues. The following table shows the course of some crucial values occurring in the computation (status after the Jacobi steps, ks denoting their step numbers):

z1	ks	g	h	d1	d2	d3	d4	d5	d6	d7	d8
0	0	0	0	15881.3	23.257	20.724	18.554	17.582	15.714	13.490	10.057
1	2	0	0	40179.8	30.659	30.139	27.458	25.297	24.814	22.787	20.449
2	6	1	1	40179.8	31.973	31.954	30.404	28.464	27.632	26.110	21.825
.	.	.	.								
6	42	1	1	40179.8	32.135	32.128	30.792	28.897	28.541	27.713	25.463
7	56	1	3	40179.8	32.135	32.128	30.792	28.897	28.541	27.713	25.573
.	.	.	.								
10	110	3	6	40179.8	32.135	32.128	30.792	28.897	28.541	27.713	25.798
11	132	4	7	40179.8	32.135	32.128	30.792	28.897	28.541	27.713	25.815
12	156	4	7	40179.8	32.135	32.128	30.792	28.897	28.541	27.713	25.822

The values indicate that the last Rayleigh quotient $d[8]$ might lead to a too optimistic discounting quotient, while our algorithm uses $d[8]$ only after $d[7]$ has been accepted which is the case after step 132.

After step 164, x_1 was within 10-digit precision, while $\|x_2 - v_2\| = 1.47_{10} - 8$, the errors of x_3, x_4, x_5 being of the same order of magnitude. That these errors are bigger than eps, is easily explained by the smallness of the corresponding eigenvalues as compared to the size of the matrix elements.

For $n = 40$, $M = {}_{10}10$, the unsharp definition of the eigenvectors v_2, v_3, v_4, ... is still more pronounced. Again x_1 is accepted after a few iteration steps, but since for the iteration vectors x_2 and following the Rayleigh quotients are already in the roundoff error level, the $d[k]$ are oscillating to such an extent that em is reduced to 1 whereupon the process terminates with only one computed eigenvector $x_1 = (x_{k1} = 0.1581138830$, $k = 1, ..., 40)$.

Acknowledgement. The author is deeply indebted to Dr. C. Reinsch of the Technische Universität München for valuable suggestions concerning the computing process, and to A. Mazzario of the Computer Science group of the Swiss Federal Instiute of Technology for assisting in streamlining the program and for performing the calculations.

References

1. Barth, W., Martin, R. S., Wilkinson, J. H.: Calculation of the eigenvalues of a symmetric tridiagonal matrix by the bisection method. Numer. Math. **9**, 386–393 (1967). Cf. II/5.
2. Bauer, F. L.: Das Verfahren der Treppeniteration und verwandte Verfahren zur Lösung algebraischer Eigenwertprobleme. Z. Angew. Math. Phys. **8**, 214–235 (1957).
3. Bowdler, H., Martin, R. S., Reinsch, C., Wilkinson, J. H.: The QL and QR algorithms for symmetric matrices. Numer. Math. **11**, 293–306 (1968). Cf. II/3.
4. Engeli, M., Ginsburg, Th., Rutishauser, H., Stiefel, E.: Refined iterative methods for computing of the solution and the eigenvalues of selfadjoint boundary value problems. Mitteilung Nr. 8 aus dem Institut für angewandte Mathematik der ETH, Zürich. Basel: Birkhäuser 1959.
5. Jennings, A.: A direct method for computing latent roots and vectors of a symmetric matrix. Proc. Cambridge Philos. Soc. **63**, 755–765 (1967).
6. Läuchli, P., Waldburger, H.: Berechnung von Beulwerten mit Hilfe von Mehrstellenoperatoren. Z. Angew. Math. Phys. **11**, 445–454 (1960).
7. Martin, R. S., Reinsch, C., Wilkinson, J. H.: Householder tridiagonalisation of a symmetric matrix. Numer. Math. **11**, 181–195 (1968). Cf. II/2.
8. Rutishauser, H.: Lineares Gleichungssystem mit symmetrischer positiv definiter Bandmatrix nach Cholesky. Computing **1**, 77–78 (1966).
9. — The Jacobi method for real symmetric matrices. Numer. Math. **9**, 1–10 (1966). Cf. II/1.
10. — Description of ALGOL 60 (Handbook of automatic computation, Vol. 1 a). Berlin-Heidelberg-New York: Springer 1967.
11. — Computational aspects of F. L. Bauer's simultaneous iteration method. Numer. Math. **13**, 4–13 (1969).
12. Schwarz, H. R.: Matrizen-Numerik. Stuttgart: Teubner 1968.
13. Stuart, G. W.: Accelerating the orthogonal iteration for the eigenvalues of a Hermitian matrix. Numer. Math. **13**, 362–376 (1969).

Reduction of the Symmetric Eigenproblem $Ax=\lambda Bx$ and Related Problems to Standard Form*

by R. S. MARTIN and J. H. WILKINSON

1. Theoretical Background

In many fields of work the solution of the eigenproblems $Ax = \lambda Bx$ and $ABx = \lambda x$ (or related problems) is required, where A and B are symmetric and B is positive definite. Each of these problems can be reduced to the standard symmetric eigenproblem by making use of the Cholesky factorization [4] of B.

For if L is defined by

$$LL^T = B \tag{1}$$

then **

$$Ax = \lambda Bx \quad \text{implies} \quad (L^{-1}AL^{-T})(L^T x) = \lambda(L^T x) \tag{2}$$

and

$$ABx = \lambda x \quad \text{implies} \quad (L^T AL)(L^T x) = \lambda(L^T x). \tag{3}$$

Hence the eigenvalues of $Ax = \lambda Bx$ are those of $Py = \lambda y$, where P is the symmetric matrix $L^{-1}AL^{-T}$, while the eigenvalues of $ABx = \lambda x$ are those of $Qy = \lambda y$, where Q is the symmetric matrix $L^T AL$. Note that in each case the eigenvector x of the original problem is transformed into $L^T x$.

There are a number of closely related problems which are defined by the equations

$$y^T AB = \lambda y^T \quad \text{or} \quad (L^{-1}y)^T(L^T AL) = \lambda(L^{-1}y)^T$$
$$BAy = \lambda y \quad \text{or} \quad (L^T AL)(L^{-1}y) = \lambda(L^{-1}y) \tag{4}$$
$$x^T BA = \lambda x^T \quad \text{or} \quad (L^T x)^T(L^T AL) = \lambda(L^T x)^T.$$

Here it is assumed throughout that it is B which is required to be positive definite (A may or may not be positive definite). It will be observed that all the problems can be reduced if we have procedures for computing $L^T AL$ and $L^{-1}AL^{-T}$. Further if we denote an eigenvector of the derived problem by z, we need only procedures for solving

$$L^T x = z \quad \text{and} \quad L^{-1}y = z \quad \text{i.e.} \quad y = Lz. \tag{5}$$

If z is normalized so that $z^T z = 1$ then x and y determined from Eqs. (5) satisfy

$$x^T Bx = 1 \quad \text{and} \quad y^T B^{-1}y = 1, \quad y^T x = 1 \tag{6}$$

and these are the normalizations that are usually required in practice.

---

* Prepublished in Numer. Math. **11**, 99—110 (1968).
** Throughout this paper we have used L^{-T} in place of the cumbersome $(L^{-1})^T$ and $(L^T)^{-1}$.

Any decomposition $B = X X^T$ may be used in a similar way, but the Cholesky decomposition is the most economical, particularly when carried out as described in Section 5. If the standard eigenproblem for the matrix B is solved by a method *which gives an accurately orthogonal system Y of eigenvectors* then we have

$$B = Y \operatorname{diag}(d_i^2) Y^T = Y D^2 Y^T \tag{7}$$

where d_i^2 are the eigenvalues (positive) of B. Hence corresponding to $A x = \lambda B x$ we may solve $(D^{-1} Y^T A Y D^{-1})(D Y^T x) = \lambda (D Y^T x)$ and similarly for the other problems. Generally this is much less efficient than the use of the Cholesky decomposition, but it may have slight advantages in very special cases (see the discussion on pp. 337—338, 344 of ref. [7]).

2. Applicability

reduc 1 may be used to reduce the eigenproblem $A x = \lambda B x$ to the standard symmetric eigenproblem $P z = \lambda z$, where $P = L^{-1} A L^{-T}$ and $B = L L^T$.

reduc 2 may be used to reduce the eigenproblems $y^T A B = \lambda y^T$, $B A y = \lambda y$, $A B x = \lambda x$, $x^T B A = \lambda x^T$ to the standard symmetric eigenproblem $Q z = \lambda z$, where $Q = L^T A L$ and $B = L L^T$.

When A and B are narrow symmetric band matrices of high order these procedures should not be used as they destroy the band form.

The derived standard symmetric eigenproblem may be solved by procedures which are included in this series [1, 3, 5, 6]. The eigenvalues of the derived standard problem are those of the original problem but the vectors are related as indicated by Eq. (2), (3), (4). The eigenvectors of the original problem may then be obtained using the procedures *rebak a* and *rebak b* described below.

rebak a provides eigenvectors x of the problems $A x = \lambda B x$, $A B x = \lambda x$, $x^T B A = \lambda x^T$ from the corresponding eigenvectors $z = L^T x$ of the derived standard symmetric problem.

rebak b provides eigenvectors y of the problems $y^T A B = \lambda y^T$, $B A y = \lambda y$ from the corresponding eigenvectors $z = L^{-1} y$ of the derived standard symmetric problem.

3. Formal Parameter List

3.1. Input to procedure *reduc 1*

n	order of matrices A and B, or negative order if L already exists.
a	elements of the symmetric matrix A given as upper-triangle of an $n \times n$ array (strict lower-triangle can be arbitrary).
b	elements of the symmetric positive definite matrix B given as upper-triangle of an $n \times n$ array (strict lower-triangle can be arbitrary).

Output of procedure *reduc 1*

a	elements of symmetric matrix $P = L^{-1} A L^{-T}$ given as lower-triangle of $n \times n$ array. The strict upper-triangle of A is still preserved but its diagonal is lost.

b sub-diagonal elements of matrix L such that $LL^T=B$ stored as strict lower-triangle of an $n \times n$ array.

dl the diagonal elements of L stored as an $n \times 1$ array.

fail   the exit used if B, possibly on account of rounding errors, is not positive definite.

3.2. Input to procedure *reduc 2*

n order of matrices A and B or negative order if L already exists.

a elements of the symmetric matrix A given as upper-triangle of an $n \times n$ array (strict lower-triangle can be arbitrary).

b elements of the symmetric positive definite matrix B given as upper-triangle of an $n \times n$ array (strict lower-triangle can be arbitrary).

Output from procedure *reduc 2*

a elements of symmetric matrix $Q=L^TAL$ given as lower-triangle of $n \times n$ array. The strict upper-triangle of A is still preserved but its diagonal is lost.

b sub-diagonal elements of matrix L such that $LL^T=B$ stored as strict lower-triangle of an $n \times n$ array.

dl the diagonal elements of L stored as an $n \times 1$ array.

fail   the exist used if B, possibly on account of rounding errors, is not positive definite.

3.3. Input to procedure *rebak a*

n order of matrices A and B.

$m1, m2$ eigenvectors $m1, \ldots, m2$ of the derived standard symmetric eigenproblem have been found.

b the sub-diagonal elements of the matrix L such that $LL^T=B$ stored as the strict lower-triangle of an $n \times n$ array (as given by *reduc 1* or *reduc 2*).

dl the diagonal elements of L stored as an $n \times 1$ array.

z an $n \times (m2 - m1 + 1)$ array containing eigenvectors $m1, \ldots, m2$ of the derived symmetric eigenproblem.

Output from procedure *rebak a*

z an $n \times (m2 - m1 + 1)$ array containing eigenvectors of the original problem; output $z = L^{-T}$ (input z).

3.4. Input to procedure *rebak b*

n order of matrices A and B.

$m1, m2$ eigenvectors $m1, \ldots, m2$ of the derived standard symmetric eigenproblem have been found.

b   the sub-diagonal elements of the matrix L such that $LL^T = B$ stored
            as the strict lower-triangle of an $n \times n$ array (as given by *reduc 1* or
            reduc 2).
dl   the diagonal elements of L stored as an $n \times 1$ array.
z   an $n \times (m2 - m1 + 1)$ array containing eigenvectors $m1, \ldots, m2$ of the
            derived symmetric eigenproblem.

Output from procedure *rebak b*

z   an $n \times (m2 - m1 + 1)$ array containing eigenvectors of the original
            problem. output $z = L$(input z).

4. ALGOL Programs

procedure *reduc1* (n) *trans*: (a, b) *result*: (dl) *exit*: $(fail)$;
value n; **integer** n; **array** a, b, dl; **label** *fail*;
comment Reduction of the general symmetric eigenvalue problem

$$A \times x = lambda \times B \times x,$$

with symmetric matrix A and symmetric positive definite matrix B,
to the equivalent standard problem $P \times z = lambda \times z$.

The upper triangle, including diagonal elements, of A and B are
given in the arrays $a[1:n, 1:n]$ and $b[1:n, 1:n]$.

L $(B = L \times LT)$ is formed in the remaining strictly lower triangle
of the array b with its diagonal elements in the array $dl[1:n]$, and the
lower triangle of the symmetric matrix P $(P = inv(L) \times A \times inv(LT))$
is formed in the lower triangle of the array a, including the diagonal
elements. Hence the diagonal elements of A are lost.

If $n < 0$, it is assumed that L has already been formed.

The procedure will fail if B, perhaps on account of rounding errors,
is not positive definite;

```
begin      integer i, j, k;
           real x, y;
           if n < 0 then n := − n
           else
           for i := 1 step 1 until n do
           for j := i step 1 until n do
           begin x := b[i, j];
                     for k := i − 1 step − 1 until 1 do
                     x := x − b[i, k] × b[j, k];
                     if i = j then
                     begin if x ≤ 0 then go to fail;
                              y := dl[i] := sqrt(x)
                     end
                     else b[j, i] := x/y
           end ji;
           comment L has been formed in the array b;
           for i := 1 step 1 until n do
           begin y := dl[i];
                     for j := i step 1 until n do
```

```
        begin x := a[i, j];
            for k := i − 1 step − 1 until 1 do
            x := x − b[i, k] × a[j, k];
            a[j, i] := x/y
        end j
    end i;
    comment The transpose of the upper triangle of inv(L) × A has been
            formed in the lower triangle of the array a;
    for j := 1 step 1 until n do
    for i := j step 1 until n do
    begin x := a[i, j];
        for k := i − 1 step − 1 until j do
        x := x − a[k, j] × b[i, k];
        for k := j − 1 step − 1 until 1 do
        x := x − a[j, k] × b[i, k];
        a[i, j] := x/dl[i]
    end ij
end reduc1;


procedure reduc2 (n) trans: (a, b) result: (dl) exit: (fail);
value n; integer n; array a, b, dl; label fail;
comment Reduction of the general symmetric eigenvalue problems
```

$$A \times B \times x = lambda \times x, \qquad yT \times A \times B = yT \times lambda,$$
$$B \times A \times y = lambda \times y, \qquad xT \times B \times A = xT \times lambda,$$

with symmetric matrix A and symmetric positive definite matrix B, to the equivalent standard problem $Q \times z = lambda \times z$.

The upper triangle, including diagonal elements, of A and B are given in the arrays $a[1:n, 1:n]$ and $b[1:n, 1:n]$.

L ($B = L \times LT$) is formed in the remaining strictly lower triangle of the array b with its diagonal elements in the array $dl[1:n]$, and the lower triangle of the symmetric matrix Q ($Q = LT \times A \times L$) is formed in the lower triangle of the array a, including the diagonal elements. Hence the diagonal elements of A are lost.

If $n < 0$, it is assumed that L has already been formed.

The procedure will fail if B, perhaps on account of rounding errors, is not positive definite;

```
begin    integer i, j, k;
         real x, y;
         if n < 0 then n := − n
         else
         for i := 1 step 1 until n do
         for j := i step 1 until n do
         begin x := b[i, j];
             for k := i − 1 step − 1 until 1 do
             x := x − b[i, k] × b[j, k];
             if i = j then
```

```
            begin if x<0 then go to fail;
                y := dl[i] := sqrt(x)
            end
            else b[j, i] := x/y
      end ji;
      comment L has been formed in the array b;
      for i := 1 step 1 until n do
      for j := 1 step 1 until i do
      begin x := a[j, i] × dl[j];
            for k := j+1 step 1 until i do
            x := x + a[k, i] × b[k, j];
            for k := i+1 step 1 until n do
            x := x + a[i, k] × b[k, j];
            a[i, j] := x
      end ji;
      comment The lower triangle of A × L has been formed in the lower
                triangle of the array a;
      for i := 1 step 1 until n do
      begin y := dl[i];
            for j := 1 step 1 until i do
            begin x := y × a[i, j];
                  for k := i+1 step 1 until n do
                  x := x + a[k, j] × b[k, i];
                  a[i, j] := x
            end j
      end i
end reduc2;


procedure rebaka (n) data: (m1, m2, b, dl) trans: (z);
value n, m1, m2;  integer n, m1, m2;  array b, dl, z;
comment This procedure performs, on the matrix of eigenvectors, Z, stored in
        the array z[1:n, m1:m2], a backward substitution LT × X = Z, over-
        writing X on Z.
            The diagonal elements of L must be stored in the array dl[1:n],
        and the remaining triangle in the strictly lower triangle of the array
        b[1:n, 1:n]. The procedures reduc1 and reduc2 leave L in this desired
        form.
            If x denotes any column of the resultant matrix X, then x satisfies
        xT × B × x = zT × z, where B = L × LT;
begin    integer i, j, k;
         real x;
         for j := m1 step 1 until m2 do
         for i := n step −1 until 1 do
```

begin $x := z[i, j];$
    for $k := i + 1$ **step** 1 **until** n **do**
    $x := x - b[k, i] \times z[k, j];$
    $z[i, j] := x/dl[i]$
end ij
end *rebaka*;

procedure *rebakb* (n) *data*: $(m1, m2, b, dl)$ *trans*: (z);
value $n, m1, m2$; **integer** $n, m1, m2$; **array** b, dl, z;
comment This procedure performs, on the matrix of eigenvectors, Z, stored in
       the array $z[1\!:\!n, m1\!:\!m2]$, a forward substitution $Y = L \times Z$, over-
       writing Y and Z.
         The diagonal elements of L must be stored in the array $dl[1\!:\!n]$,
       and the remaining triangle in the strictly lower triangle of the array
       $b[1\!:\!n, 1\!:\!n]$. The procedures *reduc1* and *reduc2* leave L in this desired
       form.
         If y denotes any column of the resultant matrix Y, then y satis-
       fies $yT \times inv(B) \times y = zT \times z$, where $B = L \times LT$;
begin   **integer** i, j, k;
            real x;
            for $j := m1$ **step** 1 **until** $m2$ **do**
            for $i := n$ **step** -1 **until** 1 **do**
            begin $x := dl[i] \times z[i, j];$
                for $k := i - 1$ **step** -1 **until** 1 **do**
                $x := x + b[i, k] \times z[k, j];$
                $z[i, j] := x$
          end ij
end *rebakb*;

5. Organisational and Notational Details

In *reduc 1* and *reduc 2* the symmetric matrices A and B are stored in $n \times n$
arrays though only the upper-triangles need be stored. The procedures *choldet 1*
and *cholsol 1* [4] could have been used but since we wish to take full advantage
of symmetry in computing the derived symmetric matrices special purpose procedures are preferable.

The matrix L such that $LL^T = B$ is determined in either case. Its strict lower-
triangle is stored in the strict lower-triangle of B and its diagonal elements are
stored in the linear array dl. Hence full information on B is retained. Since the
factorization of B may already be known (if, for example several matrices A are
associated with a fixed matrix B) provision is made for omitting the Cholesky
factorization by inputting $-n$ instead of n.

In *reduc 1* the matrix $P = L^{-1}A L^{-T}$ is formed in the two steps

$$LX = A, \qquad PL^T = X. \tag{8}$$

The matrix X is not symmetric but only half of it is required for the computation
of P. The transpose of the upper-triangle of X is computed and written on the
lower-triangle of A. The diagonal of A is therefore destroyed and must be copied

before using *reduc 1* if the residuals or Rayleigh quotients are required. The lower triangle of P is then determined and overwritten on X (see pp. 337—340 of ref. [7]).

Similarly in *reduc 2* the matrix $Q=L^T A L$ is computed in the steps

$$Y=AL, \quad Q=L^T Y. \tag{9}$$

Here only the lower-triangle of Y is required (although it is not symmetric). It is overwritten on the lower-triangle of A. The lower-triangle of Q is then formed and overwritten on Y. Again the diagonal of A is lost and a copy must be made independently if the residuals or Rayleigh quotients are required.

The derived standard symmetric eigenproblem may be solved in a variety of ways and the algorithms are given elsewhere in this series [1, 3, 5, 6]. Assuming eigenvectors of the derived problem have been found by some method and are normalized according to the Euclidean norm, then the corresponding eigenvectors of the original problem may be obtained using *rebak a* or *rebak b*, as appropriate (see 2).

6. Discussion of Numerical Properties

In general, the accuracy of the procedures described here is very high. The computed $\bar{L}$ satisfies

$$\bar{L}\bar{L}^T=B+E \tag{10}$$

and we have discussed the bounds for E in [4]. The errors made in computing $\bar{L} A \bar{L}^T$ or $\bar{L}^{-1} A \bar{L}^{-T}$ are comparatively unimportant. In general the errors in the eigenvalues are such as could be produced by very small perturbations in A and B. However, with the eigenproblem $A x=\lambda B x$ small perturbations in A and B can correspond to quite large perturbations in the eigenvalues if B is ill-conditioned with respect to inversion and this can lead to the inaccurate determination of eigenvalues.

It might be felt that in this case such a loss of accuracy is inherent in the data but this is not always true. Consider, for example, the case

$$A = \begin{bmatrix} 2 & 2 \\ 2 & 1 \end{bmatrix}, \quad B = \begin{bmatrix} 1 & 2 \\ 2 & 4+\varepsilon \end{bmatrix} \tag{11}$$

for which the eigenvalues are the roots of

$$-2-\lambda(1+2\varepsilon)+\varepsilon\,\lambda^2=0$$

i.e.

$$\lambda=[+(1+2\varepsilon)\pm(1+12\varepsilon+4\varepsilon^2)^{\frac{1}{2}}]/2\varepsilon. \tag{12}$$

As $\varepsilon\to0$ one root tends to -2 and the other to $+\infty$. It may be readily verified that the root -2 is not at all sensitive to small random changes in the elements of A and B.

Rather surprisingly, well-conditioned eigenvalues such as the -2 of the previous example are often given quite accurately, in spite of the ill-condition of B. In the above example for instance the true matrix $P=L^{-1}A L^{-T}$ is

$$\begin{bmatrix} 2 & -2/\varepsilon^{\frac{1}{2}} \\ -2/\varepsilon^{\frac{1}{2}} & 1/\varepsilon \end{bmatrix}, \tag{13}$$

and for small ε it has some very large elements. When rounding errors are involved one might easily obtain a matrix $\bar{P}$ of the form

$$\bar{P} = \begin{bmatrix} 2/(1+\varepsilon_1) & -2(1+\varepsilon_2)/(\varepsilon+\varepsilon_3)^{\frac{1}{2}} \\ -2(1+\varepsilon_2)/(\varepsilon+\varepsilon_3)^{\frac{1}{2}} & (1+\varepsilon_4)/(\varepsilon+\varepsilon_3) \end{bmatrix} \tag{14}$$

where the ε_i are of the order of machine precision. If ε itself is of this order the computed $\bar{P}$ will differ substantially from the true matrix and the large elements may well have a high relative error. However $\bar{P}$ has an eigenvalue close to -2 for all relevant values of the ε_i.

As regards the eigenvectors, there are procedures which give an eigensystem of the derived problem which is accurately orthogonal. However, when transforming back to the eigensystem of the original eigenproblem, the computed set of vectors will be ill-conditioned when B (and hence L) is ill-conditioned. This deterioration would appear to be inherent in the data.

In our experience the considerations have not been important in practice since the matrices B which arise are, in general, extremely well-conditioned.

7. Test Results

To give a formal test of the procedures described here the matrices

$$F = \begin{bmatrix} 10 & 2 & 3 & 1 & 1 \\ 2 & 12 & 1 & 2 & 1 \\ 3 & 1 & 11 & 1 & -1 \\ 1 & 2 & 1 & 9 & 1 \\ 1 & 1 & -1 & 1 & 15 \end{bmatrix}, \quad G = \begin{bmatrix} 12 & 1 & -1 & 2 & 1 \\ 1 & 14 & 1 & -1 & 1 \\ -1 & 1 & 16 & -1 & 1 \\ 2 & -1 & -1 & 12 & -1 \\ 1 & 1 & 1 & -1 & 11 \end{bmatrix}$$

were used. Since both F and G are positive definite either matrix can be used as A or B in either of the main procedures. When used in conjunction with procedures for solving the standard eigenproblems this gives quite extensive checks on the overall performance. Since both A and B are well-conditioned with respect to inversion good agreement should be achieved between computed matrices $L^{-1}A L^{-T}$, $L^{T}A L$, eigenvalues and eigenvectors obtained on different computers.

A selection from the results obtained on KDF 9, a computer using a 39 binary digit mantissa are given in Tables 1—4. The eigensystems of the derived standard eigenproblems were found using the procedures *tred 1* [3], *bisect* [1], *tridi inverse iteration* [6] and *rebak 1* [3]. Any accurate procedures for dealing with the standard eigenproblem could be used including the procedure *jacobi* [5] already published.

Table 1 gives the output matrices a obtained from *reduc 1* and *reduc 2* taking $A=F$, $B=G$.

Table 2 gives the output matrices b and vectors dl obtained from *reduc 1* and *reduc 2* taking $A=F$, $B=G$. Since the computations involved in the two procedures are identical in every detail the output matrices are identical.

Table 3 gives the eigenvalues of $F - \lambda G$ and those of $G - \lambda F$ obtained using *reduc 1*; these involve the Cholesky factorizations of G and F respectively. The product of corresponding eigenvalues is equal to unity to about eleven decimal places. It also gives the eigenvalues of $FG - \lambda I$ and $GF - \lambda I$ calculated using

Table 1

Column 1	Column 2	Column 3	Column 4	Column 5
Output matrix a from reduc 1 with A = F, B = G				
$+8.3333\,3333\,334_{10}-1;$	$+2.0000\,0000\,000_{10}+0;$	$+3.0000\,0000\,000_{10}+0;$	$+1.0000\,0000\,000_{10}+0;$	$+1.0000\,0000\,000_{10}+0;$
$+9.0279\,3771\,290_{10}-2;$	$+8.4331\,3373\,256_{10}-1;$	$+1.0000\,0000\,000_{10}+0;$	$+2.0000\,0000\,000_{10}+0;$	$+1.0000\,0000\,000_{10}+0;$
$+2.7151\,9129\,955_{10}-1;$	$-4.4710\,9421\,464_{10}-3;$	$+7.2690\,8512\,779_{10}-1;$	$+1.0000\,0000\,000_{10}+0;$	$-1.0000\,0000\,000_{10}+0;$
$-3.3434\,1703\,880_{10}-2;$	$+2.1334\,5628\,682_{10}-2;$	$+6.2497\,8723\,923_{10}-2;$	$+8.1234\,1947\,476_{10}-1;$	$+1.0000\,0000\,000_{10}+0;$
$-1.6323\,2422\,021_{10}-2;$	$+2.4171\,4784\,237_{10}-2;$	$-1.5020\,8557\,974_{10}-1;$	$+1.3606\,1578\,747_{10}-1;$	$+1.4260\,4956\,979_{10}+0;$
Output matrix a from reduc 2 with A = F, B = G				
$+1.3100\,0000\,000_{10}+2;$	$+2.0000\,0000\,000_{10}+0;$	$+3.0000\,0000\,000_{10}+0;$	$+1.0000\,0000\,000_{10}+0;$	$+1.0000\,0000\,000_{10}+0;$
$+4.4649\,6005\,171_{10}+1;$	$+1.6856\,8862\,275_{10}+2;$	$+1.0000\,0000\,000_{10}+0;$	$+2.0000\,0000\,000_{10}+0;$	$+1.0000\,0000\,000_{10}+0;$
$+3.1565\,2935\,415_{10}+1;$	$+2.6097\,6424\,615_{10}+1;$	$+1.7183\,5449\,373_{10}+2;$	$+1.0000\,0000\,000_{10}+0;$	$-1.0000\,0000\,000_{10}+0;$
$+2.8625\,9509\,072_{10}+1;$	$+1.5509\,5804\,016_{10}+1;$	$+8.7809\,0481\,507_{10}+0;$	$+1.0313\,4644\,279_{10}+2;$	$+1.6046\,1044\,072_{10}+2;$
$+2.9269\,1592\,858_{10}+1;$	$+2.2283\,8464\,931_{10}+1;$	$-1.1468\,4872\,835_{10}+0;$	$-3.9498\,1287\,698_{10}+0;$	

Table 2

Column 1	Column 2	Column 3	Column 4	Column 5
Output matrix b from reduc 1 and reduc 2 with A = F, B = G				
$+1.2000\,0000\,000_{10}+1;$	$+1.0000\,0000\,000_{10}+1;$	$-1.0000\,0000\,000_{10}+0;$	$+2.0000\,0000\,000_{10}+0;$	$+1.0000\,0000\,000_{10}+0;$
$+2.8867\,5134\,595_{10}-1;$	$+1.4000\,0000\,000_{10}+1;$	$+1.0000\,0000\,000_{10}+0;$	$-1.0000\,0000\,000_{10}+0;$	$+1.0000\,0000\,000_{10}+0;$
$-2.8867\,5134\,595_{10}-1;$	$+2.9039\,8583\,546_{10}-1;$	$+1.6000\,0000\,000_{10}+1;$	$-1.0000\,0000\,000_{10}+0;$	$+1.0000\,0000\,000_{10}+0;$
$+5.7735\,0269\,190_{10}-1;$	$-3.1273\,6936\,127_{10}-1;$	$-1.8660\,9059\,553_{10}-1;$	$+1.2000\,0000\,000_{10}+1;$	$-1.0000\,0000\,000_{10}+0;$
$+2.8867\,5134\,595_{10}-1;$	$+2.4572\,1878\,385_{10}-1;$	$+2.5433\,0089\,229_{10}-1;$	$-3.0692\,1313\,649_{10}-1;$	$+1.1000\,0000\,000_{10}+1;$
Output vector dl from reduc 1 and reduc 2 with A = F, B = G				
		$+3.4641\,0161\,514_{10}+0;$		
		$+3.7305\,0488\,093_{10}+0;$		
		$+3.9789\,8672\,143_{10}+0;$		
		$+3.3961\,8010\,923_{10}+0;$		
		$+3.2706\,8845\,017_{10}+0;$		

Table 3. *Eigenvalues*

$A - \lambda B,\ A = F,\ B = G$	$A - \lambda B,\ A = G,\ B = F$	$AB - \lambda I,\ A = F,\ B = G$	$AB - \lambda I,\ A = G,\ B = F$
$+4.3278\,7211\,020_{10} - 1;$	$+6.7008\,2644\,107_{10} - 1;$	$+7.7697\,1911\,953_{10} + 1;$	$+7.7697\,1911\,963_{10} + 1;$
$+6.6366\,2748\,402_{10} - 1;$	$+9.0148\,1958\,801_{10} - 1;$	$+1.1215\,4193\,247_{10} + 2;$	$+1.1215\,4193\,246_{10} + 2;$
$+9.4385\,9004\,670_{10} - 1;$	$+1.0594\,8027\,732_{10} - 0;$	$+1.3468\,6463\,320_{10} + 2;$	$+1.3468\,6463\,320_{10} + 2;$
$+1.1092\,8454\,002_{10} + 0;$	$+1.5067\,8940\,837_{10} + 0;$	$+1.6748\,4878\,917_{10} + 2;$	$+1.6748\,4878\,915_{10} + 2;$
$+1.4923\,5323\,254_{10} + 0;$	$+2.3106\,0432\,137_{10} + 0;$	$+2.4297\,7273\,320_{10} + 2;$	$+2.4297\,7273\,319_{10} + 2;$

Table 4. *Eigenvectors*

1st eigenvector of $F - \lambda G$	last eigenvector of $G - \lambda F$	1st eigenvector of $FG - \lambda I$ via $G = LL^T$	1st eigenvector of $FG - \lambda I$ via $F = LL^T$
$+1.3459\,0573\,962_{10} - 1;$	$-2.0458\,6718\,183_{10} - 1;$	$+2.3491\,1413\,526_{10} - 1;$	$+2.0706\,5038\,597_{10} + 0;$
$-6.1294\,7224\,718_{10} - 2;$	$+9.3172\,0977\,419_{10} - 2;$	$-4.1091\,5167\,469_{10} - 2;$	$-3.6220\,5325\,515_{10} - 1;$
$-1.5790\,2562\,211_{10} - 1;$	$+2.4002\,2507\,111_{10} - 1;$	$-3.8307\,5945\,797_{10} - 2;$	$-3.3766\,6162\,397_{10} - 1;$
$+1.0946\,5787\,725_{10} - 1;$	$-1.6639\,5354\,480_{10} - 1;$	$-2.0590\,0367\,490_{10} - 1;$	$-1.8149\,2958\,995_{10} + 0;$
$-4.1473\,0117\,966_{10} - 2;$	$+6.3041\,7653\,099_{10} - 2;$	$-7.3470\,7965\,853_{10} - 2;$	$-6.4761\,5758\,762_{10} - 1;$

reduc 2 and again using the Cholesky factorization of G and F respectively. Corresponding eigenvalues have a maximum disagreement of 1 in the eleventh decimal.

Table 4 gives the first eigenvector of $F - \lambda G$ and the last eigenvector of $G - \lambda F$. These should be the same apart from the normalization. The first is normalized so that $x^T G x = 1$ and the second so that $x^T F x = 1$. The angle between the vectors is smaller than 10^{-11} radians. It also gives the first eigenvector of $FG - \lambda I$ obtained via the factorization of G and via the factorization of F. Again the angle between the two vectors is smaller than 10^{-11} radians. These results are typical of those obtained for all five eigenvectors in each case.

The complete sets of results were subjected to extensive cross checking and proved generally to be "best possible" for a 39 digit binary computer. A second test was carried out by C. REINSCH at the Technische Hochschule, München, and confirmed the above results.

Acknowledgements. The authors wish to thank Dr. C. REINSCH of the Mathematisches Institut der Technischen Hochschule, München, for testing these algorithms and suggesting several important modifications.

The work described above has been carried out at the National Physical Laboratory.

References

1. BARTH, W., R. S. MARTIN, and J. H. WILKINSON: Calculation of the eigenvalues of a symmetric tridiagonal matrix by the method of bisection. Numer. Math. **9**, 386—393 (1967). Cf. II/5.
2. BOWDLER, H., R. S. MARTIN, C. REINSCH, and J. H. WILKINSON: The QR and QL algorithms for symmetric matrices. Numer. Math. **11**, 293—306 (1968). Cf. II/3.
3. MARTIN, R. S., C. REINSCH, and J. H. WILKINSON: Householder's tridiagonalization of a symmetric matrix. Numer. Math. **11**, 181—195 (1968). Cf. II/2.
4. — G. PETERS, and J. H. WILKINSON: Symmetric decomposition of a positive definite matrix. Numer. Math. **7**, 362—383 (1965). Cf. I/1.
5. RUTISHAUSER, H.: The Jacobi method for real symmetric matrices. Numer. Math. **9**, 1—10 (1966). Cf. II/1.
6. Wilkinson, J. H.: Calculation of the eigenvectors of a symmetric tridiagonal matrix by inverse iteration. Numer. Math. **4**, 368—376 (1962). See also: PETERS, G., and J. H. WILKINSON: The calculation of specified eigenvectors by inverse iteration. Cf. II/18.
7. — The algebraic eigenvalue problem. London: Oxford University Press 1965.

Balancing a Matrix for Calculation of Eigenvalues and Eigenvectors*

by B. N. Parlett and C. Reinsch

1. Theoretical Background

Introduction

This algorithm is based on the work of Osborne [1]. He pointed out that existing eigenvalue programs usually produce results with errors at least of order $\varepsilon\|A\|_E$, where ε is the machine precision and $\|A\|_E$ is the Euclidean (Frobenius) norm of the given matrix A**. Hence he recommends that one precede the calling of such a routine by certain diagonal similarity transformations of A designed to reduce its norm. To this end Osborne discussed the case $p=2$ of an algorithm which produces a sequence of matrices A_k, $(k=1, 2, \ldots)$ diagonally similar to A such that for an irreducible*** A

(i) $A_f = \lim\limits_{k\to\infty} A_k$ exists and is diagonally similar to A,

(ii) $\|A_f\|_p = \inf (\|D^{-1}AD\|_p)$, where D ranges over the class of all non-singular diagonal matrices,

(iii) A_f is *balanced* in $\|.\|_p$, i.e. $\|a_i\|_p = \|a^i\|_p$, $i=1, \ldots, n$ where a_i and a^i denote respectively the i-th column and i-th row of A_f,

(iv) A and $D^{-1}AD$ produce the same A_f.

The process may be organized to compute either A_f explicitly or only the corresponding diagonal transformation matrix. In the latter case, rounding errors can not cause a loss of accuracy in the eigensystem; however, existing eigenvalue programs must be modified so that they act on the matrix $D^{-1}AD$, implicitly given by A and D. This realization is preferable from a systematic point of view. In practice, however, it appears simpler to scale the given matrix explicitly without changing existing programs. No rounding errors need occur in the execution if the process is implemented on a digital computer in floating point arithmetic, and if the elements of the diagonal matrices are restricted to be exact powers of the radix base employed, usually 2, 10, or 16.

---

* Prepublished in Numer. Math. **13**, 293−304 (1969).

** The following norms will be used. Let $1\leq p<\infty$, $x=(\xi_1, \ldots, \xi_n)^T$ and $B=(\beta_{ij})$ of order n, then $\|x\|_p=(\sum|\xi_i|^p)^{1/p}$ and $\|B\|_p=(\sum\sum|\beta_{ij}|^p)^{1/p}$, where $\sum$ denotes summation over the index values $1, \ldots, n$. $\|B\|_E=\|B\|_2$.

*** For a definition of irreducible see [1].

The Algorithm

Let A_0 denote the off-diagonal part of A. For any non-singular diagonal matrix D,

$$D^{-1}AD = \operatorname{diag}(A) + D^{-1}A_0D$$

and only A_0 is affected. Assume that no row or column of A_0 identically vanishes.

From A_0 a sequence $\{A_k\}$ is formed. The term A_k differs from A_{k-1} only in one row and the corresponding column. Let $k=1, 2, \ldots$ and let i be the index of row and column modified in the step from A_{k-1} to A_k. Then, if n is the order of A_0, i is given by

$$i - 1 \equiv k - 1 \,(\operatorname{mod} n).$$

Thus, the rows are modified cyclically in their natural order. The k-th step is as follows.

(a) Let R_k and C_k denote the $\|.\|_p$ norms of row i and column i of A_{k-1}. According to the above assumption $R_k C_k \neq 0$. Hence, if β denotes the radix base, there is a unique signed integer $\sigma = \sigma_k$ such that

$$\beta^{2\sigma-1} < R_k/C_k \leq \beta^{2\sigma+1}.$$

Define $f = f_k$ by

$$f = \beta^\sigma.$$

(b) For given constant $\gamma \leq 1$ take

$$\bar{D}_k = \begin{cases} I + (f-1)\, e_i e_i^T & \text{if } (C_k f)^p + (R_k/f)^p < \gamma(C_k^p + R_k^p) \\ I & \text{otherwise}. \end{cases}$$

Here $I = (e_1, \ldots, e_n)$ is the identity matrix.

(c) Form

$$D_k = \bar{D}_k D_{k-1}; \qquad D_0 = I,$$
$$A_k = \bar{D}_k^{-1} A_{k-1} \bar{D}_k.$$

We note that with this algorithm if $\gamma = 1$ then in every step, the factor f is that integer power of the machine base β which gives maximum reduction of the contribution of the i-th row and column to $\|A_k\|_p$. If γ is slightly smaller than 1, a step is skipped if it would produce an insubstantial reduction of $\|A_{k-1}\|_p$. Iteration is terminated if for a complete cycle $\bar{D}_k = I$.

For $\gamma = 1$ and $p = 2$ the algorithm differs from Osborne's proposal only by the constraint on f given in (a). Nevertheless it shows different properties. Corresponding to the above listed points we have (under the assumption that A_0 has no null rows or columns):

(i) The process may also be applied to a reducible A and in all cases terminates after a finite number of iterations with a matrix A_f,

(ii) $\|A_f\|_p \geq \inf \|D_\beta^{-1} A D_\beta\|_p$), where D_β ranges over the class of all diagonal matrices with entries restricted to integer powers of the machine base β, and inequality may well hold,

(iii) $1/x \leq \|a^i\|_p/\|a_i\|_p \leq x, \qquad x^p = (\beta^p - \gamma)/(\gamma - \beta^{-p}), \quad$ for A_f,

(iv) A and $D_\beta^{-1} A D_\beta$ may not produce the same A_f.

For example,

$$A_0 = \begin{pmatrix} 0 & 2 & \beta^{-1} & 0 \\ 2 & 0 & 0 & \beta^{-1} \\ \beta & 0 & 0 & 2 \\ 0 & \beta & 2 & 0 \end{pmatrix}$$

and A_0^T (which is similar to A_0) remain unaltered for every $\gamma \leq 1, \beta \geq 2, 1 \leq p < \infty$, while optimum scaling would yield a symmetric A_f.

Preliminary Permutations

If row i or column i of A_0 is null then a_{ii} is an eigenvalue of A and the calculations should proceed on the submatrix obtained by deleting row and column i. In the present program such columns are moved to the left side of the matrix and such rows to the bottom by means of row and column interchanges. After this preliminary phase the matrix $P^T A P$ has the form indicated below. P is the corresponding permutation matrix.

$\times$ indicates elements which are not necessarily zero.

The submatrix Z is such that Z_0 has no null rows or columns, in particular, $low \neq hi$.

If only the eigenvalues of A are required it would be possible to continue the calculation with Z alone. However, in case the condition numbers of the eigenvalues [2, Chap. 2] are wanted or in case eigenvectors are required it is necessary to keep Z as a submatrix of $P^T A P$.

Osborne's modified algorithm is applied to Z in the following sense. The quantities f_k of step (a) are computed from the rows and columns z^i and z_i of Z_0. However, the similarity transform (c) is performed on the whole matrix. This affects X and W as well as Z.

After an eigenvalue is detected in this preliminary phase the search must begin afresh in the complementary submatrix.

A more economical approach suggests itself. In one sweep through the whole matrix it is possible to permute A into the form

$$
\begin{pmatrix}
\overline{D}_+ & \overline{X} & \overline{Y} \\
\hline
0 & \overline{Z} & \overline{W} \\
\hline
0 & 0 & \overline{D}_-
\end{pmatrix}
$$

where $\overline{D}_+$, $\overline{D}_-$ are diagonal and no row of $(\overline{Z}_0, \overline{W})$ and no column of $(\overline{X}^T, \overline{Z}_0^T)^T$ is null. $\overline{Z}_0$, however, may have null rows or columns. Thus, the modified algorithm can not be applied as described above. Moreover, this algorithm will not detect all permutations of a triangular matrix and this can be a grave defect. Consider, for example, matrices of the form

$$
\begin{pmatrix}
\alpha & 0 & 0 & 0 & \gamma & 0 \\
0 & \alpha & 0 & 0 & 0 & \gamma \\
0 & \gamma & \alpha & 0 & 0 & 0 \\
0 & 0 & \gamma & \alpha & 0 & 0 \\
0 & \delta & 0 & \gamma & \alpha & 0 \\
0 & 0 & 0 & 0 & 0 & \alpha
\end{pmatrix}.
$$

If $\overline{Z}$ is reduced to upper Hessenberg form in a manner which destroys the direct relation to lower triangular form then for example the QR algorithm will yield answers with unnecessary loss of accuracy. This variant is not recommended for general matrices.

Backtransformation of Eigenvectors

A right-hand eigenvector x and a left-hand eigenvector y of the given matrix A are derived from the corresponding right-hand eigenvector x_f and left-hand eigenvector y_f of the balanced matrix $A_f = D^{-1} P^T A P D$ according to

$$
x = P D x_f, \qquad y = P D^{-1} y_f.
$$

Both transformations are real, hence they may be applied independently to the real and imaginary parts of complex eigenvectors.

2. Applicability

Balancing is recommended if the eigenvalues and/or eigenvectors of a given non-symmetric matrix are to be determined via the procedures given in this series [3—6]. They include as major steps

(i) reduction of the matrix to upper Hessenberg form using either elementary transformations stabilized by partial pivoting, or unitary transformations,

(ii) stabilized LR transformations or QR transformations,

(iii) test for convergence involving the magnitude of certain subdiagonal elements,

(iv) inverse iteration using a triangular decomposition with partial pivoting. As is outlined elsewhere the effectiveness of these steps depends on an appropriate scaling.

The algorithm may be applied to any real or complex matrix. However, convergence may be deferred if the matrix is highly reducible. There are classes of matrices which do not need balancing; for example, normal matrices are already balanced in $\|.\|_2$. For this class of matrices it is also unnecessary to detect isolated eigenvalues by permutation of row and columns, because the eigenvalue programs given in this series will reveal such eigenvalues.

Finally, there is no point in balancing A beforehand, if an eigenvalue program is used which is invariant under scaling (e.g. vector iteration with an appropriate criterion for terminating).

Balancing is merely a device to reduce errors made during the execution of some algorithm either by rounding or by terminating the iteration. In contrast with this any inaccuracy E of the input matrix A gives rise to some inaccuracy in the results (eigenvalues and eigenvectors) which is completely independent of scaling.

Procedure *balance* balances a given real matrix in $\|.\|_1$. Procedure *balbak* performs the backtransformation of a set of right-hand eigenvectors.

3. Formal Parameter List

3.1. Input to procedure *balance*

n the order of the matrix A.

b the base of the floating point representation on the machine.

a an array $[1:n, 1:n]$ giving the elements of the matrix to be balanced.

Output from procedure *balance*

a the original matrix will be overwritten by the balanced matrix. The original may be recovered exactly.

low, hi two integers such that $a[i, j]$ is equal to zero if

         (1) $i > j$ and

         (2) $j = 1, \ldots, low - 1$ or $i = hi + 1, \ldots, n$.

d an array $[1:n]$ which contains information determining the permutations used and the scaling factors.

3.2. Input to procedure *balbak*

n order of eigenvectors.

low, hi two integers which are output of procedure *balance*.

m number of eigenvectors to be transformed.

d output vector of procedure *balance*.

z an array $[1:n, 1:m]$ with each column representing an eigenvector (or its real part, or its imaginary part) of the balanced matrix.

Output from procedure *balbak*

z the set of corresponding eigenvectors (real parts, imaginary parts) of the original matrix.

4. ALGOL Programs

procedure *balance* (*n*, *b*) *trans*: (*a*) *result*: (*low*, *hi*, *d*);

    value *n*, *b*;

    integer *n*, *b*, *low*, *hi*;

    array *a*, *d*;

comment reduce the norm of $a[1:n, 1:n]$ by exact diagonal similarity trans-
           formations stored in $d[1:n]$;

begin

    integer *i*, *j*, *k*, *l*; **real** *b2*, *c*, *f*, *g*, *r*, *s*; **Boolean** *noconv*;

    procedure *exc* (*m*); **value** *m*; **integer** *m*;

    begin

      $d[m] := j$;

      if $j \neq m$ **then**

      begin

         for $i := 1$ **step** 1 **until** k **do**

         begin $f := a[i, j]$; $a[i, j] := a[i, m]$; $a[i, m] := f$ **end**;

         for $i := l$ **step** 1 **until** n **do**

         begin $f := a[j, i]$; $a[j, i] := a[m, i]$; $a[m, i] := f$ **end**

      end $j \neq m$

    end *exc*;

    $b2 := b \times b$; $l := 1$; $k := n$;

    comment search for rows isolating an eigenvalue and push them down;

L1:

    for $j := k$ **step** -1 **until** 1 **do**

    begin

      $r := 0$;

      for $i := 1$ **step** 1 **until** $j - 1, j + 1$ **step** 1 **until** k **do**

        $r := r + abs(a[j, i])$;

      if $r = 0$ **then**

      begin *exc* (*k*); $k := k - 1$; **go to** *L1* **end**

    end j;

    comment search for columns isolating an eigenvalue and push them left;

L2:

    for $j := l$ **step** 1 **until** k **do**

    begin

      $c := 0$;

      for $i := l$ **step** 1 **until** $j - 1, j + 1$ **step** 1 **until** k **do**

        $c := c + abs(a[i, j])$;

      if $c = 0$ **then**

      begin *exc* (*l*); $l := l + 1$; **go to** *L2* **end**

    end j;

```
    comment now balance the submatrix in rows l through k;
    low := l; hi := k;
    for i := l step 1 until k do d [i] := 1;
iteration:
    noconv := false;
    for i := l step 1 until k do
    begin
        c := r := 0;
        for j := l step 1 until i − 1, i + 1 step 1 until k do
        begin c := c + abs (a [j, i]); r := r + abs (a [i, j]) end j;
        g := r/b; f := 1; s := c + r;
L3:     if c < g then begin f := f × b; c := c × b2; go to L3 end;
        g := r × b;
L4:     if c ≥ g then begin f := f/b; c := c/b2; go to L4 end;
        comment the preceding four lines may be replaced by a machine language
                procedure computing the exponent sig such that sqrt (r/(c × b)) ≤
                b↑sig < sqrt (r × b/c). Now balance;
        if (c + r)/f < 0.95 × s then
        begin
            g := 1/f; d [i] := d [i] × f; noconv := true;
            for j := l step 1 until n do a [i, j] := a [i, j] × g;
            for j := 1 step 1 until k do a [j, i] := a [j, i] × f
        end if;
        comment the j loops may be done by exponent modification in machine
                language;
    end i;
    if noconv then go to iteration
end balance;


procedure balbak (n, low, hi, m) data: (d) trans: (z);
    value n, low, hi, m;
    integer n, low, hi, m;
    array d, z;
comment Backward transformation of a set of right-hand eigenvectors of a
        balanced matrix into the eigenvectors of the original matrix from
        which the balanced matrix was derived by a call of procedure balance;
begin
    integer i, j, k; real s;
    for i := low step 1 until hi do
    begin
        s := d [i];
        comment left-hand eigenvectors are back transformed, if the foregoing
                statement is replaced by s := 1/d [i];
        for j := 1 step 1 until m do z [i, j] := z [i, j] × s
    end i:
```

```
for i := low −1 step −1 until 1, hi +1 step 1 until n do
begin
    k := d [i];
    if k ≠ i then for j := 1 step 1 until m do
    begin s := z [i, j]; z [i, j] := z [k, j]; z [k, j] := s end j
end i
end balbak;
```

5. Organisational and Notational Details

(i) Scaling factors and information concerning the preliminary permutations are stored in the array d.

Suppose that the principle submatrix in rows low through hi has been balanced. Then

$$d[j] = \begin{cases} p_j & 1 \leq j < low \\ d_j & low \leq j \leq hi \\ p_j & hi < j \leq n. \end{cases}$$

Here p_j is the index interchanged with j and d_j is the multiplier of row j after the permutations have been accomplished. The order in which the interchanges were made is n to $hi +1$ then 1 to $low -1$.

(ii) The program balances A in $\|.\|_1$ rather than $\|.\|_2$ because fewer multiplications are required. (If machine language code is used at the two places indicated in the program then no floating point multiplications are necessary.) In addition, danger of overflow (underflow) is reduced (removed) when forming C_k and R_k. Since

$$\|A\|_2 \leq \|A\|_1 \leq n \|A\|_2$$

a significant reduction in one norm implies a significant reduction in the other.

(iii) Effort has been made to prevent *intermediate* results passing outside the range of permissible numbers. Assume that positive numbers are representable in the range

$$\beta^{-\sigma} \leq z < \beta^{+\tau} \quad \text{with} \quad \sigma \geq \tau.$$

Then underflow can not occur or will be irrelevant if a result beyond the lower limit is replaced by zero. Overflow may occur in computing C_k, R_k, $C_k + R_k$, or in other rare cases. Note, however, that the scaling factors d_j may well pass out of range if the matrix is highly reducible. This short-coming will not occur if only the corresponding exponents are stored.

(iv) No ALGOL program is given for the balancing of complex matrices. A straightforward transliteration of procedure *balance* is possible using the procedure *cabs* for the computation of the modulus of a complex number [3]. It is more economical, however, to replace $|x + iy|$ by $|x| + |y|$. Backtransformation, of course, may be done by calling twice the procedure *balbak*. To this end no normalization of eigenvectors is provided.

(v) Transformation of left-hand eigenvectors is performed if the scaling factors are replaced by their reciprocals.

Table 1

$$A =
\begin{pmatrix}
6_{10}+00 & 0 & 0 & 0 & 0 & 3_{10}-04 & 0 & 1_{10}+00 & 0 \\
0 & 4_{10}+00 & 0 & 0 & 1_{10}-02 & 0 & 2_{10}-02 & 1_{10}-01 \\
1_{10}+00 & 1_{10}+02 & 7_{10}+00 & 0 & 0 & 0 & -2_{10}+00 & 2_{10}+01 \\
0 & 2_{10}+04 & 0 & 1_{10}+00 & -4_{10}+02 & 3_{10}+02 & 2_{10}+00 & -4_{10}+03 & 4_{10}+01 \\
-2_{10}+00 & -3_{10}+02 & 0 & 1_{10}-02 & 2_{10}+00 & 2_{10}+00 & 2_{10}+00 & 4_{10}+01 \\
0 & 0 & 0 & 0 & 0 & 0 & 0 & 0 & 0 \\
0 & 1_{10}+01 & 0 & 1_{10}+01 & 4_{10}-03 & 1_{10}-01 & -2_{10}-01 & 3_{10}+00
\end{pmatrix}$$

$$A_f =
\begin{pmatrix}
7 & 0.1 & 0.2 & 0 & 0 & 1 & -2 \\
0 & 4 & 1 & 3 & 1 & 0 & 20 \\
0 & 1 & 3 & 4 & 1 & 0 & 20 \\
0 & 2 & -4 & 1 & -4 & 0 & -20 \\
0 & -3 & 4 & 1 & 2 & -20 & 30 \\
0 & 0 & 0 & 0 & 0 & 6 & 20 \\
0 & 0 & 0 & 0 & 0 & 0 & 1 \\
0 & 0 & 0 & 0 & 0 & 0 & 0
\end{pmatrix}$$

$$d = (3, 10^{-3}, 10^{-2}, 10^{1}, 10^{-1}, 1, 6)$$
$$b = 10,\ low = 2,\ hi = 5,\ n = 7$$

$$B \approx
\begin{pmatrix}
4_{10}-07 & 7_{10}-12 & 6_{10}-11 & -1_{10}-07 & -1_{10}+00 & 4_{10}-09 & 7_{10}-09 & -1_{10}-10 & 9_{10}-13 & 3_{10}-11 \\
1_{10}-01 & 9_{10}-03 & -5_{10}-02 & -2_{10}-06 & -1_{10}+00 & 8_{10}-07 & 5_{10}-07 & -3_{10}-09 & 5_{10}-12 & -1_{10}-09 \\
0 & 0 & 5_{10}-07 & 0 & 0 & 0 & -5_{10}-07 & 0 & 0 & 0 \\
-4_{10}-06 & 2_{10}-10 & 9_{10}-10 & 4_{10}-06 & 1_{10}-06 & -1_{10}-06 & -4_{10}-08 & -1_{10}-08 & 6_{10}-11 & -1_{10}-09 \\
6_{10}-08 & -2_{10}-01 & -3_{10}+00 & 5_{10}-13 & 1_{10}-13 & -6_{10}-14 & 1_{10}-13 & -1_{10}-15 & 9_{10}-10 & -1_{10}-16 \\
2_{10}+01 & 1_{10}+00 & 8_{10}+00 & 1_{10}-04 & 3_{10}-05 & -2_{10}-05 & 2_{10}-05 & -2_{10}-07 & 2_{10}-01 & -3_{10}-08 \\
0 & 0 & 0 & 0 & 0 & 0 & 0 & 0 & 0 & 0 \\
0 & 0 & 0 & 0 & 0 & 0 & 0 & 0 & 0 & -1_{10}+00 \\
4_{10}-03 & 0 & 4_{10}+03 & -4_{10}-03 & -2_{10}+04 & 2_{10}+04 & 4_{10}-04 & 4_{10}+02 & -1_{10}-08 & 1_{10}-06 \\
0 & 0 & 2_{10}+03 & 0 & 0 & 0 & -4_{10}+03 & 0 & 0 & 0 \\
7_{10}+03 & 4_{10}+02 & 2_{10}+03 & 5_{10}-02 & 1_{10}-02 & -7_{10}-03 & 8_{10}-03 & -1_{10}-04 & 1_{10}+02 & -1_{10}-05 \\
2_{10}+06 & 1_{10}+02 & 7_{10}+02 & -2_{10}+06 & 2_{10}+05 & 7_{10}+04 & 1_{10}+05 & -1_{10}+03 & 2_{10}+01 & 3_{10}+02
\end{pmatrix}$$

$$n = 12$$

(vi) Rounding errors in the computed values of C_k and R_k may cause an infinite loop in procedure balance if the test for convergence is done with $\gamma = 1$. Thus, the value of γ has been reduced to equal 0.95.

6. Numerical Properties

All arithmetic operations on the matrix and the eigenvectors may be performed exactly in which case there are no rounding errors.

7. Test Results

Tests were performed at the Leibniz-Rechenzentrum der Bayerischen Akademie der Wissenschaften, München, using a TR 4 computer with radix base 16 and machine precision $\varepsilon = {}_2-35$. A second test was made on the CDC 6400 at the Department of Computer Science, University of California, Berkeley, confirming the results subject to the use of a radix base 2. In the following rounded values are listed if not otherwise specified.

Matrix A (Table 1) has three isolated eigenvalues; it was used as a formal test of the procedures described here. With $b = 10$ convergence was reached after two sweeps. The corresponding output is also listed in Table 1. To demonstrate the effect of balancing we applied the procedures *orthes* and *hqr* of this series [3, 5] to A as well as to A_f. These procedures compute the eigenvalues of a real, non-symmetric matrix by elementary orthogonal transformations. The first

Table 2. *Matrix A and A_f*

Eigenvalues λ_i	Sensitivity s_i ($\varepsilon = 2^{-35}$)	Errors in the eigenvalues computed via *orthes* and *hqr*	
		Without balancing	Balancing with $b = 10$
7	$2.04_{10}-10$	$-1.84_{10}-6$	0
6	$1.75_{10}-10$	$2.37_{10}-8$	0
$4.5895389489\ldots$	$3.49_{10}-10$	$2.59_{10}-6$	$4.10_{10}-10$
0	0	$9.09_{10}-13$	0
$-0.0771557967\ldots$	$3.65_{10}-10$	$-2.79_{10}-6$	$0.63_{10}-10$
$2.7438084239\ldots$ $\pm i\,3.8811589838\ldots$	—	$(0.88 \pm i\,2.77)_{10}-6$	$(2 \mp i\,2)_{10}-10$

column of Table 2 gives the correct eigenvalues, the second column lists the sensitivity of real eigenvalues λ_i against small perturbations in the entries of the matrix defined by

$$s_i = \sup\{|\lambda_i(A + \delta A) - \lambda_i(A)| \ : \ |\delta A| \leqq \varepsilon|A|\}.$$

This definition was chosen because it is invariant under scaling. In the absence of multiple eigenvalues

$$s_i = \varepsilon|y_i|^T |A| \, |x_i| / |y_i^T x_i| + O(\varepsilon^2)$$

with

$$A x_i = \lambda_i x_i, \qquad y_i^T A = y_i^T \lambda_i.$$

These values have to be compared with the errors in the computed eigenvalues which are also given in Table 2.

Matrix B (Table 1) was produced in a parameter study by an industrial company. Spectral radius $\approx 3.55_{10}+2$, no isolated eigenvalues. For brevity we list its entries rounded to one significant digit. Table 3 gives the 1-norms when balancing with $b=16$, 10 (rounding errors), or 2. B_f is not listed.

Table 3. $\|B\|_1$

Sweep	$b=16$	$b=10$	$b=2$
0	$3.43_{10}6$	$3.43_{10}6$	$3.43_{10}6$
1	$4.24_{10}3$	$7.68_{10}3$	$3.53_{10}3$
2	$2.77_{10}3$	$4.92_{10}3$	$2.37_{10}3$
3	$2.77_{10}3$	$4.92_{10}3$	$2.34_{10}3$
4	—	—	$2.34_{10}3$

Output vector for $b=2$ is

$$\log_2 d = (-15, -10, -7, -13, -11, -8, -9, -7, -2, 0, -1, 2).$$

Eigenvalues of B and B_f were computed using the procedures *elmhes* and *hqr* of this series [3, 5]. There are only two real eigenvalues which may be considered to be representative (Table 4).

Table 4. *Matrix B and B_f*

Eigenvalues λ_i	Sensitivity s_i ($\varepsilon = 2^{-35}$)	Errors in the eigenvalues computed via *elmhes* and *hqr*	
		Without balancing	Balancing with $b=2$
$37.2403\,7684\ldots$	$1.07_{10}-8$	$-1.56_{10}-3$	$-0.23_{10}-8$
$0.0083\,24096\ldots$	$5.84_{10}-9$	$-4.90_{10}-5$	$-0.94_{10}-9$

Matrix C of order 50 was chosen so as to illustrate the case of reducibility:

$$c_{ij} = \begin{cases} 1, 100, 1, 100, \ldots, 1, 100, 1 & \text{(Super-diagonal)} \\ 1, \ \ 0, \ 1, \ \ 0, \ \ldots, 1, \ \ 0, \ 1 & \text{(Sub-diagonal)} \\ 0 & \text{(otherwise)} . \end{cases}$$

For $b=2$ convergence was reached after 12 iterations, reducing $\|C\|_1$ from 2450 to 417.1875, (inf $(\|D^{-1}CD\|_1 = 50)$. Output vector

$$\log_2 d = (12, 12, 5, 6, 1, 3, -1, 2, -1, 2, \ldots, -1, 2, -1, 2, 0, -5, -4, -11, -11).$$

Backtransformation of right-hand eigenvectors and of left-hand eigenvectors were tested by their subsequent application to the matrix A_f with transpositions after step 1 and step 2. This gives $PDA_f D^{-1}P^T$, i.e. the original matrix.

Acknowledgements. The authors would like to thank Dr. J. H. WILKINSON for his helpful suggestions. The work of B. N. PARLETT was supported by Contract Nonr 3656 (23) with the Computing Center of the University of California, Berkeley, and that of C. REINSCH has been carried out at the Technische Hochschule München.

References

1. OSBORNE, E. E.: On pre-conditioning of matrices. J. Assoc. Comput. Mach. **7**, 338—345 (1960).
2. WILKINSON, J. H.: The algebraic eigenvalue problem. London: Oxford University Press 1965.
3. MARTIN, R. S., and J. H. WILKINSON: Similarity reduction of general matrices to Hessenberg form. Numer. Math. **12**, 349—368 (1969). Cf. II/13.
4. — The modified LR algorithm for complex Hessenberg matrices. Numer. Math. **12**, 369—376 (1969). Cf. II/16.
5. — G. PETERS, and J. H. WILKINSON: The QR algorithm for real Hessenberg matrices. Numer. Math. **14**, 219—231 (1970). Cf. II/14.
6. PETERS, G., and J. H. WILKINSON: Eigenvectors of real and complex matrices by LR and QR triangularizations. Numer. Math. **16**, 181—204 (1970). Cf. II/15.

Solution to the Eigenproblem
by a Norm Reducing Jacobi Type Method*

by P. J. EBERLEIN and J. BOOTHROYD

1. Theoretical Background

Let A be an $n \times n$ real matrix. A matrix $T = T_1 T_2 \ldots T_i \ldots$ (or T^{-1}) is constructed as a product of a sequence of two dimensional transformations T_i. From $A' = T^{-1} A T$ the eigenvalues may be read off and from T (or T^{-1}) the right (or left) vectors. Each T_i is of the form RS where R is a rotation and S a shear or complex rotation.

$$R: \quad r_{kk} = r_{mm} = \cos x; \qquad r_{km} = -\sin x = -r_{mk}; \qquad r_{ij} = \delta_{ij} \quad \text{otherwise.}$$

$$S: \quad s_{kk} = s_{mm} = \cosh y; \qquad s_{km} = s_{mk} = -\sinh y; \qquad s_{ij} = \delta_{ij} \quad \text{otherwise.}$$

The pairs (k, m) $k < m$ are chosen cyclically**. The rotational parameter x is determined by $\tan 2x = (a_{km} + a_{mk})/(a_{kk} - a_{mm})$; that solution x being chosen such that after the transformation the norm of the k-th column is greater than or equal to the norm of the m-th column. The shear parameter y is chosen to reduce the Euclidean norm of

$$A^{(i+1)} = (T_1 T_2 \ldots T_i)^{-1} A (T_1 T_2 \ldots T_i).$$

In particular:

$$\tanh y = (ED - H/2)/(G + 2(E^2 + D^2))$$

where

$$E = (a_{km} - a_{mk}), \qquad D = \cos 2x (a_{kk} - a_{mm}) + \sin 2x (a_{km} + a_{mk}),$$

$$G = \sum_{i \neq k, m} (a_{ki}^2 + a_{ik}^2 + a_{im}^2 + a_{mi}^2),$$

$$H = \cos 2x \left(2 \sum_{i \neq k, m} (a_{ki} a_{mi} - a_{ik} a_{im}) \right) - \sin 2x \left(\sum_{i \neq k, m} (a_{ki}^2 + a_{im}^2 - a_{ik}^2 - a_{mi}^2) \right).$$

---

* Prepublished in Numer. Math. 11, 1—12 (1968).

** Other patterns for the choice of (k, m) have been tried and for some matrices are clearly better. Generally over sets of matrices no pattern seemed superior. The pairs (k, m) are chosen cyclically here for simplicity only; there is no reason to believe this is the best choice of pattern.

Note that $\tanh y \leq \frac{1}{2}$. It can be shown that for this choice of $\tanh y$,

$$\|A^{(i)}\|^2 - \|A^{(i+1)}\|^2 \geq \tfrac{1}{3} |c_{km}|^2/\|A\|^2 \quad \text{where} \quad C = (c_{ij}) = A\,A* - A*\,A.$$

The final matrix A' is a normal matrix where:

$$a'_{ij} = -a'_{ji} \quad \text{and either} \quad a'_{ij} = a'_{ji} = 0 \quad \text{or} \quad a'_{ii} = a'_{jj} \quad \text{for all} \quad (i, j),$$

and

$$\|a^1\| \geq \|a^2\| \geq \cdots \geq \|a^n\|$$

where $\|a^i\|$ is the norm of the i-th column. Thus A' is block diagonal: the 1×1 blocks contain the real eigenvalues, the 2×2 blocks

$$\begin{pmatrix} a'_{jj} & a'_{j\,j+1} \\ -a'_{j\,j+1} & a'_{jj} \end{pmatrix}$$

correspond to the eigenvalues $a'_{jj} \pm i a'_{j\,j+1}$. The columns (rows) of T (T^{-1}) are right (left) eigenvectors corresponding to real eigenvalues. The right (left) eigenvectors corresponding to the complex eigenvalues $a'_{jj} \pm i a'_{j\,j+1}$ are $t^j \pm i t^{j+1} (t_j \pm i t_{j+1})$ where $t^j (t_j)$ is the j-th column (row) of $T (T^{-1})$.[*] Since the columns of A' are in order of decreasing norm, the eigenvalues appear in order of decreasing absolute value.

An annoying special form of normal block may also appear: $aI + S$ where S is skew symmetric. All roots of this block have real part a and imaginary parts equal to the roots of S. Should this case arise, we may choose $y = 0$ and $\tan 2x =$

$$2 \sum_{i=1}^{n} a_{ik} a_{im}/(\|a^k\|^2 - \|a^m\|^2)$$ where again that solution x is chosen to maximize $\|a^k\|$

in the transformed matrix [2]. We then have convergence to the form of 1×1 and 2×2 blocks described above. Note that the ALGOL program in Section 4 does not include statements to deal with this rare special case.

When A is symmetric the algorithm reduces to the (cyclic) Jacobi Method.

2. Applicability

This algorithm may be used to solve the eigenproblem for any real $n \times n$ matrix. It is not to be recommended for symmetric matrices since other methods are much faster. The appearance of multiple complex roots will slow convergence considerably; and if these roots are defective, convergence is markedly slow. An upper limit of 50 to the number of complete sweeps or iterations is contained within the procedure.

If it is desired to solve a complex matrix $C = \bar{A} + i\bar{\bar{A}}$, then the real $2n \times 2n$ matrix

$$C' = \begin{pmatrix} \bar{A} & \bar{\bar{A}} \\ -\bar{\bar{A}} & \bar{A} \end{pmatrix}$$

may be block diagonalized. If the roots of C are $\lambda_1 \lambda_2 \ldots \lambda_n$ then the roots of C' are λ_i and $\bar{\lambda}_i$, $i = 1, 2, \ldots, n$.

---

[*] The left eigenvectors are defined by $y^H A = \lambda y^H$ where H denotes conjugate transpose.

3. Formal Parameter List

n is the order of the $n \times n$ real matrix in array

a for which the eigenproblem is to be solved. At exit the real eigenvalues occupy diagonal elements of a while the real and imaginary parts of complex eigenvalues $\lambda \pm i\,\mu$ occupy respectively the diagonal and off diagonal corners of 2×2 blocks on the main diagonal so that $\lambda = a_{j,j} = a_{j+1,j+1}$ and $a_{j,j+1} = -a_{j+1,j}$, $|a_{j,j+1}| = |\mu|$.

t is an $n \times n$ array intended to receive the left or right transformation matrices or neither depending on the input value of

tmx which serves the dual purpose of input and output parameter. For input, $tmx = 0$ generates no eigenvectors and t is unaltered so that in this case, (since an actual parameter corresponding to t must be supplied) $t = a$ would be appropriate in a call of the procedure. Otherwise, values of $tmx < 0$ or $tmx > 0$ generate respectively left or right transformation matrices in t. Eigenvectors of real eigenvalues occupy corresponding rows (or columns) of the left (or right) transformation matrices. Eigenvectors corresponding to a complex eigenvalue pair $a_{j,j} \pm a_{j,j+1}$ may be formed by $t_j \pm i\,t_{j+1}$ where t_j, t_{j+1} are the corresponding rows (or columns) of the left (or right) transformation matrices.

    At exit from the procedure the absolute value of tmx records the number of iterations performed. Convergence will have occurred if $0 < tmx < 50$. Failure to converge is indicated by $tmx = 50$ or, if all transformations in one iteration are identity matrices, by $tmx < 0$.

4. ALGOL Program

procedure *eigen(n) trans*: *(tmx, a) result*: *(t)*;
value n;
integer n, tmx;
array a, t;
comment solves the eigenproblem for a real matrix $a[1:n,\ 1:n]$ by a sequence of Jacobi-like similarity transformations $T^{-1} A\,T$ where $T = T_1 T_2 \ldots$ $T_i \ldots$. Each T_i is of the form $R_i \times S_i$ with R, S given by

$$
\begin{aligned}
R: \quad & R_{k,k} = R_{m,m} = \cos(x), && R_{m,k} = -R_{k,m} = \sin(x), \\
& R_{i,i} = 1, \quad R_{i,j} = 0, && (i, j \neq k, m), \\
S: \quad & S_{k,k} = S_{m,m} = \cosh(y), && S_{k,m} = S_{m,k} = -\sinh(y), \\
& S_{i,i} = 1, \quad S_{i,j} = 0, && (i, j \neq k, m),
\end{aligned}
$$

in which x, y are determined by the elements of A_i.

    In the limiting matrix real eigenvalues occupy diagonal elements while the real and imaginary parts of complex eigenvalues occupy respectively the diagonal and off diagonal corners of 2×2 blocks centred on the main diagonal.

    An array $t[1:n,\ 1:n]$ must be provided to receive the eigenvectors, under the control of the input value of the input-output parameter tmx. For $tmx = 0$ no eigenvectors are generated, t is not used and in this case $t = a$ would be appropriate in the procedure call. Otherwise $tmx < 0$

or $tmx > 0$ generate $left(T^{-1})$ or $right(T)$ transformation matrices in t.
Eigenvectors of real eigenvalues occur as *rows* (*cols*) of $T^{-1}(T)$. Eigenvectors for a complex eigenvalue pair $a_{j,j} \pm i a_{j,j+1}$ may be formed
by $t_j \pm i t_{j+1}$ where t_j, t_{j+1} are the corresponding *rows* (*cols*) of $T^{-1}(T)$.

Iterations are limited to 50 maximum. On exit from the procedure
the absolute value of *tmx* records the number of iterations performed.
Convergence will have occured if $0 < tmx < 50$. Failure to converge is
indicated either by $tmx = 50$ or, if all transformations in one iteration
are the identity matrix, by $tmx < 0$.

The machine dependent variable *ep* is set to 10^{-8} within the procedure. *ep* should be changed appropriately before the program is used;

begin real   *eps, ep, aii, aij, aji, h, g, hj, aik, aki, aim, ami,*
                tep, tem, d, c, e, akm, amk, cx, sx, cot 2x, sig, cotx,
                cos 2x, sin 2x, te, tee, yh, den, tanhy, chy, shy,
                c1, c2, s1, s2, tki, tmi, tik, tim;
   integer *i, j, k, m, it, nless1;*   **Boolean** *mark, left, right;*
   mark := *left* := *right* := **false**;
   if *tmx* $\neq$ 0 **then**
   begin comment form identity matrix in t;
        if *tmx* < 0 **then** *left* := **true else** *right* := **true**;
        for $i := 1$ **step** 1 **until** n **do**
        begin $t[i, i] := 1.0$;
           for $j := i+1$ **step** 1 **until** n **do** $t[i, j] := t[j, i] := 0.0$
        end
   end;
   $ep :=_{10} -8$;   **comment** machine precision;
   eps := *sqrt* (*ep*);
   nless1 := $n - 1$;
   comment main loop, 50 iterations;
   for $it := 1$ **step** 1 **until** 50 **do**
   begin if *mark* **then**
        begin comment *mark* is a safety variable with value **true** in case
                    all transformations in the previous iteration were
                    omitted. As A_i failed to meet the convergence
                    test and since $A_i = A_{i+1} = \cdots = A_{50}$, exit now;
              tmx := $1 - it$; **goto** *done*
        end;
        comment compute convergence criteria;
        for $i := 1$ **step** 1 **until** *nless1* **do**
        begin $aii := a[i, i]$;
            for $j := i+1$ **step** 1 **until** n **do**
            begin $aij := a[i, j]$; $aji := a[j, i]$;
                if $abs(aij + aji) > eps$ $\lor$
                  $(abs(aij - aji) > eps \land abs(aii - a[j, j]) > eps)$
                then goto *cont*
            end

```
          end convergence test, all i, j;
          tmx := it − 1;   goto done;
          comment next transformation begins;
cont:     mark := true;
          for k := 1 step 1 until nless1 do
          for m := k+1 step 1 until n do
          begin h := g := hj := yh := 0.0;
                  for i := 1 step 1 until n do
                  begin aik := a[i, k];   aim := a[i, m];
                          te := aik × aik;   tee := aim × aim;
                          yh := yh+te − tee;
                          if i ≠ k ∧ i ≠ m then
                          begin aki := a[k, i];   ami := a[m, i];
                                  h := h+ aki × ami − aik × aim;
                                  tep := te+ ami × ami;
                                  tem := tee+ aki × aki;
                                  g := g+tep+tem;
                                  hj := hj − tep+tem
                          end if
                  end i;
                  h := h+h;   d := a[k, k] − a[m, m];
                  akm := a[k, m];   amk := a[m, k];
                  c := akm+amk;   e := akm − amk;
                  if abs (c) ≦ ep then
                  begin comment take R_i as identity matrix;
                          cx := 1.0;   sx := 0.0
                  end
                  else
                  begin comment compute elements of R_i;
                          cot2x := d/c;
                          sig := if cot2x < 0 then − 1.0 else 1.0;
                          cotx := cot2x + (sig × sqrt (1.0 + cot2x × cot2x));
                          sx := sig/sqrt (1.0 + cotx × cotx);
                          cx := sx × cotx
                  end else;
                  if yh < 0.0 then
                  begin tem := cx;   cx := sx;   sx := − tem end;
                  cos2x := cx × cx − sx × sx;
                  sin2x := 2.0 × sx × cx;
                  d := d × cos2x+c × sin2x;
                  h := h × cos2x − hj × sin2x;
                  den := g+2.0 × (e × e+d × d);
                  tanhy := (e × d − h/2.0)/den;
                  if abs (tanhy) ≦ ep then
```

```
        begin comment take Sᵢ as identity matrix;
            chy := 1.0;   shy := 0.0
    end
    else
    begin comment compute elements of Sᵢ;
            chy := 1.0/sqrt (1.0 − tanhy × tanhy);
            shy := chy × tanhy
    end else;
    comment elements of Rᵢ × Sᵢ = Tᵢ;
    c1 := chy × cx − shy × sx;
    c2 := chy × cx + shy × sx;
    s1 := chy × sx + shy × cx;
    s2 := − chy × sx + shy × cx;
    comment decide whether to apply this transformation;
    if abs (s1) > ep ∨ abs (s2) > ep then
    begin comment at least one transformation is made so;
            mark := false;
            comment transformation on the left;
            for i := 1 step 1 until n do
            begin aki := a[k, i];   ami := a[m, i];
                a[k, i] := c1 × aki + s1 × ami;
                a[m, i] := s2 × aki + c2 × ami;
                if left then
                begin comment form left vectors;
                        tki := t[k, i];   tmi := t[m, i];
                        t[k, i] := c1 × tki + s1 × tmi;
                        t[m, i] := s2 × tki + c2 × tmi
                end
            end left transformation;
            comment transformation on the right;
            for i := 1 step 1 until n do
            begin aik := a[i, k];   aim := a[i, m];
                a[i, k] := c2 × aik − s2 × aim;
                a[i, m] := − s1 × aik + c1 × aim;
                if right then
                begin comment form right vectors;
                        tik := t[i, k];   tim := t[i, m];
                        t[i, k] := c2 × tik − s2 × tim;
                        t[i, m] := − s1 × tik + c1 × tim
                end
            end right transformation
    end if
end k, m loops
```

end *it loop*;
   tmx := 50;
done: **end** *eigen*;

5. Organisational and Notational Details

The original matrix A contained in the array a is destroyed and replaced by A'. The internal value ep is machine precision and must be set for the individual machine. The convergence criterion is to exit when:

1. Either $|a_{ij} + a_{ji}| \leq eps\star$ and $|a_{ij} - a_{ji}| \leq eps$
   or $|a_{ij} + a_{ji}| \leq eps$ and $|a_{ii} - a_{jj}| \leq eps$.

2. No transformation takes place for one full sweep or iteration.

3. 50 iterations have taken place.

The transformations RS are performed as a single transformation. Within the procedure the liberal use of real local variables is used to assist the reader towards an easier understanding of the procedure.

The variable yh is the Euclidean norm of the k-th column less that of the m-th column. Since yh is needed in the computation of $\tanh y$, there is no additional cost in computing time to use it to determine whether or not to permute rows and columns. This is done in the rotation $R(k, m)$.

6. Discussion of Numerical Properties

This algorithm has been used as a Fortran program since early 1960. The accuracy of this algorithm is comparable with that of QR but it is somewhat slower. For a real matrix with some complex eigenvalues *comeig* [3] is to be preferred. For matrices, ill-conditioned in the sense that small perturbations off the diagonal produce large changes in some or all of the eigenvalues, poor results may be expected by any method. How poor the results depends on the word length and rounding procedures of the machine used as well as the matrix.

7. Test Results

The procedure was tested on the Elliott 503 (University of Tasmania) which has a 29 bit mantissa, and round off designed to force the least significant bit to one unless the number has an exact representation in 29 bits.

Example I:

$$\begin{pmatrix} 1 & 0 & 0.01 \\ 0.1 & 1 & 0 \\ 0 & 1 & 1 \end{pmatrix}.$$

The true eigenvalues are $1 + 0.1\omega$, where ω are the cube roots of unity:

$1, -\frac{1}{2}(1 \pm i\sqrt{3})$. The right vectors corresponding to a given ω are $\begin{pmatrix} 1 \\ \omega^2 \\ 10\omega \end{pmatrix}$; the left vectors corresponding to a given ω are $(1, \omega, 0.1\omega^2)$.

$\star$ $eps = \sqrt{ep}$.

Example II:

$$\begin{pmatrix}
-1 & 1 \\
-1 & 0 & 1 & & & 0 \\
-1 & 0 & 0 & 1 \\
-1 & 0 & 0 & 0 & 1 \\
-1 & 0 & 0 & 0 & 0 & 1 \\
-1 & 0 & 0 & 0 & 0 & 0 & 1 \\
-1 & 0 & 0 & 0 & 0 & 0 & 0
\end{pmatrix}.$$

The eigenvalues are $\lambda=-1$, $\pm i$, $\dfrac{\sqrt2}{2}\,(-1\pm i)$ and $\dfrac{\sqrt2}{2}\,(1\pm i)$. The right vectors corresponding to a given λ are:

$$\begin{pmatrix}
\lambda^{-6}(1+\lambda+\lambda^2+\lambda^3+\lambda^4+\lambda^5+\lambda^6) \\
\lambda^{-5}(1+\lambda+\lambda^2+\lambda^3+\lambda^4+\lambda^5) \\
\lambda^{-4}(1+\lambda+\lambda^2+\lambda^3+\lambda^4) \\
\lambda^{-3}(1+\lambda+\lambda^2+\lambda^3) \\
\lambda^{-2}(1+\lambda+\lambda^2) \\
\lambda^{-1}(1+\lambda) \\
1
\end{pmatrix}.$$

Example III:

$$\begin{pmatrix}
6 & -3 & 4 & 1 \\
4 & 2 & 4 & 0 \\
4 & -2 & 3 & 1 \\
4 & 2 & 3 & 1
\end{pmatrix}.$$

This matrix is defective with two pairs of multiple roots $3\pm\sqrt5$. The right and left vectors corresponding to $3\pm\sqrt5$ are:

$$\begin{pmatrix}
\pm\sqrt5 \\
3\pm\sqrt5 \\
2 \\
6
\end{pmatrix} \quad \text{and} \quad (5\pm\sqrt5,\ -(5\pm\sqrt5),\ \pm\tfrac{3}{2}\sqrt5,\ \tfrac{5}{2}).$$

Example IV:

$$B_{12}=\begin{bmatrix}
12 & 11 & 10 & 9 & 8 & 7 & 6 & 5 & 4 & 3 & 2 & 1 \\
11 & 11 & 10 & 9 & 8 & 7 & 6 & 5 & 4 & 3 & 2 & 1 \\
0 & 10 & 10 & 9 & 8 & 7 & 6 & 5 & 4 & 3 & 2 & 1 \\
 & 0 & 9 & 9 & 8 & 7 & 6 & 5 & 4 & 3 & 2 & 1 \\
 & & 0 & 8 & 8 & 7 & 6 & 5 & 4 & 3 & 2 & 1 \\
 & & & 0 & 7 & 7 & 6 & 5 & 4 & 3 & 2 & 1 \\
 & & & & 0 & 6 & 6 & 5 & 4 & 3 & 2 & 1 \\
 & & & & & 0 & 5 & 5 & 4 & 3 & 2 & 1 \\
 & & & & & & 0 & 4 & 4 & 3 & 2 & 1 \\
 & & & & & & & 0 & 3 & 3 & 2 & 1 \\
 & & & & & & & & 0 & 2 & 2 & 1 \\
 & & & & & & & & & 0 & 1 & 1
\end{bmatrix}.$$

The six largest eigenvalues of this matrix are well-conditioned; the six smallest roots, reciprocals of the six largest, are badly conditioned. They are:

$$32.2288\,9150\,1572^1 \qquad 6.9615\,3308\,5567^1$$
$$0.0310\,2806\,0644\,010^0 \qquad 0.1436\,4651\,9769\,219^8$$
$$20.1989\,8864\,5877^9 \qquad 3.5118\,5594\,8580\,7^4$$
$$0.0495\,0742\,9185\,276^4 \qquad 0.2847\,4972\,0558\,47^9$$
$$12.3110\,7740\,0868^3 \qquad 1.5539\,8870\,9132\,1^2$$
$$0.0812\,2765\,9240\,406^2 \qquad 0.6435\,0531\,9004\,85^2.$$

Example I

Iterations = 7   limiting *a* matrix | input *tmx* = 1   *t* matrix of right vectors

1.1000000	−0.00000004	−0.00000001	−0.27457418	0.32628057	0.18363505
0.00000007	0.95000006	0.08660255	−0.27457417	−0.00410766	−0.37438470
0.00000001	−0.08660255	0.95000006	−2.7457401	−3.2217283	1.9074973

Example II

Iterations = 25   limiting *a* matrix

.70710732	0.70710731	−0.00000001	0.00000001	−0.00000025	0.00000635	−0.00000000
.70710731	0.70710731	−0.00000003	−0.00000001	0.00002767	−0.00001939	−0.00000001
.00000000	0.00000000	−0.70710733	0.70710729	−0.00004113	0.00002821	−0.00000001
.00000001	−0.00000001	−0.70710729	−0.70710730	−0.00000371	0.00002646	0.00000001
.00001820	−0.00002767	0.00003866	0.00000371	−0.00000000	−1.0000006	0.00000000
.00000635	−0.00000670	0.00002821	−0.00002996	1.0000006	0.00000002	−0.00000000
.00000000	0.00000001	0.00000001	−0.00000001	−0.00000000	0.00000000	−1.0000004

ut *tmx* = 1   *t* matrix of right vectors

.34622465	−0.04496589	−0.48732530	−0.63619791	0.50154299	−0.32519741	−0.55850679
.55925796	−0.32159108	0.30713589	−0.53091749	0.17636683	−0.82677230	−0.00000001
.51428894	−0.66781957	−0.32910429	−0.04357935	−0.32524874	−0.50156694	−0.55850679
.23765934	−0.88083050	−0.22382486	−0.83808074	−0.00002864	0.00005058	0.00000001
.10856606	−0.83585761	0.26355542	−0.20185347	0.50158158	−0.32518170	−0.55850681
.32157260	−0.55925043	−0.53094496	−0.30709256	0.17632803	−0.82675999	0.00000002
.27660285	−0.21302890	0.10524022	−0.79446005	−0.32520958	−0.50153568	−0.55850681

Example III

Iterations = 20   limiting *a* matrix

5.2368575	−0.00000057	−0.00000000	0.00000000
0.00000057	5.2352803	0.00000000	−0.00000000
−0.00000000	−0.00000000	0.76396751	0.00004814
−0.00000000	−0.00000000	−0.00004815	0.76389683

input *tmx* = 1   *t* matrix of right vectors

24.524324	24.513097	39.721367	−39.726512
57.410200	57.417827	−13.574828	13.567806
21.932831	21.927573	−35.526797	35.533545
65.783711	65.797500	−106.59775	106.58327

Example IV

Iterations = 30   limiting *a* matrix

32.228914	−0.0000000	0.00000000	0.00000000	0.00000007
0.00000026	−0.00000047			
−0.00000000	20.198996	−0.00000000	0.00000000	0.00000015
−0.00000006	0.00000004			
0.00000000	0.00000000	12.311085	0.00000001	−0.00000001
0.00000005	−0.00000008			
0.00000000	0.00000000	−0.00000001	6.9614381	0.00000000
0.00000001	−0.00000003			
0.00000004	0.00000001	−0.00000002	0.00000000	3.5118585
0.00000002	−0.00000002			
−0.00000013	0.00000013	−0.00000002	−0.00000003	−0.00000000
−0.00000003	0.00000000			
−0.00000000	0.00000016	0.00000001	−0.00000002	−0.00000000
−0.00000000	−0.00000000			
0.00000000	−0.00000017	0.00000001	0.00000006	−0.00000000
−0.00000000	0.00000000			
0.00000007	−0.00000004	−0.00000000	0.00000011	−0.00000000
0.00001718	0.00003092			
0.00000003	0.00000002	−0.00000001	0.00000008	0.00000000
−0.00000000	0.00001197			
−0.00000025	−0.00000024	0.00000000	0.00000006	−0.00000001
0.01740461	0.04002855			
0.00000007	−0.00000011	0.00000001	0.00000006	0.00000001
−0.04002855	0.01740462			

input *tmx* = 1   *t* matrix of right vectors

−1.1304541	1.4647447	0.96462393	−0.36366056	−0.04976648
0.00000991	−0.00000035			
−1.0953784	1.3922290	0.88626980	−0.31142200	−0.03559550
−0.00009015	−0.00003981			
−0.67555710	0.52563006	−0.04761536	0.30793704	0.13042125
0.00064890	0.00068319			
−0.31472127	−0.18964936	−0.76364382	0.71104979	0.19464190
−0.00305003	−0.00760883			
−0.11630503	−0.41446365	−0.66680570	0.21080323	−0.19501896
0.00216424	0.06466296			
−0.03457479	−0.31883221	−0.11641063	−0.63659660	−0.58288118
0.12550564	−0.43812094			
−0.00824098	−0.15941442	0.27218661	−0.75711972	−0.02818475
−1.4643408	2.3847407			
−0.00154842	−0.05681473	0.30681210	−0.09969277	0.97569242
10.163236	−10.330029			
−0.00022250	−0.01454144	0.17134508	0.45841542	0.73799196
−49.308840	34.813889			
−0.00002118	−0.00256930	0.05774069	0.44984927	−0.58346281
166.59953	−87.635749			
−0.00000517	−0.00028453	0.01129611	0.18767679	−1.0477529
−357.44853	151.24431			
0.00000373	−0.00001377	0.00099936	0.03148461	−0.41712133
369.42972	−138.88363			

Example IV

0.00000040	−0.00000025	−0.00000001	−0.00000035	0.00000006
0.00000013	−0.00000007	0.00000017	0.00000001	0.00000060
0.00000008	−0.00000002	−0.00000003	−0.00000001	0.00000016
0.00000003	−0.00000000	0.00000001	−0.00000002	0.00000001
0.00000001	−0.00000005	−0.00000005	−0.00000000	−0.00000000
1.5539860	−0.00000001	−0.00000001	0.00000002	−0.00000000
−0.00000001	0.64361736	0.00000000	0.00000000	−0.00000000
0.00000001	0.00000000	0.28679831	0.00000000	−0.00000000
−0.00000000	−0.00000000	−0.00000000	0.13422201	−0.04521722
0.00000001	−0.00000000	−0.00000000	0.04521722	0.13422221
−0.00000000	−0.00000000	−0.00000000	−0.00001719	0.00000000
0.00000000	−0.00000000	0.00000000	0.00003092	−0.00001196

−0.00398794	−0.00026110	0.00002444	−0.00000004	0.00000827
−0.00142175	0.00014531	−0.00006580	−0.00005691	−0.00003498
0.02772201	0.00438249	−0.00077315	0.00043477	−0.00027915
0.01903180	−0.00468438	0.00420396	−0.00020836	0.00426024
−0.15376892	−0.05868937	0.01381945	−0.02096011	−0.01581126
−0.15279446	0.09071978	−0.15169285	0.17136122	−0.08693305
0.63818915	0.58807817	0.03996928	−0.43956307	1.1796285
0.81745572	−1.1713467	3.0741529	−2.8901258	−4.5497475
−1.7619778	−3.9199497	−8.3416046	34.174849	−2.1049114
−2.7322872	9.4503328	−22.131802	−162.85057	82.780507
2.4274939	13.038671	142.29261	413.03358	−292.88811
4.3818747	−36.586100	−199.51235	−458.16058	362.22509

A second test (including the case $tmx = -1$) was made at the Leibniz-Rechenzentrum of the Bavarian Academy of Sciences confirming the results.

References

1. EBERLEIN, P. J.: A Jacobi-like method for the automatic computation of eigenvalues and eigenvectors of an arbitrary matrix. J. Soc. Indust. Appl. Math.
   10, 74—88 (1962).
2. — Algorithms for the scaling of inner products with applications to linear least
   squares and the eigenproblem. Submitted to the SIAM J. Numer. Anal.
3. — Solution to the complex eigenproblem by a norm reducing Jacobi type method.
   Numer. Math. **14**, 232—245 (1970). Cf. II/17.

Similarity Reduction of a General Matrix to Hessenberg Form*

by R. S. Martin and J. H. Wilkinson

1. Theoretical Background

With several algorithms for finding the eigensystem of a matrix the volume of work is greatly reduced if the matrix A is first transformed to upper-Hessenberg form, i.e. to a matrix H such that $h_{ij}=0$ $(i>j+1)$. The reduction may be achieved in a stable manner by the use of either stabilized elementary matrices or elementary unitary matrices [2].

The reduction by *stabilized elementary* matrices takes place in $n-2$ major steps; immediately before the r-th step the matrix $A_1=A$ has been reduced to A_r which is of upper-Hessenberg form in its first $r-1$ columns. For real matrices the r-th step is then as follows.

(i) Determine the maximum of the quantities $|a_{ir}^{(r)}|$ $(i=r+1, \ldots, n)$. (When this is not unique we take the first of these maximum values.) If the maximum is zero the r-th step is complete. Otherwise denote this maximum element by $|a_{(r+1)',r}^{(r)}|$ and proceed as follows.

(ii) Interchange rows $(r+1)'$ and $(r+1)$ and columns $(r+1)'$ and $(r+1)$.

(iii) For each value of i from $r+2$ to n:

Compute $n_{i,r+1}=a_{ir}^{(r)}/a_{r+1,r}^{(r)}$, subtract $n_{i,r+1}\times$row $r+1$ from row i and add $n_{i,r+1}\times$column i to column $r+1$. (Notice that if $n_{i,r+1}=0$ these last two operations can be skipped.)

We may express the relationship between A_r and A_{r+1} in the form

$$A_{r+1}=N_{r+1}^{-1}I_{r+1,(r+1)'}\,A_r I_{r+1,(r+1)'}N_{r+1},\tag{1}$$

where $I_{r+1,(r+1)'}$ is an elementary permutation matrix and N_{r+1} is an elementary matrix with

$$(N_{r+1})_{i,r+1}=n_{i,r+1}\ (i=r+2,\ldots,n)\quad\text{and}\quad (N_{r+1})_{ij}=\delta_{ij}\ \text{otherwise.}\tag{2}$$

This is the basis of the procedure *elmhes*.

When the matrix A_1 is complex the same procedure carried out in complex arithmetic gives a stable reduction, but it is somewhat simpler to determine the maximum of the quantities $|\mathscr{R}(a_{ij}^{(r)})|+|\mathscr{I}(a_{ij}^{(r)})|$ rather than the maximum of the moduli. This ensures that $|n_{i,r+1}|\leq 2^{\frac{1}{2}}$ which is quite adequate.

---

* Prepublished in Numer. Math. **12**, 349—368 (1968).

The row and column interchanges used in this reduction effectively determine a permutation matrix P such that when $\tilde{A}_1 = P A_1 P^T$ is reduced no row interchanges are required. It further determines a unit lower triangular matrix N with e_1 as its first column such that

$$\tilde{A}_1 N = N H \tag{2a}$$

where H is of upper-Hessenberg form. From Eq. (2a) N and H can be determined directly column by column by equating in succession columns 1, 2, ..., n of the two sides and the interchanges can be determined at the same time. If innerproducts are accumulated without intermediate rounding of results, very many fewer rounding errors are involved in this realization. (For a detailed discussion see [2].) This is the basis of the procedure *dirhes*. We give the procedures only for the case when A is real. Although it has the advantage over *elmhes* as regards rounding errors it has the weakness that one cannot readily take advantage of zero elements in N and these will be common if A_1 is sparse. Its effectiveness depends, of course, on the efficiency of the computer when accumulating innerproducts in double-precision.

The reduction by *unitary transformations* also takes place in $n - 2$ major steps and again immediately before the r-th step A_1 has been reduced to A_r which is of upper-Hessenberg form in its first $r - 1$ columns. We consider only these when A_1 (and hence all A_r) is real, since when A_1 is complex it is recommended that stabilized elementary matrices be used. The matrix A_{r+1} is derived from A_r via the relation

$$A_{r+1} = P_r A_r P_r. \tag{3}$$

The orthogonal matrix P_r is of the form

$$P_r = I - u_r u_r^T / H_r, \tag{4}$$

where

$$u_r^T = [0, \ldots, 0, a_{r+1,r}^{(r)} \pm \sigma_r^{\frac{1}{2}}, a_{r+2,r}^{(r)}, \ldots, a_{nr}^{(r)}], \tag{5}$$

$$\sigma_r = (a_{r+1,r}^{(r)})^2 + (a_{r+2,r}^{(r)})^2 + \cdots + (a_{nr}^{(r)})^2, \tag{6}$$

$$H_r = \sigma_r \pm a_{r+1,r}^{(r)} \sigma_r^{\frac{1}{2}}. \tag{7}$$

It is evident that this leaves the zeros already produced in the first $r - 1$ columns of A_r unaffected. The computation of A_{r+1} takes place in the steps

$$B_{r+1} = P_r A_r = A_r - u_r (u_r^T A_r) / H_r, \tag{8}$$

$$A_{r+1} = B_{r+1} P_r = B_{r+1} - (B_{r+1} u_r / H_r) u_r^T. \tag{9}$$

In practice it is advantageous to "prepare" a general matrix before attempting to compute its eigensystem. It is envisaged that this will normally be done and a description of the algorithm is given in a further paper in this series [3]. This has the effect of reducing a matrix A to the form B given by

$$B = D P A P^{-1} D^{-1}, \tag{10}$$

where P is a permutation matrix, D is a positive diagonal matrix and B is of the form

$$B = \begin{bmatrix} B_{11} & B_{12} & B_{13} \\ 0 & B_{22} & B_{23} \\ 0 & 0 & B_{33} \end{bmatrix}. \tag{11}$$

The matrices D and P are determined so that B_{11} and B_{33} are upper-triangular while B_{22} is a square matrix such that corresponding rows and columns have approximately equal l_1 norms. The eigenvalues of A are therefore the diagonal elements of B_{11} and B_{33} together with the eigenvalues of B_{22}. If the matrix B_{22} lies on the intersection of rows and columns k to l it is clear that the reduction to Hessenberg form of the complete matrix involves only operations on rows and columns k to l and the procedures have been framed in this form.

2. Applicability

Each of the procedures *elmhes, dirhes, comhes* and *orthes* is designed to reduce a general square matrix A to a matrix of upper Hessenberg form by means of a sequence of similarity transformations. The procedures will, in general, be more effective if A has already been prepared by means of a "balancing" procedure [3] though they can be used on a matrix which has not been prepared in this way.

It is assumed that the matrix is in the form given in (11) and described in section 1, with the matrix B_{22} occupying rows and columns k to l. If the matrix has not been prepared one has only to take $k = 1$ and $l = n$ when B_{22} becomes the full matrix.

elmhes is used to reduce a real matrix to upper-Hessenberg form by real stabilized elementary similarity transformations.

dirhes performs the same reduction as *elmhes* but it reduces rounding errors by accumulating inner-products wherever possible.

comhes is used to reduce a complex matrix to upper-Hessenberg form by stabilized elementary similarity transformations.

orthes is used to reduce a real matrix to upper Hessenberg form by orthogonal similarity transformations.

The solution of the eigenproblems of Hessenberg matrices of real and complex elements is discussed in further papers in this series [4, 5].

When the eigenvectors of a reduced Hessenberg matrix have been computed by any method then the eigenvectors of the original matrix can be found by using the appropriate procedure of the set *elmbak, dirbak, combak* and *ortbak*.

elmbak, dirbak, combak and *ortbak* are to be used to determine r eigenvectors of a matrix A from corresponding eigenvectors of the Hessenberg matrix derived from it by the procedures *elmhes, dirhes, comhes* and *orthes* respectively.

3. Formal Parameter List

3.1 (a) Input to procedure *elmhes*

n order of the full matrix A.

k, l parameters output by a procedure for preparing A. (See [3].) If A is not prepared in this way $k = 1$, $l = n$.

a the $n \times n$ matrix A, normally in prepared form [3].

Output from procedure *elmhes*

a an $n \times n$ array consisting partly of the derived upper Hessenberg matrix; the quantity $n_{i,r+1}$ involved in the reduction is stored in the $[i, r]$ element.

int   an integer array describing the row and column interchanges involved
        in the reduction.

3.1 (b) Input to procedure *elmbak*

k, l   parameters output by a procedure preparing A for reduction to Hessen-
        berg form [3].

r   the number of eigenvectors of the Hessenberg matrix which have been
        determined.

a   the $n \times n$ array output by *elmhes* which includes the row and column
        multipliers used in the reduction.

int   an integer array describing the row and column interchanges used in
        elmhes.

z   an $n \times r$ array giving r eigenvectors of the derived Hessenberg matrix.
        corresponding to a real eigenvalue there is one column of this array;
        corresponding to a complex conjugate pair of eigenvalues $\lambda \pm i\mu$ the
        matrix z contains two consecutive columns a and b such that $a \pm ib$ are
        the vectors corresponding to $\lambda \pm i\mu$ respectively.

Output from procedure *elmbak*

z   an $n \times r$ array giving r eigenvectors of the matrix reduced by *elmhes* in
        an analogous form to that used in the input z. The output vectors are
        not normalized since usually they will be the eigenvectors of the pre-
        pared version of the original matrix and will need further back trans-
        formation to give eigenvectors of the original matrix (see [3]). Moreover
        complex conjugate pairs would need another form of normalization.

3.2 (a) Input to procedure *dirhes*

n   order of full matrix A

k, l   parameters output by a procedure for preparing A. (See [3].) If A is not
        prepared in this way $k = 1, l = n$.

a   the $n \times n$ matrix A, normally in prepared form [3].

Output from procedure *dirhes*.

a   an $n \times n$ array consisting partly of the derived upper Hessenberg matrix;
        the element $n_{i, r+1}$ of N is stored in the $[i, r]$ position.

int   an integer array describing the row and column interchanges involved
        in the reduction.

3.2 (b) Input to procedure *dirbak*

k, l   parameters output by a procedure preparing A for reduction to Hessen-
        berg form [3].

r   the number of eigenvectors of the Hessenberg matrix which have been
        determined.

a   the $n \times n$ array output by *dirhes* which includes the elements of N.

int   an integer array describing the interchanges used in *dirhes*.

z   an $n \times r$ array giving r eigenvectors of the derived Hessenberg matrix. Corresponding to a real eigenvalue there is one column of this array; corresponding to a complex conjugate pair of eigenvalues $\lambda \pm i\mu$ the matrix z contains two consecutive columns a and b such that $a \pm ib$ are the vectors corresponding to $\lambda \pm i\mu$ respectively.

Output from procedure *dirbak*

z   an $n \times r$ array giving r eigenvectors of the matrix reduced by *dirhes* in an analogous form to that used in the input z. The output vectors are not normalized since usually they will be the eigenvectors of the prepared version of the original matrix and will need further back transformation to give eigenvectors of the original matrix (see [3]).

3.3 (a) Input to procedure *comhes*

n   order of the full complex matrix A.

k, l   parameters output by a procedure for preparing A [3]. If A is not prepared in this way then $k = 1$, $l = n$.

ar, ai   two $n \times n$ arrays, ar giving the matrix of real parts of A and ai the matrix of imaginary parts.

Output from procedure *comhes*

ar, ai   two $n \times n$ arrays of which ar contains the matrix of real parts and ai contains the matrix of imaginary parts of the derived Hessenberg matrix. The multipliers used in the reduction are also stored in these arrays with $\mathcal{R}(n_{i, r+1})$ in ar $[i, r]$ and $\mathcal{I}(n_{i, r+1})$ in ai $[i, r]$.

int   an integer array describing the row and column interchanges involved in the reduction.

3.3 (b) Input to procedure *combak*

k, l   parameters output by the procedure used for preparing A [3].

r   the number of eigenvectors of the derived complex Hessenberg matrix which have been determined.

ar, ai   the arrays output by *comhes* which include the real and imaginary parts respectively of the row and column multipliers used in the reduction.

int   an integer array describing the row and column interchanges used in *comhes*.

zr, zi   two $n \times r$ arrays of which the r columns of zr and zi are the real parts and imaginary parts respectively of r eigenvectors of the derived Hessenberg matrix. They are not normalized since usually they will belong to the prepared version of the matrix and will need further back transformation to give vectors of the original matrix (see [3]).

Output from procedure *combak*

zr, zi   two $n \times r$ arrays giving the real and imaginary parts of the eigenvectors of the matrix reduced by *comhes*.

3.4 (a) Input to procedure *orthes*

n order of the full real matrix A.

k, l parameters output by a procedure preparing A for reduction to Hessenberg form [3]. If A has not been prepared then $k = 1$, $l = n$.

tol   a machine dependent constant. Should be set equal to eta/macheps where eta is the smallest number representable in the computer and macheps is the smallest positive number for which $1 + \text{macheps} \neq 1$.

a the $n \times n$ real matrix A normally in prepared form.

Output from procedure *orthes*

a the upper part of this $n \times n$ array gives the elements of the derived real upper-Hessenberg matrix. The remaining elements give information on the orthogonal transformations for use in *ortbak*.

d a linear array giving the remaining information on the orthogonal transformations to be used in *ortbak*.

3.4 (b) Input to procedure *ortbak*

k, l parameters output by the procedure used for preparing A [3].

r number of eigenvectors of the derived Hessenberg matrix which have been found.

a the $n \times n$ array output by *orthes* which includes information on the orthogonal transformations.

d a linear array output by *orthes* giving the remaining information on the orthogonal transformations.

z an $n \times r$ array giving r eigenvectors of the derived Hessenberg matrix. Corresponding to a real eigenvalue there is one column of this array; corresponding to a complex conjugate pair of eigenvalues $\lambda \pm i\mu$ the matrix z contains two consecutive columns a and b such that $a \pm ib$ are the vectors corresponding to $\lambda \pm i\mu$ respectively.

Output from procedure *ortbak*

z an $n \times r$ array giving r eigenvectors of the matrix reduced by *orthes* in a form analogous to that used in the input z. The output vectors are not normalized since usually they will belong to the prepared version of the original matrix and will need further back transformation to give vectors of the original matrix. (See [3].)

d the original information in d is destroyed.

4. ALGOL Procedures

procedure *elmhes* (n, k, l) *trans*: (a) *result*: (int);
value n, k, l; **integer** n, k, l; **array** a; **integer array** int;
comment Given the unsymmetric matrix, A, stored in the array $a[1:n, 1:n]$, this procedure reduces the sub-matrix of order $l - k + 1$, which starts at the element $a[k, k]$ and finishes at the element $a[l, l]$, to Hessenberg form, H, by non-orthogonal elementary transformations. The matrix H is overwritten on A with details of the transformations stored in the remaining triangle under H and in the array $int[k:l]$;

```
begin      integer i, j, la, m;
           real x, y;
           la := l − 1;
           for m := k + 1 step 1 until la do
           begin i := m; x := 0;
                   for j := m step 1 until l do
                   if abs (a[j, m − 1]) > abs (x) then
                   begin x := a[j, m − 1]; i := j
                   end;
                   int [m] := i;
                   if i ≠ m then
                   begin comment interchange rows and columns of array a;
                           for j := m − 1 step 1 until n do
                           begin y := a[i, j]; a[i, j] := a[m, j]; a[m, j] := y
                           end j;
                           for j := 1 step 1 until l do
                           begin y := a[j, i]; a[j, i] := a[j, m]; a[j, m] := y
                           end j
                   end interchange;
                   if x ≠ 0 then
                   for i := m + 1 step 1 until l do
                   begin y := a[i, m − 1];
                           if y ≠ 0 then
                           begin y := a[i, m − 1] := y/x;
                                   for j := m step 1 until n do
                                   a[i, j] := a[i, j] − y × a[m, j];
                                   for j := 1 step 1 until l do
                                   a[j, m] := a[j, m] + y × a[j, i]
                           end
                   end i
           end m
end elmhes;
```

procedure elmbak (k, l) data: (r, a, int) trans: (z);
value k, l, r; integer k, l, r; array a, z; integer array int;
comment Given r eigenvectors, stored as columns in the array z[1:n, 1:r], of
        the Hessenberg matrix, H, which was formed by elmhes, and details
        of the transformations as left by elmhes below H and in the array
        int[k:l], 1 ≤ k ≤ l ≤ n, this procedure forms the eigenvectors of the
        matrix A and overwrites them on the given vectors;

```
begin      integer i, j, m;
           real x;
           k := k + 1;
           for m := l − 1 step −1 until k do
           begin for i := m + 1 step 1 until l do
```

```
                    begin x := a[i, m − 1];
                          if x ≠ 0 then
                          for j := 1 step 1 until r do
                          z[i, j] := z[i, j] + x × z[m, j]
                    end i;
                    i := int[m];
                    if i ≠ m then
                    for j := 1 step 1 until r do
                    begin x := z[i, j]; z[i, j] := z[m, j]; z[m, j] := x
                    end j
              end m
end elmbak;
```

procedure *dirhes* (n, k, l) *trans*: (a) *result*: (int);
value n, k, l; **integer** n, k, l; **array** a; **integer array** int;
comment Given the unsymmetric matrix, A, stored in the array $a[1:n, 1:n]$, this procedure reduces the sub-matrix of order $l − k + 1$, which starts at the element $a[k, k]$ and finishes at the element $a[l, l]$, to Hessenberg form, H, by the direct method $(AN=NH)$. The matrix H is overwritten on A with details of the transformations (N) stored in the remaining triangle under H and in the array $int[k:l]$. The code procedure *innerprod* is used. See Contribution I/2;

```
begin      integer i, j, m;
           real x, y, z;
           for j := k + 1 step 1 until l do
           begin m := j; x := 0;
                   for i := j step 1 until l do
                   if abs (a[i, j − 1]) > abs (x) then
                   begin x := a[i, j − 1]; m := i
                   end i;
                   int[j] := m;
                   if m ≠ j then
                   begin comment interchange rows and columns of a;
                           for i := k step 1 until n do
                           begin y := a[m, i]; a[m, i] := a[j, i]; a[j, i] := y
                           end i;
                           for i := 1 step 1 until l do
                           begin y := a[i, m]; a[i, m] := a[i, j]; a[i, j] := y
                           end i
                   end interchange;
                   if x ≠ 0 then
                   for i := j + 1 step 1 until l do
                   a[i, j − 1] := a[i, j − 1]/x;
                   for i := 1 step 1 until l do
```

begin $y := a[i, j]$;
   if $x \neq 0$ **then** $innerprod (j + 1, 1, l, y, 0, a[i, m],$
$$a[m, j - 1], m, y, z)$$
      else $z := 0$;
   $innerprod (k + 1, 1,$ **if** $i \leq j$ **then** $i - 1$ **else** $j, -y, -z,$
$$a[i, m - 1], a[m, j], m, y, z);$$
   $a[i, j] := -y$
   end i
end j
end *dirhes*;

procedure *dirbak* (k, l) *data*: (r, a, int) *trans*: (z);
value k, l, r; **integer** k, l, r; **array** a, z; **integer array** *int*;
comment Given r eigenvectors, stored as columns in the array $z[1:n, 1:r]$, of
   the Hessenberg matrix, H, which was formed by *dirhes*, and details
   of the transformations as left by *dirhes* below H and in the array
   $int[k:l]$, $1 \leq k \leq l \leq n$, this procedure forms the eigenvectors of the
   matrix A and overwrites them on the given vectors. The code procedure *innerprod* is used. See Contribution I/2;
begin   **integer** i, j, m;
   real x;
   $k := k + 2$;
   comment form Nz;
   for $j := 1$ **step** 1 **until** r **do**
   for $i := l$ **step** -1 **until** k **do**
   $innerprod (k - 1, 1, i - 1, z[i, j], 0, a[i, m - 1], z[m, j], m, z[i, j], x)$;
   $k := k - 1$;
   comment interchange where necessary;
   for $i := l$ **step** -1 **until** k **do**
   begin $m := int[i]$;
      if $m \neq i$ **then**
      for $j := 1$ **step** 1 **until** r **do**
      begin $x := z[m, j]$; $z[m, j] := z[i, j]$; $z[i, j] := x$
      end j
   end i
end *dirbak*;

procedure *comhes* (n, k, l) *trans*: (ar, ai) *result*: (int);
value n, k, l; **integer** n, k, l; **array** ar, ai; **integer array** *int*;
comment Given the complex unsymmetric matrix, A, stored in the arrays
   $ar[1:n, 1:n]$ and $ai[1:n, 1:n]$, this procedure reduces the sub-matrix
   of order $l - k + 1$, which starts at the elements $ar[k, k]$ and $ai[k, k]$
   and finishes at the elements $ar[l, l]$ and $ai[l, l]$, to Hessenberg form,
   H, by non-orthogonal elementary transformations. The matrix H is
   overwritten on A with details of the transformations stored in the
   remaining triangle under H and in the array $int[k:l]$. The procedure
   cdiv is used, see Appendix;

```
begin      integer i, j, la, m;
           real xr, xi, yr, yi;
           la := l − 1;
           for m := k + 1 step 1 until la do
           begin i := m;  xr := xi := 0;
                 for j := m step 1 until l do
                 if abs (ar[j, m − 1]) + abs (ai[j, m − 1]) > abs (xr) + abs (xi) then
                 begin xr := ar[j, m − 1];  xi := ai[j, m − 1];  i := j
                 end;
                 int [m] := i;
                 if i ≠ m then
                 begin comment interchange rows and columns of arrays ar
                                 and ai;
                       for j := m − 1 step 1 until n do
                       begin yr := ar[i, j];  ar[i, j] := ar[m, j];  ar[m, j] := yr;
                             yi := ai[i, j];  ai[i, j] := ai[m, j];  ai[m, j] := yi
                       end j;
                       for j := 1 step 1 until l do
                       begin yr := ar[j, i];  ar[j, i] := ar[j, m];  ar[j, m] := yr;
                             yi := ai[j, i];  ai[j, i] := ai[j, m];  ai[j, m] := yi
                       end j
                 end interchange;
                 if xr ≠ 0 ∨ xi ≠ 0 then
                 for i := m + 1 step 1 until l do
                 begin yr := ar[i, m − 1];  yi := ai[i, m − 1];
                       if yr ≠ 0 ∨ yi ≠ 0 then
                       begin cdiv (yr, yi, xr, xi, yr, yi);
                             ar[i, m − 1] := yr;  ai[i, m − 1] := yi;
                             for j := m step 1 until n do
                             begin ar[i, j] := ar[i, j] − yr × ar[m, j]
                                                          + yi × ai[m, j];
                                   ai[i, j] := ai[i, j] − yr × ai[m, j]
                                                          − yi × ar[m, j]
                             end j;
                             for j := 1 step 1 until l do
                             begin ar[j, m] := ar[j, m] + yr × ar[j, i]
                                                          − yi × ai[j, i];
                                   ai[j, m] := ai[j, m] + yr × ai[j, i]
                                                          + yi × ar[j, i]
                             end j
                       end
                 end
           end i
           end m
end comhes;


procedure combak (k, l) data: (r, ar, ai, int) trans: (zr, zi);
value k, l, r;  integer k, l, r;  array ar, ai, zr, zi;  integer array int;
```

comment Given r eigenvectors, stored as columns in the arrays $zr[1:n, 1:r]$ and $zi[1:n, 1:r]$, of the complex Hessenberg matrix, H, which was formed by *comhes*, and details of the transformations as left by *comhes* below H and in the array $int[k:l]$, $1 \leq k \leq l \leq n$, this procedure forms the eigenvectors of the matrix A and overwrites them on the given vectors;

begin **integer** i, j, m;

         real xr, xi;

         $k := k+1$;

         for $m := l-1$ **step** -1 **until** k **do**

         begin for $i := m+1$ **step** 1 **until** l **do**

               begin $xr := ar[i, m-1]$; $xi := ai[i, m-1]$;

                    if $xr \neq 0 \lor xi \neq 0$ **then**

                    for $j := 1$ **step** 1 **until** r **do**

                    begin $zr[i, j] := zr[i, j] + xr \times zr[m, j] - xi \times zi[m, j]$;

                            $zi[i, j] := zi[i, j] + xr \times zi[m, j] + xi \times zr[m, j]$

                    end j

               end i;

               $i := int[m]$;

               if $i \neq m$ **then**

               for $j := 1$ **step** 1 **until** r **do**

               begin $xr := zr[i, j]$; $zr[i, j] := zr[m, j]$; $zr[m, j] := xr$;

                    $xi := zi[i, j]$; $zi[i, j] := zi[m, j]$; $zi[m, j] := xi$

               end j

         end m

end *combak*;

procedure *orthes* (n, k, l) *data*: (tol) *trans*: (a) *result*: (d);

value n, k, l, tol; **integer** n, k, l; **real** tol; **array** a, d;

comment Given the unsymmetric matrix, A, stored in the array $a[1:n, 1:n]$, this procedure reduces the sub-matrix of order $l-k+1$, which starts at the element $a[k, k]$ and finishes at the element $a[l, l]$, to Hessenberg form, H, by orthogonal transformations. The matrix H is overwritten on A with details of the transformations stored in the remaining triangle under H and in the array $d[k:l]$. *tol* is a tolerance for checking if the transformation is valid;

begin **integer** i, j, m, la;

         real f, g, h;

         $la := l-1$;

         for $m := k+1$ **step** 1 **until** la **do**

         begin $h := 0$;

                for $i := l$ **step** -1 **until** m **do**

                begin $f := d[i] := a[i, m-1]$; $h := h+f\uparrow 2$

                end i;

                if $h \leq tol$ **then**

```
              begin g := 0; go to skip
              end;
              g := if f ≥ 0 then − sqrt (h) else sqrt (h);
              h := h − f × g; d[m] := f − g;
              comment form  (I − (u × uT)/h) × A ;
              for j := m step 1 until n do
              begin f := 0;
                        for i := l step −1 until m do
                        f := f + d[i] × a[i, j];
                        f := f/h;
                        for i := m step 1 until l do
                        a[i, j] := a[i, j] − f × d[i]
              end j;
              comment form  (I − (u × uT)/h) × A × (I − (u × uT)/h);
              for i := 1 step 1 until l do
              begin f := 0;
                        for j := l step −1 until m do
                        f := f + d[j] × a[i, j];
                        f := f/h;
                        for j := m step 1 until l do
                        a[i, j] := a[i, j] − f × d[j]
              end i;
        skip:  a[m, m − 1] := g
        end m
end orthes;


procedure ortbak (k, l) data : (r, a) trans : (d, z);
value k, l, r; integer k, l, r; array a, d, z;
comment Given r eigenvectors, stored as columns in the array z[1:n, 1:r], of
          the Hessenberg matrix, H, which was formed by orthes, and details
          of the transformations as left by orthes below H and in the array d[k:l],
          1 ≤ k ≤ l ≤ n, this procedure forms the eigenvectors of the matrix A
          and overwrites them on the given vectors. The array d is used as a
          temporary store and is not restored to its original form;
begin     integer i, j, m, ma;
          real g, h;
          for m := l − 2 step −1 until k do
          begin ma := m + 1; h := a[ma, m];
                    if h = 0 then go to skip;
                    h := h × d[ma];
                    for i := m + 2 step 1 until l do
                    d[i] := a[i, m];
                    for j := 1 step 1 until r do
```

```
begin g := 0;
        for i := ma step 1 until l do
        g := g + d[i] × z[i, j];
        g := g/h;
        for i := ma step 1 until l do
        z[i, j] := z[i, j] + g × d[i]
    end j;
skip:
    end m
end ortbak;
```

5. Organisational and Notational Details

In *elmhes* a form of *pivoting* is used before each elementary similarity transformation is performed. The pivoting is done in such a way that all row and column multipliers are bounded in modulus by unity. This is in the interests of numerical stability. For example with the matrix

$$\begin{bmatrix} 1 & 2 & 3 \\ 10^{-10} & 1 & 4 \\ 1 & 1 & 3 \end{bmatrix} \tag{12}$$

it would be fatal on, say, an 8-digit decimal computer to perform the reduction without pivoting. In fact this would lead to

$$\begin{bmatrix} 1 & 3 \times 10^{10} & 3 \\ 10^{-10} & 4 \times 10^{10} & 4 \\ 0 & -4 \times 10^{20} & -4 \times 10^{10} \end{bmatrix} \tag{13}$$

and it is easy to show that exactly the same matrix would be obtained for example, if the element a_{32} takes any value from -5.10^{12} to $+5.10^{12}$. Pivoting immediately deals with this difficulty.

However, it should be appreciated that by means of a diagonal similarity transformation any pivotal selection may be induced. For example the matrix (12) above may be transformed into

$$\begin{bmatrix} -1 & 2 \times 10^{-11} & 3 \\ 10 & 1 & 4 \times 10^{11} \\ 1 & 10^{-11} & 3 \end{bmatrix}. \tag{14}$$

The element $|a_{21}|$ is now the largest of the elements $|a_{21}|$ and $|a_{31}|$ and the reduced matrix is

$$\begin{bmatrix} 1 & 3 \times 10^{-1} & 3 \\ 10 & 4 \times 10^{10} & 4 \times 10^{11} \\ 0 & -4 \times 10^{9} & -4 \times 10^{10} \end{bmatrix}. \tag{15}$$

The transformation is just as fatal as that giving (13). Clearly the selection of pivots we have used can only be justified if the original matrix is *prepared* in some way. This is discussed in [3]. In addition to preparing the matrix so that it is suitable for the pivotal strategy we have adopted, the preparatory procedure also detects whether any of the eigenvalues of the original matrix are isolated and produces a prepared matrix of the type indicated in (11).

The pivoting procedure in *dirhes* is directly related to that in *elmhes* and indeed, apart from rounding errors, the two procedures are identical. The procedure *dirhes* uses the procedure *innerprod* (see Contribution I/2).

In *comhes*, for economy of effort, we have not selected the maximum of the elements $|a_{r+1,r}|, \ldots, |a_{n,r}|$ but have used the simpler selection $\max\limits_{i=r+1}^{n}(|\mathscr{R}a_{i,r}| + |\mathscr{I}a_{i,r}|)$. This ensures that all multipliers are bounded by $2^{\frac{1}{2}}$ rather than unity. The procedure *comhes* uses the procedure *cdiv*, see Appendix.

The complex matrix A is stored by means of two independent $n \times n$ arrays giving $\mathscr{R}(A)$ and $\mathscr{I}(A)$ respectively. This differs from the storage convention used earlier for complex arrays.

During the reductions in *elmhes* and *comhes* each multiplier is determined so as to lead to the annihilation of an element of the current matrix. The computed multiplier is stored in the position previously occupied by the annihilated element. Analogous storage strategies are used in *dirhes*.

Further storage locations are needed to indicate the positions of the pivotal rows in each of the major steps. If the matrix were not "prepared" these would be $n-2$ major steps. A prepared matrix however is already partly in triangular form as is indicated in (11) and only $k-l-1$ major steps, and therefore $k-l-1$ row and column interchanges are required.

In *orthes* precautions must be taken to ensure that underflow during the computation of σ_r does not result in severe departure of the transformation from orthogonality. This was discussed in detail in [1] and the same precautions have been adopted here.

There is not quite sufficient space to store full details of the orthogonal transformations in the positions occupied by the zeros they introduce. The information is supplemented by the storage of $a_{r+1,r}^{(r)} \pm \sigma_r^{\frac{1}{2}}$ at the r-th step in the linear array $d[i]$.

Where the orders of magnitude of elements of a matrix change rapidly as one moves in a direction parallel to the principal diagonal the matrix should be presented so that the largest elements are in the top left-hand corner. This may involve reflecting the matrix in the secondary diagonal (c.f. the argument in [1]; there such matrices were presented in the opposite way since the reduction started with the last row and column).

In the procedures *elmbak*, *dirbak* and *ortbak* provision must be made for the fact that although A is real it may have some complex conjugate pairs of roots and vectors. It is convenient that the complete set of eigenvectors should be stored in one $n \times n$ array whatever the distribution of real and complex roots. Corresponding to a complex pair of roots $\lambda \pm i\mu$ two consecutive columns u and v are stored in the eigenvector array and $u+iv$ gives the vector corresponding to $\lambda + i\mu$.

6. Discussion of Numerical Properties

All four of the reductions described here are, in general, numerically stable in the sense that the derived Hessenberg matrix is usually exactly similar to a matrix which is close to the original. For the procedure *orthes* the result can be expressed in very categorical terms. For computation with a t-digit mantissa the computed A_{n-1} is exactly orthogonally similar to some matrix $A_1 + E$ where

$$\|E\|_F \leq K_1 n^2 2^{-t} \|A_1\|_F \tag{16}$$

(here $\|\cdot\|_F$ denotes the Frobenius norm and K_1 is of order unity). The errors introduced in the individual eigenvalues depend, of course, on their inherent sensitivities.

With *dirhes* we have

$$\tilde{A}_1 N = N H + E \tag{17}$$

where

$$|e_{ij}| \leqq 2^{-t} \max |h_{ij}|. \tag{18}$$

This may be written in the form

$$N^{-1}(\tilde{A}_1 - E N^{-1}) N = H \tag{19}$$

so that H is exactly similar to $\tilde{A}_1 - E N^{-1}$. Although $\|E N^{-1}\|$ can be much larger than $\|E\|$ this is not generally true and *dirhes* is particularly satisfactory in practice.

It is not possible to make such unequivocal statements about the stability of *elmhes* and *comhes*. Probably the simplest way of describing the stability is the following. If N_{r+1} is the *computed* transformation involved in Eq. (1) then

$$A_{r+1} = N_{r+1}^{-1} I_{r+1, (r+1)'} (A_r + E_r) I_{r+1, (r+1)'} N_{r+1}, \tag{20}$$

where A_{r+1} and A_r are the computed matrices. E_r certainly satisfies the relation

$$\|E_r\|_2 \leq K_2 n 2^{-t} \max_r, \tag{21}$$

where $\max_r$ is the maximum attained by the moduli of all elements of the successive matrices involved in passing from A_r to A_{r+1}.

Any attempt to give a rigorous *a priori* bound for a perturbation in A_1 which is equivalent to all the errors made in passing from A_1 to A_{n-1} tends to be misleading since it has to cover extremely remote contingencies. In general, when pivoting is used there is no tendency for the elements of successive A_r to increase, and $\max_r$ remains of the order of magnitude of $\max |a_{ij}^{(1)}|$. Also the sensitivity of individual eigenvalues tends to remain roughly constant as we pass from A_1 to A_{n-1}. The net effect is that *elmhes* and *comhes* are about as stable as *orthes* in practice, though it is possible to construct matrices for which this is untrue (for a detailed discussion see [2] Chapter 6).

The procedures *elmbak*, *dirbak*, *combak* and *ortbak* are all very stable and in general even if this part of the computation could be performed without rounding errors there would be little improvement in the accuracy of the computed vectors.

7. Test Results

As a purely formal test of *elmhes* and *orthes* the matrix

$$\begin{bmatrix} 1 & 2 & 3 & 5 \\ 2 & 4 & 1 & 6 \\ 1 & 2 & -1 & 3 \\ 2 & 0 & 1 & 3 \end{bmatrix}$$

was reduced using KDF9, which has a 39 binary digit mantissa. The output arrays obtained using *elmhes* is given in Table 1 and that obtained using *orthes* is given in Table 2. Here and in later tables the separating line between the upper-Hessenberg matrix and the transformation matrix is indicated.

As a formal test of *comhes* the matrix

$$\begin{bmatrix} 1+i & 3+i & 2-i & 1+i \\ 3+i & 2-i & 1+i & 3+i \\ 4+i & 5+i & 6-i & 2+i \\ 1+i & 3-i & 3-i & 2-i \end{bmatrix}$$

was reduced and the output arrays are given in Table 3.

Table 1. *Output array a*

$+1.0000000000_{10}+0;$	$+8.5000000000_{10}+0;$	$+5.3214285714_{10}+0;$	$+3.0000000000_{10}+0;$
$+2.0000000000_{10}+0;$	$+1.0500000000_{10}+1;$	$+6.1071428571_{10}+0;$	$+1.0000000000_{10}+0;$
$+5.0000000000_{10}-1;$	$-7.0000000000_{10}+0;$	$-3.0000000000_{10}+0;$	$+0.0000000000$;
$+1.0000000000_{10}+0;$	$+1.0714285714_{10}-1;$	$+1.6071428571_{10}-1;$	$-1.5000000000_{10}+0;$

Output array int 2; 4;

Table 2. *Output array a*

$+1.0000000000_{10}+0;$	$-5.6666666667_{10}+0;$	$+1.8633410513_{10}+0;$	$+1.5546218238_{10}+0;$
$-3.0000000000_{10}+0;$	$+7.2222222219_{10}+0;$	$-2.8262510226_{10}+0;$	$+9.8128611579_{10}-1;$
$+1.0000000000_{10}+0;$	$+3.3591592128_{10}+0;$	$+2.3839046924_{10}-1;$	$+1.5401167031_{10}+0;$
$+2.0000000000_{10}+0;$	$+3.3555555555_{10}+0;$	$+2.0678336980_{10}-1;$	$-1.4606126914_{10}+0;$

Relevant part of vector d $+5.0000000000_{10}+0;$ $-3.5147147684_{10}+0;$

Table 3. *Output array ar*

$+1.0000000000_{10}+0;$	$+4.3529411764_{10}+0;$	$-2.2228897566_{10}-1;$	$+3.0000000000_{10}+0;$
$+4.0000000000_{10}+0;$	$+1.0176470588_{10}+1;$	$-4.8813457494_{10}-1;$	$+5.0000000000_{10}+0;$
$+7.6470588235_{10}-1;$	$+3.2491349480_{10}+0;$	$-9.6379410881_{10}-1;$	$+1.7058823529_{10}+0;$
$+2.9411764705_{10}-1;$	$-6.3292279963_{10}-1;$	$+1.6065541256_{10}+0;$	$+7.8732352058_{10}-1;$

Output array ai

$+1.0000000000_{10}+0;$	$+4.1176470588_{10}-1;$	$-1.6623610694_{10}+0;$	$+1.0000000000_{10}+0;$
$+1.0000000000_{10}+0;$	$+7.0588235294_{10}-1;$	$-3.0153199159_{10}+0;$	$+1.0000000000_{10}+0;$
$+5.8823529411_{10}-2;$	$-3.5328719723_{10}+0;$	$-1.4235152757_{10}+0;$	$-2.1764705882_{10}+0;$
$+1.7647058823_{10}-1;$	$-6.7647942325_{10}-1;$	$+1.0615484720_{10}+0;$	$-2.2823670772_{10}+0;$

Output array int 3; 4;

The arrays obtained on other computers should agree with those given in Tables 1, 2, 3 almost to working accuracy.

In order to illustrate the nature of the "stability" of the reductions the matrix

$$
\begin{bmatrix}
2+2i & 2+2i & 2 & 0 & -6-6i & -6 \\
8+8i & 8+8i & 16 & 24+24i & 0 & -8 \\
36i & 72i & 18+18i & 108i & 36i & 0 \\
0 & 6+6i & 6 & 2+2i & 2+2i & 2 \\
-24-24i & 0 & 8 & 8+8i & 8+8i & 16 \\
-108i & -36i & 0 & 36i & 72i & 18+18i
\end{bmatrix} = A + iB \quad (22)
$$

Table 4. *Output array ar*

$+2.0000000000_{10}+0;$	$-9.6296296296_{10}+0;$	$-4.4444444434_{10}-1;$	$+1.7000000000_{10}+0;$
$+0.0000000000 \quad ;$	$+3.6666666666_{10}+1;$	$+2.0800000000_{10}+1;$	$-9.0000000000_{10}-1;$
$-3.3333333334_{10}-1;$	$+8.8888888891_{10}+0;$	$-1.0666666666_{10}+1;$	$+1.5000000000_{10}+0;$
$+0.0000000000 \quad ;$	$+1.5347723092_{10}-13;$	$+1.1641532182_{10}-10;$	$+5.5400000000_{10}+1;$
$+2.2222222223_{10}-1;$	$+3.3333333309_{10}-2;$	$+7.5000000000_{10}-1;$	$-4.5149999998_{10}+0;$
$-7.4074074074_{10}-2;$	$-5.1111111109_{10}-1;$	$+1.7500000000_{10}-1;$	$-3.2131405748_{10}-1;$

$+3.7622770992_{10}+0;$	$-6.0000000000_{10}+0;$
$-2.2012713052_{10}+1;$	$+0.0000000000 \quad ;$
$-1.8343927543_{10}+1;$	$+0.0000000000 \quad ;$
$+6.5426337404_{10}+1;$	$+3.6000000001_{10}+1;$
$-1.2889677673_{10}+1;$	$-5.2000000001_{10}+0;$
$-8.1684242416_{10}+0;$	$-1.4510322326_{10}+1;$

Output array ai

$+2.0000000000_{10}+0;$	$-3.6379788070_{10}-12;$	$+2.5465851649_{10}-11;$	$+1.4000000000_{10}+0;$
$-1.0800000000_{10}+2;$	$+3.6666666667_{10}+1;$	$+2.0799999998_{10}+1;$	$-2.4300000000_{10}+1;$
$+0.0000000000 \quad ;$	$+8.8888888893_{10}+0;$	$-1.0666666667_{10}+1;$	$+8.5500000000_{10}+1;$
$+0.0000000000 \quad ;$	$+5.1159076974_{10}-14;$	$+5.8207660913_{10}-11;$	$+5.2799999998_{10}+1;$
$-2.2222222222_{10}-1;$	$-3.3333333347_{10}-2;$	$+2.5000000000_{10}-2;$	$-1.3549999997_{10}+0;$
$+7.4074074074_{10}-2;$	$+5.1111111111_{10}-1;$	$-2.5000000000_{10}-2;$	$+3.0573212572_{10}-1;$

$+9.3491590530_{10}-2;$	$-6.0000000000_{10}+0;$
$+1.2865387861_{10}+1;$	$+7.2000000000_{10}+1;$
$+1.0072115655_{10}+2;$	$+6.0000000001_{10}+1;$
$+8.7439050456_{10}+1;$	$+3.6000000000_{10}+1;$
$-1.1697339258_{10}+1;$	$-3.4000000000_{10}+0;$
$-1.1855498885_{10}+1;$	$-1.3102660741_{10}+1;$

Eigenvalues	Real part	Imaginary part	Output array int
	$+5.0486596842_{10}+1;$	$+5.0486596842_{10}+1;$	6
	$-5.3153974198_{10}+0;$	$-5.3153974197_{10}+0;$	3
	$-1.7171199422_{10}+1;$	$-1.7171199422_{10}+1;$	6
	$+5.0486596842_{10}+1;$	$+5.0486596842_{10}+1;$	6
	$-5.3153974196_{10}+0;$	$-5.3153974197_{10}+0;$	
	$-1.7171199422_{10}+1;$	$-1.7171199422_{10}+1;$	

was reduced using *comhes*. This matrix has three pairs of complex roots *and all its elementary divisors are linear*. Accordingly the corresponding Hessenberg matrix must have at least one zero sub-diagonal element if exact computation is used. (In fact a_{43} should be exactly zero.) In practice when rounding errors are involved this means that the multipliers used in the third major step of the reduction are *entirely machine dependent*.

The reduced matrix obtained on KDF9 is given in Table 4. It will be seen that the computed element a_{43} is of order 10^{-10} (in fact a_{43} is $2^{-33}+i2^{-34}$) and the multipliers in the third step are entirely determined by rounding errors. The Hessenberg matrices obtained on different computers will therefore agree, in general, only in the first three rows and columns. *However, the eigenvalues and eigenvectors should be preserved very accurately.* The eigenvalues of the reduced matrix as found by procedure *comlr* [4] are given in Table 4.

The real matrix

$$\begin{bmatrix} A & -B \\ B & A \end{bmatrix} \tag{23}$$

derived from the matrix (22) above was also reduced using *elmhes* and *orthes*. The roots of this augmented real matrix also consist entirely of equal pairs since they are the roots of $A+iB$ together with their complex conjugates. With both *elmhes*

Table 5. *Eigenvalues using elmhes*

$-1.71711994216_{10}+1;$	$+1.71711994219_{10}+1;$
$-1.71711994216_{10}+1;$	$-1.71711994219_{10}+1;$
$+5.04865968409_{10}+1;$	$+5.04865968416_{10}+1;$
$+5.04865968409_{10}+1;$	$-5.04865968416_{10}+1;$
$+5.04865968422_{10}+1;$	$+5.04865968415_{10}+1;$
$+5.04865968422_{10}+1;$	$-5.04865968415_{10}+1;$
$-1.71711994222_{10}+1;$	$+1.71711994217_{10}+1;$
$-1.71711994222_{10}+1;$	$-1.71711994217_{10}+1;$
$-5.31539741965_{10}+0;$	$+5.31539741986_{10}+0;$
$-5.31539741965_{10}+0;$	$-5.31539741986_{10}+0;$
$-5.31539741973_{10}+0;$	$+5.31539741981_{10}+0;$
$-5.31539741973_{10}+0;$	$-5.31539741981_{10}+0;$

Eigenvalues using orthes

Real part	Imaginary part
$+5.04865968419_{10}+1;$	$+5.04865968418_{10}+1;$
$+5.04865968419_{10}+1;$	$-5.04865968418_{10}+1;$
$+5.04865968423_{10}+1;$	$+5.04865968417_{10}+1;$
$+5.04865968423_{10}+1;$	$-5.04865968417_{10}+1;$
$-1.71711994219_{10}+1;$	$+1.71711994219_{10}+1;$
$-1.71711994219_{10}+1;$	$-1.71711994219_{10}+1;$
$-1.71711994219_{10}+1;$	$+1.71711994218_{10}+1;$
$-1.71711994219_{10}+1;$	$-1.71711994218_{10}+1;$
$-5.31539741954_{10}+0;$	$+5.31539741994_{10}+0;$
$-5.31539741954_{10}+0;$	$-5.31539741994_{10}+0;$
$-5.31539741989_{10}+0;$	$+5.31539741973_{10}+0;$
$-5.31539741989_{10}+0;$	$-5.31539741973_{10}+0;$

and *orthes* the element a_{76} of the Hessenberg matrix should be zero; *elmhes* gave $-1.5\ldots\times10^{-8}$ and *orthes* $-2.6\ldots\times10^{-8}$.

The eigenvalues of the Hessenberg matrix given by *elmhes* and *orthes* are given in Table 5 (both sets were computed by the procedure *hqr* [5]) for comparison with those obtained by the reduction of $A+iB$. In spite of the fact that a_{76} is of order 10^{-8} instead of zero the errors in the eigenvalues are all less than 10^{-9}. With matrices of this kind the "zero" element in H may turn out to be surprisingly "large" without the eigenvalues being disturbed to a comparable extent.

A second test run with these matrices was made by C. REINSCH at the Technische Hochschule Munich, confirming the given results.

The procedures *elmbak*, *dirbak*, *combak* and *ortbak* were included in this paper for convenience since they are independent of the method used to find the eigensystem of the Hessenberg matrices. However, they cannot be used until eigenvectors of the Hessenberg matrices have been determined, and accordingly test results will be given in connexion with procedures to be published later in the series. (See contribution II/18.)

Appendix

The procedures for performing the fundamental complex arithmetic operations in complex arithmetic will be used repeatedly in the algorithms in this series. For convenience these procedures are assembled in this appendix. It was considered not worth while to have procedures for the addition, subtraction and multiplication of complex numbers but procedures are given for finding the modulus of a complex number, and the square root of a complex number and for dividing one complex number by another. In each of these procedures special precautions are taken in order to avoid unnecessary overflow and underflow at intermediate stages of the computation. The procedures themselves are self explanatory and are as follows.

ALGOL Procedures

real procedure *cabs* (xr, xi);
value xr, xi;   **real** xr, xi;
comment modulus of the complex number $xr + i \times xi$, $cabs := sqrt(xr\uparrow2 + xi\uparrow2)$;
begin real h;
 $xr := abs(xr)$; $xi := abs(xi)$;
 if $xi > xr$ **then begin** $h := xr$; $xr := xi$; $xi := h$ **end**;
 $cabs :=$ **if** $xi = 0$ **then** xr **else** $xr \times sqrt(1 + (xi/xr)\uparrow2)$
end *cabs*;

procedure *cdiv* (xr, xi, yr, yi) *result*: (zr, zi);
value xr, xi, yr, yi;   **real** xr, xi, yr, yi, zr, zi;
comment complex division $zr + i \times zi := (xr + i \times xi)/(yr + i \times yi)$. This procedure should not be called with $yr = yi = 0$;
begin real h;
 if $abs(yr) > abs(yi)$
 then begin $h := yi/yr$; $yr := h \times yi + yr$; $zr := (xr + h \times xi)/yr$;
$$zi := (xi - h \times xr)/yr$$
  end

```
    else begin h := yr/yi; yi := h × yr + yi; zr := (h × xr + xi)/yi;
                                              zi := (h × xi − xr)/yi
        end
end cdiv;


procedure csqrt (xr, xi) result: (yr, yi);
value xr, xi; real xr, xi, yr, yi;
comment complex square root yr + i × yi := sqrt(xr + i × xi). The procedure
        cabs is used;
begin real h;
    h := sqrt((abs (xr) + cabs (xr, xi))/2);
    if xi ≠ 0 then xi := xi/(2 × h);
    if xr ≥ 0 then xr := h else
        if xi ≥ 0 then begin xr := xi; xi := h end
        else begin xr := − xi; xi := − h end;
    yr := xr; yi := xi
end csqrt;
```

Acknowledgements. The work of R. S. MARTIN and J. H. WILKINSON has been carried out at the National Physical Laboratory. The authors wish to acknowledge a number of helpful suggestions received from Dr. C. REINSCH of the Technische Hochschule Munich, Dr. J. VARAH of Stanford University and Professor B. PARLETT of Berkeley University.

References

1. MARTIN, R. S., C. REINSCH, and J. H. WILKINSON: Householder's tridiagonalization of a symmetric matrix. Numer. Math. **11**, 181−195 (1968). Cf. II/2.
2. WILKINSON, J. H.: The algebraic eigenvalue problem. London: Oxford University Press 1965.
3. PARLETT, B. N., and C. REINSCH: Balancing a matrix for calculation of eigenvalues and eigenvectors. Numer. Math. **13**, 293−304 (1969). Cf. II/11.
4. MARTIN, R. S., and J. H. WILKINSON: The modified LR algorithm for complex Hessenberg matrices. Numer. Math. **12**, 369—376 (1968). Cf. II/16.
5. −, G. PETERS, and J. H. WILKINSON: The QR algorithm for real Hessenberg matrices. Numer. Math. **14**, 219−231 (1970). Cf. II/14.

The QR Algorithm for Real Hessenberg Matrices*

by R. S. MARTIN, G. PETERS and J. H. WILKINSON

1. Theoretical Background

The QR algorithm of Francis [1] and Kublanovskaya [4] with shifts of origin is described by the relations

$$Q_s(A_s - k_s I) = R_s, \quad A_{s+1} = R_s Q_s^T + k_s I, \quad \text{giving} \quad A_{s+1} = Q_s A_s Q_s^T, \qquad (1)$$

where Q_s is orthogonal, R_s is upper triangular and k_s is the shift of origin. When the initial matrix A_1 is of upper Hessenberg form then it is easy to show that this is true of all A_s. The volume of work involved in a QR step is far less if the matrix is of Hessenberg form, and since there are several stable ways of reducing a general matrix to this form [3, 5, 8], the QR algorithm is invariably used after such a reduction.

Parlett [6] has shown that if all k_s are taken to be zero then, in general, A_s tends to a form in which $a_{i+1,i}^{(s)} a_{i+2,i+1}^{(s)} = 0$ $(i = 1, \ldots, n-2)$ and hence all the eigenvalues are either isolated on the diagonal or are the eigenvalues of a 2×2 diagonal submatrix. In order to achieve rapid convergence it is essential that origin shifts should be used and that ultimately each shift should be close to an eigenvalue of A_1 (and hence of all A_s).

However, even when A_1 is real some of the eigenvalues may be complex, and if the transformation (1) is carried out with a complex value of k_s then A_{s+1} is, in general, a complex matrix. In theory this difficulty could be overcome by performing two steps of (1) with shifts of k_s and $\bar{k}_s$ respectively. With exact computation A_{s+2} would then be real. However, although the QR transformation is stable in that it preserves eigenvalues, one finds in practice that elements of A_{s+2} may have substantial imaginary components and neglect of these can introduce serious errors (see the discussion on pp. 512—514 of Ref. [8]).

Francis [1] described an economical method of performing two steps of the QR algorithm, either with two real values of k_s and k_{s+1} or with complex conjugate values, without using complex arithmetic. It is based on the observation that if B is non-singular and

$$BQ = QH, \qquad (2)$$

with unitary Q and upper-Hessenberg H, then the whole of Q and H are determined by the first column of Q, the determination being unique if H has positive sub-

---

* Prepublished in Numer. Math. **14**, 219—231 (1970).

diagonal elements. Now if two steps of QR are performed we have

$$A_{s+2}=Q_{s+1}Q_s A_s Q_s^T Q_{s+1}^T \quad \text{giving} \quad A_s(Q_s^T Q_{s+1}^T)=(Q_s^T Q_{s+1}^T) A_{s+2} \qquad (3)$$

and

$$(Q_s^T Q_{s+1}^T)(R_{s+1} R_s)=(A_s-k_s I)(A_s-k_{s+1} I). \qquad (4)$$

Writing

$$Q_{s+1}Q_s=Q, \qquad R_{s+1} R_s=R, \qquad (A_s-k_s I)(A_s-k_{s+1} I)=M \qquad (5)$$

we have

$$A_s Q^T=Q^T A_{s+2}, \qquad R=QM. \qquad (6)$$

Suppose then a matrix H is derived from A_s (by any method) such that

$$A_s \tilde{Q}^T=\tilde{Q}^T H \quad \text{or} \quad \tilde{Q} A_s \tilde{Q}^T=H, \qquad (7)$$

where $\tilde{Q}$ is orthogonal and H is upper-Hessenberg, then if $\tilde{Q}^T$ has the same first column as Q^T (i.e., $\tilde{Q}$ has the same first row as Q) we must have

$$\tilde{Q}=Q \quad \text{and} \quad A_{s+2}=H. \qquad (8)$$

Now from (6) Q is the matrix which triangularises the matrix product M; this product is a real matrix if k_s and k_{s+1} are both real or are complex conjugates. Hence Q is real. Householder has shown that any real matrix can be triangularised by successive premultiplication by real elementary Hermitian matrices $P_1, P_2, \dots, P_{n-1}$, where

$$P_r=I-2w_r w_r^T \qquad (9)$$

and w_r is a unit vector with zeros for its first $r-1$ components [2, 8]. The first row of the matrix $P_{n-1}\dots P_2 P_1$ is clearly the first row of P_1 itself. Hence we have only to produce a $\tilde{Q}$ satisfying Eq. (7) and having for its first row the first row of P_1.

Now in triangularizing any matrix the first factor, P_1, is determined only by the first column of that matrix and the first column of M, written as a row, is of the form

$$(p_1, q_1, r_1, 0, \dots, 0) \qquad (10)$$

where

$$p_1=a_{11}^2-a_{11}(k_1+k_2)+k_1 k_2+a_{12}a_{21},$$
$$q_1=a_{21}(a_{11}+a_{22}-k_1-k_2), \qquad (11)$$
$$r_1=a_{32}a_{21}.$$

Because of this the vector w_1 associated with P_1 has only three non-zero elements, the first three. The matrix $P_1 A_s P_1^T$ is therefore of the form illustrated when $n=7$ by

$$\begin{bmatrix} X & X & X & X & X & X & X \\ X & X & X & X & X & X & X \\ \underline{X} & X & X & X & X & X & X \\ \underline{X} & \underline{X} & X & X & X & X & X \\ & & X & X & X & X \\ & & & X & X & X \\ & & & & X & X \end{bmatrix}, \qquad (12)$$

i.e. it is an upper-Hessenberg matrix with three additional elements, the underlined elements in (12). Now Householder [2, 5] has shown that any real matrix K

can be reduced by an orthogonal similarity transformation to upper-Hessenberg form and the transformation can be written in the form

$$P_{n-1} \dots P_3 P_2 K P_2^T P_3^T \dots P_{n-1}^T \tag{13}$$

where the P_r are of the type defined above. In particular we may take K to be the matrix $P_1 A_s P_1^T$. Hence we have

$$P_{n-1} \dots P_3 P_2 P_1 A_s P_1^T P_2^T P_3^T \dots P_{n-1}^T = H, \tag{14}$$

where H is upper-Hessenberg. However, the first row of $P_{n-1} \dots P_2 P_1$ is the first row of P_1 and hence the whole matrix must be the matrix Q which would be determined by two steps of the QR algorithm, and H must be A_{s+2}. This is the two-step QR algorithm of Frances [1, 8] and clearly no complex numbers occur at any stage.

We have now only to consider the Householder similarity reduction to Hessenberg form of a matrix of the form illustrated in (12). It is easy to see that at each stage in the Householder reduction the current matrix is of a special form which is illustrated by

$$
\begin{bmatrix}
X & X & X & X & X & X & X & X \\
X & X & X & X & X & X & X & X \\
  & X & X & X & X & X & X & X \\
  &   & X & X & X & X & X & X \\
  &   & \underline{X} & X & X & X & X & X \\
  &   & \underline{X} & \underline{X} & X & X & X & X \\
  &   &   &   & X & X & X & X \\
  &   &   &   &   & X & X & X
\end{bmatrix}. \tag{15}
$$

This shows the configuration for a matrix of order 8 immediately before transformation with P_4. It is a Hessenberg matrix with three additional elements in positions $(5, 3)$, $(6, 3)$, $(6, 4)$. For this reason the matrix P_r used in the Householder reduction has a w_r of the form

$$(0, 0, \dots, 0, \alpha_r, \beta_r, \gamma_r, 0, \dots, 0), \tag{16}$$

i.e. it has non-zero elements only in positions r, $r+1$, $r+2$, these elements being determined by the elements $(r, r-1)$, $(r+1, r-1)$, $(r+2, r-1)$ of the current matrix. The preliminary step, pre-multiplication and post-multiplication with P_1 is therefore seen to be of precisely the same form as the steps in the Householder reduction.

The shifts of origin at each stage are taken to be the two roots of the 2×2 matrix in the bottom right-hand corner of the current A_s, giving

$$k_s + k_{s+1} = a^{(s)}_{n-1, n-1} + a^{(s)}_{nn}, \quad k_s k_{s+1} = a^{(s)}_{n-1, n-1} a^{(s)}_{nn} - a^{(s)}_{n-1, n} a^{(s)}_{n, n-1} \star. \tag{17}$$

$\star$ The proper combination of (11) and (17) is described in Contribution II/15.

Using these shifts of origin, convergence is usually very fast, and a stage is reached at which either $a^{(s)}_{n,n-1}$ or $a^{(s)}_{n-1,n-2}$ is "negligible". If the former of these occurs then $a^{(s)}_{nn}$ may be regarded as an eigenvalue and deflation to a Hessenberg matrix of order $n-1$ may be achieved by dropping the last row and column; if $a^{(s)}_{n-1,n-2}$ is "negligible" then the two eigenvalues of the 2×2 matrix in the bottom right-hand corner may be regarded as eigenvalues of $A^{(s)}$ and deflation to a Hessenberg matrix of order $n-2$ may be achieved by dropping the last two rows and columns.

After each two-step iteration all subdiagonal elements are examined to see if any of them is "negligible". If so the eigenproblem for the current matrix splits into that for two or more smaller Hessenberg matrices, and iteration continues with the submatrix in the bottom right-hand corner. It may happen that while no individual sub-diagonal element is sufficiently small to be regarded as negligible, the product of two consecutive elements may be small enough to allow us to work with a submatrix. This device was introduced by Francis [1] and a detailed exposition of it has been given by Wilkinson in Ref. [8], pp. 526—528.

Failure of the process to converge is a comparatively rare event and its emergence would appear to be the result of a quite exceptional combination of circumstances. An example of such a failure is provided by matrices of the type

$$\begin{bmatrix} 0 & 0 & 0 & 0 & 1 \\ 1 & 0 & 0 & 0 & 0 \\ & 1 & 0 & 0 & 0 \\ & & 1 & 0 & 0 \\ & & & 1 & 0 \end{bmatrix}. \tag{18}$$

Here the shifts of origin given by our strategy are both zero, and since the matrix is orthogonal, it is invariant with respect to the QR algorithm without shifts. However, if one double step is performed with any shifts of origin which are loosely related to the norm of the matrix, convergence is very rapid. Accordingly after ten iterations have been taken without determining an eigenvalue, the usual shifts k_1 and k_2 are replaced for the next iteration by shifts defined by

$$k_1+k_2=1.5\,(|a_{n,n-1}|+|a_{n-1,n-2}|), \qquad k_1 k_2=(|a_{n,n-1}|+|a_{n-1,n-2}|)^2. \tag{19}$$

No deep significance should be attributed to the factor 1.5; it was included since it appeared to diminish the probability of the choice being "unfortunate". This strategy is used again after 20 unsuccessful iterations. If 30 unsuccessful iterations are needed then a failure indication is given.

2. Applicability

The procedure *hqr* may be used to find all the eigenvalues of a real upper Hessenberg matrix. Normally such a matrix will have been produced from a full real matrix by one of the procedure *elmhes*, *dirhes* or *orthes*. It is assumed that the original matrix is balanced in the sense that the $l1$ norms of corresponding rows and columns are roughly of the same order of magnitude [7].

3. Formal Parameter List

Input to procedure *hqr*

n order of the real upper Hessenberg matrix H.

macheps a machine constant equal to the smallest positive machine number for which $1 + macheps \neq 1$.

h an $n \times n$ array giving the elements of H.

Output from procedure *hqr*

h the matrix H is destroyed in the process.

wr   an $n \times 1$ array giving the real parts of the n eigenvalues of H.

wi   an $n \times 1$ array giving the imaginary parts of the n eigenvalues of H.

cnt   an $n \times 1$ integer array giving the number of iterations for the eigenvalues. If two eigenvalues are found simultaneously, i.e. a 2×2 matrix splits off, then the number of iterations is given with a positive sign for the first of the pair and with a negative sign for the second.

fail   exit used when more than 30 successive iterations are used without an eigenvalue being located.

4. ALGOL Procedures

procedure *hqr* $(n, macheps)$ *trans*: (h) *result*: (wr, wi, cnt) *exit*: $(fail)$;
value n, *macheps*; **integer** n; **real** *macheps*; **array** h, wr, wi;
integer array *cnt*; **label** *fail*;

      comment Finds the eigenvalues of a real upper Hessenberg matrix, H, stored in the array $h[1:n, 1:n]$, and stores the real parts in the array $wr[1:n]$ and the imaginary parts in the array $wi[1:n]$. *macheps* is the relative machine precision. The procedure fails if any eigenvalue takes more than 30 iterations;

  begin   **integer** i, j, k, l, m, na, its;
            real $p, q, r, s, t, w, x, y, z$;
            Boolean *notlast*;

            $t := 0$;
 nextw: **if** $n = 0$ **then goto** *fin*;
            $its := 0$; $na := n - 1$;
            comment look for single small sub-diagonal element;
 nextit: **for** $l := n$ **step** -1 **until** 2 **do**
            if $abs(h[l, l-1]) \leq macheps \times (abs(h[l-1, l-1])$
                                $+ abs(h[l, l]))$ **then goto** *cont1*;
            $l := 1$;
 cont1: $x := h[n, n]$;
            if $l = n$ **then goto** *onew*;
            $y := h[na, na]$; $w := h[n, na] \times h[na, n]$;
            if $l = na$ **then goto** *twow*;
            if $its = 30$ **then goto** *fail*;
            if $its = 10 \; \vee \; its = 20$ **then**

begin comment form exceptional shift;
$\quad t := t + x$;
$\quad$ **for** $i := 1$ **step** 1 **until** n **do** $h[i, i] := h[i, i] - x$;
$\quad s := abs\,(h[n, na]) + abs\,(h[na, n-2])$;
$\quad x := y := 0.75 \times s$; $w := -0.4375 \times s\uparrow 2$
end;
$its := its + 1$;

comment look for two consecutive small sub-diagonal
$\qquad\qquad$ elements;
for $m := n - 2$ **step** -1 **until** l **do**
begin
$\quad z := h[m, m]$; $r := x - z$; $s := y - z$;
$\quad p := (r \times s - w)/h[m+1, m] + h[m, m+1]$;
$\quad q := h[m+1, m+1] - z - r - s$;
$\quad r := h[m+2, m+1])$;
$\quad s := abs\,(p) + abs\,(q) + abs\,(r)$;
$\quad p := p/s$; $q := q/s$; $r := r/s$;
$\quad$ **if** $m = l$ **then goto** $cont2$;
$\quad$ **if** $abs\,(h[m, m-1]) \times (abs\,(q) + abs\,(r)) \leqq$
$\qquad macheps \times abs\,(p) \times (abs\,(h[m-1, m-1]) + abs\,(z)$
$\qquad + abs\,(h[m+1, m+1]))$ **then goto** $cont2$
end m;

$cont2$:$\qquad$ **for** $i := m + 2$ **step** 1 **until** n **do** $h[i, i-2] := 0$;
$\qquad\qquad$ **for** $i := m + 3$ **step** 1 **until** n **do** $h[i, i-3] := 0$;
$\qquad\qquad$ **comment** double QR step involving rows l to n and
$\qquad\qquad\qquad\qquad$ columns m to n;
$\qquad\qquad$ **for** $k := m$ **step** 1 **until** na **do**
$\qquad\qquad$ **begin** $notlast := k \neq na$;
$\qquad\qquad\qquad$ **if** $k \neq m$ **then**
$\qquad\qquad\qquad$ **begin** $p := h[k, k-1]$; $q := h[k+1, k-1]$;
$\qquad\qquad\qquad\qquad r :=$ **if** $notlast$ **then** $h[k+2, k-1]$ **else** 0;
$\qquad\qquad\qquad\qquad x := abs\,(p) + abs\,(q) + abs\,(r)$;
$\qquad\qquad\qquad\qquad$ **if** $x = 0$ **then goto** $cont3$;
$\qquad\qquad\qquad\qquad p := p/x$; $q := q/x$; $r := r/x$
$\qquad\qquad\qquad$ **end**;
$\qquad\qquad\qquad s := sqrt\,(p\uparrow 2 + q\uparrow 2 + r\uparrow 2)$;
$\qquad\qquad\qquad$ **if** $p < 0$ **then** $s := -s$;
$\qquad\qquad\qquad$ **if** $k \neq m$ **then** $h[k, k-1] := -s \times x$
$\qquad\qquad\qquad$ **else if** $l \neq m$ **then** $h[k, k-1] := -h[k, k-1]$;
$\qquad\qquad\qquad p := p + s$; $x := p/s$; $y := q/s$; $z := r/s$;
$\qquad\qquad\qquad q := q/p$; $r := r/p$;
$\qquad\qquad\qquad$ **comment** row modification;
$\qquad\qquad\qquad$ **for** $j := k$ **step** 1 **until** n **do**
$\qquad\qquad\qquad$ **begin** $p := h[k, j] + q \times h[k+1, j]$;
$\qquad\qquad\qquad\qquad$ **if** $notlast$ **then**

$$\textbf{begin } p := p + r \times h[k+2, j];$$
$$h[k+2, j] := h[k+2, j] - p \times z$$
$$\textbf{end};$$
$$h[k+1, j] := h[k+1, j] - p \times y;$$
$$h[k, j] := h[k, j] - p \times x$$
$$\textbf{end } j;$$
$$j := \textbf{if } k+3 < n \textbf{ then } k+3 \textbf{ else } n;$$
$$\textbf{comment column modification};$$
$$\textbf{for } i := l \textbf{ step } 1 \textbf{ until } j \textbf{ do}$$
$$\textbf{begin } p := x \times h[i, k] + y \times h[i, k+1];$$
$$\textbf{if } notlast \textbf{ then}$$
$$\textbf{begin } p := p + z \times h[i, k+2];$$
$$h[i, k+2] := h[i, k+2] - p \times r$$
$$\textbf{end};$$
$$h[i, k+1] := h[i, k+1] - p \times q;$$
$$h[i, k] := h[i, k] - p$$
$$\textbf{end } i;$$

cont3:

$$\textbf{end } k;$$
$$\textbf{goto } nextit;$$
$$\textbf{comment one root found};$$

onew: $\quad wr[n] := x + t; \ wi[n] := 0; \ cnt[n] := its;$
$$n := na; \ \textbf{goto } nextw;$$
$$\textbf{comment two roots found};$$

twow: $\quad p := (y - x)/2; \ q := p{\uparrow}2 + w; \ y := sqrt(abs(q));$
$$cnt[n] := -its; \ cnt[na] := its; \ x := x + t;$$
$$\textbf{if } q > 0 \textbf{ then}$$
$$\textbf{begin comment real pair};$$
$$\textbf{if } p < 0 \textbf{ then } y := -y;$$
$$y := p + y; \ wr[na] := x + y; \ wr[n] := x - w/y;$$
$$wi[na] := wi[n] := 0$$
$$\textbf{end}$$
$$\textbf{else}$$
$$\textbf{begin comment complex pair};$$
$$wr[na] := wr[n] := x + p; \ wi[na] := y;$$
$$wi[n] := -y$$
$$\textbf{end};$$
$$n := n - 2; \ \textbf{goto } nextw;$$

fin:
$$\textbf{end } hqr;$$

5. Organisational and Notational Details

The Householder transformation at each stage is determined by three elements p_s, q_s, r_s. For the preliminary step these elements are derived using Eqs. (11); in all subsequent steps p_s, q_s, r_s are the elements $(s, s-1), (s+1, s-1), (s+2, s-1)$ of the current matrix. In order to avoid unnecessary underflow and overflow the p_s, q_s and r_s are normalised by dividing them by $|p_s|+|q_s|+|r_s|$. (The transformation depends only on the ratios of p_s, q_s and r_s.)

The number of arithmetic operations in the transformation with the matrix $I-2w_s w_s^T$ is further reduced by expressing $2w_s w_s^T$ in the form $u_s v_s^T$, where u_s and v_s are both parallel to w_s and their non-zero elements are given respectively by

$$(p_s \pm \sigma_s)/(\pm \sigma_s), \ q_s/(\pm \sigma_s), \ r_s/(\pm \sigma_s) \tag{20}$$

and

$$1, \ q_s/(p_s \pm \sigma_s), \ r_s/(p_s \pm \sigma_s) \tag{21}$$

with

$$\sigma_s^2 = p_s^2 + q_s^2 + r_s^2. \tag{22}$$

Before each iteration the matrix is searched for negligible subdiagonal elements. If the last negligible element is in position $(l, l-1)$ then attention is focused on the submatrix in the rows and columns l to n. If none of the subdiagonals is negligible then l is taken to be unity. It is not easy to devise a perfectly satisfactory test for negligible elements; we have used the criterion

$$|a_{l, l-1}| \leq macheps \, (|a_{l-1, l-1}| + |a_{l, l}|) \tag{23}$$

which tests whether $a_{l, l-1}$ is negligible compared with the local diagonal elements.

After determining l the submatrix in rows l to n is examined to see if any two consecutive subdiagonal elements are small enough to make it possible to work with an even smaller submatrix. We wish to establish the criterion to be satisfied if we are to start at row m. To do this we compute the elements p_m, q_m, r_m defined by

$$\begin{aligned}
p_m &= a_{mm}^2 - a_{mm}(k_1 + k_2) + k_1 k_2 + a_{m, m+1} a_{m+1, m} \\
q_m &= a_{m+1, m}(a_{mm} + a_{m+1, m+1} - k_1 - k_2) \\
r_m &= a_{m+2, m+1} a_{m+1, m}
\end{aligned} \tag{24}$$

and from them determine the appropriate elementary Hermitian matrix P_m which has a w_m with the usual three non-zero elements. If this is applied to the full matrix it will generate non-zero elements in positions $(m+1, m-1)$, $(m+2, m-1)$. The required condition is that these elements should be negligible. To obtain a satisfactory criterion which is not unnecessarily complicated we merely use approximations for these elements which are valid in the case when they will turn out to be negligible. This leads to the criterion

$$|a_{m, m-1}| \, (|q_m| + |r_m|)/|p_m| \leq macheps \, (|a_{m+1, m+1}| + |a_{m, m}| + |a_{m-1, m-1}|), \tag{25}$$

where we have effectively tested that the elements which would appear in positions $(m+1, m-1)$, $(m+2, m-1)$ are negligible compared with the three local diagonal elements $h_{m+1, m+1}$, $h_{m, m}$ and $h_{m-1, m-1}$. Condition (25) is used in the form

$$|a_{m, m-1}| \, (|q_m| + |r_m|) \leq macheps \, |p_m| \, (|a_{m+1, m+1}| + |a_{m, m}| + |a_{m-1, m-1}|). \tag{26}$$

In practice we take m to be the largest integer $(\geq l)$ for which condition (26) is satisfied. The double-step Francis iteration then uses only transformations with matrices P_m, P_{m+1}, ..., P_{n-1} and these are applied only to the submatrix in rows and columns l to n.

6. Discussion of Numerical Properties

The QR algorithm ultimately reduces the Hessenberg matrix A_1 to a Hessenberg matrix A_s for which $a^{(s)}_{i+1,i}\, a^{(s)}_{i+2,i+1} = 0$ $(i = 1, \ldots, n-2)$. The computed matrix A_s is always exactly orthogonally similar to a matrix $A_1 + E$ where

$$\|E\|_F \leq K \; macheps \; n\, p \, \|A_1\|_F. \tag{27}$$

Here $\|\cdot\|_F$ denotes the Frobenius norm, p is the total number of iterations and K is a constant of order unity. In practice p very rarely exceeds $2n$ (the average number of iterations per eigenvalue for all matrices solved by QR at the National Physical Laboratory is about 1.7). The right hand side of (27) is therefore of the order of magnitude of $macheps \; n^2 \|A_1\|_F$. In practice the statistical distribution of errors seems to ensure that $macheps \; n \|A_1\|_F$ is rarely exceeded. The accuracy of individual eigenvalues depends, in general, as it must, on their inherent sensitivity to perturbations in A_1 but it is clear that the QR algorithm is a very stable method.

7. Test Results

The algorithm has been used extensively at the National Physical Laboratory over the last nine years. (It was communicated to J. H. Wilkinson privately by J. G. Francis long before publication. Although the procedure has been progressively refined over this period the changes are not important as far as the majority of matrices are concerned.) It has proved to be the most effective of the procedures we have used for general matrices. We give now the results obtained on KDF 9 (a machine using a 39 binary digit mantissa) for a number of simple matrices designed to test various features of the procedure.

To test the *ad hoc* device and to deal with cycling, the matrix

$$\begin{bmatrix} 0 & 0 & 0 & 0 & 0 & 0 & 0 & 1 \\ 1 & 0 & 0 & 0 & 0 & 0 & 0 & 0 \\ 0 & 1 & 0 & 0 & 0 & 0 & 0 & 0 \\ 0 & 0 & 1 & 0 & 0 & 0 & 0 & 0 \\ 0 & 0 & 0 & 1 & 0 & 0 & 0 & 0 \\ 0 & 0 & 0 & 0 & 1 & 0 & 0 & 0 \\ 0 & 0 & 0 & 0 & 0 & 1 & 0 & 0 \\ 0 & 0 & 0 & 0 & 0 & 0 & 1 & 0 \end{bmatrix}$$

was solved. The first ten iterations leave the matrix unchanged. The exceptional shifts are then used for one iteration and all eight eigenvalues were found in 14 more iterations. A wide variety of exceptional shifts were tried and this showed that virtually any "reasonable" values broke up the cycling. The computed eigenvalues are given in Table 1 together with the number of iterations (NB., an entry of $-k$ preceded by an entry of k shows that k iterations were required to find a pair of eigenvalues.)

Table 1

Eigenvalues		
Real parts	Imaginary parts	Iterations
$-1.0000\,0000\,001_{10} + 0;$	$+0.0000\,0000\,000\qquad;$	0
$-7.0710\,6781\,190_{10} - 1;$	$+7.0710\,6781\,187_{10} - 1;$	3
$-7.0710\,6781\,190_{10} - 1;$	$-7.0710\,6781\,187_{10} - 1;$	-3
$-5.4569\,6821\,064_{10} - 12;$	$+1.0000\,0000\,000_{10} + 0;$	2
$-5.4569\,6821\,064_{10} - 12;$	$-1.0000\,0000\,000_{10} + 0;$	-2
$+9.9999\,9999\,998_{10} - 1;$	$+0.0000\,0000\,000\qquad;$	2
$+7.0710\,6781\,192_{10} - 1;$	$+7.0710\,6781\,185_{10} - 1;$	17
$+7.0710\,6781\,192_{10} - 1;$	$-7.0710\,6781\,185_{10} - 1;$	-17
	Total	24

To test the use of criterion (26) the matrix

$$\begin{bmatrix}
3 & 2 & 1 & 2 & 1 & 4 & 1 & 2 \\
2 & 1 & 3 & 1 & 2 & 2 & 1 & 4 \\
 & 3 & 1 & 2 & 1 & 2 & 1 & 3 \\
 & & 1 & 1 & 2 & 1 & 3 & 1 \\
 & & & 10^{-7} & 3 & 1 & 4 & 2 \\
 & & & & 10^{-6} & 2 & 1 & 4 \\
 & & & & & 1 & 2 & 3 \\
 & & & & & & 3 & 2
\end{bmatrix}$$

was solved. The elements (5, 4) and (6, 5) are both "fairly" small and iteration did, in fact, start at row 5. We would expect that four of the eigenvalues of the leading principal submatrix of order five would agree with four of the eigenvalues of the full matrix to working accuracy; a similar remark applies to three of the eigenvalues of the final principal submatrix of order four. The computed eigenvalues of the full matrix and the two relevant submatrices are given in Table 2 and they confirm these observations.

The third test matrix

$$\begin{bmatrix}
1 & 2 & 3 & 4 \\
10^{-3} & 3\cdot 10^{-3} & 2\cdot 10^{-3} & 10^{-3} \\
 & 10^{-6} & 3\cdot 10^{-6} & 2\cdot 10^{-6} \\
 & & 10^{-9} & 2\cdot 10^{-9}
\end{bmatrix}$$

has rows which have rapidly decreasing norms. It was solved twice, once in the given form and once after applying the balancing procedure described by Parlett and Reinsch [7]. The balanced matrix produced by that procedure was

$$\begin{bmatrix}
1_{10}0 & 6.25_{10}-2 & 3.6621\,0937\,500_{10}-4 & 9.5367\,4316\,406_{10}-7 \\
3.2_{10}-2 & 3.0_{10}-3 & 7.8125_{10}-6 & 7.6293\,9453\,126_{10}-9 \\
 & 2.56_{10}-4 & 3.0_{10}-6 & 3.90625_{10}-9 \\
 & & 5.12_{10}-7 & 2.0_{10}-9
\end{bmatrix}$$

Table 2

Eigenvalues of full matrix	Iterations	Eigenvalues of 5×5 matrix	Iterations	Eigenvalues of 4×4 matrix	Iterations
$+5.6181\,8555\,238_{10} +0$;	0	$+5.6181\,8555\,240_{10} +0$;	0	$+2.9999\,9990\,475_{10} +0$;	0
$-2.4295\,1247\,406_{10} +0$;	0	$-2.4295\,1247\,407_{10} +0$;	0	$+3.5424\,7302\,818_{10} -1$;	0
$+1.9449\,5965\,031_{10} +0$;	3	$+1.9449\,5965\,029_{10} +0$;	3	$+1.1666\,5894\,308_{10} -6$;	0
$+8.6636\,7190\,136_{10} -1$;	-3	$+8.6636\,7190\,130_{10} -1$;	-3	$+5.6457\,5162\,584_{10} +0$;	7
$+2.9999\,9998\,601_{10} +0$;	0	$+3.0000\,0008\,124_{10} +0$;	1		Total 7
$+3.5424\,7302\,818_{10} -1$;	1		Total 4		
$+1.1666\,5894\,308_{10} -6$;	0				
$+5.6457\,5162\,584_{10} +0$;	7				
	Total 11				

Table 3

Eigenvalues (unbalanced)	Iterations	Eigenvalues (balanced)	Iterations
$+1.0020\,0200\,098_{10} +0$;	0	$+1.0020\,0200\,099_{10} +0$;	0
$+9.9699\,4963\,657_{10} -4$;	0	$+9.9699\,4963\,685_{10} -4$;	0
$+4.0052\,9353\,527_{10} -6$;	2	$+4.0052\,9353\,525_{10} -6$;	2
$+7.4976\,5325\,656_{10} -10$;	-2	$+7.4976\,5324\,706_{10} -10$;	-2
	Total 2		Total 2

The results for the two matrices are given in Table 3. The first incorrect figure is underlined. It will be seen that balancing is of little relevance for a matrix of this kind though the splitting criterion is important.

The final matrix was chosen to illustrate the performance on an ill conditioned eigenproblem. It is the matrix of order 13 given by

$$\begin{bmatrix}
13 & 12 & 11 & \ldots & 3 & 2 & 1 \\
12 & 12 & 11 & \ldots & 3 & 2 & 1 \\
   & 11 & 11 & \ldots & 3 & 2 & 1 \\
   &   & . & \ldots & 3 & 2 & 1 \\
   &   &   & \ldots & . & . & . \\
   &   &   &   & 2 & 2 & 1 \\
   &   &   &   &   & 1 & 1
\end{bmatrix}.$$

The eigenvalues are such that $\lambda_i = 1/(\lambda_{14-i})$. The larger eigenvalues are very well conditioned and the smaller ones very ill-conditioned. The results are given in Table 4 and the first incorrect figure is underlined. The errors are such as can be caused by changes of one unit in the last digits (i.e. the 39th significant binary digit) of the data.

Table 4

Eigenvalues	Iterations
$+ 3.5613\,8612\,027_{10} + 1;$	0
$+ 2.3037\,5322\,906_{10} + 1;$	2
$+ 1.4629\,7821\,337_{10} + 1;$	-2
$+ 8.7537\,7956\,658_{10} + 0;$	0
$+ 4.7673\,0653\,278_{10} + 0;$	2
$+ 2.2989\,1123\,271_{10} + 0;$	1
$+ 9.9999\,9999\,418_{10} - 1;$	-1
$+ 2.7734\,3688\,695_{10} - 2;$	1
$+ 4.4163\,1575\,355_{10} - 2;$	-1
$+ 6.7831\,7802\,945_{10} - 2;$	2
$+ 1.1435\,2899\,315_{10} - 1;$	3
$+ 2.0975\,6535\,265_{10} - 1;$	2
$+ 4.3498\,8302\,529_{10} - 1;$	6
Total	19

Other examples have been given in [5] and these illustrate the use of *hqr* in connexion with procedures for reduction to Hessenberg form.

The results obtained with the examples given here have been confirmed by Mrs. E. Mann at the Technische Hochschule Munich.

Acknowledgements. The authors wish to thank Dr. C. Reinsch of the Technische Hochschule München for many helpful suggestions and for testing several versions of the procedure. They also wish to thank Mrs. E. Mann of the Technische Hochschule Munich for confirming the test results. The work described above has been carried out at the National Physical Laboratory.

References

1. Francis, J. C. F.: The QR transformation — a unitary analogue to the LR transformation. Comput. J. 4, 265—271 and 332—345 (1961/62).
2. Householder, A. S.: Unitary triangularization of a non-symmetric matrix. J. Assoc. Comput. Mach. 5, 339—342 (1958).
3. — Bauer, F. L.: On certain methods for expanding the characteristic polynomial. Numer. Math. 1, 29—37 (1959).
4. Kublanovskaya, V. N.: On some algorithms for the solution of the complete eigenvalue problem. Ž. Vyčisl. Mat. i Mat. Fiz. 1, 555—570 (1961).
5. Martin, R. S., Wilkinson, J. H.: Similarity reduction of a general matrix to Hessenberg form. Numer. Math. 12, 349—368 (1968). Cf. II/13.
6. Parlett, B. N.: Global convergence of the basic QR algorithm on Hessenberg matrices. Math. Comp. 22, 803—817 (1968).
7. — Reinsch, C.: Balancing a matrix for calculation of eigenvalues and eigenvectors. Numer. Math. 13, 293—304 (1969). Cf. II/11.
8. Wilkinson, J. H.: The algebraic eigenvalue problem. London: Oxford University Press 1965.

Eigenvectors of Real and Complex Matrices
by LR and QR triangularizations[*]

by G. Peters and J. H. Wilkinson

1. Theoretical Background

In a recent paper [4] the triangularization of complex Hessenberg matrices using the LR algorithm was described. Denoting the Hessenberg matrix by H and the final triangular matrix by T we have

$$P^{-1}HP = T, \tag{1}$$

where P is the product of all the transformation matrices used in the execution of the LR algorithm. In practice H will almost invariably have been derived from a general complex matrix A using the procedure *comhes* [3] and hence for some nonsingular S we have

$$P^{-1}S^{-1}ASP = T. \tag{2}$$

The eigenvalues of A lie along the diagonal of T and we may write $t_{ii} = \lambda_i$; the ordering of the λ_i is, in general, of an arbitrary nature. The eigenvector x_i of T corresponding to λ_i is given by

$$
\begin{aligned}
x_{ji} &= 0 \quad (j = i+1, \ldots, n) \\
x_{ii} &= 1 \\
x_{ji} &= -\left(\sum_{k=j+1}^{i} t_{jk} x_{ki}\right) \bigg/ (\lambda_j - \lambda_i) \quad (j = i-1, \ldots, 1)
\end{aligned}
\tag{3}
$$

and the corresponding eigenvector of A is SPx_i. If the product SP of all the transformations is retained this provides a method of computing the eigenvectors of A. This technique was first used on the computer ACE at the National Physical Laboratory and provides the most efficient program known to us *when the complete eigensystem is required*.

Analogous procedures may be used in connexion with the Francis QR algorithm for real Hessenberg matrices [2, 8]. A program based on this algorithm was also used on the computer, ACE and a sophisticated version has been produced by

---

[*] Prepublished in Numer. Math. **16**, 181—204 (1970).

Parlett; more recently Dekker has produced an ALGOL procedure [1] for the case when the eigenvalues are real and plans one for the complex case.

The final matrix produced by the QR algorithm given in [5] is not, in general strictly triangular but may contain 2×2 blocks on the diagonal. Each 2×2 block corresponds either to a real pair of eigenvalues or a complex conjugate pair. For a 2×2 block in rows and columns r and $r+1$ corresponding to a real pair, true triangular form can be established by a similarity transformation involving a real rotation in the $(r, r+1)$ plane. True triangular form cannot be established for 2×2 blocks corresponding to complex conjugate eigenvalues without moving into the complex field.

However, the eigenvectors can still be determined by a simple back substitution process. For a real eigenvalue λ_i the components of the vector x_i are given by the Eqs. (3) except that corresponding to a 2×2 block in row j and $j-1$ the two components x_{ji} and $x_{j-1, i}$ are given by

$$
(t_{j-1, j-1} - \lambda_i) x_{j-1, i} + t_{j-1, j} x_{ji} = - \sum_{k=j+1}^{i} t_{j-1, k} x_{ki},
$$

$$
t_{j, j-1} x_{j-1, i} + (t_{jj} - \lambda_i) x_{ji} = - \sum_{k=j+1}^{i} t_{jk} x_{ki}.
$$

(4)

This is a 2×2 system of equations with real coefficients and right hand side.

For a complex conjugate pair we need determine only the vector corresponding to the eigenvalue with positive imaginary part, since the other vector is merely the complex conjugate. Taking x_{ii} to be $1+0i$ the component $x_{i-1, i}$ satisfies the equations

$$
(t_{i-1, i-1} - \lambda_i) x_{i-1, i} + t_{i-1, i} = 0
$$

$$
t_{i, i-1} x_{i-1, i} + (t_{ii} - \lambda_i) = 0
$$

(5)

and may be determined using the equation with the "larger" coefficients (for the more stable determination). The remaining components are given by Eqs. (3) and (4) though it should be appreciated that the diagonal coefficients are now complex.

Again in practice the real Hessenberg form H will usually have been derived from a full matrix, A, using one of the three procedures *elmhes*, *dirhes* or *orthes*. If the product S of the transformations involved in the reduction to Hessenberg form is derived before starting the QR algorithm, this may then be multiplied by the matrices involved in the QR quasi-triangularization. The eigenvectors of A can then be found from those for the quasi-triangular matrix.

2. Applicability

comlr2 may be used to find *all* the eigenvalues *and* eigenvectors of a complex matrix A which has been reduced to Hessenberg form H by procedure *comhes*. When only a few selected eigenvectors are required it is more efficient to use *comlr* [4] and complex inverse iteration with the matrix H [9].

hqr2 may be used to find *all* the eigenvalues *and* eigenvectors (real and/or complex) of a real matrix A which has been reduced to real Hessenberg form H

by procedures *elmhes, dirhes* or *orthes*. Before entering *hqr2* the product of the matrices reducing A to H must be determined using *elmtrans, dirtrans* or *ortrans* respectively. If the latter is used then at the point in the procedure *hqr2* when the QR reduction is complete, the orthogonal matrix X which reduces A to real quasi-triangular form is available. This matrix is sometimes a requirement in its own right and the procedure may be modified in the obvious way to provide X as an output.

The recommended program for dealing with a general complex matrix A is first to balance it using a complex variant of the procedure *balance* [6], then to reduce it to a complex Hessenberg matrix H by procedure *comhes* [3] and then use *comlr2* to find the eigenvalues and eigenvectors of the balanced A. The eigenvectors of the original matrix A can then be recovered by the complex analogue of procedure *balbak* [6].

The recommended program for dealing with a general real matrix A is first to balance it using procedure *balance* [6], then to reduce it to real Hessenberg form by procedures *elmhes, dirhes* or *orthes* [3]. The transformation matrix is then derived using *elmtrans, dirtrans* or *ortrans* and the eigenvalues and eigenvectors of the balanced A are found by entering *hqr2*. The eigenvectors of the original A can then be recovered by using *balbak* [6].

3. Formal Parameter List

3.1. Input to procedure *comlr2*

n the order of the complex Hessenberg matrix H.

low, upp   two integers produced by the complex analogue of *balance* [6]. If balancing has not been used then set $low = 1$, $upp = n$. The Hessenberg matrix is effectively upper-triangular as far as its rows 1 to $low - 1$ and rows $upp + 1$ to n are concerned, so that $low - 1 + n - upp$ eigenvalues are already isolated.

macheps the smallest number on the computer for which $1 + macheps > 1$.

int   an $n \times 1$ integer array produced by *comhes*. If the primary matrix is already of Hessenberg form then set $int[i] = i$.

hr, hi   two $n \times n$ arrays, *hr* giving the matrix of real parts of H and *hi* the matrix of imaginary parts stored in the relevant locations. The remaining locations of *hr* and *hi* store the real and imaginary parts of the multipliers used in *comhes*. The latter procedure provides an *hr* and *hi* in the required form. If the primary matrix is already of Hessenberg form the remaining locations of *hr* and *hi* must be set to zero.

Output from procedure *comlr2*

hr, hi   the strict upper triangles of these two $n \times n$ arrays contain the components of the eigenvectors of the triangular matrix produced by the LR algorithm.

wr, wi   two $n \times 1$ arrays giving the real and imaginary parts of the eigenvalues.

vr, vi	two $n \times n$ arrays giving the real and imaginary parts of un-normalized eigenvectors of the original full matrix used as input to *comhes*.
fail	exit used if more than 30 iterations are used to find any one eigenvalue.

3.2. Input to procedure *elmtrans*

n	order of the Hessenberg matrix H.
low, upp	integers produced by procedure *balance* if used. Otherwise $low = 1$, $upp = n$.
int	an $n \times 1$ integer array produced by *elmhes*
h	an $n \times n$ array produced by *elmhes* containing the Hessenberg matrix H and the multipliers used in producing it from the general matrix A.

Output from procedure *elmtrans*

v	the $n \times n$ array defining the similarity reduction from A to H.

Input and output parameters for *dirtrans* are analogous to those for *elmtrans* though the procedures are not identical.

Input to procedure *ortrans*

n	order of the Hessenberg matrix H.
low, upp	integers produced by procedure *balance* if used. Otherwise $low = 1$, $upp = n$.
d	an $n \times 1$ array produced by *orthes* giving information on the orthogonal reduction from A to H.
h	an $n \times n$ array produced by *orthes* giving the upper Hessenberg matrix H and the remaining information on the orthogonal reduction.

Output from procedure *ortrans*

d	the input array is destroyed by the procedure
v	an $n \times n$ array defining the similarity reduction from A to H.

3.3. Input to procedure *hqr2*

n	order of the Hessenberg matrix H.
low, upp	integers produced by *balance* if used. Otherwise $low = 1$, $upp = n$.
macheps	the smallest number on the computer for which $1 + macheps > 1$.
h	an $n \times n$ array containing the matrix H in the relevant part.
vecs	an $n \times n$ array containing the matrix defining the similarity transformation from A to H. (This is produced by *elmtrans*, *dirtrans* or *ortrans*.) If H is the primary matrix then *vecs* must be set to the identity matrix.

Output from procedure *hqr2*

h	the upper part of this $n \times n$ array contains the eigenvectors of the quasi-triangular matrix produced by the QR algorithm.

wr, wi two $n \times 1$ arrays giving the real and imaginary parts of the eigenvalues.

int an $n \times 1$ integer array giving the number of iterations for the eigenvalues. If two eigenvalues are found simultaneously as a pair, then the number of iterations is given with a positive sign for the first and a negative sign for the second.

vecs an $n \times n$ array giving the unnormalized eigenvectors of the original full matrix A (unless H was the primary matrix). If the i-th eigenvalue is real the i-th column of *vecs* is the corresponding real eigenvector. If eigenvalues i and $i+1$ are a complex pair the columns i and $i+1$ give the real and imaginary part of the eigenvector corresponding to the eigenvalue with positive imaginary part.

fail exit used when 30 iterations fail to isolate the current eigenvalue.

4. ALGOL Programs

procedure *comlr2* $(n, low, upp, macheps)$ *data*: (int) *trans*: (hr, hi)
    result: (wr, wi, vr, vi) *exit*: $(fail)$;

value $n, low, upp, macheps$;

integer n, low, upp;

real *macheps*;

array hr, hi, wr, wi, vr, vi;

integer array *int*;

label *fail*;

comment Finds the eigenvalues and eigenvectors of a complex matrix which has been reduced by procedure *comhes* to upper Hessenberg form, H, stored in the arrays $hr[1:n, 1:n]$ and $hi[1:n, 1:n]$. The real and imaginary parts of the eigenvalues are formed in the arrays $wr[1:n]$ and $wi[1:n]$ respectively and the un-normalised eigenvectors are formed as columns of the arrays $vr[1:n, 1:n]$ and $vi[1:n, 1:n]$. *low* and *upp* are two integers produced in balancing where eigenvalues are isolated in positions 1 to $low-1$ and $upp+1$ to n. If balancing is not used $low=1$, $upp=n$. *macheps* is the relative machine precision. The procedure fails if any eigenvalue takes more than 30 iterations. The procedures *cdiv* and *csqrt* are used, see [3];

begin
    integer i, j, k, m, its, en;
    real $sr, si, tr, ti, xr, xi, yr, yi, zr, zi, norm$;
    $tr := ti := 0$;
    for $i := 1$ **step** 1 **until** n **do**
    begin
        for $j := 1$ **step** 1 **until** n **do** $vr[i, j] := vi[i, j] := 0$;
        $vr[i, i] := 1$
    end i;

```
for i := upp − 1 step −1 until low + 1 do
begin
    j := int [i];
    for k: = i + 1 step 1 until upp do
    begin
        vr [k, i] := hr [k, i − 1]; vi [k, i] := hi [k, i − 1]
    end k;
    if i ≠ j then
    begin
        for k := i step 1 until upp do
        begin
            vr [i, k] := vr [j, k]; vi [i, k] := vi [j, k];
            vr [j, k] := vi [j, k] := 0
        end k;
        vr [j, i] := 1
    end i ≠ j
end i;
for i := 1 step 1 until low − 1, upp + 1 step 1 until n do
begin
    wr [i] := hr [i, i]; wi [i] := hi [i, i]
end isolated roots;
en := upp;
nextw:
    if en < low then goto fin;
    its := 0;
    comment look for single small sub-diagonal element;
nextit:
    for k := en step −1 until low + 1 do
    if abs (hr [k, k − 1]) + abs (hi [k, k − 1]) ≦ macheps × (abs (hr [k − 1, k − 1]) +
            abs (hi [k − 1, k − 1]) + abs (hr [k, k]) + abs (hi [k, k])) then goto cont1;
    k := low;
cont1:
    if k = en then goto root;
    if its = 30 then goto fail;
    comment form shift;
    if its = 10 ∨ its = 20 then
    begin
        sr := abs (hr [en, en − 1]) + abs (hr [en − 1, en − 2]);
        si := abs (hi [en, en − 1]) + abs (hi [en − 1, en − 2])
    end
    else
    begin
        sr := hr [en, en]; si := hi [en, en];
        xr := hr [en − 1, en] × hr [en, en − 1] − hi [en − 1, en] × hi [en, en − 1];
        xi := hr [en − 1, en] × hi [en, en − 1] + hi [en − 1, en] × hr [en, en − 1];
        if xr ≠ 0 ∨ xi ≠ 0 then
```

```
    begin
        yr := (hr [en − 1, en − 1] − sr)/2;  yi := (hi [en − 1, en − 1] − si)/2;
        csqrt (yr↑2 − yi↑2 + xr, 2 × yr × yi + xi, zr, zi);
        if yr × zr + yi × zi < 0 then
        begin
            zr := − zr;  zi := − zi
        end;
        cdiv (xr, xi, yr + zr, yi + zi, xr, xi);
        sr := sr − xr;  si := si − xi
    end
end;
for i := low step 1 until en do
begin
    hr [i, i] := hr [i, i] − sr;  hi [i, i] := hi [i, i] − si
end i;
tr := tr + sr;  ti := ti + si;  its := its + 1;  j := k + 1;
comment look for two consecutive small sub-diagonal elements;
xr := abs (hr [en − 1, en − 1]) + abs (hi [en − 1, en − 1]);
yr := abs (hr [en, en − 1]) + abs (hi [en, en − 1]);
zr := abs (hr [en, en]) + abs (hi [en, en]);
for m := en − 1 step − 1 until j do
begin
    yi := yr;  yr := abs (hr [m, m − 1]) + abs (hi [m, m − 1]);
    xi := zr;  zr := xr;
    xr := abs (hr [m − 1, m − 1]) + abs (hi [m − 1, m − 1]);
    if yr ≤ macheps × zr/yi × (zr + xr + xi) then goto cont2
end m;
m := k;
comment triangular decomposition H = L × R;
cont2:
    for i := m + 1 step 1 until en do
    begin
        xr := hr [i − 1, i − 1];  xi := hi [i − 1, i − 1];
        yr := hr [i, i − 1];  yi := hi [i, i − 1];
        if abs (xr) + abs (xi) < abs (yr) + abs (yi) then
        begin
            comment interchange rows of hr and hi;
            for j := i − 1 step 1 until n do
            begin
                zr := hr [i − 1, j];  hr [i − 1, j] := hr [i, j];  hr [i, j] := zr;
                zi := hi [i − 1, j];  hi [i − 1, j] := hi [i, j];  hi [i, j] := zi
            end j;
            cdiv (xr, xi, yr, yi, zr, zi);  wr [i] := 1
        end
        else
```

```
      begin
          cdiv (yr, yi, xr, xi, zr, zi); wr [i] := — 1
      end;
      hr [i, i — 1] := zr; hi [i, i — 1] := zi;
      for j := i step 1 until n do
      begin
          hr [i, j] := hr [i, j] — zr × hr [i — 1, j] + zi × hi [i — 1, j];
          hi [i, j] := hi [i, j] — zr × hi [i — 1, j] — zi × hr [i — 1, j]
      end j
  end i;
  comment composition R × L = H;
  for j := m + 1 step 1 until en do
  begin
      xr := hr [j, j — 1]; xi := hi [j, j — 1]; hr [j, j — 1] := hi [j, j — 1] := 0;
      comment interchange columns of hr, hi, vr, and vi, if necessary;
      if wr [j] > 0 then
      begin
          for i := 1 step 1 until j do
          begin
              zr := hr [i, j — 1]; hr [i, j — 1] := hr [i, j]; hr [i, j] := zr;
              zi := hi [i, j — 1]; hi [i, j — 1] := hi [i, j]; hi [i, j] := zi
          end;
          for i := low step 1 until upp do
          begin
              zr := vr [i, j — 1]; vr [i, j — 1] := vr [i, j]; vr [i, j] := zr;
              zi := vi [i, j — 1]; vi [i, j — 1] := vi [i, j]; vi [i, j] := zi
          end
      end interchange columns;
      for i := 1 step 1 until j do
      begin
          hr [i, j — 1] := hr [i, j — 1] + xr × hr [i, j] — xi × hi [i, j];
          hi [i, j — 1] := hi [i, j — 1] + xr × hi [i, j] + xi × hr [i, j]
      end i;
      for i := low step 1 until upp do
      begin
          vr [i, j — 1] := vr [i, j — 1] + xr × vr [i, j] — xi × vi [i, j];
          vi [i, j — 1] := vi [i, j — 1] + xr × vi [i, j] + xi × vr [i, j]
      end accumulate transformations
  end j;
  goto nextit;
  comment a root found;
root:
  wr [en] := hr [en, en] + tr; wi [en] := hi [en, en] + ti; en := en — 1;
  goto nextw;
```

```
    comment all roots found;
fin:
    norm := 0.0;
    for i := 1 step 1 until n do
    begin
        norm: = norm + abs (wr [i]) + abs (wi [i]);
        for j := i + 1 step 1 until n do
        norm := norm + abs (hr [i, j]) + abs (hi [i, j])
    end;
    comment Backsubstitute to find vectors of upper triangular form;
    for en := n step −1 until 2 do
    begin
        xr := wr [en]; xi := wi [en];
        for i := en −1 step −1 until 1 do
        begin
            zr := hr [i, en]; zi := hi [i, en];
            for j := i +1 step 1 until en −1 do
            begin
                zr := zr + hr [i, j] × hr [j, en] − hi [i, j] × hi [j, en];
                zi := zi + hr [i, j] × hi [j, en] + hi [i, j] × hr [j, en]
            end;
            yr := xr − wr [i]; yi := xi − wi [i];
            if yr = 0.0 ∧ yi = 0.0 then yr := macheps × norm;
            cdiv (zr, zi, yr, yi, hr [i, en], hi [i, en])
        end
    end backsub;

    comment multiply by transformation matrix to give vectors of original full
                matrix;
    for i := 1 step 1 until low − 1, upp + 1 step 1 until n do
    for j := i +1 step 1 until n do
    begin
        vr [i, j] := hr [i, j]; vi [i, j] := hi [i, j]
    end vectors of isolated roots;
    for j := n step −1 until low do
    for i := low step 1 until upp do
    begin
        zr := vr [i, j]; zi := vi [i, j];
        m := if upp < j then upp else j − 1;
        for k := low step 1 until m do
        begin
            zr := zr + vr [i, k] × hr [k, j] − vi [i, k] × hi [k, j];
            zi := zi + vr [i, k] × hi [k, j] + vi [i, k] × hr [k, j]
        end;
        vr [i, j] := zr; vi [i, j] := zi
    end
end comlr2;
```

```
procedure elmtrans (n, low, upp) data: (int, h) result: (v);
value n, low, upp;
integer n, low, upp;
integer array int;
array h, v;
comment form the matrix of accumulated transformations in the array v [1:n,
          1:n] from the information left by procedure elmhes below the upper
          Hessenberg matrix, H, in the array h [1:n, 1:n] and in the integer
          array int [1:n];
begin
   integer i, j, k;
   for i := 1 step 1 until n do
   begin
      for j := 1 step 1 until n do v [i, j] := 0.0;
      v [i, i] := 1.0
   end i;
   for i := upp −1 step −1 until low +1 do
   begin
      j := int [i];
      for k := i +1 step 1 until upp do v [k, i] := h [k, i −1];
      if i ≠ j then
      begin
         for k := i step 1 until upp do
         begin
            v [i, k] := v [j, k]; v [j, k] := 0.0
         end k;
         v [j, i] := 1.0
      end i ≠ j
   end i
end elmtrans;


procedure dirtrans (n, low, upp) data: (int, h) result: (v);
value n, low, upp;
integer n, low, upp;
integer array int;
array h, v;
comment form the matrix of accumulated transformations in the array v [1:n,
          1:n] from the information left by procedure dirhes below the upper
          Hessenberg matrix, H, in the array h [1:n, 1:n] and in the integer
          array int [1:n];
begin
   integer i, j, m; real x;
   for i := 1 step 1 until n do
   begin
      for j := 1 step 1 until n do v [i, j] := 0.0;
      v [i, i] := 1.0
   end i;
```

```
    for i := upp step −1 until low +1 do
    begin
        for j := low +1 step 1 until i −1 do v [i, j] := h [i, j −1];
        m := int [i];
        if m ≠ i then
        for j := low +1 step 1 until upp do
        begin
            x := v [m, j];  v [m, j] := v [i, j];  v [i, j] := x
        end m ≠ i
    end i
end dirtrans;


procedure ortrans (n, low, upp) data : (h) trans : (d) result : (v);
value n, low, upp;
integer n, low, upp;
array h, d, v;
comment form the matrix of accumulated transformations in the array v [1:n,
        1:n] from the information left by procedure orthes below the upper
        Hessenberg matrix, H, in the array h [1:n, 1:n] and in the array
        d [1:n]. The contents of the latter are destroyed;
begin
    integer i, j, k, m;  real x, y;
    for i := 1 step 1 until n do
    begin
        for j := 1 step 1 until n do v [i, j] := 0.0;
        v [i, i] := 1.0
    end i;
    for k := upp −2 step −1 until low do
    begin
        m := k +1;  y := h [m, k];
        if y = 0.0 then goto skip;
        y := y × d [m];
        for i := k +2 step 1 until upp do d [i] := h [i, k];
        for j := m step 1 until upp do
        begin
            x := 0.0;
            for i := m step 1 until upp do
            x := x + d [i] × v [i, j];
            x := x/y;
            for i := m step 1 until upp do
            v [i, j] := v [i, j] + x × d [i]
        end j;
skip:
    end k
end ortrans;
```

procedure $hqr2$ $(n, low, upp, macheps)$ $trans$: $(h, vecs)$ $result$: (wr, wi, cnt)
    $exit$: $(fail)$;

value $n, low, upp, macheps$;

integer n, low, upp ;

real $macheps$;

array $h, vecs, wr, wi$;

integer array cnt ;

label $fail$;

comment Finds the eigenvalues and eigenvectors of a real matrix which has
    been reduced to upper Hessenberg form in the array $h[1:n, 1:n]$ with
    the accumulated transformations stored in the array $vecs[1:n, 1:n]$.
    The real and imaginary parts of the eigenvalues are formed in the
    arrays $wr, wi[1:n]$ and the eigenvectors are formed in the array
    $vecs[1:n, 1:n]$ where only one complex vector, corresponding to the
    root with positive imaginary part, is formed for a complex pair. low
    and upp are two integers produced in balancing where eigenvalues
    are isolated in positions 1 to $low - 1$ and $upp + 1$ to n. If balancing
    is not used $low = 1$, $upp = n$. $macheps$ is the relative machine precision.
    The procedure fails if any eigenvalue takes more than 30 iterations.
    The non-local procedure $cdiv$ is used, see [3];

begin
  integer $i, j, k, l, m, na, its, en$;
  real $p, q, r, s, t, w, x, y, z, ra, sa, vr, vi, norm$;
  Boolean $notlast$;
  for $i := 1$ **step** 1 **until** $low - 1$, $upp + 1$ **step** 1 **until** n **do**
  begin
    $wr[i] := h[i, i]$; $wi[i] := 0.0$; $cnt[i] := 0$
  end $isolated$ $roots$;
  $en := upp$; $t := 0.0$;
$nextw$:
  if $en < low$ **then goto** fin ;
  $its := 0$; $na := en - 1$;
  comment look for single small sub-diagonal element ;
$nextit$:
  for $l := en$ **step** -1 **until** $low + 1$ **do**
  if $abs(h[l, l-1]) \leq macheps \times (abs(h[l-1, l-1]) + abs(h[l, l]))$ **then goto** $cont1$;
  $l := low$;
$cont1$:
  $x := h[en, en]$; **if** $l = en$ **then goto** $onew$;
  $y := h[na, na]$; $w := h[en, na] \times h[na, en]$; **if** $l = na$ **then goto** $twow$;
  if $its = 30$ **then**
  begin
    $cnt[en] := 31$; **goto** $fail$
  end ;
  if $its = 10 \vee its = 20$ **then**

begin comment form exceptional shift;
    $t := t + x$;
    for $i := low$ **step** 1 **until** en **do** $h[i, i] := h[i, i] - x$;
    $s := abs(h[en, na]) + abs(h[na, en - 2])$; $x := y := 0.75 \times s$;
    $w := -0.4375 \times s\uparrow 2$
end;
$its := its + 1$;

comment look for two consecutive small sub-diagonal elements;
for $m := en - 2$ **step** -1 **until** l **do**
begin
    $z := h[m, m]$; $r := x - z$; $s := y - z$;
    $p := (r \times s - w)/h[m+1, m] + h[m, m+1]$;
    $q := h[m+1, m+1] - z - r - s$; $r := h[m+2, m+1]$;
    $s := abs(p) + abs(q) + abs(r)$; $p := p/s$; $q := q/s$; $r := r/s$;
    if $m = l$ **then goto** *cont2*;
    if $abs(h[m, m-1]) \times (abs(q) + abs(r)) \leq$
        $macheps \times abs(p) \times (abs(h[m-1, m-1]) + abs(z) + abs(h[m+1, m+1]))$
    then goto *cont2*
end m;
cont2:
    for $i := m + 2$ **step** 1 **until** en **do** $h[i, i-2] := 0.0$;
    for $i := m + 3$ **step** 1 **until** en **do** $h[i, i-3] := 0.0$;
    comment double QR step involving rows l to en and columns m to en of the
            complete array;
    for $k := m$ **step** 1 **until** na **do**
    begin
        $notlast := k \neq na$;
        if $k \neq m$ **then**
        begin
            $p := h[k, k-1]$; $q := h[k+1, k-1]$;
            $r :=$ **if** $notlast$ **then** $h[k+2, k-1]$ **else** 0;
            $x := abs(p) + abs(q) + abs(r)$; **if** $x = 0$ **then goto** *cont3*;
            $p := p/x$; $q := q/x$; $r := r/x$
        end;
        $s := sqrt(p\uparrow 2 + q\uparrow 2 + r\uparrow 2)$;
        if $p < 0$ **then** $s := -s$;
        if $k \neq m$ **then** $h[k, k-1] := -s \times x$
        else if $l \neq m$ **then** $h[k, k-1] := -h[k, k-1]$;
        $p := p + s$; $x := p/s$; $y := q/s$; $z := r/s$; $q := q/p$; $r := r/p$;
        comment row modification;
        for $j := k$ **step** 1 **until** n **do**
        begin
            $p := h[k, j] + q \times h[k+1, j]$;
            if $notlast$ **then**
            begin
                $p := p + r \times h[k+2, j]$; $h[k+2, j] := h[k+2, j] - p \times z$

```
            end;
            h[k+1, j] := h[k+1, j] − p×y;  h[k, j] := h[k, j] − p×x
        end j;
        j := if k+3 < en then k+3 else en;
        comment column modification;
        for i := 1 step 1 until j do
        begin
            p := x×h[i, k] + y×h[i, k+1];
            if notlast then
            begin
                p := p + z×h[i, k+2];  h[i, k+2] := h[i, k+2] − p×r
            end;
            h[i, k+1] := h[i, k+1] − p×q;  h[i, k] := h[i, k] − p
        end i;
        comment accumulate transformations;
        for i := low step 1 until upp do
        begin
            p := x×vecs[i, k] + y×vecs[i, k+1];
            if notlast then
            begin
                p := p + z×vecs[i, k+2];  vecs[i, k+2] := vecs[i, k+2] − p×r
            end;
            vecs[i, k+1] := vecs[i, k+1] − p×q;
            vecs[i, k] := vecs[i, k] − p
        end;
cont3:
    end k;
    goto nextit;
    comment one root found;
onew:
    wr[en] := h[en, en] := x+t;  wi[en] := 0.0;
    cnt[en] := its;  en := na;  goto nextw;
    comment two roots found;
twow:
    p := (y−x)/2;  q := p↑2 + w;  z := sqrt(abs(q));
    x := h[en, en] := x+t;  h[na, na] := y+t;
    cnt[en] := −its;  cnt[na] := its;
    if q > 0 then
    begin comment real pair;
        z := if p < 0.0 then p − z else p + z;
        wr[na] := x+z;  wr[en] := s := x − w/z;  wi[na] := wi[en] := 0.0;
        x := h[en, na];  r := sqrt(x↑2 + z↑2);  p := x/r;  q := z/r;
        for j := na step 1 until n do
        begin
            z := h[na, j];  h[na, j] := q×z + p×h[en, j];
            h[en, j] := q×h[en, j] − p×z
        end row modification;
```

```
for i := 1 step 1 until en do
begin
    z := h[i, na]; h[i, na] := q × z + p × h[i, en];
    h[i, en] := q × h[i, en] − p × z
end column modification;
for i := low step 1 until upp do
begin
    z := vecs[i, na]; vecs[i, na] := q × z + p × vecs[i, en];
    vecs[i, en] := q × vecs[i, en] − p × z
end accumulate
end pair of real roots
else
begin comment complex pair;
    wr[na] := wr[en] := x + p; wi[na] := z; wi[en] := − z
end two roots found;
en := en − 2; goto nextw;

comment all roots found, now backsubstitute;
```
fin:
```
    norm := 0.0; k := 1;
    for i := 1 step 1 until n do
    begin
        for j := k step 1 until n do norm := norm + abs(h[i, j]);
        k := i
    end norm;
    comment backsubstitution;
    for en := n step −1 until 1 do
    begin
        p := wr[en]; q := wi[en]; na := en − 1;
        if q = 0.0 then
        begin comment real vector;
            m := en; h[en, en] := 1.0;
            for i := na step −1 until 1 do
            begin
                w := h[i, i] − p; r := h[i, en];
                for j := m step 1 until na do r := r + h[i, j] × h[j, en];
                if wi[i] < 0.0 then
                begin
                    z := w; s := r
                end
                else
                begin
                    m := i;
                    if wi[i] = 0.0 then
                        h[i, en] := −r/(if w ≠ 0.0 then w else macheps × norm)
                    else
                    begin comment solve
```
$$\begin{bmatrix} w & x \\ y & z \end{bmatrix} \begin{bmatrix} h[i, en] \\ h[i+1, en] \end{bmatrix} = \begin{bmatrix} -r \\ -s \end{bmatrix};$$

$x := h[i, i+1]; \; y := h[i+1, i];$
$q := (wr[i] - p)\uparrow 2 + wi[i]\uparrow 2;$
$h[i, en] := t := (x \times s - z \times r)/q;$
$h[i+1, en] := \textbf{if } abs(x) > abs(z) \textbf{ then } (-r - w \times t)/x$
$\qquad\qquad\qquad\qquad\qquad\qquad \textbf{else } (-s - y \times t)/z$

end $wi[i] > 0.0$
end $wi[i] \geqq 0.0$
end i

end *real vector*

else

if $q < 0.0$ **then**

begin comment complex vector associated with $lambda = p - i \times q;$
$m := na;$
if $abs(h[en, na]) > abs(h[na, en])$ **then**
begin
$\quad h[na, na] := -(h[en, en] - p)/h[en, na];$
$\quad h[na, en] := -q/h[en, na]$
end
else $cdiv(-h[na, en], 0.0, h[na, na] - p, q, h[na, na], h[na, en]);$
$h[en, na] := 1.0; \; h[en, en] := 0.0;$
for $i := na - 1$ **step** -1 **until** 1 **do**
begin
$\quad w := h[i, i] - p; \; ra := h[i, en]; \; sa := 0.0;$
$\quad \textbf{for } j := m \textbf{ step } 1 \textbf{ until } na \textbf{ do}$
$\quad \textbf{begin}$
$\quad\quad ra := ra + h[i, j] \times h[j, na];$
$\quad\quad sa := sa + h[i, j] \times h[j, en]$
$\quad \textbf{end};$
$\quad \textbf{if } wi[i] < 0.0 \textbf{ then}$
$\quad \textbf{begin}$
$\quad\quad z := w; \; r := ra; \; s := sa$
$\quad \textbf{end}$
$\quad \textbf{else}$
$\quad \textbf{begin}$
$\quad\quad m := i;$
$\quad\quad \textbf{if } wi[i] = 0.0 \textbf{ then } cdiv(-ra, -sa, w, q, h[i, na], h[i, en])$
$\quad\quad \textbf{else}$
$\quad\quad \textbf{begin comment} \text{ solve complex equations}$

$$\begin{bmatrix} w + q \times i & x \\ y & z + q \times i \end{bmatrix} \begin{bmatrix} h[i, na] + h[i, en] \times i \\ h[i+1, na] + h[i+1, en] \times i \end{bmatrix} = \begin{bmatrix} -ra - sa \times i \\ -r - s \times i \end{bmatrix};$$

$\quad\quad\quad x := h[i, i+1]; \; y := h[i+1, i];$
$\quad\quad\quad vr := (wr[i] - p)\uparrow 2 + wi[i]\uparrow 2 - q\uparrow 2;$
$\quad\quad\quad vi := (wr[i] - p) \times 2.0 \times q;$
$\quad\quad\quad \textbf{if } vr = 0.0 \wedge vi = 0.0 \textbf{ then}$
$\quad\quad\quad\quad vr := macheps \times norm$
$\quad\quad\quad\quad\quad \times (abs(w) + abs(q) + abs(x) + abs(y) + abs(z));$

```
                    cdiv (x × r − z × ra + q × sa, x × s − z × sa − q × ra, vr, vi,
                            h [i, na], h [i, en]);
                    if abs (x) > abs (z) + abs (q) then
                    begin
                        h [i + 1, na] := (−ra − w × h [i, na] + q × h [i, en])/x;
                        h [i + 1, en] := (−sa − w × h [i, en] − q × h [i, na])/x
                    end
                    else cdiv (−r − y × h [i, na], −s − y × h [i, en], z, q,
                            h [i + 1, na], h [i + 1, en])
                end wi [i] > 0.0
            end wi [i] ≧ 0.0
        end i
    end complex vector
end backsubstitution;

comment vectors of isolated roots;
for i := 1 step 1 until low − 1, upp + 1 step 1 until n do
    for j := i + 1 step 1 until n do vecs [i, j] := h [i, j];

comment multiply by transformation matrix to give vectors of original full
            matrix;
for j := n step −1 until low do
begin
    m := if j ≦ upp then j else upp; l := j − 1;
    if wi [j] < 0.0 then
    begin
        for i := low step 1 until upp do
        begin
            y := z := 0.0;
            for k := low step 1 until m do
            begin
                y := y + vecs [i, k] × h [k, l]; z := z + vecs [i, k] × h [k, j]
            end;
            vecs [i, l] := y; vecs [i, j] := z
        end i
    end
    else
    if wi [j] = 0.0 then
        for i := low step 1 until upp do
        begin
            z := 0.0;
            for k := low step 1 until m do
                z := z + vecs [i, k] × h [k, j];
            vecs [i, j] := z
        end i
    end j
end hqr2;
```

5. Organisational and Notational Details

The LR reduction proceeds essentially as in *comlr* [4], but since the resulting triangular matrix is used to find the vectors, each transformation must operate on the full $n \times n$ matrix even when the eigenvalues could be found from a number of Hessenberg matrices of lower order. The procedure *comlr2* first finds the transformation matrix S used to reduce the original general matrix to Hessenberg form (no multiplications are required) and then updates S as each LR iteration is performed.

The eigenvector of the triangular matrix T corresponding to λ_i has zero components in position $i+1$ to n. Hence when deriving the corresponding eigenvector Px_i of A we do not need the last $n-i$ columns of P. Accordingly if the vectors are found in the order $Px_n, \ldots, Px_2, Px_1$, each can be overwritten on the corresponding column of P.

If $\lambda_j = \lambda_i$ $(j < i)$ then a zero divisor occurs when computing x_{ji} from Eq. (3). Such a zero divisor is replaced by *macheps* $\times$ *norm* where

$$norm = \sum\sum \left(|\Re(t_{ij})| + |\Im(t_{ij})| \right). \tag{6}$$

When the equal eigenvalues correspond to linear divisors this ensures, even when rounding errors are taken into account, that we obtain the requisite number of independent eigenvectors. It should be appreciated that with exact computation the numerator of the expression for x_{ji} would also be zero.

When the equal eigenvalues correspond to a non-linear divisor, the numerator of the expression for x_{ji} will be non-zero, even with exact computation. If we were to replace the $\lambda_j - \lambda_i$ by an infinitesimally small quantity we would effectively obtain an infinite multiple of the eigenvector corresponding to λ_j, i.e. after normalization we would merely obtain the same vectors from λ_i and λ_j. In practice the solution is complicated by the fact that if rounding errors have occurred the computed λ_i and λ_j will not be equal; in fact since eigenvalues corresponding to non-linear divisors are very sensitive the computed values will usually differ quite substantially.

Corresponding to a quadratic divisor we will usually find that the computed v_i and v_j make an angle with each other of the order of magnitude of $2^{-t/2}$ (for a machine with a t-digit binary mantissa) and $v_i - v_j$ will give an approximation to a vector of second grade. Similar remarks apply for divisors of higher degree. Two code procedures *cdiv* and *csqrt* [3] are used.

Again *hqr2* follows roughly the same lines as *hqr* [5] but before entry, the matrix S of the transformation reducing the original A to Hessenberg form must be produced using *elmtrans*, *dirtrans* or *ortrans* corresponding to the use of *elmhes*, *dirhes* or *orthes* in the reduction. If the primary matrix is itself upper-Hessenberg then *hqr2* must be provided with the identity matrix in lieu of this transformation matrix. As each QR iteration is performed S is updated. After isolating a pair of real eigenvalues in rows r and $r+1$ a rotation is performed in plane $(r, r+1)$, the angle being chosen to make $h_{r+1, r}$ equal to zero.

There is an additional feature in *hqr2* which is now also included in *hqr*. This is designed to overcome a shortcoming in the original form of the latter. The first step of procedure *hqr* is to determine the three non-zero elements of the first column of $(H - k_1 I)(H - k_2 I)$ where k_1 and k_2 are the shifts. The formulae for these elements are better transformed according to the equations

$$(h_{11} - k_1)(h_{11} - k_2) + h_{12}h_{21}$$
$$= h_{21}\left[((h_{nn} - h_{11})(h_{n-1,n-1} - h_{11}) - h_{n,n-1}h_{n-1,n})/h_{21} + h_{12}\right]$$
$$h_{21}(h_{11} - k_2 + h_{22} - k_1) = h_{21}\left(h_{22} - h_{11} - (h_{nn} - h_{11}) - (h_{n-1,n-1} - h_{11})\right)$$
$$h_{21}h_{32} = h_{21}(h_{32})$$

(7)

and since only the ratios are of interest, the factor h_{21} can be omitted. That this formulation is superior to that originally used in *hqr* is evident if we consider the tridiagonal matrix defined by

$$h_{ii} = 1, \qquad h_{i+1,i} - h_{i,i+1} - e.$$

(8)

If ε^2 is smaller than *macheps* the old formulation gave the values ε^2, 0, ε^2 instead of 0, 0, ε^2. This does not harm the eigenvalues but convergence is jeopardized. As an additional precaution after ten iterations have been performed without locating an eigenvalue H is replaced by $H - h_{nn}I$ before starting the double-Francis step.

Similar considerations arise when computing the vectors of the quasi-triangular matrix T to those discussed in connexion with *comlr2*. Corresponding to a complex pair of eigenvalues only one eigenvector is determined (that corresponding to the eigenvalue with positive imaginary part) and the real and imaginary parts of the vector are stored as two columns of the eigenvector array.

6. Discussion of Numerical Properties

As far as the LR and QR reductions are concerned the remarks given for *comlr* [4] and *hqr* [5] apply without modification. The computed vectors should give very small residuals even when, as a result of sensitivity, the values and vectors are fairly inaccurate. This is because the algorithms guarantee that each computed vector and value is exact for a matrix which is very close to the original matrix.

7. Test Results

The two algorithms were used on KDF (39 binary digit mantissa) to find the eigenvalues and eigenvectors of a number of test matrices. Most of the test matrices are of Hessenberg form so as to provide tests of *comlr2* and *hqr2* which are independent of other procedures. Procedures which are not essentially different from these have been used over a very long period at the National Physical Laboratory, often on matrices of quite high order and have given consistently good results.

7.1. Tests on comlr2

The matrix of Table 1 has a well-conditioned eigenvalue problem and has simple eigenvalues; it provides a straightforward test of the algorithm. Results obtained on other computers should agree with the quoted results to within machine accuracy. The eigenvectors in this and later examples are normalized for convenience. The published procedures do not include normalization because usually the vectors will be transformed to give the vectors corresponding to the full matrix from which the Hessenberg form was derived. The residuals corresponding to the normalised vectors are all less than 10^{-10}.

Table 1

Real part of matrix	Imaginary part of matrix
$\begin{bmatrix} 2 & 1 & 1 & -1 \\ 1 & 2 & 0 & 2 \\ & 2 & 1 & 3 \\ & & 4 & 1 \end{bmatrix}$	$\begin{bmatrix} 1 & 2 & -1 & 1 \\ 2 & 1 & 1 & 0 \\ & 3 & 2 & 1 \\ & & 1 & 3 \end{bmatrix}$

Eigenvalues

Real parts	Imaginary parts
$+3.4250\,6739\,820_{10}-1$	$-2.2078\,0890\,098_{10}+0$
$-2.3101\,4123\,972_{10}+0$	$+2.5966\,0306\,664_{10}+0$
$+2.7313\,9648\,762_{10}+0$	$+2.3407\,2215\,071_{10}+0$
$+5.2362\,3801\,229_{10}+0$	$+4.2704\,8368\,366_{10}+0$

Eigenvectors

v_1 real	v_1 imaginary	v_2 real	v_2 imaginary
$-9.5653\,5377\,924_{10}-1$	$-1.0156\,2354\,788_{10}-2$	$+4.2687\,2844\,242_{10}-1$	$-1.7751\,7856\,560_{10}-2$
$+1.0000\,0000\,000_{10}+0$	$+0.0000\,0000\,000$	$-4.7947\,9818\,465_{10}-1$	$-1.8536\,4426\,192_{10}-1$
$-4.3074\,4833\,806_{10}-1$	$+3.9212\,5597\,416_{10}-1$	$-8.0258\,5993\,661_{10}-1$	$+9.9797\,2650\,739_{10}-2$
$-1.6457\,2377\,905_{10}-1$	$-4.2691\,8537\,943_{10}-1$	$+1.0000\,0000\,000_{10}+0$	$+0.0000\,0000\,000$

v_3 real	v_3 imaginary	v_4 real	v_4 imaginary
$+1.0000\,0000\,000_{10}+0$	$+0.0000\,0000\,000$	$+2.8381\,7639\,694_{10}-1$	$+1.7881\,1316\,481_{10}-2$
$+4.6222\,0295\,055_{10}-1$	$+1.6411\,4613\,473_{10}-1$	$+5.5424\,4481\,908_{10}-1$	$-9.6903\,8376\,463_{10}-2$
$-2.9682\,8088\,600_{10}-1$	$-5.6928\,8524\,047_{10}-2$	$+1.0000\,0000\,000_{10}+0$	$+0.0000\,0000\,000$
$-4.6944\,7324\,807_{10}-1$	$-4.8171\,5035\,708_{10}-1$	$+9.3126\,7051\,967_{10}-1$	$-4.3236\,3795\,722_{10}-2$

The matrix of Table 2 has linear divisors but one double eigenvalue. Two independent eigenvectors are given corresponding to the double root. The procedure can be used with *low* = 1, *upp* = 3 or with *low* = 2, *upp* = 3 and gives identical results for the two cases. All results were exact.

Table 2

Real part of matrix	Imaginary part of matrix	Eigenvalues
$\begin{bmatrix} 1 & -1 & 2 \\ & 0 & 2 \\ & -1 & 3 \end{bmatrix}$	$\begin{bmatrix} 1 & -1 & 2 \\ & 1 & 0 \\ & 0 & 1 \end{bmatrix}$	$1+i$ $1+i$ $2+i$

Eigenvectors

v_1	v_2	v_3
$1+0i$	0	$1+0i$
0	$1+0i$	$0.5-0.5i$
0	$0.5+0i$	$0.5-0.5i$

The matrix of Table 3 has two pairs of quadratic divisors. The eigenvalues have about half the figures correct, as is to be expected. The eigenvectors corresponding to equal eigenvalues agree to about half the word-length. The difference between two such vectors when correspondingly normalised is a vector of grade two, accurate to about half the word length. This is well illustrated in the table by the difference $v_1 - v_2$. The maximum residual is 4×10^{-11} in spite of errors of 10^{-6} in the results.

Table 3

Real part of matrix	Imaginary part of matrix
$\begin{bmatrix} -4 & -5 & -2 & 0 \\ 1 & 0 & 0 & 0 \\ & 1 & 0 & 0 \\ & & 1 & 0 \end{bmatrix}$	$\begin{bmatrix} -2 & -6 & -6 & -2 \\ 0 & 0 & 0 & 0 \\ & 0 & 0 & 0 \\ & & 0 & 0 \end{bmatrix}$

Eigenvalues

Real parts	Imaginary parts
$-1.0000\,0119\,397_{10}+0$	$-1.0000\,0037\,299_{10}+0$
$-9.9999\,8806\,028_{10}-1$	$-9.9999\,9627\,000_{10}-1$
$-1.0000\,0165\,028_{10}+0$	$-1.9151\,4772\,872_{10}-6$
$-9.9999\,8349\,722_{10}-1$	$+1.9151\,3879\,547_{10}-6$

Eigenvectors

v_1 real	v_1 imaginary	v_2 real	v_2 imaginary
$+1.0000\,0000\,000_{10}+0$	$+0.0000\,0000\,000$	$+1.0000\,0000\,000_{10}+0$	$+0.0000\,0000\,000$
$-4.9999\,9813\,507_{10}-1$	$+4.9999\,9403\,019_{10}-1$	$-5.0000\,0186\,501_{10}-1$	$+5.0000\,0596\,989_{10}-1$
$+4.1049\,1015\,728_{10}-7$	$-4.9999\,9216\,523_{10}-1$	$-4.1048\,6172\,928_{10}-7$	$-5.0000\,0783\,488_{10}-1$
$+2.4999\,9104\,523_{10}-1$	$+2.4999\,9720\,260_{10}-1$	$+2.5000\,0895\,480_{10}-1$	$+2.5000\,0279\,751_{10}-1$

v_3 real	v_3 imaginary	v_4 real	v_4 imaginary
$+1.0000\,0000\,000_{10}+0$	$+0.0000\,0000\,000$	$-9.9999\,5049\,169_{10}-1$	$+5.7453\,9365\,890_{10}-6$
$-9.9999\,8349\,722_{10}-1$	$+1.9151\,4240\,547_{10}-6$	$+9.9999\,6699\,446_{10}-1$	$-3.8302\,7218\,016_{10}-6$
$+9.9999\,6699\,438_{10}-1$	$-3.8302\,7633\,405_{10}-6$	$-9.9999\,8349\,726_{10}-1$	$+1.9151\,4248\,255_{10}-6$
$-9.9999\,5049\,153_{10}-1$	$+5.7454\,0480\,570_{10}-6$	$+1.0000\,0000\,000_{10}+0$	$+0.0000\,0000\,000$

(v_1-v_2) real	(v_1-v_2) imaginary	(v_1-v_2) normalised	
0	0	0	
$+3.72994_{10}-7$	$-1.19397_{10}-6$	$(1+i)^2/3$	to six
$+8.20977_{10}-7$	$+1.56696\,5_{10}-6$	$-2(1+i)/3$	decimals
$-1.79095\,7_{10}-6$	$-5.59491_{10}-7$	1	

The final test matrix is a full complex matrix and the eigenproblem was solved using *comhes* and *comlr2*. The results are given in Table 4. All residuals are less than 10^{-10}.

Table 4

Real part of matrix				Imaginary part of matrix			
1	2	3	1	3	1	2	1
3	1	2	4	4	2	1	3
2	1	3	5	3	5	1	2
1	3	1	5	2	1	4	3

Eigenvalues

Real parts	Imaginary parts
$+9.7836\,5812\,748_{10}+0$	$+9.3225\,1422\,480_{10}+0$
$-3.3710\,0978\,501_{10}+0$	$-7.7045\,3986\,959_{10}-1$
$+2.2216\,8234\,772_{10}+0$	$+1.8489\,9335\,975_{10}+0$
$+1.3656\,6930\,995_{10}+0$	$-1.4010\,5359\,747_{10}+0$

Eigenvectors

v_1 real	v_1 imaginary	v_2 real	v_2 imaginary
$+6.3233\,7764\,500_{10}-1$	$-1.4378\,0794\,462_{10}-2$	$-5.0609\,6461\,966_{10}-1$	$+5.8345\,1962\,595_{10}-1$
$+8.7375\,8591\,710_{10}-1$	$+8.1057\,8078\,216_{10}-3$	$+1.0000\,0000\,000_{10}+0$	$+0.0000\,0000\,000$
$+1.0000\,0000\,000_{10}+0$	$+0.0000\,0000\,000$	$+5.1831\,9483\,943_{10}-1$	$-7.1465\,7072\,196_{10}-1$
$+9.4371\,7588\,374_{10}-1$	$+3.7984\,6348\,035_{10}-2$	$-5.5348\,4898\,199_{10}-1$	$+1.8756\,3320\,202_{10}-2$

v_3 real	v_3 imaginary	v_4 real	v_4 imaginary
$-7.9662\,0817\,424_{10}-1$	$+3.0498\,0785\,942_{10}-1$	$-8.4659\,8704\,792_{10}-4$	$+7.3020\,3551\,249_{10}-1$
$-1.7883\,4394\,745_{10}-1$	$+4.2972\,4147\,740_{10}-1$	$-8.7941\,7015\,790_{10}-2$	$-3.8790\,1799\,596_{10}-1$
$-2.5281\,4311\,199_{10}-1$	$+3.8173\,4049\,516_{10}-2$	$+1.0000\,0000\,000_{10}+0$	$+0.0000\,0000\,000$
$+1.0000\,0000\,000_{10}+0$	$+0.0000\,0000\,000$	$-4.3208\,9374\,798_{10}-1$	$-4.3342\,6369\,341_{10}-1$

7.2. Tests on hqr2

The matrix of Table 5 has a well-conditioned eigenvalue problem and simple eigenvalues, and again provides a straightforward test.

Table 5

Matrix				Eigenvalues	
				Real parts	Imaginary parts
3	1	2	5	$+6.0779\,0591\,265_{10}+0$	$+0.0000\,0000\,000$
2	1	3	-1	$-3.1955\,2891\,925_{10}+0$	$+0.0000\,0000\,000$
0	4	1	1	$+1.5588\,1150\,332_{10}+0$	$+2.0015\,1592\,083_{10}+0$
0	0	2	1	$+1.5588\,1150\,332_{10}+0$	$-2.0015\,1592\,083_{10}+0$

Eigenvectors

v_1 real	v_2 real	v_3 real	v_3 imaginary
$+1.0000\,0000\,000_{10}+0$	$+2.1195\,8893\,230_{10}-1$	$+1.0000\,0000\,000_{10}+0$	$+0.0000\,0000\,000$
$+7.0117\,3512\,001_{10}-1$	$-9.2970\,7744\,406_{10}-1$	$-6.6545\,4727\,311_{10}-2$	$-3.4571\,4918\,782_{10}-1$
$+5.9877\,6353\,665_{10}-1$	$+1.0000\,0000\,000_{10}+0$	$-4.7651\,7960\,726_{10}-1$	$+4.2077\,9033\,315_{10}-2$
$+2.3583\,5938\,664_{10}-1$	$-4.7669\,7941\,656_{10}-1$	$-8.4321\,4205\,020_{10}-2$	$+4.5261\,5006\,593_{10}-1$

The matrix of Table 6 has a double eigenvalue with linear divisors. Two independent eigenvectors are produced. It can be used with $low = 1$, $upp = 3$ or $low = 2$, $upp = 3$ and gives identical results in the two cases.

Table 6

Matrix			Eigenvalues		Eigenvectors	
2	4	4	2	1	1.00	0
0	3	1	4	0	0.25	1.0
0	1	3	2	0	0.25	−1.0

The matrix of Table 7 has two pairs of quadratic divisors, all the eigenvalues being complex. As for the corresponding example for *comlr2*, the eigenvalues have about half the figures correct, the "two eigenvectors" agree to about half word length and the difference between them gives a vector of grade two accurate to about half word length. Nevertheless, the maximum residual is only 7×10^{-11}.

Table 7

Matrix				Eigenvalues	
				Real parts	Imaginary parts
2	−3	2	−1	$+4.9999\,9008\,420_{10} - 1$	$+8.6602\,7888\,416_{10} - 1$
1	0	0	0	$+4.9999\,9008\,420_{10} - 1$	$-8.6602\,7888\,416_{10} - 1$
	1	0	0	$+5.0000\,0991\,538_{10} - 1$	$+8.6602\,2919\,134_{10} - 1$
		1	0	$+5.0000\,0991\,538_{10} - 1$	$-8.6602\,2919\,134_{10} - 1$

Eigenvectors

v_1 real	v_1 imaginary	v_3 real	v_3 imaginary
$+1.0000\,0000\,000_{10} + 0$	$+0.0000\,0000\,000$	$-9.9999\,5032\,063_{10} - 1$	$+6.3030\,4295\,975_{10} - 6$
$+4.9999\,7352\,445_{10} - 1$	$-8.6602\,5020\,176_{10} - 1$	$-4.9999\,4704\,950_{10} - 1$	$+8.6602\,4636\,562_{10} - 1$
$-5.0000\,1983\,124_{10} - 1$	$-8.6602\,0434\,467_{10} - 1$	$+5.0000\,0991\,557_{10} - 1$	$+8.6602\,2919\,137_{10} - 1$
$-9.9999\,5032\,025_{10} - 1$	$+6.3031\,0928\,537_{10} - 6$	$+1.0000\,0000\,000_{10} - 0$	$+0.0000\,0000\,000$

The final test is with a full matrix. It was solved using *hqr2* combined with *elmhes*, *dirhes* and *orthes* and *elmtrans*, *dirtrans* and *ortrans* respectively and the three sets of results were in close agreement. All residuals were less than 10^{-10}. The results are given in Table 8.

Table 8

Matrix				Eigenvalues
3	1	2	5	$+1.0591\,9821\,216_{10} + 1$
2	1	3	7	$-2.3663\,0331\,169_{10} + 0$
3	1	2	4	$+1.9134\,6321\,403_{10} - 1$
4	1	3	2	$-4.1702\,5131\,203_{10} - 1$

Eigenvectors

v_1	v_2	v_3	v_4
$+8.5798428341 1_{10}-1$	$+3.6016263951 7_{10}-1$	$-2.9505033518 2_{10}-1$	$-1.1208317067 1_{10}-2$
$+1.0000000000 0_{10}+0$	$+1.0000000000 0_{10}+0$	$+1.0000000000 0_{10}+0$	$+1.0000000000 0_{10}+0$
$+7.8345839182 3_{10}-1$	$+9.6050802709 7_{10}-2$	$+1.0637549393 3_{10}-1$	$-2.4113619293 9_{10}-1$
$+7.8937691131 6_{10}-1$	$-6.2496871411 5_{10}-1$	$-7.6811355716 0_{10}-2$	$-9.5885702603 9_{10}-2$

A second test run was made by E. Mann of the Rechencentrum der Technischen Hochschule München and confirmed these results.

Acknowledgements. This work was carried out at the National Physical Laboratory. The authors wish to thank E. Mann of the Rechenzentrum der Technischen Hochschule München for performing test runs with earlier versions, and C. Reinsch for suggesting substantial improvements in the published algorithms.

References

1. Dekker, T. J., Hoffman, W.: Algol 60 procedures in linear algebra. Mathematical Centre Tracts, 23. Mathematisch Centrum Amsterdam (1968).
2. Francis, J. G. F.: The QR transformation—a unitary analogue to the LR transformation. Comput. J. **4**, 265–271, 332–345 (1961/62).
3. Martin, R. S., Wilkinson, J. H.: Similarity reduction of a general matrix to Hessenberg form. Numer. Math. **12**, 349–368 (1968). Cf. II/13.
4. — — The modified LR algorithm for complex Hessenberg matrices. Numer. Math. **12**, 369–376 (1968). Cf. II/16.
5. — Peters, G., Wilkinson, J. H.: The QR algorithm for real Hessenberg matrices. Numer. Math. **14**, 219–231 (1970). Cf. II/14.
6. Parlett, B. N., Reinsch, C.: Balancing a matrix for calculation of eigenvalues and eigenvectors. Numer. Math. **13**, 293–304 (1969). Cf. II/11.
7. Rutishauser, H.: Solution of eigenvalue problems with the LR transformation. Nat. Bur Standards Appl. Math. Ser. **49**, 47–81 (1958).
8. Wilkinson, J. H.: The algebraic eigenvalue problem. London: Oxford University Press 1965.
9. Peters, G., Wilkinson, J. H.: The calculation of specified eigenvectors by inverse iteration. Cf. II/18.

The Modified LR Algorithm
for Complex Hessenberg Matrices [*]

by R. S. Martin and J. H. Wilkinson

1.

The LR algorithm of Rutishauser [3] is based on the observation that if

$$A = LR \tag{1}$$

where L is unit lower-triangular and R is upper-triangular then B defined by

$$B = L^{-1}AL = RL \tag{2}$$

is similar to A. Hence if we write

$$A_s = L_s R_s, \qquad R_s L_s = A_{s+1}, \tag{3}$$

a sequence of matrices is obtained each of which is similar to A_1. Rutishauser showed [3, 4, 5] if A_1 has roots of distinct moduli then, in general A_s tends to upper triangular form, the diagonal elements tending to the roots arranged in order of decreasing modulus. If A_1 has some roots of equal modulus then A_s does not tend to strictly triangular form but rather to block-triangular form. Corresponding to a block of k roots of the same modulus there is a diagonal block of order k which does not tend to a limit, but its roots tend to the k eigenvalues.

In its classical form the LR algorithm involves too much work and is often unstable. If the original matrix A_1 is of upper-Hessenberg form then L_1 has only one line of non-zero elements below the diagonal and A_2 is therefore of upper-Hessenberg form. Hence all A_s are of upper-Hessenberg form. This greatly reduces the volume of computation.

Stability may be introduced by using partial pivoting [4] in the triangular factorization of each of the matrices. This is best performed in the following manner. A_1 is reduced to upper triangular form R_1 in $n-1$ major steps, in the r-th of which the current matrix is pre-multiplied first by $I_{r, r'}$ (a permutation matrix which interchanges rows r and r') and then by N_r, an elementary matrix which is equal to the identity matrix apart from the sub-diagonal elements of its r-th column. The non-zero sub-diagonals of N_r are chosen so as to annihilate the sub-diagonal elements of the r-th column of the current transformed matrix and r' is

---

[*] Prepublished in Numer. Math. **12**, 369—376 (1968).

chosen so that all elements of $|N_r|$ are bounded by unity. Hence we have

$$N_{n-1}I_{n-1,(n-1)'} \cdots N_2I_{2,2'}N_1I_{1,1'}A_1=R_1 \tag{4}$$

and A_2 is defined by

$$A_2= R_1I_{1,1'}N_1^{-1}I_{2,2'}N_2^{-1} \cdots I_{n-1,(n-1)'}N_{n-1}^{-1}. \tag{5}$$

It is obvious that A_2 is similar to A_1 and in the case when no row interchanges are used (i.e. $r'=r\ (r=1, \ldots, n-1)$), A_2 is the matrix given by the classical LR algorithm. This modified version of the LR algorithm also preserves upper-Hessenberg form and there is the same economy of effort as with the classical algorithm.

Unfortunately it is no longer possible to show that the A_s converge to upper triangular form with this modified version. It is easy to see, for example, that each of the matrices

$$\begin{bmatrix} 1 & 2 \\ 3 & 0 \end{bmatrix} \text{ and } \begin{bmatrix} 1 & 3 \\ 2 & 0 \end{bmatrix} \tag{6}$$

is transformed into the other by the modified algorithm and hence the algorithm does not converge although the eigenvalues have distinct moduli.

In practice this is not as important as it appears because the LR algorithm is normally used with 'origin' shifts. The classical algorithm then becomes

$$A_s-k_sI=L_sR_s, \quad R_sL_s+k_sI=A_{s+1} \tag{7}$$

and A_{s+1} is again similar to A_s. There is no need to add back the k_sI and if this is omitted. A_{s+1} is similar to $A_1-\sum_{i=1}^{s}k_iI$. The version with origin shifts can be modified to give numerical stability in exactly the same way as the original algorithm. The values of k_s should be chosen so as to accelerate convergence and in practice extremely satisfactory results have been obtained by taking k_s to be the root of the 2×2 matrix

$$\begin{bmatrix} a_{n-1,n-1}^{(s)} & a_{n-1,n}^{(s)} \\ a_{n,n-1}^{(s)} & a_{nn}^{(s)} \end{bmatrix} \tag{8}$$

nearer to $a_{nn}^{(s)}$. In general, this root will be complex but since in this paper we are concerned only with the case when A_1 is a complex matrix this causes no inconvenience. When A_1 is real we recommend the use of the Francis double QR algorithm [1, 4, 8]. With this choice of k_s the simple oscillations observed with the matrices of (6) disappear, since convergence occurs in one iteration for any matrix of order two.

Since we recommend that the LR algorithm be used only on upper-Hessenberg matrices, only the $n-1$ subdiagonal elements need to be examined. If at any stage a sub-diagonal element $a_{r+1,r}^{(s)}$ is regarded as negligible then the eigenvalues of A_s are those of the leading principal minor of order r and the trailing principal minor of order $n-r$; both these matrices are of upper-Hessenberg form.

It is difficult to devise an entirely satisfactory criterion for determining when a sub-diagonal element is negligible. If the original matrix is balanced (see the discussion in [6]), a reasonable criterion is that the sub-diagonal element is small compared with the two neighbouring diagonal elements. For economy of computation the criterion has been taken to be

$$m(a^{(s)}_{r+1,r}) \leqq \text{macheps} \; [m(a^{(s)}_{r+1,r+1}) + m(a^{(s)}_{rr})], \tag{9}$$

where

$$m(x) \equiv |\mathscr{R}(x)| + |\mathscr{I}(x)| \tag{10}$$

and *macheps* is the machine precision. When $a^{(s)}_{n,n-1}$ is negligible $a^{(s)}_{nn}$ may be regarded as an eigenvalue of A_s and the last row and column may be deleted to give a Hessenberg matrix of order $n-1$.

When a single sub-diagonal $a^{(s)}_{r+1,r}$ is zero (or 'negligible') the leading principal sub-matrix of order r is effectively decoupled from the lower part of the matrix. A similar decoupling is effected when two consecutive elements are 'small' even though, individually, neither is small enough to effect a decoupling. This observation was first made by FRANCIS [1] in connexion with the QR algorithm and has been discussed in detail in Chapter 8 of reference [4]. The criterion adopted in this procedure is again based on the assumption that the original matrix is balanced and for economy of computation has been chosen to be

$$m(a^{(s)}_{r+1,r}) m(a^{(s)}_{r,r-1}) \leqq \text{macheps} \; m(a^{(s)}_{r,r}) [m(a^{(s)}_{r+1,r+1}) + m(a^{(s)}_{rr}) + m(a^{(s)}_{r-1,r-1})]. \tag{11}$$

When this criterion is satisfied the matrix A_s is subjected to transformation only in rows and columns r to n; in the notation of (4) and (5) we have

$$A_{s+1} = N_{n-1} I_{n-1,(n-1)'} \cdots N_{r+1} I_{r+1,(r+1)'} N_r I_{r,r} A_s I_{r,r} N_r^{-1} I_{r+1,(r+1)'} N_{r+1}^{-1}$$
$$\cdots I_{n-1,(n-1)'} N_{n-1}^{-1}. \tag{12}$$

The convergence properties of the modified algorithm with the above origin shifts are difficult to analyze theoretically but experience with it has been very satisfactory. It may be thought that row interchanges ultimately die out and hence the later behaviour is covered by the analysis of the classical algorithm. This is far from true in practice, and usually at the stage when an eigenvalue is crystallizing out at the bottom of the matrix, interchanges are still taking place as frequently in the upper part of the matrix as they were in the initial stages. It should be appreciated that when a matrix has a number of unequal roots of the same modulus the use of origin shifts destroys this property. The algorithm therefore gives convergence to true diagonal form rather than block diagonal form.

It would appear that failure to converge or the need for an excessive number of iterations arises only as the result of very exceptional circumstances. The following empirical device has accordingly been added to the procedure. When the number of iterations for any one root has reached ten or twenty then the shift used in the next iteration is taken to be

$$(|\mathscr{R}a_{n,n-1}| + |\mathscr{R}a_{n-1,n-2}|) + i(|\mathscr{I}a_{n,n-1}| + |\mathscr{I}a_{n-1,n-2}|). \tag{13}$$

The choice of this exceptional shift has no very deep significance. All that is needed is some reasonable value for the shift which will break up an established cycle. Since cycling is improbable in any case, it is virtually certain that it will not be established a second time. The empirical device has proved very effective in breaking up cycling.

2. Applicability

The procedure *comlr* may be used to find all the eigenvalues of a complex upper-Hessenberg matrix. Normally such a matrix will have been produced from a full complex matrix by the procedure *comhes*. It is assumed that the matrix is 'balanced' in the sense that the $l1$ norms of corresponding rows and columns are roughly of the same orders of magnitude [6].

3. Formal Parameter List

Input to procedure *comlr*.

n order of the complex upper Hessenberg matrix H.

macheps a machine constant equal to the smallest positive machine number for which $1 + macheps \neq 1$.

hr an $n \times n$ array storing the real parts of the elements of H.

hi an $n \times n$ array storing the imaginary parts of the elements of H.

Output from procedure *comlr*.

hr, hi the given matrix is destroyed in the process.

wr an $n \times 1$ array giving the real parts of the n eigenvalues of H.

wi an $n \times 1$ array giving the imaginary parts of the n eigenvalues of H.

fail   exit used if more than 30 iterations are required to find any one eigenvalue.

4. ALGOL Program

procedure *comlr* $(n, macheps)$ *trans*: (hr, hi) *result*: (wr, wi) *exit*: $(fail)$;
value n, *macheps*; **integer** n; **real** *macheps*; **array** hr, hi, wr, wi; **label** *fail*;
comment Finds the eigenvalues of a complex upper Hessenberg matrix, H, stored in the arrays $hr[1:n, 1:n]$ and $hi[1:n, 1:n]$, and stores the real parts in the array $wr[1:n]$ and the complex parts in the array $wi[1:n]$. *macheps* is the relative machine precision. The procedure fails if any eigenvalue takes more than 30 iterations. The procedures *cdiv* and *csqrt* are used. See [7];
begin integer i, j, l, m, its;
      real $sr, si, tr, ti, xr, xi, yr, yi, zr, zi$;
      $tr := ti := 0$;
nextw: **if** $n = 0$ **then go to** *fin*;
      $its := 0$;
      comment look for single small sub-diagonal element;
nextit: **for** $l := n$ **step** -1 **until** 2 **do**
      if $abs(hr[l, l-1]) + abs(hi[l, l-1]) \leq macheps \times$
            $(abs(hr[l-1, l-1]) + abs(hi[l-1, l-1]) + abs(hr[l, l])$
                      $+ abs(hi[l, l]))$ **then go to** *cont1*;
      $l := 1$;
cont1: **if** $l = n$ **then go to** *root*;
      if $its = 30$ **then go to** *fail*;
      comment form shift;
      if $its = 10 \lor its = 20$ **then**

```
begin sr := abs (hr[n, n − 1]) + abs (hr[n − 1, n − 2]);
      si := abs (hi[n, n − 1]) + abs (hi[n − 1, n − 2])
end
else
begin sr := hr[n, n]; si := hi[n, n];
      xr := hr[n − 1, n] × hr[n, n − 1] − hi[n − 1, n] × hi[n, n − 1];
      xi := hr[n − 1, n] × hi[n, n − 1] + hi[n − 1, n] × hr[n, n − 1];
      if xr ≠ 0 ∨ xi ≠ 0 then
      begin yr := (hr[n − 1, n − 1] − sr)/2; yi := (hi[n − 1, n − 1] − si)/2;
            csqrt (yr↑2 − yi↑2 + xr, 2 × yr × yi + xi, zr, zi);
            if yr × zr + yi × zi < 0 then
            begin zr := − zr; zi := − zi
            end;
            cdiv (xr, xi, yr + zr, yi + zi, xr, xi);
            sr := sr − xr; si := si − xi
      end
end;
for i := 1 step 1 until n do
begin hr[i, i] := hr[i, i] − sr; hi[i, i] := hi[i, i] − si
end i;
tr := tr + sr; ti := ti + si; its := its + 1; j := l + 1;
comment look for two consecutive small sub-diagonal elements;
xr := abs (hr[n − 1, n − 1]) + abs (hi[n − 1, n − 1]);
yr := abs (hr[n, n − 1]) + abs (hi[n, n − 1]);
zr := abs (hr[n, n]) + abs (hi[n, n]);
for m := n − 1 step − 1 until j do
begin yi := yr; yr := abs (hr[m, m − 1]) + abs (hi[m, m − 1]);
      xi := zr; zr := xr;
      xr := abs (hr[m − 1, m − 1]) + abs (hi[m − 1, m − 1]);
      if yr ≤ macheps × zr/yi × (zr + xr + xi) then go to cont2
end m;
m := l;
comment triangular decomposition H = L × R;
cont2: for i := m + 1 step 1 until n do
begin xr := hr[i − 1, i − 1]; xi := hi[i − 1, i − 1];
      yr := hr[i, i − 1]; yi := hi[i, i − 1];
      if abs (xr) + abs (xi) < abs (yr) + abs (yi) then
      begin comment interchange rows of hr and hi;
            for j := i − 1 step 1 until n do
            begin zr := hr[i − 1, j]; hr[i − 1, j] := hr[i, j]; hr[i, j] := zr;
                  zi := hi[i − 1, j]; hi[i − 1, j] := hi[i, j]; hi[i, j] := zi
            end j;
            cdiv (xr, xi, yr, yi, zr, zi); wr[i] := 1
      end
      else
```

```
      begin cdiv (yr, yi, xr, xi, zr, zi); wr[i] := −1
      end;
      hr[i, i − 1] := zr; hi[i, i − 1] := zi;
      for j := i step 1 until n do
      begin hr[i, j] := hr[i, j] − zr × hr[i − 1, j] + zi × hi[i − 1, j];
            hi[i, j] := hi[i, j] − zr × hi[i − 1, j] − zi × hr[i − 1, j]
      end j
  end i;
  comment composition R × L = H;
  for j := m + 1 step 1 until n do
  begin xr := hr[j, j − 1]; xi := hi[j, j − 1]; hr[j, j − 1] := hi[j, j − 1] := 0;
        comment interchange columns of hr and hi, if necessary;
        if wr[j] > 0 then
        for i := l step 1 until j do
        begin zr := hr[i, j − 1]; hr[i, j − 1] := hr[i, j]; hr[i, j] := zr;
              zi := hi[i, j − 1]; hi[i, j − 1] := hi[i, j]; hi[i, j] := zi
        end i;
        for i := l step 1 until j do
        begin hr[i, j − 1] := hr[i, j − 1] + xr × hr[i, j] − xi × hi[i, j];
              hi[i, j − 1] := hi[i, j − 1] + xr × hi[i, j] + xi × hr[i, j]
        end i
  end j;
  go to nextit;
  comment a root found;
root: wr[n] := hr[n, n] + tr; wi[n] := hi[n, n] + ti; n := n − 1; go to nextw;
fin:
end comlr;
```

5. Organisational and Notational Details

The complex matrix is stored as two $n \times n$ arrays, one containing the real parts of H and the other containing the imaginary parts. H is destroyed in the process of obtaining the eigenvalues.

Since H and all the transforms are of Hessenberg form, the triangular factorizations are particularly simple. If we consider Eqs. (4) and (5) in the case when A_1 is upper Hessenberg, only $\frac{1}{2}n^2$ (complex) multiplications are required to produce R_1 and another $\frac{1}{2}n^2$ to complete the transform. At each stage in the production of R_1, r' is either r or $r+1$, i.e. one has only to decide whether or not to interchange rows r and $r+1$. Information on these interchanges is stored in the array $wr[i]$. Similarly only one multiplier is involved in each major step of the triangularization. Each of these is stored in the position occupied by the zero it produces. Two code procedures *cdiv* and *csqrt* [7] are used.

6. Discussion of Numerical Properties

Any attempt to give a rigorous *a priori* bound for the errors introduced in s steps of the modified LR algorithm is likely to be misleading since it must cater for extremely remote contingencies. It is more informative to consider one LR step. If R_s is the computed upper-triangular matrix derived from H_s then R_s is the exact matrix for $H_s + E_s$, where E_s certainly satisfies the relation

$$\|E_s\|_2 \le k_1 2^{-t} n^{\frac{3}{2}} \|H_s\|_2, \tag{14}$$

k_1 being of order unity. The factor $n^{\frac{3}{2}}$ is an extreme upper estimate and it is rare for a factor even as large as $n^{\frac{1}{2}}$ to be necessary. Similarly the computed H_{s+1} satisfies the relation

$$T_s(H_s + E_s)T_s^{-1} = H_{s+1} + F_s \tag{15}$$

where

$$\|F_s\|_2 \le k_2 2^{-t} n^{\frac{3}{2}} \|H_{s+1}\|_2. \tag{16}$$

If the sensitivity of each eigenvalue remained constant throughout all the transformation the effect of the errors made in s steps would be no greater than that made by a perturbation in H_1 bounded by $k_3 2^{-t} n^{\frac{3}{2}} s \|H_1\|_2$. Although the possibility that the sensitivity of individual eigenvalues might be increased at an intermediate stage cannot be completely ruled out it would appear that in practice the use of interchanges makes it extremely unlikely.

7. Test Results

The algorithm was used on KDF 9 to find the eigenvalues of a complex Hessenberg matrix $A + iB$ of order seven with

$$A = \begin{bmatrix}
2 & 1 & 6 & 3 & 5 & 2 & 1 \\
1 & 1 & 3 & 5 & 1 & 7 & 2 \\
  & 3 & 1 & 6 & 2 & 1 & 2 \\
  &   & 2 & 3 & 1 & 2 & 3 \\
  &   &   & 2 & 2 & 3 & 1 \\
  &   &   &   & 1 & 2 & 2 \\
  &   &   &   &   & 2 & 1
\end{bmatrix}$$

$$B = \begin{bmatrix}
4 & 1 & 2 & 3 & 5 & 6 & 1 \\
2 & 3 & 1 & -4 & 1 & 2 & 3 \\
  & -2 & 1 & 3 & 1 & 4 & 1 \\
  &   & 3 & 1 & 2 & 2 & 1 \\
  &   &   & -1 & 2 & 1 & 3 \\
  &   &   &   & -1 & 1 & 2 \\
  &   &   &   &   & -2 & 1
\end{bmatrix}.$$

For comparison the algorithm was then used to find the eigenvalues of $X + iY$ where X and Y are defined by

$$x_{ij} = b_{8-j,\,8-i}, \qquad y_{ij} = a_{8-j,\,8-i}.$$

Clearly if $\lambda + i\mu$ is an eigenvalue of $A + iB$ then $\mu + i\lambda$ is an eigenvalue of $X + iY$, though the two calculations have no common features. The two computed sets of eigenvalues are given in the Table and it will be seen that the maximum disagreement is less than $10^{-11} \|A\|_\infty$, a very satisfactory result for computation with a 39 binary digit mantissa. The average number of iterations per eigenvalue is

Table

Real Part	Imaginary Part	Number of Iterations
	Eigenvalues of $A + iB$	
$+ 8.2674\ 0340\ 990_{10} + 0;$	$+ 3.7046\ 7442\ 049_{10} + 0;$	4
$- 2.7081\ 1509\ 774_{10} + 0;$	$- 2.8029\ 1696\ 409_{10} + 0;$	5
$- 8.0109\ 2562\ 924_{10} - 1;$	$+ 4.9693\ 6407\ 474_{10} + 0;$	4
$+ 5.3137\ 8016\ 733_{10} + 0;$	$+ 1.2242\ 0710\ 121_{10} + 0;$	3
$+ 2.0808\ 8120\ 722_{10} + 0;$	$+ 1.9340\ 7686\ 266_{10} + 0;$	2
$+ 9.9492\ 5700\ 448_{10} - 1;$	$+ 3.1688\ 3486\ 630_{10} + 0;$	1
$- 1.1477\ 8282\ 403_{10} + 0;$	$+ 8.0175\ 9638\ 742_{10} - 1;$	0
	Eigenvalues of $X + iY$	
$+ 3.7046\ 7442\ 048_{10} + 0;$	$+ 8.2674\ 0340\ 976_{10} + 0;$	6
$- 2.8029\ 1696\ 406_{10} + 0;$	$- 2.7081\ 1509\ 749_{10} + 0;$	3
$+ 1.2242\ 0710\ 126_{10} + 0;$	$+ 5.3137\ 8016\ 740_{10} - 0;$	4
$+ 8.0175\ 9638\ 773_{10} - 1;$	$- 1.1477\ 8282\ 402_{10} + 0;$	3
$+ 4.9693\ 6407\ 477_{10} + 0;$	$- 8.0109\ 2562\ 982_{10} - 1;$	3
$+ 3.1688\ 3486\ 629_{10} + 0;$	$+ 9.9492\ 5700\ 441_{10} - 1;$	1
$+ 1.9340\ 7686\ 267_{10} + 0;$	$+ 2.0808\ 8120\ 724_{10} + 0;$	0

about three for both matrices. This average has been typical for eigenvalues of matrices of orders 5 to 50 which have been computed.

The above results have been confirmed by Dr. C. REINSCH at the Technische Hochschule München.

Further examples of the use of the procedure are given in [7], in which the reduction of a general complex matrix to upper-Hessenberg form is described.

Acknowledgements. The authors are indebted to Dr. C. REINSCH of the Technische Hochschule München who suggested a number of important improvements in the algorithm and tested successive versions of the procedure.

The work described above has been carried out at the National Physical Laboratory.

References

1. FRANCIS, J. G. F.: The QR transformation — a unitary analogue to the LR transformation. Comput. J. **4**, 265—271 and 332—345 (1961/62).
2. PARLETT, B. N.: The development and use of methods of LR type. SIAM Rev. **6**, 275—295 (1964).
3. RUTISHAUSER, H.: Solution of eigenvalue problems with the LR transformation. Nat. Bur. Standards Appl. Math. Ser. **49**, 47—81 (1958).
4. WILKINSON, J. H.: The algebraic eigenvalue problem. London: Oxford University Press 1965.
5. — Convergence of the LR, QR and related algorithms. Comput. J. **8**, 77—84 (1965).
6. PARLETT, B. N., and C. REINSCH: Balancing a matrix for calculation of eigenvalues and eigenvectors. Numer. Math. **13**, 293—304 (1969). Cf. II/11.
7. MARTIN, R. S., and J. H. WILKINSON: Similarity reduction of a general matrix to Hessenberg form. Numer. Math. **12**, 349—368 (1968). Cf. II/13.
8. —, G. PETERS and J. H. WILKINSON: The QR algorithm for real Hessenberg matrices. Numer. Math. **14**, 219—231 (1970). Cf. II/14.

Solution to the Complex Eigenproblem by a Norm Reducing Jacobi Type Method[*]

by P. J. EBERLEIN

1. Theoretical Background

Let $C = (c_{ij}) = A + iZ$ be a complex $n \times n$ matrix having real part $A = (a_{ij})$ and imaginary part $Z = (z_{ij})$. We construct a complex matrix $W = T + iU = W_1 W_2 \ldots W_k$ as a product of non-singular two dimensional transformations W_j such that the off diagonal elements of $W^{-1} C W = \tilde{C}$ are arbitrarily small[1]. The diagonal elements of $\tilde{C}$ are now approximations to the eigenvalues and the columns of W are approximations to the corresponding right eigenvectors.

The program given here is a generalization of an earlier Jacobi type program for real matrices appearing in [1]. As in [1] each W_j is expressed as a product RS, where R now is a complex rotation defined by:

$R: \quad r_{kk} = r_{mm} = \cos x; \qquad r_{km} = -r_{mk}^* = -e^{i\alpha} \sin x;\text{[2]} \qquad r_{ij} = \delta_{ij} \text{ otherwise};$

and S is a complex shear:

$S: \quad s_{kk} = s_{mm} = \cosh y; \qquad s_{km} = s_{mk}^* = -i e^{i\beta} \sinh y; \qquad s_{ij} = \delta_{ij} \text{ otherwise}.$

As before the pairs (k, m) are arbitrarily chosen in cyclic order.

In the earlier paper [1] the choice of the single rotational parameter x reduces to that of the usual Jacobi parameter when the matrix is symmetric. Here in the complex case the parameters α and x are chosen so that if the matrix is either Hermitian or skew-Hermitian then we obtain the usual Jacobi parameters. In particular if

$$|c_{km} + c_{mk}^*|^2 + [\operatorname{Re}(c_{kk} - c_{mm})]^2 \geq |c_{km} - c_{mk}^*|^2 + [\operatorname{Im}(c_{kk} - c_{mm})]^2$$

then we choose α and x by (1.1) below; otherwise by (1.2).

(1.1) Choose α so that

$$\operatorname{Im}(c_{km} - c_{mk}) \cos \alpha - \operatorname{Re}(c_{km} + c_{mk}) \sin \alpha = 0,$$

and so that $\cos \alpha \operatorname{Re}(c_{km} + c_{mk}) \geq 0$.

x is chosen so that

$$\tan 2x = |c_{km} + c_{km}^*| / \operatorname{Re}(c_{kk} - c_{mm}); \quad \text{and so that} \quad \sin x \operatorname{Re}(c_{kk} - c_{mm}) \geq 0,$$

i.e., that smaller angle is chosen.

---

[*] Prepublished in Numer. Math. **14**, 232—245 (1970).
[1] Note that the "ones" above the diagonal of a matrix in Jordan canonical form may be taken as "epsilons".
[2] The star * denotes complex conjugate.

(1.2)   Choose α so that

$$\text{Re}\,(c_{km}-c_{mk})\cos\alpha+\text{Im}\,(c_{km}+c_{mk})\sin\alpha=0,$$

and so that $\cos\alpha\,\text{Im}\,(c_{km}+c_{mk})\geq0$.

x is chosen so that

$$\tan 2x=|c_{km}-c_{mk}^*|/\text{Im}\,(c_{kk}-c_{mm}).$$

Again the smaller angle is chosen.

The transformations $W_i=R_iS_i$ are performed together in order to reduce the number of multiplications involved. To discuss the choice of the shear parameters, let us suppose that c_{ij} are the already rotated elements and denote by $C'=(c'_{ij})$ the elements of $C'=S^{-1}CS$ for a single shear S in the (k, m) plane. The parameters y and β of S are chosen to decrease $\|C'\|_E$ thus making the matrix C' closer to normal than C [3]:

$$\cos\beta=-\text{Im}\,[(CC^*-C^*C)_{km}]/|(C^*C-CC^*)_{km}|;\,[4]$$

(1.3) $\sin\beta=\text{Re}\,[(CC^*-C^*C)_{km}]/|(C^*C-CC^*)_{km}|;$

$$\tanh y=|(C^*C-CC^*)_{km}|/(G+2(|c_{kk}-c_{mm}|^2+|c_{km}\,e^{-i\beta}+c_{mk}\,e^{i\beta}|^2))$$

where

$$G=\sum_{i\neq k,\,m}(|c_{ki}|^2+|c_{ik}|^2+|c_{mi}|^2+|c_{im}|^2).$$

The shear parameters β and y so chosen reduce the Euclidean norm of

$$C^{(i+1)}=W_i^{-1}C^{(i)}W_i.$$

It is shown [2] that

$$\|C^{(i)}\|_E^2-\|C^{(i+1)}\|_E^2\geq\tfrac{1}{3}|(C^{(i)}*C^{(i)}-C^{(i)}C^{(i)}*)_{km}|^2/\|C^{(i)}\|^2;$$

if the matrix is normal then $\beta=y=0$ and $S=I$ and $\|C^{(i)}\|_E^2$ is unchanged.

The details of the computation of the rotated elements in (1.3) are discussed in Section 5.

2. Applicability

This algorithm may be used to solve the eigenproblem for any complex $n\times n$ matrix. If the matrix is normal then other methods are clearly faster; however, if the matrix is Hermitian or skew-Hermitian, the algorithm reduces to the usual Jacobi method.

Since the publication of [1] correspondence has suggested the desirability of a procedure for complex matrices. Also it was clear that the slow convergence of the real procedure for real matrices with pairs of complex roots made the early program prohibitive for large matrices. The complex procedure given here is generally recommended over the real procedure. Empirically we have found that the real procedure is competitive with the complex procedure only for real

---

3 Note the Euclidean norm is invariant under the rotations.
4 The star denotes conjugate transpose.

matrices all of whose roots are real. The complex program does not fail, as does the real procedure, in the complex defective case; however convergence is linear.

Left vectors, or both right and left vectors may be obtained by making the obvious changes to the program.

3. Formal Parameter List

n is the order of the $n \times n$ complex matrix $a + iz$.

a, z are $n \times n$ arrays, the real and imaginary parts of the matrix $a + iz$ for which the eigenproblem is to be solved. These matrices are destroyed during the procedure; upon exit the real and imaginary parts of the eigenvalues occupy the diagonal elements of a and z respectively. These will appear in order of decreasing magnitude.

t, u are the $n \times n$ arrays reserved to receive the real and imaginary parts of the eigenvectors which will occupy the columns of $W = t + iu$.

4. ALGOL Program

procedure comeig (n) trans: (a, z) result: (t, u);

value n;

integer n;

array a, z, t, u;

comment solves the eigenproblem for the complex matrix $a + iz = C$ ($a[1:n, 1:n]$, $z[1:n, 1:n]$) by finding a sequence of similarity transformations $W_1 W_2 \ldots = W$ such that $W^{-1}CW$ is approximately diagonal. Arrays $t[1:n, 1:n]$ and $u[1:n, 1:n]$ must be provided to receive the eigenvectors $W = t + iu$. The maximum number of iterations, or sweeps, is limited to 35. A machine dependent variable eps is set to 10^{-14} within the program. This variable should be changed appropriately when the program is used;

begin real eps, tau, tem, tep, max, hj, hr, hi, g, te, tee, br, bi, er, ei, dr, di, root, root1, root2, ca, sa, cb, sb, cx, sx, sh, ch, eta, tse, ind, nc, s, c, cot2x, d, de, sig, cotx, cos2a, sin2a, tanh, c1r, c2r, s1r, s2r, c1i, c2i, s1i, s2i, isw, b, e, zim, zik, aik, aim, tim, tik, uim, uik, ami, aki, zki, zmi;

integer i, j, k, m, it; **Boolean** mark; **array** en$[1:n]$;

eps: $= {}_{10}-14$;

mark: $=$ **false**;

comment put identity in t and zero in u ;

for $i := 1$ **step** 1 **until** n **do**

   begin $t[i, i] := 1.0$; $u[i, i] := 0.0$;

                  for $j := i+1$ **step** 1 **until** n **do**

                     $t[i, j] := t[j, i] := u[i, j] := u[j, i] := 0.0$

      end;

comment safety loop;

for $it := 1$ **step** 1 **until** 35 **do**

```
begin if mark then goto done;
      comment convergence criteria;
      tau := 0.0 ;
      for k := 1 step 1 until n do
      begin tem := 0.0 ;
            for i := 1 step 1 until n do
                if i ⧸= k then tem := abs (a[i, k]) + abs (z[i, k]) + tem;
            tau := tau + tem ;
            en [k] := tem + abs (a [k, k]) + abs (z [k, k])
      end;
      comment interchange rows and columns ;
      for k := 1 step 1 until n − 1 do
      begin max := en [k]; i := k ;
            for j := k + 1 step 1 until n do
            if en[j] > max then
            begin max := en[j]; i := j
            end;
            if i ⧸= k then
            begin en[i] := en[k];
                  for j := 1 step 1 until n do
                  begin tep := a[k, j]; a[k, j] := a[i, j];
                        a[i, j] := tep; tep := z[k, j];
                        z[k, j] := z[i, j]; z[i, j] := tep
                  end j;
                  for j := 1 step 1 until n do
                  begin tep := a[j, k]; a[j, k] := a[j, i];
                        a[j, i] := tep; tep := z[j, k];
                        z[j, k] := z[j, i]; z[j, i] := tep;
                        tep := t[j, k]; t[j, k] := t[j, i];
                        t[j, i] := tep; tep := u[j, k];
                        u[j, k] := u[j, i]; u[j, i] := tep;
                  end j
            end if
      end k;
      if tau < 100.0 × eps then goto done;
      comment begin sweep;
      mark := true;
      for k := 1 step 1 until n − 1 do
      for m := k + 1 step 1 until n do
      begin hj := hr := hi := g := 0.0 ;
```

```
for i := 1 step 1 until n do
if   i ≠ k ∧ i ≠ m then
begin hr := hr + a[k, i] × a[m, i] + z[k, i] × z[m, i]
               − a[i, k] × a[i, m] − z[i, k] × z[i, m];
      hi := hi + z[k, i] × a[m, i] − a[k, i] × z[m, i]
               − a[i, k] × z[i, m] + z[i, k] × a[i, m];
      te := a[i, k] × a[i, k] + z[i, k] × z[i, k]
               + a[m, i] × a[m, i] + z[m, i] × z[m, i];
      tee := a[i, m] × a[i, m] + z[i, m] × z[i, m]
               + a[k, i] × a[k, i] + z[k, i] × z[k, i];
      g  := g + te + tee; hj := hj − te + tee;
end;
br := a[k, m] + a[m, k];  bi := z[k, m] + z[m, k];
er := a[k, m] − a[m, k];  ei := z[k, m] − z[m, k];
dr := a[k, k] − a[m, m];  di := z[k, k] − z[m, m];
te := br × br + ei × ei + dr × dr;
tee := bi × bi + er × er + di × di;
if te ≥ tee then
begin isw := 1.0;  c := br;  s := ei;  d := dr;
      dc := di;  root2 := sqrt (te)
end
```
else
```
      begin isw := − 1.0;  c := bi;  s := − er;  d := di;
            de := dr;  root2 := sqrt (tee)
      end;
root1 := sqrt (s × s + c × c);
sig := if d ≥ 0.0 then 1.0 else − 1.0;
sa := 0.0;  ca := if c ≥ 0.0 then 1.0 else − 1.0;
if root1 < eps then
      begin sx := sa := 0.0;  cx := ca := 1.0;
            e := if isw > 0.0 then er else ei;
            b := if isw > 0.0 then bi else − br;
            nd := d × d + de × de;
            goto enter1
      end;
if abs (s) > eps then
      begin ca := c/root1;  sa := s/root1 end;
cot2x := d/root1;
cotx := cot2x + (sig × sqrt (1.0 + cot2x × cot2x));
sx := sig/sqrt (1.0 + cotx × cotx);  cx := sx × cotx;
```

comment find rotated elements;

$eta := (er \times br + bi \times ei)/root1;$

$tse := (br \times bi - er \times ei)/root1;$

$te := sig \times (-root1 \times de + tse \times d)/root2;$

$tee := (d \times de + root1 \times tse)/root2;$

$nd := root2 \times root2 + tee \times tee;$

$tee := hj \times cx \times sx; \ cos2a := ca \times ca - sa \times sa;$

$sin2a := 2.0 \times ca \times sa; \ tem := hr \times cos2a + hi \times sin2a;$

$tep := hi \times cos2a - hr \times sin2a;$

$hr := cx \times cx \times hr - sx \times sx \times tem - ca \times tee;$

$hi := cx \times cx \times hi + sx \times sx \times tep - sa \times tee;$

$b := isw \times te \times ca + eta \times sa;$

$e := ca \times eta - isw \times te \times sa;$

enter1: $s := hr - sig \times root2 \times e;$

$c := hi - sig \times root2 \times b;$

$root := sqrt(c \times c + s \times s);$

if $root < eps$ **then**

begin $cb := ch := 1.0; \ sb := sh := 0.0;$ **goto** *trans*

end;

$cb := -c/root; \ sb := s/root;$

$tee := cb \times b - e \times sb; \ nc := tee \times tee;$

$tanh := root/(g + 2.0 \times (nc + nd));$

$ch := 1.0/sqrt(1.0 - tanh \times tanh);$

$sh := ch \times tanh;$

comment prepare for transformation;

trans: $tem := sx \times sh \times (sa \times cb - sb \times ca);$

$c1r := cx \times ch - tem; \ c2r := cx \times ch + tem;$

$c1i := c2i := -sx \times sh \times (ca \times cb + sa \times sb);$

$tep := sx \times ch \times ca; \ tem := cx \times sh \times sb;$

$s1r := tep - tem; \ s2r := -tep - tem;$

$tep := sx \times ch \times sa; \ tem := cx \times sh \times cb;$

$s1i := tep + tem; \ s2i := tep - tem;$

comment decide whether to make transformation;

$tem := sqrt(s1r \times s1r + s1i \times s1i);$

$tep := sqrt(s2r \times s2r + s2i \times s2i);$

if $tem > eps$ v $tep > eps$ **then**

begin $mark :=$ **false**;

comment transformation on left;

for $i := 1$ **step** 1 **until** n **do**

```
        begin aki := a[k, i];  ami := a[m, i];
              zki := z[k, i];  zmi := z[m, i];
              a[k, i] := c1r × aki − c1i × zki
                        + s1r × ami − s1i × zmi;
              z[k, i] := c1r × zki + c1i × aki
                        + s1r × zmi + s1i × ami;
              a[m, i] := s2r × aki − s2i × zki
                        + c2r × ami − c2i × zmi;
              z[m, i] := s2r × zki + s2i × aki
                        + c2r × zmi + c2i × ami;
        end;
        comment transformation on right;
        for i := 1 step 1 until n do
        begin aik := a[i, k];  aim := a[i, m];
              zik := z[i, k];  zim := z[i, m];
              a[i, k] := c2r × aik − c2i × zik
                        − s2r × aim + s2i × zim;
              z[i, k] := c2r × zik + c2i × aik
                        − s2r × zim − s2i × aim;
              a[i, m] := − s1r × aik + s1i × zik
                        + c1r × aim − c1i × zim;
              z[i, m] := − s1r × zik − s1i × aik
                        + c1r × zim + c1i × aim;
              tik := t[i, k];  tim := t[i, m];
              uik := u[i, k];  uim := u[i, m];
              t[i, k] := c2r × tik − c2i × uik
                        − s2r × tim + s2i × uim;
              u[i, k] := c2r × uik + c2i × tik
                        − s2r × uim − s2i × tim;
              t[i, m] := − s1r × tik + s1i × uik
                        + c1r × tim − c1i × uim;
              u[i, m] := − s1r × uik − s1i × tik
                        + c1r × uim + c1i × tim;
        end;
      end if;
    end km loops;
  end it loop;
done: end comeig;
```

5. Organisational and Notational Details

The original matrix $C = a + iz$ is replaced by $\tilde{C} = W^{-1} C W$, an approximately diagonal matrix. The internal variable *eps* is machine precision and must be set appropriately for the individual machine. Contrary to what might be expected, convergence is faster when *eps* is chosen sufficiently small; the shear parameter must be used to its full extent in order that the iterates come sufficiently close to a normal matrix to allow convergence. The convergence criteria (we assume that the norm of the matrix does not differ by more than a few orders of magnitude from one) is to exit whenever one of the following occurs:

1. $\sum_{i \neq j} |c_{ij}| < 100 \, eps \,^5$.

2. No transformation occurs for one full sweep.

3. 35 iterations have taken place.

The columns of the matrix are permuted once each sweep as necessary so that upon exit, if $\lambda_i = \bar{\lambda}_i + i \bar{\bar{\lambda}}_i$. Then $|\bar{\lambda}_1| + |\bar{\bar{\lambda}}_1| \geq |\bar{\lambda}_2| + |\bar{\bar{\lambda}}_2| \geq \cdots \geq |\bar{\lambda}_n| + |\bar{\bar{\lambda}}_n|$.

This procedure finds the right vectors only. This choice was made because the complex arithmetic is clumsy. The program is easily altered so that left vectors, or both right and left vectors may be obtained.

Since the $W_i = R_i S_i$ are performed together the elements β and y of the shear transformation are computed in terms of the already rotated elements: we let $C' = R^{-1} C R = (c'_{ij})$ whether (1.1) or (1.2) is used. We include here the formulae we use in finding the necessary elements of C' since their derivation is by no means obvious in the program. For the two cases we have:

$$d = \operatorname{Re}(c_{kk} - c_{mm}) \qquad\qquad d = \operatorname{Im}(c_{kk} - c_{mm})$$
$$de = \operatorname{Im}(c_{kk} - c_{mm}) \qquad\qquad de = \operatorname{Re}(c_{kk} - c_{mm})$$
$$root1 = |c_{km} + c^*_{mk}| \qquad\qquad root1 = |c_{km} - c^*_{mk}|$$
$$(5.1) \quad isw = +1 \qquad\qquad (5.2) \quad isw = -1$$
$$c = \operatorname{Re}(c_{km} + c_{mk}) \qquad\qquad c = \operatorname{Im}(c_{km} + c_{mk})$$
$$s = \operatorname{Im}(c_{km} - c_{mk}) \qquad\qquad s = -\operatorname{Re}(c_{km} - c_{mk})$$
$$root2 = \sqrt{d^2 + root1^2} \qquad\qquad root2 = \sqrt{d^2 + root1^2}$$

Now for either case we define, if $root1 \neq 0$,

$$\cos \alpha = c/root1, \quad \sin \alpha = s/root1, \quad \tan 2x = root1/d,$$

$$sig = \begin{matrix} +1 \\ -1 \end{matrix} \quad \text{as} \quad d \begin{matrix} \geq 0 \\ < 0 \end{matrix}.$$

$$eta = (|c_{km}|^2 - |c_{mk}|^2)/root1$$
$$(5.3) \qquad tse = 2 \operatorname{Im}(c_{km} c_{mk})/root1$$
$$T = sig(-root1 \times de + tse \times d)/root2$$
$$TT = (d \times de + root1 \times tse)/root2$$
$$B = T \times \cos \alpha + eta \times \sin \alpha$$
$$E = eta \times \cos \alpha - T \times \sin \alpha.$$

In the first case we have $(c'_{kk} - c'_{mm}) = sig(root2 + iTT)$; and in the second $(c'_{kk} - c'_{mm}) = sig(TT + i \, root2)$. Hence in either case we have

$$|c'_{kk} - c'_{mm}|^2 = root2^2 + (TT)^2, \quad \text{which we need for (1.3)}.$$

---

5 The factor of 100 may be chosen larger without changing the rate of convergence.

To find $(C'C*'-C*'C')_{km}$ we define $H = \sum\limits_{i \neq k, m} (c_{ki} c^*_{mi} - c^*_{ik} c_{im})$, and directly find

$$H' = hr + i\, hi = \cos^2 x\, H - e^{2i\alpha} H^* \sin^2 x - \tfrac{1}{2} \sin 2x\, hj \cdot e^{i\alpha}$$

where

$$hj = \sum\limits_{i \neq k, m} (|c_{ki}|^2 + |c_{im}|^2 - |c_{ik}|^2 - |c_{mi}|^2);$$

G is invariant under R and stands as defined in (1.3). Using the above definition of hr and hi, we have in either of the two cases

$$\mathrm{Re}\left((C'C*' - C*'C')_{km}\right) = hr - sig \times root2 \times E,$$

(5.4) $\qquad \mathrm{Im}\left((C'C*' - C*C')_{km}\right) = hi - sig \times root2 \times B, \qquad$ and

$$|c'_{km} e^{-i\beta} + c'_{mk} e^{i\beta}|^2 = (B \cos \beta - E \sin \beta)^2.$$

Suppose now $root1 = 0$; we do not rotate, i.e. $\cos x = \cos \alpha = 1$ and $\sin x = \sin \alpha = 0$, hr and hi are just the real and imaginary parts of H. We may still use (5.4) if we set

(1)	(2)
$B = \mathrm{Im}(c_{km} + c_{mk})$	$B = \mathrm{Re}(c_{km} + c_{mk})$
$E = \mathrm{Re}(c_{km} - c_{mk})$	$E = \mathrm{Im}(c_{km} - c_{mk}).$
(when $isw = 1$)	(when $isw = -1$)

6. Discussion of Numerical Properties

The general remarks made in [1] hold for this procedure.

7. Test Results

The procedure was tested on the CDC 6400 (S.U.N.Y. at Buffalo) which has a 48 bit mantissa and on the IBM 360-50 (University of Rochester) using the "long word" which has 14 hexadecimal digits.

Example I

$$\begin{pmatrix} -1 & 1 & & & & & \\ -1 & 0 & 1 & & & & \\ -1 & 0 & 0 & 1 & & 0 & \\ -1 & 0 & 0 & 0 & 1 & & \\ -1 & 0 & 0 & 0 & 0 & 1 & \\ -1 & 0 & 0 & 0 & 0 & 0 & 1 \\ -1 & 0 & 0 & 0 & 0 & 0 & 1 \\ -1 & 0 & 0 & 0 & 0 & 0 & 0 \end{pmatrix}$$

The eigenvalues of this matrix are -1, $\pm i$, $\sqrt{2}/2(-1 \pm i)$ and $\sqrt{2}/2(1 \pm i)$; The right vector corresponding to -1 is $(1, 0, 1, 0, 1, 0, 1)$.

This example is given in [1] where the real procedure took 25 sweeps before convergence took place. For the complex procedure given here convergence took place after only 8 sweeps.

Example II

$$\begin{pmatrix} 15 & 11 & 6 & -9 & -15 \\ 1 & 3 & 9 & -3 & -8 \\ 7 & 6 & 6 & -3 & -11 \\ 7 & 7 & 5 & -3 & -11 \\ 17 & 12 & 5 & -10 & -16 \end{pmatrix}.$$

This matrix is defective with eigenvalues -1, and doubleroots $\frac{1}{2}(3 \pm i \sqrt{51})$. The real program [1] does not converge for this matrix but the complex procedure given here does at least to half a word length.

Example III

$$B_{13} = \begin{bmatrix} 13 & 12 & 11 & \ldots & \ldots & 2 & 1 \\ 12 & 12 & 11 & 10 & \ldots & 2 & 1 \\ 0 & 11 & 11 & 10 & \ldots & 2 & 1 \\ 0 & 0 & 10 & 10 & \ldots & 2 & 1 \\ \cdot & \cdot & \cdot & \cdot & \cdot & \cdot & \cdot \\ 0 & 0 & \ldots & & & 2 & 2 & 1 \\ 0 & 0 & \ldots & & & 0 & 1 & 1 \end{bmatrix}.$$

This matrix has six well conditioned roots, and their reciprocals, which are poorly conditioned. The thirteenth root, well conditioned, is one.

They are (rounded):   1.

35.61386 12008	0.02807 89548
23.03753 22899	0.04340 74270
14.62978 21335	0.06835 37178
8.75377 95670	0.11423 63698
4.76730 65328	0.20976 20518
2.29891 12322	0.43498 85224

The right vector associated with the eigenvalue one is:

$$\left(\frac{1}{43680}, 0, -\frac{1}{3640}, 0, \frac{1}{384}, 0, -\frac{1}{48}, 0, \frac{1}{8}, 0, -\frac{1}{2}, 0, 1 \right).$$

Example IV

$$\begin{pmatrix} 1+2i & 3+4i & 21+22i & 23+24i & 41+42i \\ 43+44i & 13+14i & 15+16i & 33+34i & 35+36i \\ 5+6i & 7+8i & 25+26i & 27+28i & 45+46i \\ 37+48i & 17+18i & 19+20i & 37+38i & 39+40i \\ 9+10i & 11+12i & 29+30i & 31+32i & 49+50i \end{pmatrix}.$$

This is a well conditioned complex matrix with unknown exact roots.

Example V

$$\begin{bmatrix} 5i & 1+2i & 2+3i & -3+6i & 6 \\ -1+2i & 6i & 4+5i & -3-2i & 5 \\ -2+3i & -4+5i & 7i & +3 & 2 \\ 3+6i & 3-2i & -3 & -5i & 2+i \\ -6 & -5 & -2 & -2+i & 2i \end{bmatrix}$$

This skew-Hermitian matrix in real form is skew symmetric and the real program [1] fails. As for example IV the roots are unknown.

The actual results obtained for these matrices (U. of R. IBM 360-50 with $eps = 10^{-14}$) are given below. The vectors were normalized outside the procedure.

Example I

Iteration number

$tau = +1.20000\,00000\,00000' + 01$ 0

$tau = +1.23066\,84573\,69602' + 01$ 1

$tau = +3.60671\,61862\,27806' + 00$ 2

$tau = +7.44677\,14753\,83664' - 01$ 3

$tau = +8.66522\,29564\,55960' - 02$ 4

$tau = +2.33068\,28366\,74229' - 03$ 5

$tau = +1.59919\,52962\,01154' - 06$ 6

$tau = +7.20201\,92566\,13738' - 12$ 7

$tau = +8.10365\,15784\,68963' - 16$ 8

$Re(\lambda_i)$	$Im(\lambda_i)$
$+7.07106\,78118\,65344' - 01$	$-7.07106\,78118\,65301' - 01$
$+7.07106\,78118\,65338' - 01$	$+7.07106\,78118\,65299' - 01$
$-7.07106\,78118\,65338' - 01$	$-7.07106\,78118\,65290' - 01$
$-7.07106\,78118\,65315' - 01$	$+7.07106\,78118\,65290' - 01$
$-9.99999\,99999\,99824' - 01$	$-2.04975\,70945\,87373' - 16$
$+1.14687\,38108\,10045' - 15$	$-9.99999\,99999\,99728' - 01$
$+8.13000\,11835\,25542' - 16$	$+9.99999\,99999\,99728' - 01$

The approximate eigenvector corresponding to $\lambda = -1$ is:

$(+1.00000\,00000\,00000' + 00$	$+ \qquad\qquad\qquad 0i)$
$(-5.98719\,61276\,57948' - 15$	$+ -3.13525\,17466\,58850' - 15i)$
$(+9.99999\,99999\,99968' - 01$	$+1.54240\,46514\,44486' - 16i)$
$(-6.64742\,85795\,64983' - 16$	$+1.59402\,48389\,08926' - 15i)$
$(+9.99999\,99999\,99936' - 01$	$-2.00512\,60468\,77833' - 15i)$
$(-2.17566\,90259\,60504' - 16$	$+7.57092\,69240\,88728' - 16i)$
$(+9.99999\,99999\,99984' - 01$	$+4.62721\,39543\,33459' - 16i)$

Example II. The quadratic convergence exhibited in example I is not present for this defective matrix. Convergence is linear and after 35 sweeps $tau = 1.7 \times_{10} - 7$. The real and imaginary parts of the eigenvalues obtained are as follows.

$Re(\lambda_i)$	$Im(\lambda_i)$
$+1.50000\,04047\,91030' + 00$	$+3.57071\,42586\,29270' + 00$
$+1.49999\,98730\,82816' + 00$	$-3.57071\,46850\,18749' + 00$
$+1.50000\,01269\,16637' + 00$	$-3.57071\,37435\,23373' + 00$
$+1.49999\,95952\,08434' + 00$	$+3.57071\,41699\,12738' + 00$
$-9.99999\,99999\,99344' - 01$	$+9.83671\,15171\,43487' - 15$

Example III. Convergence is slow, but ultimately quadratic for this ill-conditioned matrix.

	Iteration number
$tau = +4.42000\,00000\,00000' + 02$	0
$tau = +1.94896\,86244\,00691' + 02$	1
$\vdots$	
$tau = +9.76651\,95692\,58384' - 01$	23
$tau = +7.00200\,16021\,21469' - 01$	24
$tau = +5.32647\,02772\,29778' - 01$	$\vdots$
$tau = +3.33257\,62098\,09161' - 01$	
$tau = +2.19668\,41491\,14417' - 01$	
$tau = +1.20931\,37833\,50892' - 01$	
$tau = +4.81534\,84148\,48147' - 02$	
$tau = +9.44738\,21526\,14893' - 03$	
$tau = +2.09693\,32462\,38059' - 04$	31
$tau = +3.76999\,22047\,00596' - 09$	32
$tau = +6.12567\,69029\,63803' - 14$	33

$Re(\lambda_i)$	$Im(\lambda_i)$
$+3.56138\,61200\,82135' + 01$	$+3.29022\,92211\,93522' - 15$
$+2.30375\,32289\,94105' + 01$	$-2.17539\,11269\,16770' - 14$
$+1.46297\,82133\,52061' + 01$	$+1.16413\,32815\,93331' - 14$
$+8.75377\,95670\,20512' + 00$	$-9.90569\,98619\,23577' - 15$
$+4.76730\,65327\,94814' + 00$	$-2.93213\,77662\,41567' - 14$
$+2.29891\,12322\,32686' + 00$	$+1.21094\,52046\,96430' - 13$
$+9.99999\,99994\,52672' - 01$	$+3.11979\,69359\,58598' - 12$
$+4.34988\,52503\,09144' - 01$	$-3.40749\,00868\,62150' - 10$
$+2.09762\,03389\,05566' - 01$	$+5.59286\,86126\,60224' - 09$
$+1.14236\,26673\,02596' - 01$	$-2.21076\,95079\,85426' - 08$
$+6.83545\,12912\,37060' - 02$	$+1.71587\,89409\,37890' - 08$
$+4.34060\,63030\,43178' - 02$	$+2.17243\,89703\,67746' - 08$
$+2.80796\,42115\,05525' - 02$	$-2.20308\,14978\,42396' - 08$

The approximate eigenvector corresponding to the eigenvalue 1 is:

$+2.17013\,88880\,52055' - 05$ $+7.23400\,17224\,55552' - 16$

$+1.34884\,09740\,21067' - 14$ $-8.71969\,02505\,94967' - 16$

$-2.60416\,66659\,51648' - 04$ $-4.55609\,90477\,83963' - 15$

$-1.45625\,99480\,81003' - 13$ $+9.28305\,91932\,48772' - 15$

$+2.60416\,66661\,04733' - 03$ $+3.17788\,04457\,30703' - 14$

$+1.12544\,54734\,04106' - 12$ $-6.56698\,25639\,60713' - 14$

$-2.08333\,33329\,94071' - 02$ $-1.93400\,11258\,41413' - 13$

$-6.76814\,08217\,46133' - 12$ $+3.90047\,52292\,35965' - 13$

$+1.24999\,99998\,64248' - 01$ $+7.74073\,58618\,49072' - 13$

$+2.71005\,73824\,72533' - 11$ $-1.55426\,73403\,73056' - 12$

$-4.99999\,99997\,28573' - 01$ $-1.55182\,79506\,28571' - 12$

$-5.42868\,55910\,32183' - 11$ $+3.10692\,82087\,17494' - 12$

$+1.00000\,00000\,00000' + 00$

The first digit known to be in error is underlined.

Example IV.

	Iteration number
$tau = +1.02000\,00000\,00000' + 03$	0
$tau = +2.42287\,73820\,23283' + 02$	1
$tau = +8.14773\,83395\,19814' + 01$	2
$tau = +3.17717\,42118\,82053' + 01$	3
$tau = +1.06935\,13707\,40482' + 01$	$\vdots$
$tau = +1.87322\,41109\,31331' + 00$	
$tau = +5.36583\,75535\,98266' - 02$	
$tau = +3.99647\,53349\,39394' - 05$	
$tau = +4.92498\,86690\,38457' - 11$	
$tau = +3.67301\,62274\,59524' - 14$	9

$Re(\lambda_i)$	$Im(\lambda_i)$
$+1.27386\,67077\,30661' + 02$	$+1.32278\,20320\,01201' + 02$
$-9.45998\,40218\,91152' + 00$	$+7.28018\,58369\,23632' + 00$
$+7.07331\,32488\,23640' + 00$	$-9.55838\,90370\,45312' + 00$
$+1.05851\,29313\,37603' - 14$	$+1.91511\,77468\,57339' - 14$
$-2.01340\,53877\,18155' - 15$	$-1.37504\,65667\,64295' - 14$

Example V

	Iteration number
$tau = +1.00000\,00000\,00000' + 02$	0
$tau = +4.76312\,51466\,56103' + 01$	1
$tau = +9.41633\,41427\,69200' + 00$	2
$tau = +1.96085\,02333\,2847' - 01$	3
$tau = +3.07330\,88718\,68031' - 04$	4
$tau = +7.98424\,17913\,25072' - 17$	5

$Re(\lambda_i)$	$Im(\lambda_i)$
$+4.26484\,13949\,50907' - 16$	$+1.78171\,16774\,09606' + 01$
$+2.14310\,09488\,95139' - 16$	$-1.07336\,41878\,00040' + 01$
$+1.98916\,15695\,13088' - 16$	$+9.31585\,83897\,72048' + 00$
$+2.19296\,05043\,06690' - 16$	$-2.99912\,28321\,55645' + 00$
$-4.61944\,19859\,98074' - 17$	$+1.59978\,95462\,87834' + 00$

This work was partially supported by NSF grants GP-8954 GP 13287.

The author would like to thank Mrs. Elizabeth Reineman, University of Rochester and Mr. G. Shadaksharappa, State University of New York at Buffalo, for their programming assistance. She is also indebted to Mr. John Boothroyd of the University, of Tasmania for a helpful correspondence.

References

1. Eberlein, P. J., Boothroyd, J.: Solution to the eigenproblem by a norm-reducing Jacobi-type method. Numer. Math. **11**, 1, 1—12 (1968). Cf. II/12.
2. — A Jacobi-like method for the automatic computation of eigenvalues and eigenvectors of an arbitrary matrix. J. Soc. Indust. Appl. Math. **10**, 1, 74—88 (1962).

The Calculation of Specified Eigenvectors by Inverse Iteration

by G. PETERS and J. H. WILKINSON

1. Theoretical Background

When an approximation μ is known to an eigenvalue of a matrix A, inverse iteration provides an efficient algorithm for computing the corresponding eigenvector. It consists essentially of the determination of a sequence of vectors x_r defined by

$$(A - \mu I)\, x_{r+1} = k_r\, x_r \quad (r = 0, 1, \ldots),\tag{1}$$

where k_r is chosen so that $\|x_{r+1}\| = 1$ in some norm and x_0 is an arbitrary unit vector.

For convenience in this discussion we assume that μ is an approximation to λ_1 and that $\|A\| = 1$ since it makes it easier to recognize when quantities may be regarded as small. In the algorithms presented here we assume that μ has been found by a reliable algorithm and that it is an exact eigenvalue of some matrix $A + E$ where

$$\|E\| = \varepsilon,\tag{2}$$

ε being usually of the order of the machine precision. The eigenvalue will therefore *usually* be a good eigenvalue in that η defined by

$$\mu - \lambda_1 = \eta\tag{3}$$

will be 'small', but if λ_1 is an ill-conditioned eigenvalue η will not be of the order of the machine precision.

Discussion of the algorithm is complicated by the fact that x_{r+1} is defined by an almost singular system of linear equations so that the computed x_{r+1} differs substantially from the exact solution. However, in this section we shall ignore the effect of rounding errors.

If A is non-defective and u_i $(i = 1, \ldots, n)$ is a complete set of normalized eigenvectors, we may write

$$x_0 = \sum_{i=1}^{n} \alpha_i\, u_i,\tag{4}$$

and

$$x_r = C_r \left[\alpha_1\, u_1 + (-\eta)^r \sum_{2}^{n} \alpha_i\, u_i/(\lambda_i - \mu)^r \right]\tag{5}$$

where C_r is the appropriate normalising factor. If η is small compared with $\lambda_i - \mu$ $(i = 2, \ldots, n)$, x_r tends rapidly to u_1. Indeed if α_1 is not small and η is of the order of machine precision the first iterated vector x_1 may well be correct almost to working accuracy and this will often be the case in our algorithms.

Equation (5) is a little deceptive in that it implies a rapid improvement *with each iteration*. The use of the vectors u_i as a basis is, in this respect, unsatisfactory. This becomes evident when we consider the case of a defective matrix. We assume for simplicity that λ_1 corresponds to a cubic elementary divisor and that all other eigenvalues are well separated from λ_1 and correspond to linear divisors. If u_1, u_2, u_3 are principal vectors such that

$$(A - \lambda_1 I) u_1 = 0, \quad (A - \lambda_1 I) u_2 = u_1, \quad (A - \lambda_1 I) u_3 = u_2 \tag{6}$$

and if (4) now gives x_0 in terms of these vectors and the remaining eigenvectors, then we have

$$x_1 = k_1 \left[- u_1 (\alpha_1/\eta + \alpha_2/\eta^2 + \alpha_3/\eta^3) - u_2 (\alpha_2/\eta + \alpha_3/\eta^2) - u_3 (\alpha_3/\eta) \right.$$
$$\left. + \sum_4^n \alpha_i u_i/(\lambda_i - \mu) \right] \tag{7}$$

or effectively

$$x_1 = u_1 (\alpha_3 + \alpha_2 \eta + \alpha_1 \eta^2) + u_2 (\alpha_3 \eta + \alpha_2 \eta^2) + u_3 (\alpha_3 \eta^2) + O(\eta^3). \tag{8}$$

In one iteration u_1 has become very much the dominant component provided α_3 is not "small". Notice that it is α_3 which is relevant and not α_1. After the second iteration we have effectively

$$x_2 = u_1 (\alpha_3 + \tfrac{2}{3} \alpha_2 \eta + \tfrac{1}{3} \alpha_1 \eta^2) + u_2 (\tfrac{2}{3} \alpha_3 \eta + \tfrac{1}{3} \alpha_2 \eta^2) + u_3 (\tfrac{1}{3} \alpha_3 \eta^2) + O(\eta^4). \tag{9}$$

In this iteration only a marginal improvement is made and the same is true of all subsequent iterations.

When A is non-defective but λ_1 is a very ill-conditioned eigenvalue, the behaviour is of the type we have described for a defective matrix. (See [9].)

The simplest assessment of the 'accuracy' of an approximate normalized eigenvector is provided by a calculation of the residual r defined by

$$r = A x - \mu x. \tag{10}$$

If $\|x\|_2 = 1$, equation (10) may be written in the form

$$(A - r x^H) x = \mu x. \tag{11}$$

Since $\|r x^H\|_2 = \|r\|_2$ we see that when r is small, x and μ are exact for a matrix "close to A". Returning to one step of inverse iteration, if we write

$$(A - \mu I) y = x \quad (\|x\|_2 = 1) \tag{12}$$

then, defining w by

$$w = y/\|y\|_2, \quad \text{i.e. } \|w\|_2 = 1, \tag{13}$$

we have

$$(A - \mu I) w = x/\|y\|_2. \tag{14}$$

Hence the larger the norm of y, the smaller the residual. We may refer to $\|y\|_2$ as the 'growth' factor. For the cubic elementary divisor discussed above the growth factor in the first iteration is of the order $1/\eta^3$ while in all subsequent iterations it is of order $1/\eta$ in spite of the fact that the vector is slowly improving. Hence provided α_3 is not small the residual after one iteration is far smaller than at subsequent stages. Since $(A + E - \mu I)$ is exactly singular, $\|(A - \mu I)^{-1}\| \geq 1/\|E\| = 1/\varepsilon$ and hence there is always at least one vector x for which the growth factor is as large as $1/\varepsilon$ and for which consequently the residual is as small as ε. (This fact was pointed out by Varah [10].) For non-normal matrices there is much to be said for choosing the initial vector in such a way that the full growth occurs in one iteration, thus ensuring a small residual. This is the only simple way we have of recognising a satisfactory performance. For well conditioned eigenvalues (and therefore for all eigenvalues of normal matrices) there is no loss in performing more than one iteration and subsequent iterations offset an unfortunate choice of the initial vector.

The above discussion might give the impression that the situation as regards residuals is actually more favourable in the defective case. This is not so; if λ_1 is well conditioned, η will be of the order of ε, while if λ_1 corresponds to a cubic elementary divisor it will be of the order of $\varepsilon^{\frac{1}{3}}$.

When λ_1 is an eigenvalue of multiplicity s corresponding to linear elementary divisors (5) is replaced by

$$x_r = C_r \left[\sum_1^s \alpha_i u_i + (-\eta)^r \sum_{s+1}^n \alpha_i u_i / (\lambda_i - \mu)^r \right] \tag{15}$$

so that x_r tends to an eigenvector which is dependent on the initial vector. The choice of a different initial vector will give a different eigenvector in the subspace.

In general one will wish to determine s vectors which give full "digital information" about the invariant subspace. With exact computation this could be achieved by orthogonalising each iterated vector with respect to all earlier accepted vectors thus obtaining ultimately a set of s eigenvectors which are orthogonal to working accuracy. The fact that $(A - \mu I)$ is nearly singular, so that we are solving very ill-conditioned sets of linear equations means in practice that the situation is not quite so straightforward.

If detailed attention is given to all the above problems algorithms for inverse iteration become very sophisticated and are likely to contain many parameters. This makes them a little tiresome to use. In our experience multiple eigenvalues and severe clusters of eigenvalues are common with real symmetric matrices and it seems worthwhile to deal rather carefully with this case. For non-Hermitian matrices the problem of ill-conditioned eigenvalues is quite common and the derogatory case is of less importance. The algorithms given here reflect this experience.

In order to minimise the computation in inverse iteration, it is desirable to work with some condensed form of the matrix. For real symmetric matrices it is assumed that A has been reduced to tridiagonal form and for real and complex non-Hermitian matrices that it has been reduced to upper Hessenberg form.

2. Applicability

tristurm may be used to find the eigenvalues of a symmetric tridiagonal matrix lying between two given values and the corresponding eigenvectors. The orthogonality of the computed eigenvectors corresponding to close eigenvalues is controlled as discussed in section 5.1. *tristurm* should not be used unless less than (say) 25% of the eigenvalues and eigenvectors are required since otherwise *tql2* [2] and *imtql2* [7] are more efficient. When the tridiagonal matrix has been determined from a full matrix A using *tred1* or *tred3* the eigenvectors of A may be recovered by using *trbak1* or *trbak3* respectively [4].

invit may be used to find eigenvectors of a real upper Hessenberg matrix H from prescribed eigenvalues, which may be real or complex conjugates. It is assumed that the eigenvalues have been found using *hqr* [3] and accordingly when the Hessenberg matrix 'splits' as the result of negligible sub-diagonal elements, the affiliation of each eigenvalue is known. When H has been derived from a general matrix A by procedure *elmhes*, *dirhes* or *orthes* [5] the eigenvectors of A may be recovered by using *elmbak*, *dirbak* or *ortbak* [5] respectively. If more than (say) 25% of the eigenvectors are required, *hqr2* [8] is to be preferred.

cxinvit may be used to find eigenvectors of a complex upper Hessenberg matrix H from prescribed eigenvalues. It is assumed that the eigenvalues have been found using *comlr* [6] and accordingly when the Hessenberg matrix splits as the result of negligible sub-diagonal elements the affiliation of each eigenvalue is known. When H has been derived from a general complex matrix A by procedure *comhes* [5] the eigenvectors of A may be recovered using *combak* [5]. If more than (say) 25% of the eigenvectors are required *comlr2* [8] is to be preferred.

3. Formal Parameter List

3.1.   Input to procedure *tristurm*

n	order of the symmetric tridiagonal matrix T.
lb, ub	all eigenvalues of T between lb and ub are determined.
macheps	the smallest machine number for which $1 + macheps > 1$.
eps1	the error to be tolerated in the eigenvalue of smallest modulus; if not positive, the quantity $macheps \ (|lb| + |ub|)$ is used instead, so that $eps1 = 0$ is admissible.
c	the diagonal elements of T.
b	the sub-diagonal elements of T with $b[1] = 0$.
beta	the squares of the subdiagonal elements of T with $beta[1] = 0$.
m	the procedure fails if m on entry turns out to be less than the number of eigenvalues in the interval. A call with $m = -1$ will determine this number.

Output from procedure *tristurm*

b	the subdiagonal elements of T with negligible elements replaced by zero.
beta	the squares of the output elements b.
m	the actual number of eigenvalues in the interval.
root	an array of m components giving the m eigenvalues in the given interval, not always in ascending sequence.

vec   an $n \times m$ array giving the m corresponding eigenvectors.

count   an integer array giving the number of iterations for each vector.

fail   failure exit used when

(i) the input m is less than the output m; *root*, *vec*, and *count* are not used.

(ii) more than five iterations are required for any eigenvector. The corresponding component of *count* is set equal to 6. All m eigenvalues and the preceding eigenvectors are correctly determined.

3.2.   Input to procedure *invit*

n   order of the upper Hessenberg matrix A.

macheps   the smallest machine number for which $1 + macheps > 1$.

c   an $n \times 1$ Boolean array indicating the eigenvalues for which the eigenvectors are required. r as used below is equal to the total number of eigenvalues involved, a complex conjugate pair contributing 2 to this count.

a $n \times n$ array giving the matrix A.

wr, wi   two $n \times 1$ arrays giving the real and imaginary parts of the eigenvalues of A; these must be as given by *hqr* [3] since the eigenvalues must be ordered so as to have the correct affiliation (see 5.2).

             Output from procedure *invit*

wr   the real parts of the selected eigenvalues perturbed, where necessary, in the case of close eigenvalues.

z   an $n \times r$ array giving non-redundant information on the r eigenvectors. The eigenvectors $u \pm iv$ corresponding to a complex conjugate pair of eigenvalues $\lambda \pm i\mu$, are represented by two consecutive columns u and v in this array.

3.3.   Input to procedure *cxinvit*

n   order of the complex Hessenberg matrix A.

macheps   the smallest machine number for which $1 + macheps > 1$.

c   an $n \times 1$ Boolean array indicating the eigenvalues for which the eigenvectors are required. r as used below is the number of components which are true.

ar, ai   two $n \times n$ arrays giving the real and imaginary parts of the matrix A.

wr, wi   two $n \times 1$ arrays giving the real and imaginary parts of the eigenvalues of A; these must be as given by *comlr* [6] since eigenvalues must be ordered to have the correct affiliations (see 5.2, 5.3).

             Output from procedure *cxinvit*

wr   the real parts of the selected eigenvalues, perturbed where necessary in the case of close eigenvalues.

zr, zi   two $n \times r$ arrays giving the real and imaginary parts of the r eigenvectors.

4. ALGOL Procedures

procedure *tristurm* $(n, lb, ub, macheps, eps1)$ *data*: (c) *trans*: $(b, beta, m)$
         result: $(root, vec, count)$ *exit*: $(fail)$;
value $n, lb, ub, macheps, eps1$; **integer** n, m; **real** $lb, ub, macheps, eps1$;
array $c, b, beta, root, vec$; **integer array** $count$; **label** $fail$;
comment c is the diagonal, b the sub-diagonal and *beta* the squared subdiagonal
        of a symmetric tridiagonal matrix of order n. The eigenvalues which
        are less than ub and not less than lb are calculated by the method of
        bisection and stored in the vector $root[1:m]$. The procedure fails if m
        on entry is less than the number of eigenvalues required and on exit m
        gives the actual number of eigenvalues found. The corresponding
        eigenvectors are calculated by inverse iteration and are stored in the
        array $vec[1:n, 1:m]$, normalised so that the sum of squares is 1, with
        the number of iterations stored in the vector $count[1:m]$. The procedure
        fails if any vector has not been accepted after 5 iterations. Elements
        of b regarded as negligible and the corresponding *beta* are replaced by
        zero. *macheps* is the relative machine precision and *eps1* should be
        equal to the error to be tolerated in the smallest eigenvalue;
begin
    integer $i, p, q, r, m1, m2$;
    real *norm*;

integer procedure *sturmcnt* $(p, q, d, e, f, lambda)$;
value $p, q, lambda$; **integer** p, q; **real** *lambda*;
array d, e, f;
begin
    integer *count*, i;
    real x;
    $count := 0$; $x := 1.0$;
    for $i := p$ **step** 1 **until** q **do**
    begin
        $x := d[i] - lambda - ($**if** $x \neq 0.0$ **then** $f[i]/x$ **else** $abs(e[i])/macheps)$;
        if $x < 0.0$ **then** $count := count + 1$
    end;
    $sturmcnt := count$
end *sturmcnt*;

    if $eps1 \leq 0.0$ **then** $eps1 := (abs(lb) + abs(ub)) \times macheps$;
    comment look for small sub-diagonal entries;
    for $i := 2$ **step** 1 **until** n **do**
    if $abs(b[i]) \leq macheps \times (abs(c[i]) + abs(c[i-1]))$ **then** $b[i] := beta[i] := 0.0$;
    $r := m$;
    $m := sturmcnt(1, n, c, b, beta, ub) - sturmcnt(1, n, c, b, beta, lb)$;
    if $m > r$ **then goto** *fail*;
    $q := 0$; $r := 1$;
nextp:
    $p := q + 1$;

```
      for q := p step 1 until n − 1 do
      if abs (b[q + 1]) = 0.0 then goto sub;
      q := n;
sub:
   if p = q then
   begin
      if lb ≤ c[p] ∧ c[p] < ub then
      begin
         for i := 1 step 1 until n do vec[i, r] := 0.0;
         root[r] := c[p]; vec[p, r] := 1.0;
         count[r] := 0; r := r + 1
      end isolated root
   end
   else
   begin
      m1 := sturmcnt (p, q, c, b, beta, lb) + 1;
      m2 := sturmcnt (p, q, c, b, beta, ub);
      if m1 ≤ m2 then
      begin
         integer j, k, s, its, group;
         real x0, x1, xu, u, v, bi, eps2, eps3, eps4;
         array x, wu[m1:m2], d, e, f, y, z[p:q];
         Boolean array int[p:q];

         comment find roots by bisection;
         x0 := ub;
         for i := m1 step 1 until m2 do
         begin
            x[i] := ub; wu[i] := lb
         end;
         comment loop for k-th eigenvalue;
         for k := m2 step −1 until m1 do
         begin
            xu := lb;
            for i := k step −1 until m1 do
            if xu < wu[i] then
            begin
               xu := wu[i]; goto contin
            end i;
contin:
            if x0 > x[k] then x0 := x[k];
            for x1 := (xu + x0) × 0.5 while x0 − xu > 2.0 × macheps ×
               (abs (xu) + abs (x0)) + eps1 do
            begin
               s := sturmcnt (p, q, c, b, beta, x1);
               if s < k then
```

```
            begin
               if s < m1 then xu := wu[m1] := x1
               else
               begin
                  xu := wu[s + 1] := x1;
                     if x[s] > x1 then x[s] := x1
               end
            end
            else x0 := x1
         end x1;
         x[k] := (x0 + xu) × 0.5
      end k;
      comment find vectors by inverse iteration;
      norm := abs(c[p]);
      for i := p + 1 step 1 until q do norm := norm + abs(c[i]) + abs(b[i]);
      comment eps2 is the criterion for grouping,
                  eps3 replaces zero pivots and equal roots are
                     modified by eps3,
                  eps4 is taken very small to avoid overflow;
      eps2 := norm × 10 − 3; eps3 := macheps × norm;
      eps4 := eps3 × (q − p + 1);
      group := 0; s := p;
      for k := m1 step 1 until m2 do
      begin
         its := 1; root[r] := x1 := x[k];
         comment look for close or coincident roots;
         if k ≠ m1 then
         begin
            group := if x1 − x0 < eps2 then group + 1 else 0;
               if x1 ≤ x0 then x1 := x0 + eps3
         end;
         u := eps4/sqrt(q − p + 1);
         for i := p step 1 until q do z[i] := u;
         comment elimination with interchanges;
         u := c[p] − x1; v := b[p + 1];
         for i := p + 1 step 1 until q do
         begin
            bi := b[i];
            int[i] := abs(bi) ≥ abs(u);
            if int[i] then
            begin
               y[i] := xu := u/bi;
               d[i − 1] := bi; e[i − 1] := c[i] − x1;
               f[i − 1] := if i ≠ q then b[i + 1] else 0.0;
               u := v − xu × e[i − 1]; v := − xu × f[i − 1]
            end
            else
```

```
            begin
              y[i] := xu := bi/u;
              d[i-1] := u;  e[i-1] := v;  f[i-1] := 0.0;
              u := c[i] - x1 - xu × v;
              if i ≠ q then v := b[i+1]
            end
          end i;
          d[q] := if u ≠ 0.0 then u else eps3;
          e[q] := f[q] := 0.0;
          comment backsubstitution;
```

newz:

```
          for i := q step −1 until p do
          begin
            z[i] := (z[i] - u × e[i] - v × f[i])/d[i];
            v := u;  u := z[i]
          end backsub;
          for j := r − group step 1 until r − 1 do
          begin
            comment orthogonalise with respect to previous members of
                    group;
            xu := 0.0;
            for i := p step 1 until q do xu := xu + z[i] × vec[i, j];
            for i := p step 1 until q do z[i] := z[i] − xu × vec[i, j]
          end orthogonalise;
          norm := 0.0;
          for i := p step 1 until q do norm := norm + abs(z[i]);
          comment forward substitution;
          if norm < 1.0 then
          begin
            if its = 5 then begin count[r] := 6; goto fail end;
            if norm = 0.0 then
            begin
              z[s] := eps4;
              s := if s ≠ q then s + 1 else p
            end null vector
            else
            begin
              xu := eps4/norm;
              for i := p step 1 until q do z[i] := z[i] × xu
            end;
            for i := p + 1 step 1 until q do
            if int[i] then
            begin
              u := z[i − 1];  z[i − 1] := z[i];
              z[i] := u − y[i] × z[i]
```

```
          end
          else z[i] := z[i]−y[i]×z[i−1];
          its := its +1; goto newz
        end forwardsub;
        comment normalise so that sum of squares is 1 and expand to full
                  order;
        u := 0.0;
        for i := p step 1 until q do u := u +z[i]↑2;
        xu := 1.0/sqrt(u);
        for i := p step 1 until q do vec[i, r] := z[i]×xu;
        for i := p −1 step −1 until 1, q +1 step 1 until n do
            vec[i, r] := 0.0;
        count[r] := its; r := r +1; x0 := x1
      end k-th vector
    end m1 ≤ m2
  end p ≠ q;
  if q < n then goto nextp
end tristurm;

procedure invit(n, macheps, c) data: (a, wi) trans: (wr) result: (z);
value n, macheps; integer n; real macheps; array a, wr, wi, z;
Boolean array c;
comment Finds the r eigenvectors selected by the Boolean array c[1:n] of a real
          upper Hessenberg matrix, stored in the array a[1:n, 1:n], given the
          real and imaginary parts of the eigenvalues in the arrays wr, wi[1:n].
          The eigenvectors are formed in the array z[1:n, 1:r] where only one
          complex vector, corresponding to the eigenvalue with positive imag-
          inary part, is formed for a complex pair. Any vector which has not
          been accepted is set to zero. macheps is the relative machine precision.
          Local arrays (n +2) ×n and 2n are used. Uses the procedures cabs
          and cdiv;
begin

  procedure guessvec (n, j, eps, u);
  value n, eps, j; integer n, j; real eps; array u;
  begin
    integer i; real x, y;
    x := sqrt(n); y := eps/(x +1); u[1] := eps;
    for i := 2 step 1 until n do u[i] := y;
    u[j] := u[j]−eps×x
  end guessvec;

  integer i, j, k, m, s, uk, its;
  real w, x, y, rlambda, ilambda, eps3, norm, growtol, normv;
  array b[1:n +2, 1:n], u, v[1:n];
  uk := 0; s := 1;
  for k := 1 step 1 until n do
  if c[k] then
```

```
    begin
        if uk < k then
        begin
            for uk := k step 1 until n − 1 do
                if a[uk + 1, uk] = 0.0 then goto split;
            uk := n;
split:
                norm := 0.0;  m := 1;
                for i := 1 step 1 until uk do
                begin
                    x := 0.0;
                    for j := m step 1 until uk do x := x + abs (a[i, j]);
                    if x > norm then norm := x;
                    m := i
                end norm of leading uk × uk matrix;
                comment eps3 replaces zero pivot in decomposition and close roots are
                        modified by eps3, growtol is the criterion for the growth;
                eps3 := macheps × norm;  growtol := 1.0/sqrt (uk)
        end uk;
        if norm = 0.0 then
        begin
            for i := 1 step 1 until n do z [i, s] := 0.0;
            z [k, s] := 1.0;
            s := s + 1;  goto next k
        end null matrix;
        rlambda := wr[k];  ilambda := wi[k];
        comment perturb eigenvalue if it is close to any previous eigenvalue;
test:
        for i := k − 1 step −1 until 1 do
        if c[i] ∧ abs (wr[i] − rlambda) < eps3 ∧ abs (wi[i] − ilambda) < eps3 then
        begin
            rlambda := rlambda + eps3;
            goto test
        end;
        wr[k] := rlambda;
        comment form upper Hessenberg B = A − rlambda × I and initial real
                vector u;
        m := 1;
        for i := 1 step 1 until uk do
        begin
            for j := m step 1 until uk do b[i, j] := a[i, j];
            b[i, i] := b[i, i] − rlambda;  m := i;  u[i] := eps3
        end;
        its := 0;
        if ilambda = 0.0 then
        begin
```

comment real eigenvalue.
        Decomposition with interchanges, replacing zero
        pivots by $eps3$;
for $i := 2$ **step** 1 **until** uk **do**
begin
    $m := i - 1$;
    if $abs(b[i, m]) > abs(b[m, m])$ **then**
    for $j := m$ **step** 1 **until** uk **do**
    begin
        $y := b[i, j]$; $b[i, j] := b[m, j]$; $b[m, j] := y$
    end *interchange*;
    if $b[m, m] = 0.0$ **then** $b[m, m] := eps3$;
    $x := b[i, m]/b[m, m]$;
    if $x \neq 0.0$ **then**
    for $j := i$ **step** 1 **until** uk **do** $b[i, j] := b[i, j] - x \times b[m, j]$
end *decomposition*;
if $b[uk, uk] = 0.0$ **then** $b[uk, uk] := eps3$;

back:

for $i := uk$ **step** -1 **until** 1 **do**
begin
    $y := u[i]$;
    for $j := i + 1$ **step** 1 **until** uk **do** $y := y - b[i, j] \times u[j]$;
    $u[i] := y/b[i, i]$
end *backsub*;
$its := its + 1$; $norm := normv := 0.0$;
for $i := 1$ **step** 1 **until** uk **do**
begin
    $x := abs(u[i])$;
    if $normv < x$ **then**
    begin
        $normv := x$; $j := i$
    end;
    $norm := norm + x$
end;
if $norm \geq growtol$ **then**
begin comment accept vector;
    $x := 1/u[j]$;
    for $i := 1$ **step** 1 **until** uk **do** $z[i, s] := u[i] \times x$;
    $j := uk + 1$;
end
else
if $its < uk$ **then**
begin
    $guessvec(uk, uk - its + 1, eps3, u)$; **goto** *back*
end *else set vector to zero*
else $j := 1$;

```
            for i := j step 1 until n do z[i, s] := 0.0;
            s := s + 1
        end real case
    else
    if ilambda > 0.0 then
    begin comment complex eigenvalue;
            for i := 1 step 1 until uk do v[i] := 0.0;
            comment triangular decomposition. Store imaginary parts in the lower
                    triangle starting at b[3, 1];
            b[3, 1] := − ilambda;
            for i := uk + 2 step −1 until 4 do b[i, 1] := 0.0;
            for i := 2 step 1 until uk do
            begin
                m := i − 1;  w := b[i, m];
                x := b[m, m] ↑2 + b[i + 1, m] ↑2;
                if w ↑2 > x then
                begin
                    x := b[m, m]/w;  y := b[i + 1, m]/w;
                    b[m, m] := w;  b[i + 1, m] := 0.0;
                    for j := i step 1 until uk do
                    begin
                        w := b[i, j];  b[i, j] := b[m, j] − x×w;  b[m, j] := w;
                        b[j + 2, i] := b[j + 2, m] − y×w;  b[j + 2, m] := 0.0
                    end;
                    b[i + 2, m] := − ilambda;
                    b[i, i] := b[i, i] − y×ilambda;
                    b[i + 2, i] := b[i + 2, i] + x×ilambda
                end
                else
                begin
                    if x = 0.0 then
                    begin
                        b[m, m] := eps3;  b[i + 1, m] := 0.0;  x := eps3 ↑2
                    end;
                    w := w/x;  x := b[m, m]×w;  y := −b[i + 1, m]×w;
                    for j := i step 1 until uk do
                    begin
                        b[i, j] := b[i, j] − x×b[m, j] + y×b[j + 2, m];
                        b[j + 2, i] := − x×b[j + 2, m] − y×b[m, j]
                    end;
                    b[i + 2, i] := b[i + 2, i] − ilambda
                end
            end cdecomposition;
            if b[uk, uk] = 0.0 ∧ b[uk + 2, uk] = 0.0 then b[uk, uk] := eps3;
cback:
            for i := uk step −1 until 1 do
```

```
    begin
        x := u[i]; y := v[i];
        for j := i+1 step 1 until uk do
        begin
            x := x − b[i, j] × u[j] + b[j + 2, i] × v[j];
            y := y − b[i, j] × v[j] − b[j + 2, i] × u[j]
        end;
        cdiv (x, y, b[i, i], b[i + 2, i], u[i], v[i])
    end cbacksub;
    its := its +1; norm := normv := 0.0;
    for i := 1 step 1 until uk do
    begin
        x := cabs (u[i], v[i]);
        if normv < x then
        begin
            normv := x; j := i
        end;
        norm := norm + x
    end;
    m := s +1;
    if norm ≥ growtol then
    begin comment accept complex vector;
        x := u[j]; y := v[j];
        for i := 1 step 1 until uk do
            cdiv (u[i], v[i], x, y, z[i, s], z[i, m]);
        j := uk +1
    end
    else
    if its < uk then
    begin
        guessvec (uk, uk − its + 1, eps3, u);
        for i := 1 step 1 until uk do v[i] := 0.0;
        goto cback
    end else set vector to zero
    else j := 1;
    for i := j step 1 until n do z[i, s] := z[i, m] := 0.0;
    s := s +2
    end complex case;
nextk:
    end k
end invit;


procedure cxinvit (n, macheps, c) data: (ar, ai, wi) trans: (wr) result: (zr, zi);
value n, macheps; integer n; real macheps; array ar, ai, wr, wi, zr, zi;
Boolean array c;
comment Given a complex upper Hessenberg matrix in the arrays ar, ai[1:n, 1:n]
        and its eigenvalues in wr, wi[1:n] this procedure finds the r eigen-
```

vectors selected by the Boolean array $c[1:n]$ and stores them in
arrays zr, $zi[1:n, 1:r]$. Any vector which has not been accepted is set
to zero. *macheps* is the relative machine precision. Uses the procedures
cabs and *cdiv*;

begin
    integer i, j, k, uk, its, m, s;
    real $x, y, eps3, rlambda, ilambda, norm, normv, growtol$;
    array $br, bi[1:n, 1:n], u, v[1:n]$;

    procedure *guessvec* (n, j, eps, u);
    value n, j, eps; **integer** n, j; **real** eps; **array** u;
    begin
        integer i; **real** x, y;
        $x := sqrt(n)$; $y := eps/(x+1)$; $u[1] := eps$;
        for $i := 2$ **step** 1 **until** n **do** $u[i] := y$;
        $u[j] := u[j] - eps$
    end *guessvec*;

    $uk := 0$; $s := 1$;
    for $k := 1$ **step** 1 **until** n **do**
    if $c[k]$ **then**
    begin
        if $uk < k$ **then**
        begin
            for $uk := k$ **step** 1 **until** $n - 1$ **do**
            if $ar[uk+1, uk] = 0.0$ **and** $ai[uk+1, uk] = 0.0$ **then goto** *split*;
            $uk := n$;
split:
            $norm := 0.0$; $m := 1$;
            for $i := 1$ **step** 1 **until** uk **do**
            begin
                $x := 0.0$;
                for $j := m$ **step** 1 **until** uk **do** $x := x + cabs(ar[i, j], ai[i, j])$;
                if $x > norm$ **then** $norm := x$;
                $m := i$;
            end *norm of leading* $uk \times uk$ *matrix*;
            comment *eps3* replaces zero pivots in decomposition and close roots
                        are modified by *eps3*,
                           growtol is the criterion for the growth;
            $eps3 := macheps \times norm$; $growtol := 1.0/sqrt(uk)$
        end;
        $rlambda := wr[k]$; $ilambda := wi[k]$;
        comment perturb eigenvalue if it is close to any previous eigenvalue;
test:
        for $i := k - 1$ **step** -1 **until** 1 **do**
        if $c[i] \wedge abs(wr[i] - rlambda) < eps3 \wedge abs(wi[i] - ilambda) < eps3$ **then**
        begin
            $rlambda := rlambda + eps3$;

```
        goto test
    end;
    wr[k] := rlambda;
    comment generate B = A − lambda × I, the matrix to be reduced, and
               initial vector u + iv;
    m := 1;
    for i := 1 step 1 until uk do
    begin
        for j := m step 1 until uk do
        begin
            br[i, j] := ar[i, j]; bi[i, j] := ai[i, j]
        end j;
        br[i, i] := br[i, i] − rlambda; bi[i, i] := bi[i, i] − ilambda;
        m := i; u[i] := eps3; v[i] := 0.0
    end i;
    comment reduce B to LU form;
    for i := 2 step 1 until uk do
    begin
        m := i − 1;
        if cabs(br[i, m], bi[i, m]) > cabs(br[m, m], bi[m, m]) then
        for j := m step 1 until uk do
        begin
            y := − br[i, j]; br[i, j] := br[m, j]; br[m, j] := y;
            y := bi[i, j]; bi[i, j] := bi[m, j]; bi[m, j] := y
        end interchange;
        if br[m, m] = 0.0 ∧ bi[m, m] = 0.0 then br[m, m] := eps3;
        cdiv(br[i, m], bi[i, m], br[m, m], bi[m, m], x, y);
        if x ≠ 0.0 ∨ y ≠ 0.0 then
        for j := i step 1 until uk do
        begin
            br[i, j] := br[i, j] − x × br[m, j] + y × bi[m, j];
            bi[i, j] := bi[i, j] − x × bi[m, j] − y × br[m, j]
        end j
    end complex decomposition;
    if br[uk, uk] = 0.0 ∧ bi[uk, uk] = 0.0 then br[uk, uk] := eps3;
    its := 0;
back:
    for i := uk step −1 until 1 do
    begin
        x := u[i]; y := v[i];
        for j := i + 1 step 1 until uk do
        begin
            x := x − br[i, j] × u[j] + bi[i, j] × v[j];
            y := y − br[i, j] × v[j] − bi[i, j] × u[j]
        end j;
        cdiv(x, y, br[i, i], bi[i, i], u[i], v[i])
```

```
    end complex backsub;
    its := its +1;
    comment normalize u + i × v;
    norm := normv := 0.0;
    for i := 1 step 1 until uk do
    begin
        x := cabs (u[i], v[i]);
        if normv < x then
        begin
            normv := x; j := i
        end;
        norm := norm + x
    end i;
    if norm ≥ growtol then
    begin
        x := u[j]; y := v[j];
        for i := 1 step 1 until uk do cdiv (u[i], v[i], x, y, zr[i, s], zi[i, s]);
        j := uk +1
    end
    else
    if its < uk then
    begin
        guessvec (uk, uk − its +1, eps3, u);
        for i := 1 step 1 until uk do v[i] := 0.0;
        goto back
    end
    else j := 1;
    for i := j step 1 until n do zr[i, s] := zi[i, s] := 0.0;
    s := s +1
    end k
end cxinvit;
```

5. Organisational and Notational Details

5.1. Procedure *tristurm*. The procedure finds eigenvalues as well as eigen-vectors for reasons described below. The input parameter m should not be an underestimate of the number of eigenvalues in the given range or *tristurm* will fail. A call with $m = -1$ will determine this number via the failure exit. A test is first made to see if any off diagonal element is negligible compared with the two neighbouring diagonal elements and if so the matrix is split into the direct sum of smaller tridiagonal matrices. The eigenvalues of each of these matrices in the required interval are then determined by bisection as in *bisect* [1]. Since the eigenvector computation assumes that the eigenvalues are correct almost to working accuracy the parameter *eps1* cannot be set at a comparatively high value as it could in *bisect* when low accuracy was acceptable.

The corresponding eigenvectors are found independently for each of the tridiagonal sub-matrices C. Eigenvectors corresponding to different sub-matrices are accordingly exactly orthogonal. The first attempt to compute any eigenvector uses only half an iteration. $(C - \mu I)$ is factorized into $L \times U$ and the equation $U x = \alpha e$ is then solved. α is taken to be small so as to diminish the possibility of overflow. If acceptable growth takes place then the normalized x is accepted as an eigenvector. If not, up to four further full iterations are allowed. Most frequently the first half iteration is successful.

Eigenvalues within a submatrix which differ by less than $10^{-3} \|C\|_\infty$ are grouped together and after each iteration the current vector is orthogonalised with respect to earlier computed vectors belonging to the group. If orthogonalization leaves a null vector, iteration is continued with $k e_i$, where e_i is a column of the unit matrix, the columns of the unit matrix being taken in turn. (There seems little point in making a sophisticated choice of the alternative starting vectors.) The criterion for grouping eigenvalues is a subjective matter. The larger the tolerance the more accurate will be the orthogonality of members in different groups. For tridiagonal matrices having elements and eigenvalues which vary enormously in size the tolerance for grouping should be made smaller since otherwise all the smaller eigenvalues may be grouped together, and as a result a good deal of time may be spent in unnecessary reorthogonalization.

5.2. Procedure *invit*. It is assumed that the eigenvalues have been found by some procedure which determines the affiliation of each eigenvalue in the case when the Hessenberg matrix H splits. Moreover, complex conjugate pairs must be stored in consecutive components of *wr*, *wi*, with the positive component of *wi* first. (This would be true if the eigenvalues were found by *hqr* [3].) The required eigenvectors are specified by the Boolean array c, where components corresponding to a complex conjugate pair must agree.

The eigenvector corresponding to λ_k is found by inverse iteration using a truncated Hessenberg matrix H_k of order uk; the value of uk corresponds to the first point beyond k (if any) at which H splits. Components $uk + 1$ to n are set to zero. Eigenvectors are accepted only when an acceptable growth factor is achieved in one half iteration, and the initial vector is changed, if necessary, until one is found for which this is true. The first trial vector is αe, α being chosen as in *tristurm*. If this fails multiples of the columns of the orthogonal matrix $\beta v v^T - I$ are tried in turn, where

$$v^T = (uk^{\frac{1}{2}} + 1, 1, \ldots, 1), \quad \beta = 1/(uk + uk^{\frac{1}{2}}).$$

If none of these gives acceptable growth in one iteration a zero "eigenvector" is given as an indication of failure.

Inverse iteration is performed by solving the relevant equations using Gaussian elimination with partial pivoting. When λ_k is complex the solution is performed in complex arithmetic using an auxiliary $(n + 2) \times n$ array, b; advantage is taken of the fact that the relevant matrix is of Hessenberg form and the $n \times n$ triangle in the bottom left-hand corner of b is used to hold the imaginary parts of the initial, and reduced elements. When λ_i is real the upper $n \times n$ submatrix of the array b is used in the orthodox way.

Zero pivots in the Gaussian elimination are replaced by $eps3 = macheps \times norm$ and close eigenvalues are artificially separated by $eps3$. Only one eigenvector is found for a complex conjugate pair of eigenvalues, that corresponding to the eigenvalue with positive imaginary part. No attempt is made to orthogonalize eigenvectors corresponding to multiple and close eigenvalues. This is again a somewhat arbitrary decision but it should be remembered that the eigenvectors corresponding to close roots might indeed be almost parallel.

5.3. Procedure *cxinvit* is closely analogous to *invit*, though the logic is simpler since we do not have to distinguish between real and complex eigenvalues. The procedure *comlr* [6] provides eigenvalues with known affiliations.

6. Numerical Properties

Inverse iteration is capable of giving eigenvectors of almost the maximum accuracy attainable with the precision of computation, in that each computed eigenvector v_i is exact for some $A + E_i$ where E_i is of the order of magnitude of rounding errors in A. Inaccuracy in v_i therefore springs only from its inherent sensitivity. This is because the computed solution of $(A - \mu I) y = x$, satisfies exactly $(A + F - \mu I) y = x$, where F, which is a function of x, is of the order of magnitude of rounding errors in A; this is true however ill-conditioned $A - \mu I$ may be.

The main weakness of *tristurm* is that the orthogonality of the computed vectors is generally poorer than that of those given by *jacobi* or *tred2* and *tql2*. Nevertheless the individual eigenvectors are just as accurate. In order to guarantee that the eigenvectors are orthogonal to working accuracy it is necessary to increase the tolerance for grouping eigenvalues to something in the neighbourhood of $\|A\|$ itself. The amount of work involved in the orthogonalizations is then very substantial unless very few vectors are involved.

A similar criticism applies to *invit* and *cxinvit*. The eigenvectors corresponding to coincident or close eigenvalues may not give full digital information about the relevant subspaces.

Since it was assumed that μ is an eigenvalue of some matrix $A + E$, i.e. that $A + E - \mu I$ is exactly singular, it is evident that $\|(A + F - \mu I)^{-1}\| \geq 1/\|F - E\| \geq 1/(\|F\| + \|E\|)$ and hence, even when rounding errors are taken into account, there must still be an x for which y shows acceptable growth. However, since F is, in general, different for each x, it is difficult to make useful predictions about the effect of a succession of iterations; or about the relation between the vectors y computed from related vectors x. Moreover, when μ is associated with pathologically close or coincident eigenvalues a very small change in μ may lead to a complete change in the vector y corresponding to a given x.

7. Test Results

7.1. Procedure *tristurm*

To test the splitting device the matrices defined by

$$a_{ii} = n, \qquad a_{ij} = 1 \qquad (i \neq j)$$

with $n = 10$ and $n = 16$ were used having $n - 1$ eigenvalues equal to $n - 1$ and one eigenvalue equal to $2n - 1$. In each case the tridiagonal matrix produced by a call of procedure *tred1* [4] split into $n-2$ matrices of order 1 and one matrix of order 2. The $n-1$ vectors of the tridiagonal matrix corresponding to the multiple root were therefore exactly orthogonal and were obtained without computation.

To test the grouping device the tridiagonal matrix W of order 21 with

$$w_{i,\,i+1} = w_{i+1,\,i} = 1, \qquad w_{ii} = |11 - i|$$

was used. The roots between 10 and 11 were requested. This consists of two roots which are pathologically close. The computed eigenvalues and eigenvectors on KDF 9, a computer with a 39 binary digit mantissa are given in Table 1. The vectors are almost orthogonal and span the correct subspace.

Table 1

$\lambda_1 = +1.0746\,1941\,830_{10} + 1$ eigenvector 1	$\lambda_2 = +1.0746\,1941\,830_{10} + 1$ eigenvector 2
$+1.6895\,1108\,095_{10} - 1$	$+1.0000\,0000\,000_{10} + 0$
$+1.2607\,0334\,145_{10} - 1$	$+7.4619\,4184\,527_{10} - 1$
$+5.1192\,1759\,501_{10} - 2$	$+3.0299\,9943\,015_{10} - 1$
$+1.4513\,3215\,796_{10} - 2$	$+8.5902\,4948\,829_{10} - 2$
$+3.1775\,4484\,499_{10} - 3$	$+1.8807\,4819\,901_{10} - 2$
$+5.6792\,3197\,902_{10} - 4$	$+3.3614\,6508\,081_{10} - 3$
$+8.5852\,0493\,028_{10} - 5$	$+5.0814\,7443\,057_{10} - 4$
$+1.1251\,3158\,436_{10} - 5$	$+6.6594\,5820\,395_{10} - 5$
$+1.3027\,4618\,116_{10} - 6$	$+7.7054\,5921\,608_{10} - 6$
$+1.4267\,3373\,398_{10} - 7$	$+7.9719\,8770\,282_{10} - 7$
$+8.7694\,3661\,958_{10} - 8$	$+6.2533\,0361\,644_{10} - 8$
$+7.9962\,5459\,971_{10} - 7$	$-1.2686\,8386\,356_{10} - 7$
$+7.7055\,2878\,576_{10} - 6$	$-1.3006\,7873\,011_{10} - 6$
$+6.6594\,3437\,284_{10} - 5$	$-1.1250\,7821\,145_{10} - 5$
$+5.0814\,7107\,364_{10} - 4$	$-8.5851\,7259\,343_{10} - 5$
$+3.3614\,6463\,421_{10} - 3$	$-5.6792\,2787\,111_{10} - 4$
$+1.8807\,4813\,552_{10} - 2$	$-3.1775\,4431\,849_{10} - 3$
$+8.5902\,4939\,027_{10} - 2$	$-1.4513\,3209\,191_{10} - 2$
$+3.0299\,9941\,540_{10} - 1$	$-5.1192\,1752\,974_{10} - 2$
$+7.4619\,4182\,932_{10} - 1$	$-1.2607\,0333\,974_{10} - 1$
$+1.0000\,0000\,000_{10} + 0$	$-1.6895\,1108\,773_{10} - 1$

7.2. Procedure *invit*

The matrix W was also used to test *invit* and again the two largest eigenvalues were selected. The procedure separated the eigenvalues and gave two independent vectors, providing almost full digital information on the invariant subspace. See Table 2.

As a second test the Frank matrix H of order 13 was used. This is an upper Hessenberg matrix defined by

$$h_{ij} = 14 - j \quad (i \leq j), \qquad h_{i+1,\,i} = 13 - i.$$

The two smallest eigenvalues were found. These are very ill-conditioned and hence although they are exact eigenvalues of a neighbouring matrix they are not very accurate. Nevertheless each eigenvector was determined by the first 'half' iteration and was of maximum accuracy having regard to its condition. See Table 3.

Table 2

$\lambda_{20} = +1.0746\,1941\,827_{10}+1$ eigenvector 20	$\lambda_{21} = +1.0746\,1941\,829_{10}+1$ eigenvector 21
$+1.0000\,0000\,000_{10}+0$	$+6.0168\,0298\,969_{10}-1$
$+7.4619\,4182\,781_{10}-1$	$+4.4897\,0339\,088_{10}-1$
$+3.0299\,9941\,367_{10}-1$	$+1.8230\,9095\,419_{10}-1$
$+8.5902\,4937\,670_{10}-2$	$+5.1685\,8382\,136_{10}-2$
$+1.8807\,4812\,580_{10}-2$	$+1.1316\,0910\,033_{10}-2$
$+3.3614\,6456\,432_{10}-3$	$+2.0225\,2704\,707_{10}-3$
$+5.0814\,7059\,559_{10}-4$	$+3.0574\,2116\,613_{10}-4$
$+6.6594\,3653\,746_{10}-5$	$+4.0068\,6087\,536_{10}-5$
$+7.7060\,1863\,151_{10}-6$	$+4.6370\,7469\,096_{10}-6$
$+8.0416\,2473\,939_{10}-7$	$+4.8811\,4192\,560_{10}-7$
$+1.3169\,7515\,498_{10}-7$	$+1.2014\,8271\,355_{10}-7$
$+6.1127\,7122\,510_{10}-7$	$+8.0298\,9720\,354_{10}-7$
$+5.8261\,2054\,148_{10}-6$	$+7.7059\,1272\,828_{10}-6$
$+5.0345\,2969\,869_{10}-5$	$+6.6594\,3866\,160_{10}-5$
$+3.8415\,8518\,629_{10}-4$	$+5.0814\,7104\,751_{10}-4$
$+2.5412\,6294\,740_{10}-3$	$+3.3614\,6462\,278_{10}-3$
$+1.4218\,4320\,100_{10}-2$	$+1.8807\,4813\,403_{10}-2$
$+6.4942\,1765\,040_{10}-2$	$+8.5902\,4938\,832_{10}-2$
$+2.2906\,7571\,975_{10}-1$	$+3.0299\,9941\,518_{10}-1$
$+5.6412\,1857\,247_{10}-1$	$+7.4619\,4182\,915_{10}-1$
$+7.5599\,8733\,628_{10}-1$	$+1.0000\,0000\,000_{10}+0$

Table 3

$\lambda_1 = +2.7755\,3987\,291_{10}-2$ eigenvector 1	$\lambda_2 = +4.4119\,4423\,277_{10}-2$ eigenvector 2
$+4.2792\,5794\,055_{10}-11$	$-2.2722\,5077\,213_{10}-11$
$-1.3936\,1241\,100_{10}-9$	$+5.5260\,0937\,792_{10}-10$
$+3.0278\,0088\,463_{10}-8$	$-5.8253\,8031\,647_{10}-9$
$-5.0833\,3303\,903_{10}-7$	$-1.1584\,1897\,038_{10}-8$
$+6.8975\,2171\,663_{10}-6$	$+1.5713\,4904\,714_{10}-6$
$-7.6780\,6578\,297_{10}-5$	$-3.1681\,3976\,535_{10}-5$
$+7.0146\,3733\,572_{10}-4$	$+4.0147\,4709\,225_{10}-4$
$-5.2072\,6524\,397_{10}-3$	$-3.6716\,7116\,269_{10}-3$
$+3.0767\,0997\,672_{10}-2$	$+2.4951\,1499\,438_{10}-2$
$-1.3967\,8066\,480_{10}-1$	$-1.2447\,9436\,263_{10}-1$
$+4.5875\,2082\,989_{10}-1$	$+4.3479\,4099\,107_{10}-1$
$-9.7224\,4601\,275_{10}-1$	$-9.5588\,0557\,676_{10}-1$
$+1.0000\,0000\,000_{10}-0$	$+1.0000\,0000\,000_{10}+0$

7.3. Procedure *cxinvit*

As a formal test of *cxinvit* the matrix $A+iB$ was run.

$$
\text{(A)} \qquad\qquad\qquad \text{(B)}
$$

$$
\begin{bmatrix} 1 & 2 & 3 & 1 \\ 3 & 1 & 2 & 4 \\ 2 & 1 & 3 & 5 \\ 1 & 3 & 1 & 5 \end{bmatrix} \qquad \begin{bmatrix} 3 & 1 & 2 & 1 \\ 4 & 2 & 1 & 3 \\ 3 & 5 & 1 & 2 \\ 2 & 1 & 4 & 3 \end{bmatrix}.
$$

The two smallest eigenvalues and their corresponding vectors were found. See Table 4.

Table 4

$\lambda_1 = +2.2216823477110_{10} + 0$ real eigenvector 1	$+1.8489933597410_{10} + 0\,i$ imaginary eigenvector 1
$-7.966208174355_{10} - 1$	$+3.049807859411_{10} - 1$
$-1.788343947385_{10} - 1$	$+4.297241477425_{10} - 1$
$-2.528143111945_{10} - 1$	$+3.817340494945_{10} - 2$
$+1.00000000000005_{10} + 0$	$+0.00000000000$

$\lambda_2 = +1.3656693099515_{10} + 0$ real eigenvector 2	$-1.4010535974710_{10} + 0\,i$ imaginary eigenvector 2
$-8.4659871456325_{10} - 4$	$+7.302035512645_{10} - 1$
$-8.7941701579025_{10} - 2$	$-3.879017996145_{10} - 1$
$+1.00000000000005_{10} + 0$	$+0.00000000000$
$-4.3208937480425_{10} - 1$	$-4.334263693435_{10} - 1$

Acknowledgements. The authors are indebted to Dr. C. Reinsch of the Technische Hochschule München who suggested a number of important improvements and tested the procedures. The work described above has been carried out at the National Physical Laboratory.

References

1. Barth, W., Martin, R. S., Wilkinson, J. H.: Calculation of the eigenvalues of a symmetric tridiagonal matrix by the method of bisection. Numer. Math. **9**, 386–393 (1967). Cf. II/5.
2. Bowdler, Hilary, Martin, R. S., Reinsch, C., Wilkinson, J. H.: The QR and QL algorithms for symmetric matrices. Numer. Math. **11**, 293–306 (1968). Cf. II/3.
3. Martin, R. S., Peters, G., Wilkinson, J. H.: The QR algorithm for real Hessenberg matrices. Numer. Math. **14**, 219–231 (1970). Cf. II/14.
4. — Reinsch, C., Wilkinson, J. H.: Householder's tridiagonalization of a symmetric matrix. Numer. Math. **11**, 181–195 (1968). Cf. II/2.
5. — Wilkinson, J. H.: Similarity reduction of a general matrix to Hessenberg form. Numer. Math. **12**, 349–368 (1968). Cf. II/13.
6. — — The modified LR algorithm for complex Hessenberg matrices. Numer. Math. **12**, 369–376 (1968). Cf. II/16.
7. — — The implicit QL algorithm. Numer. Math. **12**, 377–383 (1968). Cf. II/4.
8. Peters, G., Wilkinson, J. H.: Eigenvectors of real and complex matrices by LR and QR triangularizations. Numer. Math. **16**, 181–204 (1970). Cf. II/15.
9. — — Theoretical and practical properties of inverse iteration. (To be published.)
10. Varah, J.: Ph. D. Thesis. Stanford University (1967).

Die Grundlehren der mathematischen Wissenschaften in Einzeldarstellungen mit besonderer Berücksichtigung der Anwendungsgebiete

Eine Auswahl

Druck der Universitätsdruckerei H. Stürtz AG, Würzburg